Nanostructure Science and Technology

Series Editor

David J. Lockwood, FRSC
National Research Council of Canada
Ottawa, ON, Canada

Nanostructure science and technology now forms a common thread that runs through all physical and materials sciences and is emerging in industrial applications as nanotechnology. The breadth of the subject material is demonstrated by the fact that it covers and intertwines many of the traditional areas of physics, chemistry, biology, and medicine. Within each main topic in this field there can be many subfields. For example, the electrical properties of nanostructured materials is a topic that can cover electron transport in semiconductor quantum dots, self-assembled molecular nanostructures, carbon nanotubes, chemically tailored hybrid magnetic-semiconductor nanostructures, colloidal quantum dots, nanostructured superconductors, nanocrystalline electronic junctions, etc. Obviously, no one book can cope with such a diversity of subject matter. The nanostructured material system is, however, of increasing significance in our technology-dominated economy and this suggests the need for a series of books to cover recent developments.

The scope of the series is designed to cover as much of the subject matter as possible – from physics and chemistry to biology and medicine, and from basic science to applications. At present, the most significant subject areas are concentrated in basic science and mainly within physics and chemistry, but as time goes by more importance will inevitably be given to subjects in applied science and will also include biology and medicine. The series will naturally accommodate this flow of developments in the sciences and technology of nanostructures and maintain its topicality by virtue of its broad emphasis. It is important that emerging areas in the biological and medical sciences, for example, not be ignored as, despite their diversity, developments in this field are often interlinked. The series will maintain the required cohesiveness from a judicious mix of edited volumes and monographs that while covering subfields in depth will also contain more general and interdisciplinary texts.

Thus the series is planned to cover in a coherent fashion the developments in basic research from the distinct viewpoints of physics, chemistry, biology, and materials science and also the engineering technologies emerging from this research. Each volume will also reflect this flow from science to technology. As time goes by, the earlier series volumes will then serve as reference texts to subsequent volumes.

More information about this series at https://link.springer.com/bookseries/6331

Warwick Bowen · Frank Vollmer · Reuven Gordon
Editors

Single Molecule Sensing Beyond Fluorescence

Springer

Editors
Warwick Bowen
School of Mathematics and Physics
University of Queensland
Brisbane, QLD, Australia

Frank Vollmer
Department of Physics and Astronomy
University of Exeter
Exeter, UK

Reuven Gordon
Department of Computer and Electrical Engineering
University of Victoria
Victoria, BC, Canada

ISSN 1571-5744 ISSN 2197-7976 (electronic)
Nanostructure Science and Technology
ISBN 978-3-030-90338-1 ISBN 978-3-030-90339-8 (eBook)
https://doi.org/10.1007/978-3-030-90339-8

This Springer imprint is published by the registered company Springer Nature Switzerland AG
The registered company address is: Gewerbestrasse 11, 6330 Cham, Switzerland

At quite uncertain times and places
The atoms left their heavenly path,
And by tortuitous embraces
Engendered all that being hath.

And though they seem to cling together
And form "associations" here,
Yet, late or soon, they burst their tether,
And through the depths of space career.

James Clerk Maxwell
From the poem Molecular Evolution
Nature 8, 473 (1873).

Almost all aspects of life are engineered at the molecular level, and without understanding molecules we can only have a very sketchy understanding of life itself.

Francis Crick, What Mad Pursuit, Basic Books, 1988.

But, to reconcile the high durability of the hereditary substance with its minute size, we had to evade the tendency to disorder by 'inventing the molecule', in fact, an unusually large molecule which has to be a masterpiece of highly differentiated order, safeguarded by the conjuring rod of quantum theory.

Erwin Schrödinger, What is Life?, Cambridge University Press, 1944.

Preface

The importance of biomolecules can hardly be overstated. They are the fundamental building blocks of life, driving active processes in cells ranging from gene transcription to catalysis and cell division. To understand a biomolecule is to understand its structure, its dynamics, and how it interacts with its environment. However, because biomolecules generally consist of many thousands of atoms and interact with complex environments, *ab initio* models are fraught with uncertainty. As such, direct measurements of single-molecule behaviours are vital. In the midst of the COVID19 pandemic this importance has perhaps never been more evident, with the behaviour of spike proteins on the virus surface playing a key role in both the function of the virus and the development of targeted vaccines.

Experimental methods to resolve single-molecule dynamics are faced with the challenge of size-scales. Biomolecules tend to be small, of size below a few nanometers, and are therefore difficult to observe. For many decades, the primary approach to overcome this challenge has been to attach bright fluorescent labels to the molecule, which act like a beacon lighting the molecule up. This has proved tremendously successful, as illustrated by the award of the 2014 Nobel Prize in Chemistry. However, fluorescent techniques have significant drawbacks. The label can alter both the physical and chemical environment of the molecule, changing its behaviour. The fluorescence can bleach, blink or quench, degrading the measurement and limiting the time over which it can be made. This motivates the development of alternative techniques that do not rely on labels and can therefore go beyond the capabilities of fluorescence. Techniques to sense and visualise molecules are particularly important in biology where movements of motor proteins and conformational (shape) changes of enzymes occur on length and timescales often difficult to access with fluorescence measurement techniques.

The past decade has seen an explosion of activity in unlabelled single-molecule biosensing. The purpose of this book is to provide an overview of this activity, current capabilities, and future possibilities. It is a particularly exciting time for such a book. Sensor systems have started to emerge that enable the detection of molecules even down to small organic molecules and single atomic ions. The emerging sensor systems often use interferometry to probe complex biomolecules such as nanoscale proteins with exceedingly high sensitivity. Some sensors enable the detection of microsecond molecular dynamics such as the conformational (shape) changes of enzymes; others measure current changes when a single DNA strand passes through a synthetic or protein nanopore. These electrical sensor signals can resolve the DNA sequence down to single nucleotides. The making of molecular movies is starting to become a reality, by operating optical and electrical sensors at their ultimate (fundamental) detection limits, and by utilising multiple detection channels to access the biomolecular information. Exquisitely precise tools to manipulate molecules using electrical and optical forces are also being developed, providing the means to deliver molecules to the sensor, to control their position and orientation while sensing and, thereby, to gain a great deal more information about them.

Single-molecule sensing transcends disciplines, benefiting from the sustained efforts of physicists, engineers, chemists and biologists, as well as researchers in allied fields. We aim for this book to serve as a proper introduction to the field of single-molecule sensing beyond fluorescence. The book introduces the reader to some of the most exciting and emerging single-molecule sensing technologies that are currently investigated around the world. It is the editors hope that the book will inspire new ideas and fruitful collaborations, stimulating further advances in the field. Our goal when preparing the book has been to provide a balanced overview of the fundamental physics, the engineering principles and the applications of the label-free single-molecule sensors, in areas such as biology and chemistry, highlighting some of the pioneering single-molecule sensing techniques that are being developed by our colleagues. We are immensely grateful to all authors for their high quality contributions to the book, and of course for driving our field forward with so many impressive achievements over the last decade. The book serves as a testament to the immense scientific breadth, collaborative spirit and research excellence that we are all deeply grateful to be a part of.

As one might understand, with a book of this nature a tension exists between wide coverage and substantive treatments. In developing the book we deliberately chose not to seek to be exhaustive, but rather focus on key advances in optical and nanopore-based single-molecule sensing techniques. For instance, we do not discuss quite remarkable progress towards room-temperature structural imaging using X-ray lasers or electron microscopy, nor do we discuss progress on quantum probes of biomolecules such as nitrogen-vacancy centres in diamond. We also do not include a substantial treatment of surface-enhanced Raman spectroscopy (SERS), which is perhaps the most well established of unlabelled single-molecule sensing technologies but which is well treated in existing literature.[1]

The book is an edited volume of chapters from prominent researchers in the field of unlabelled single-molecule sensing. It is separated into four parts, each focused on a different form of single-molecule sensing. Part 1 treats optical single-molecule sensors that use light scattering, optical cavities and nonlinear effects; Part 2 treats optomechanical single-molecule sensors, where light interacts with a mechanical

[1] For instance, see *Surface-Enhanced Raman Scattering: Physics and Applications*, Eds. Katrin Kneipp, Martin Moskovits & Harald Kneipp, Springer, 2006.

degree of freedom such as a vibrational mode of the molecule; Part 3 treats single-molecule sensors that employ optical or radio-frequency molecular traps; and Part 4 treats single-molecule sensors based on nanopores with both optical and electrical readout. Together, this comprises a wide spectrum of recent advances in unlabelled single-molecule sensors. In the future, we envisage that these sensing technologies will be combined into multi-modal technologies capable of unravel the deepest mysteries of biomolecular dynamics, advancing our understanding of the building blocks of life and transforming biomedical and pharmaceutical capabilities.

Brisbane, Australia
Exeter, UK
Victoria, Canada
May 2021

Warwick Bowen
Frank Vollmer
Reuven Gordon

Contents

Part I Optical Sensing

1 **Interferometric Biosensing** 5
Nicolas P. Mauranyapin, Larnii Booth, Igor Marinkovic, and Warwick P. Bowen

2 **Optoplasmonic Whispering Gallery Mode Sensors for Single Molecule Characterization: A Practical Guide** 37
Sivaraman Subramanian, Kulathunga Mudalige Kalani Perera, Srikanth Pedireddy, and Frank Vollmer

3 **Nonlinear Optical Microcavities Towards Single-Molecule Sensing** 97
Wenjing Liu, Shui-Jing Tang, and Yun-Feng Xiao

Part II Optomechanical Sensing

4 **Optomechanical Sensing** 127
Wenyan Yu, Wei C. Jiang, Qiang Lin, and Tao Lu

5 **Quantum Optical Theories of Molecular Optomechanics** 163
Mohsen Kamandar Dezfouli and Stephen Hughes

Part III Biomolecular Manipulation

6 **Dielectrophoresis of Single Molecules** 207
Binoy Paulose Nadappuram, Christopher T. Ertsgaard, Joshua B. Edel, and Sang-Hyun Oh

7 **Optical Trapping of Single Molecules** 233
Joshua Kolbow, Nathan Lindquist, and Sang-Hyun Oh

8 **Applications of Trapping to Protein Analysis and Interactions** 249
Reuven Gordon

9 Towards Single-Molecule Chiral Sensing and Separation 271
Riley Sanders, Yaoran Liu, and Yuebing Zheng

Part IV Nanopores

10 Experimental Approaches to Solid-State Nanopores 297
Y. M. Nuwan D. Y. Bandara, Jugal Saharia, Min Jun Kim, Scott Renkes, and George Alexandrakis

11 Challenges in Protein Sequencing Using 2-D MoS_2 Nanopores 343
Adrien Nicolaï and Patrick Senet

12 Single-Molecule Ionic and Optical Sensing with Nanoapertures 367
Wayne Yang and Cees Dekker

13 Self-induced Back-Action Actuated Nanopore Electrophoresis (SANE) Sensing 389
Scott Renkes, Sai Santosh Sasank Peri, Muhammad Usman Raza, Jon Weidanz, Min Jun Kim, and George Alexandrakis

Index 413

Editors and Contributors

About the Editors

Warwick Bowen earned a Ph.D. in experimental physics from the Australian National University in 2004 for his work on quantum information technologies. He is currently a professor of physics at the University of Queensland. His research interests centre on the development of precise measurement techniques based on quantum science, and their application in areas such as biological imaging, precision metrology and advanced sensing.

Frank Vollmer obtained his Ph.D. in 'Physics & Biology' from the Rockefeller University in NYC, USA, in 2004. He was Rowland Fellow at Harvard University from 2004 to 2009, Scholar-in-Residence at the Wyss Institute at Harvard, Group Leader (untenured Associate Professor) at the Max Planck Institute for the Science of Light in Germany, and Instructor in Medicine at Brigham and Women's Hospital/Harvard Medical School where he directed a satellite laboratory. Since 2016 he is a Professor in Biophysics at the School of Physics, University of Exeter, UK. His research centres on biosensing with optical microcavities.

Reuven Gordon earned his Ph.D. in Physics from Cambridge (UK) in 2003. He holds a Professor position at the University of Victoria (Canada) and he is the former Canada Research Chair of Nanoplasmonics (10-year term). He is a Fellow of the Optical Society of America, the Society of Photo-Optical Instrumentation Engineers and the Institute of Electrical and Electronics Engineers. His research interests are in the area of nanostructured metals for enhanced light-matter interactions.

Contributors

George Alexandrakis Department of Bioengineering, University of Texas at Arlington, Arlington, TX, USA

Y. M. Nuwan D. Y. Bandara Research School of Physics, Australian National University, Canberra, Australia

Larnii Booth School of Mathematics and Physics, The University of Queensland, Brisbane, Australia

Warwick P. Bowen School of Mathematics and Physics, The University of Queensland, Brisbane, Australia

Cees Dekker Kavli Institute of Nanoscience, Delft University of Technology, Delft, The Netherlands

Joshua B. Edel Imperial College London, London, UK

Christopher T. Ertsgaard University of Minnesota, Minneapolis, USA

Reuven Gordon University of Victoria, Victoria, BC, Canada

Stephen Hughes Department of Physics, Engineering Physics and Astronomy, Queen's University, Kingston, ON, Canada

Wei C. Jiang Science and Technology Division, Corning Research and Development Corporation, Corning, NY, USA

Kulathunga Mudalige Kalani Perera Living Systems Institute, University of Exeter, Exeter, UK

Mohsen Kamandar Dezfouli Advanced Electronics and Photonics Research Centre, National Research Council Canada, Ottawa, ON, Canada

Min Jun Kim Department of Mechanical Engineering, Southern Methodist University, Dallas, TX, USA

Joshua Kolbow University of Minnesota, Minneapolis, MN, USA

Qiang Lin Institute of Optics, University of Rochester, Rochester, NY, USA; Department of Electrical and Computer Engineering, University of Rochester, Rochester, NY, USA

Nathan Lindquist Bethel University, Minneapolis, MN, USA

Wenjing Liu State Key Laboratory for Mesoscopic Physics and Frontiers Science Center for Nano-optoelectronics, School of Physics, Peking University, Beijing, China

Yaoran Liu Department of Electrical and Computer Engineering, The University of Texas at Austin, Austin, USA

Tao Lu Department of ECE, University of Victoria, Victoria, Canada

Igor Marinkovic School of Mathematics and Physics, The University of Queensland, Brisbane, Australia

Nicolas P. Mauranyapin School of Mathematics and Physics, The University of Queensland, Brisbane, Australia

Adrien Nicolaï Laboratoire Interdisciplinaire Carnot de Bourgogne, UMR 6303 CNRS-Université de Bourgogne Franche-Comté, Dijon Cedex, France

Sang-Hyun Oh University of Minnesota, Minneapolis, MN, USA

Binoy Paulose Nadappuram Imperial College London, London, UK

Srikanth Pedireddy Living Systems Institute, University of Exeter, Exeter, UK

Sai Santosh Sasank Peri Department of Electrical Engineering, University of Texas at Arlington, Arlington, TX, USA

Muhammad Usman Raza Department of Electrical Engineering, University of Texas at Arlington, Arlington, TX, USA

Scott Renkes Department of Bioengineering, University of Texas at Arlington, Arlington, TX, USA

Jugal Saharia Department of Mechanical Engineerings, Southern Methodist University, Dallas, TX, USA

Riley Sanders Walker Department of Mechanical Engineering, The University of Texas at Austin, Austin, USA

Patrick Senet Laboratoire Interdisciplinaire Carnot de Bourgogne, UMR 6303 CNRS-Université de Bourgogne Franche-Comté, Dijon Cedex, France

Sivaraman Subramanian Living Systems Institute, University of Exeter, Exeter, UK

Shui-Jing Tang Collaborative Innovation Center of Extreme Optics, Taiyuan, Shanxi, China

Frank Vollmer Living Systems Institute, University of Exeter, Exeter, UK

Jon Weidanz Department of Biology, University of Texas at Arlington, Arlington, TX, USA

Yun-Feng Xiao Peking University Yangtze Delta Institute of Optoelectronics, Nantong, Jiangsu, China

Wayne Yang Kavli Institute of Nanoscience, Delft University of Technology, Delft, The Netherlands

Wenyan Yu Department of ECE, University of Victoria, Victoria, Canada

Yuebing Zheng Walker Department of Mechanical Engineering, Texas Materials Institute, Department of Electrical and Computer Engineering, and Department of Biomedical Engineering, The University of Texas at Austin, Austin, USA

Part I
Optical Sensing

In optical sensing, one detects the interaction of light with a molecule. The interaction with a single molecule is a relatively weak optical interaction because most molecules and biomolecules are less than 10 nm in diameter and are essentially non-absorbing at the probing wavelength. The following three chapters discuss the emerging and most exciting optical techniques that are being developed to detect the weak optical interaction of a single molecule.

Chapter 1 discusses the interferometric single-molecule sensing techniques which make use of the polarisability of molecules and biomolecules for detection. The interferometric phase shift of the light scattered from a biomolecule in close proximity to the partially reflecting surface of a flat microscope cover-glass is detected in interferometric scattering microscopy, iSCAT. iSCAT achieves contrasts on the order of 0.4%/MDa of molecular mass for illumination in the near-ultraviolet. It is not only possible to localise larger proteins on the iSCAT microscope but also to estimate their molecular mass in single-molecule mass photometry. Other important single-molecule interferometric sensing approaches modulate the input light to increase the sensitivity in heterodyne detection. An optical nanofiber is used to collect the light scattered from a single molecule located in the proximity of the fibre which is immersed in an aqueous solution. The probing light is focused onto the molecule located at a sufficient distance from the nanofiber to reduce the background scattering and then modulated, to detect the molecule from a beat note signal that results from interfering the scattered light with a frequency-shifted local oscillator. The intensity of the beat note is proportional to the scattered light intensity. This heterodyne detection scheme achieves a high single-molecule sensitivity at the shot-noise limit. Single bovine serum albumin protein (BSA) which has a molecular mass of approx. 66 kDa was detected at a time resolution of up to 200 Hz, sufficiently fast to follow the varying amplitude of the beat note signal of the slowly diffusing protein. Chapter 1 provides a scholarly introduction to the various homodyne and heterodyne interferometric biosensing techniques, an in-depth analysis of their noise sources and detection limits, and exciting examples for their applications in single-molecule biosensing.

Chapter 2 discusses optical microcavities that enhance the interaction of light with a molecule. Molecules are placed within the evanescent field of the whispering gallery mode (WGM) microcavity. The WGM microcavity is an approx. 100 um diameter glass microbead that confines light by near-total internal reflection. After water immersion, the glass microbead confines the light for prolonged time achieving a high Q factor of 10^{6-7} and a Finesse of approx. 10^4 in the visible to near-infrared. Approximately 300 femtometers of effective path-length change can be detected with the WGM glass micro-interferometer, by tracking the WGM resonance wavelength shifts. The sensitivity is sufficient to detect single Influenza A, Adenovirus, and SARS-CoV2 Coronavirus particles which are all about 100 nm in diameter and have a refractive index of approximately n=1.5. Even though the virus particles' refractive index contrast is reduced in water and they interact with only about 4% of the circulating light intensity via the evanescent tail of the WGM, the single virions and their Brownian motions are detected with a good signal-to-noise ratio >10.

In single-molecule detection, the molecule is best placed at the location of the highest WGM intensity, thereby experiencing the smallest mode volume V and maximal Q/V for a high detection sensitivity. The optoplasmonic WGM sensors achieve exactly this, by attaching a plasmonic nanoparticle such as a gold nanorod at the surface of the glass microsphere, where the WGM excites the localised plasmon resonance (LSPR) of the nanorod. The plasmonic near-field enhancement at the tip of the gold nanorod is on the order of 10^3. Different from the all-dielectric WGM microcavity, the highest field intensity of the optoplasmonic WGM is now located outside of the glass microsphere, at the tips of the gold nanorods which are immersed in aqueous solution. The gold nanorods provide the binding sites for various chemical groups (thiols, amines). Alternatively, the gold surface can be chemically modified with ligand molecules to provide specific binding and interaction sites for detecting molecules in solution. The ohmic losses of the metal nanoparticles introduce an extra decay channel for the intracavity photons, slightly degrading the quality Q factor by a factor of 1.3–1.5. Nevertheless, the large intensity enhancement factor due to the LSPR and the corresponding suppression of the mode volume greatly enhance the optoplasmonic sensor's sensitivity, practically by a factor of 1000–10000 as compared to the all-dielectric WGM sensor. The optoplasmonic WGM sensors have detected small organic molecules such as cysteamine (77 Da), amino acids such as cysteine (121 Da) and Glycine (75 Da), neurotransmitters such as GABA (103 Da) and dopamine (153 Da), and even single atomic ions such as Zn^{2+} (65 Da), all in aqueous solution. For larger molecules such as proteins, the WGM resonance shift signal further depends on the overlap of the protein molecule with the LSPR hotspots which have decay lengths on the order of 10 nm or less. Fortuitously, the LSPR decay lengths approximately match the size of a typical protein such as BSA, thereby creating a partial near-field overlap with the protein when it is attached to the nanorod. Conformational (shape) changes of the immobilised protein are detected at a microsecond time resolution. The corresponding intra-protein distance changes of a few Angstroms would be difficult to detect with the fluorescence-based rulers such as fluorescence-resonance energy transfer (FRET).

Chapter 3 discusses the nonlinear interactions of light with the analyte molecules which are enhanced by the high Q optical microcavities. The high Q factor and Finesse of the microcavity are the important metrics for the generation of the new frequencies of light. The intensity of the light that is generated by a nonlinear optical process is typically many orders of magnitude less as compared to the intensity of the excitation beam. However, the nonlinear processes generate new optical signals at well-defined frequencies away from the frequency of the excitation beam, and this greatly aides the detection of the nonlinear optical interaction, possibly down to single molecules. In nonlinear (single) molecule sensing it is the analyte molecules that generate the nonlinear optical signals. Because of their high Q factors and Finesses, optical microcavities such as the WGM microbead or microtoroid give access to a host of nonlinear signals, the generation of which would otherwise require much higher optical powers. The WGM glass microspheres and toroids have been used to generate Raman and Brillouin Scattering, Sum Frequency and Harmonic Generation, Four-Wave Mixing, and the generation of Optical Frequency Combs. In nanoparticle and molecule sensing, WGM microcavities have emerged as a powerful platform technology to investigate the Raman scattering of single nanoparticles placed in the evanescent field of the WGM. In molecule sensing, the Raman scattering of Rhodamine 6G molecules has been detected, for a layer of the dye coated onto the WGM cavity. In combination with plasmonic and other near-field enhancers, the surface-enhanced Raman sensing (SERS) on WGM-type cavities shows great promise for molecule detection, as it was demonstrated with R6G on the WGM-type liquid core optical ring resonators (LCORR). The stimulated Raman scattering and lasing have been demonstrated with organic molecules coated onto the WGM microcavities. Stimulated scattering is a great way to enhance molecular signals. Already, Raman lasers have demonstrated exceedingly high sensitivity for the detection of nanoparticles down to about 12 nm diameter. Obtaining the Raman spectrum of molecules is a particularly powerful approach for identifying the molecule. Developing the spectroscopic methods for identifying the analyte molecules on WGM sensors is a very important goal. Otherwise, the specific detection of molecules would have to rely on detecting receptor molecules and their specific interactions with the analyte molecule. Surface-enhanced sensing and Sum Frequency generation are other important approaches for developing powerful and specific molecular sensors. In precision sensing, the well-known dual-comb spectroscopy is already a powerful tool in the absorption spectroscopy of gases. The frequency comb spectroscopy can be realised on optical microcavities, ingeniously taking advantage of the Four-Wave Mixing in the silica or silicon nitride cavity which can generate the octave-spanning combs. The frequency comb techniques promise to become a powerful approach for precision sensing of molecules and biomolecules on optical microcavities. Chapter 3 gives a comprehensive and exciting overview of all of the nonlinear sensing techniques that are emerging for the optical microcavities and discusses the potential emerging applications in biosensing and single-molecule sensing.

Chapter 1
Interferometric Biosensing

Nicolas P. Mauranyapin, Larnii Booth, Igor Marinkovic, and Warwick P. Bowen

Abstract Interferometric biosensors characterise biological specimens via the effects they have on the interference between two optical fields. Generally, one field is arranged to interact with the specimen, perhaps propagating through it and incurring a phase shift, or scattering from it. Interference with the second field can both allow phase shifts to be resolved, and amplify weak scattering signals to measurable levels. While widely used in biological imaging, recent experiments have shown that interferometric sensors are also capable of probing the size and dynamics of single protein molecules. This chapter reviews this recent progress, outlines the concepts and techniques involved, and discusses areas in which such single-molecule interferometric biosensors have been applied.

1.1 Introduction

Interferometric techniques are widely used to sense and image biological specimens. In general, they rely on relative phase shifts introduced by the specimen between two optical beam paths and the subsequent interference of the light in the two paths. This is shown conceptually in Fig. 1.1. The phase shifts are introduced by the elastic dipole interaction between the specimen and the light [1]. For bulk media this interaction is responsible for the refractive index of the media, which both introduces phase shifts on the transmitted field and causes some fraction of the field to be reflected. For sub-wavelength particles such as biomolecules, or for specimens with sub-wavelength features, the interaction causes dipole scattering.

The most commonly used interferometric bio-imaging techniques are various forms of interferometric microscopy, including phase contrast microscopy [2] and differential interference contrast microscopy [3]. In the former of these methods,

N. P. Mauranyapin (✉) · L. Booth · I. Marinkovic · W. P. Bowen
School of Mathematics and Physics, The University of Queensland, Brisbane, Australia
e-mail: n.mauranyapin@uq.edu.au

W. P. Bowen
e-mail: w.bowen@uq.edu.au

W. Bowen et al. (eds.), *Single Molecule Sensing Beyond Fluorescence*,
Nanostructure Science and Technology,
https://doi.org/10.1007/978-3-030-90339-8_1

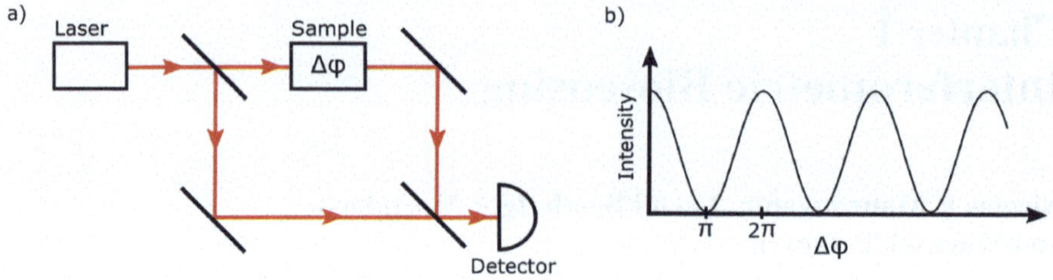

Fig. 1.1 Mach-Zehnder interferometer. Light is split into two arms by a beamsplitter. One arm is phase shifted by $\Delta\phi$. When the two arms are recombined, they constructively or destructively interfere depending on the phase difference, the intensity of the light at the output therefore depends on $\Delta\phi$

shown in Fig. 1.2, a mask is used to reduce the range of angles for which the specimen is illuminated. Light scattered from the specimen is not constrained to this narrow range of angles, introducing a second beam path. The interference between then two beam paths is arranged to be constructive using a phase shifting ring to shift the phase of the transmitted illumination field. Differential interference contrast microscopy works similarly, but uses two spatially displaced paths with orthogonal polarisations rather than controlling the range of illumination angles. Light interference is also used, for example, in optical coherence tomography [4], where low-coherence light is used in a white-light interferometer configuration; in refractive index sensing of biofluids, often enhanced by plasmonic resonances [5, 6]; and in a variety of other microscopy applications such as synthetic aperture microscopy in three dimensions [7].

All of the above mentioned techniques are limited in resolution by the Abbe diffraction limit, and used predominantly to image or sense extended biological specimens, rather than to observe or characterise single molecules. More recent developments have seen the same principles of interferometric biosensing applied to single molecules, as described in this chapter. Since the elastic dipole interaction is energy conserving, it has a high scattering cross section relative to other nonlinear scattering mechanisms such as Raman scattering or two-photon scattering [8]. This offers the prospect of high signal-to-noise and measurement speed while using low optical intensities, and allows for operation without labels or contrast agents. A significant challenge presented by the same energy conversing nature of the interaction is that background scattered light has the same wavelength as the light scattered from the specimen. It is therefore generally not possible to achieve the high specificity available to other techniques such as fluorescence and Raman scattering.

Even with the limitations presented by non-specific background scatter, we aim to convince you in this chapter that interferometric single molecule sensing has an important role to play in noninvasively observing single molecule systems and processes, and has the potential to shed new light on the dynamics and function of nanoscale systems. The ability to avoid labels is an important advantage, since they have been known to perturb function [9], and in some situations cannot be

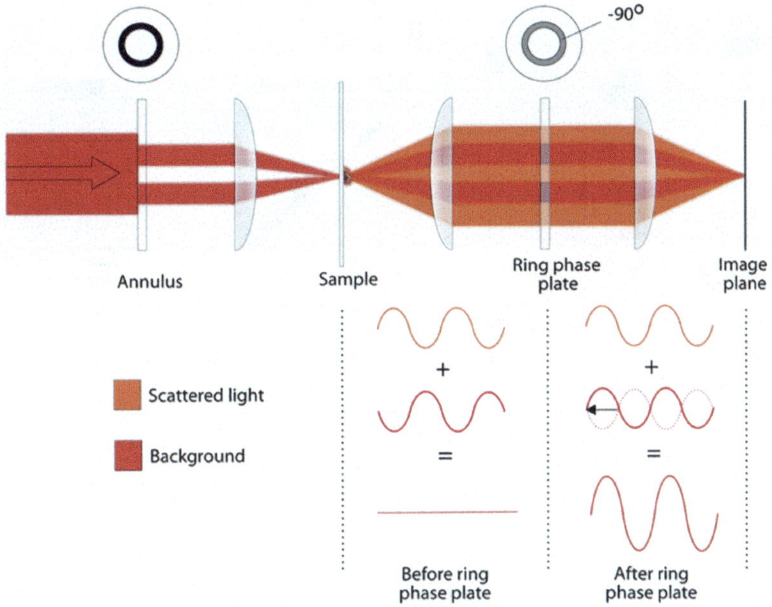

Fig. 1.2 Phase contrast microscope. Input light (red) is shone through an annulus and focused on the sample. The input light is phase shifted with a ring phase plate, making it constructively interfere with the scattered light (orange), enhancing the signal from the sample

easily introduced. Perhaps more importantly, much of the dynamics of biological machines occurs at kilohertz to gigahertz speeds. For instance, the myosin motors responsible for muscle cell contraction and nutrient transport along actin filaments can have stepping frequencies up to a few kilohertz [10], enzymes such as ATPase and carbonic anhydrase have frequencies ranging from hundreds-of-hertz to a megahertz [11, 12], while biomolecular conformational changes can occur over picosecond timescales [13]. Conventional fluorescence-based single-molecule microscopy techniques such as STORM (Stochastic Optical Reconstruction Microscopy), STED (Stimulated Emission Depletion Microscopy), FRET (Fluorescence Resonant Energy Transfer microscopy), and PALM (Photo Activated Localization Microscopy) struggle to reach these timescales and frequencies, since the quality fluorescent labels they use typically emit only around ten thousand photons per second [14, 15]. By contrast, it is becoming clear that these frequencies and timescales may be obtainable using dipole scattering, providing a new set of eyes to look upon many important nanoscale processes.

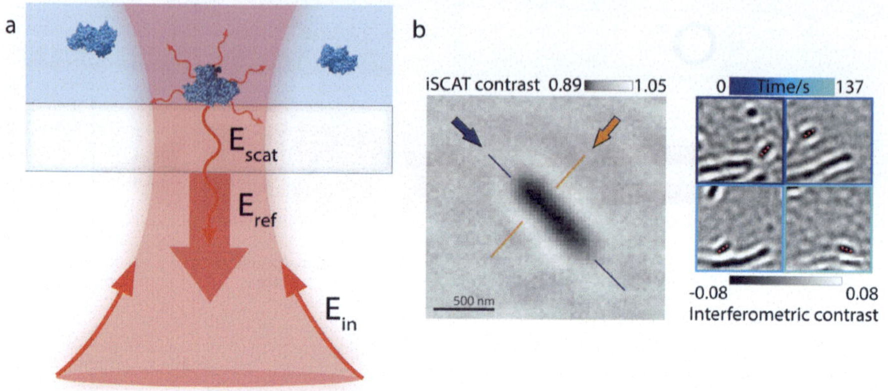

Fig. 1.3 **a** iSCAT (Interferometric scattering microscopy), excitation light E_{in} is partially reflected(E_{ref}) and interferes with field E_{scat} scattered of a molecule, producing visible contrast even for single molecules. **b** Examples of single molecule imaging using iSCAT. Left: Label-free imaging of microtubule with 8 nm resolution. Right: Actin motion tracked over 137 s. Figures adapted from Ref. [18] with permission and [19] under Creative Commons CC-BY license

1.2 Single Molecule Sensing

As discussed above, interferometric single molecule sensing relies on the scattering interaction between the molecule and light. For typical molecular size-scales, much smaller than the optical wavelength, the scattering is well approximated by dipole scattering (see details in Sect. 1.3.1). This scattering is detected by interference with a reference optical field. Two broad strategies are generally adopted in interferometric single molecule sensing. The first uses a reference field of the same frequency as the scattered field. This is commonly referred to as *homodyne detection*. The second uses a shifted reference frequency, and is termed *heterodyne detection*. These approaches each have there own advantages, the primary ones being that homodyne is more naturally compatible with imaging on a camera, and heterodyne is less sensitive to environmental and laser noise (see Sect. 1.3.2.2). In this section we introduce two important example implementations of each strategy, *interferometric SCATTering microscopy* (iSCAT) [16] and *dark-field heterodyne biosensing* [17]. These two techniques are illustrated in Figs. 1.3 and 1.4.

1.2.1 Interferometric Scattering Microscopy (iSCAT)

In iSCAT a biomolecule solution is prepared on a microscope slide. When the sample is then illuminated through the microscope, light scatters both from biomolecules and from the interface between the microscope slide and the solution in which the biomolecules are contained (see Fig. 1.3a). The iSCAT microscope images the interference between these two fields on a camera. As discussed further in Sect. 1.3.2.2, the few-micron-scale pathlength difference between the two fields results in a highly stable interference pattern. Even so, the variations in measured intensity across the camera are typically much larger than the signal from a biomolecule. To overcome this, a difference imaging technique is employed, monitoring the change in intensity distribution when a biomolecule arrives close to the microscope slide. As can been seen in Fig. 1.3b, this allows single biomolecules such as bovine serum albumin (BSA, Stokes radius 3.5 nm) to be resolved and to be distinguished from aggregates [16]. iSCAT has been shown to allow imaging and tracking single proteins [20], as well as observation of the motion of single protein filaments such as actin [19] (Fig. 1.3b). It also allows measurements of single molecule structures such as the length of microtubules with sub 5 nm precision [21].

Single molecule mass photometry is an important application of iSCAT. It has been shown experimentally that, to good approximation, the amplitude of scattered light from protein molecules is linearly proportional to the molecular mass [16, 22]. The contrast of the interference pattern collected in iSCAT can then be directly used to determine mass. This method has been used to determine the mass of single proteins with a mass resolution of 19 kDa and a precision of 1 kDa [23]. The ability to nondestructively determine the mass of single molecules is a useful tool for a range of applications, discussed further in Sect. 1.4.

1.2.2 Dark-Field Heterodyne Biosensing

In dark-field heterodyne biosensing, the light scattered from a single biomolecule in solution is collected using a high numerical aperture lens or an optical near-field collection device. Figure 1.4 shows the latter scenario, where an optical nanofibre is used for collection [17]. Optical nanofibres are optical fibres that have been drawn down to a few hundred nanometre radius, so that their optical mode extends into the medium around the fibre. This allows them to efficiently collect dipole scattered light. The collected field is interfered with light at a shifted frequency—typically shifted by a few tens of megahertz using an acousto-optic modulator—and then detected using a high bandwidth photodetector. The interference creates a beat-note on the detected photocurrent at the difference frequency between the two fields. The intensity of this beat note is proportional to the scattered light intensity, providing information about the presence of a biomolecule, its size, and its position. Any light the leaks into the fibre from the probe field used to illuminate the biomolecule contributes background to the measurement. A dark-field illumination strategy is employed to minimise this leakage. In the case of a nanofibre, a dark field configuration is achieved by positioning the fibre so that the direction of propagation of the probe light is perpendicular to the fibre axis.

Reference [17] showed that shot noise limited performance was possible in dark-field heterodyne biosensors (discussed further in Sect. 1.3.2.2), and showed that this allowed the detection of single BSA molecules with 100 Hz measurement bandwidth

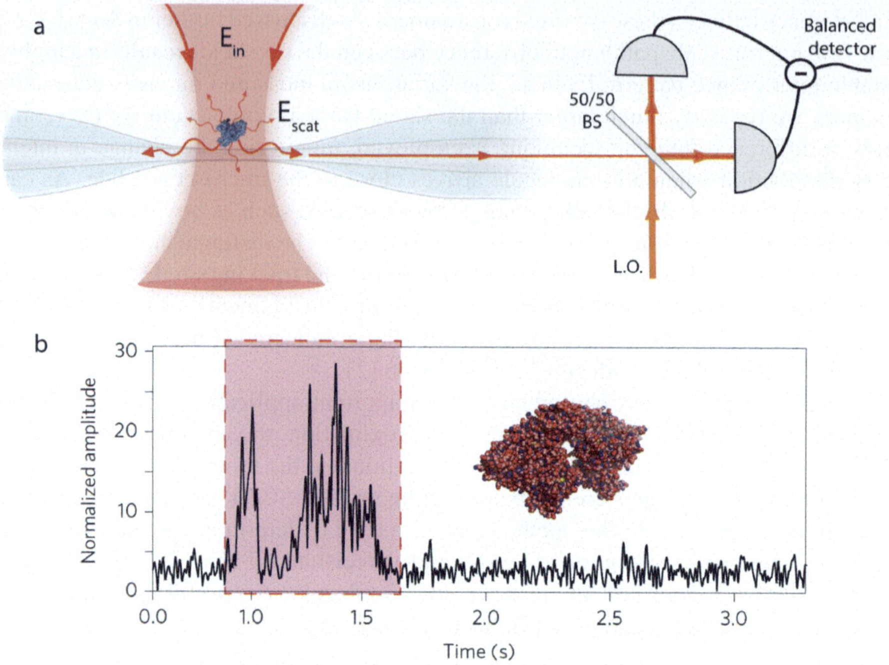

Fig. 1.4 **a** Evanescent optical nanofiber, where light scattered from the particle (E_{scat}) is evanescently coupled into the nanofibre. The collected scattered signal is interfered with a frequency shifted local oscillator and collected on a balanced detector, **b** Time trace of the detection of a single BSA molecule, with a detection event highlighted. Figure adapted from Ref. [17] with permission

using probe intensities four orders of magnitude below the photodamage threshold for biological samples (see Fig. 1.4b) [17]. Reference [17] further showed that the motion of these biomolecules within the near-field of the sensor could be track with a resolution down to 1 nm with a bandwidth 100 Hz.

In another example, Mitra et al. used dark-field heterodyne to detect viruses and impurities inside a nanofluidic channel (see Fig. 1.5) [24]. In this case, the probe beam was shaped into an annulus (see Fig. 1.5) using a diffractive element, so that only the scattered light could pass through a filter prior to photodetection. The probe beam was further suppressed by using a back-scattering configuration, where only the scattered light back-scattered in the opposite direction to the probe field is detected. The scattered light was combined with a frequency shifted reference field before photodetection. Reference [24] showed that their sensor was capable of detecting single human and bacterial viruses as small as 24 nm in radius, and 20–30 nm sized impurities in virus samples [24].

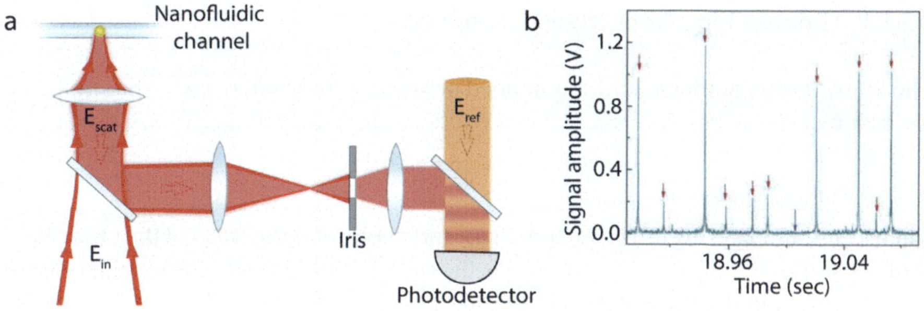

Fig. 1.5 Mitra et al. [24] use dark field heterodyne to image single molecules flowing inside a nanofluidic channel. **a** Input light (E_{in}) is filtered from the scattered light with an iris and the scattered light is combined with a frequency shifted reference beam. **b** Detection events for HIV AT2WT wild type virus (smaller spikes) and 75 nm polystyrene beads (larger spikes). Figure adapted from Ref. [24] with permission

1.3 Signal to Noise

The signal to noise (or SNR) is a key parameter for all biosensors, determining the smallest resolvable signals in a given measurement time. For a signal n, the SNR can be generally defined as

$$SNR \equiv \frac{\langle n \rangle^2}{Var[n]}, \tag{1.1}$$

where $\langle n \rangle$ and $Var[n]$ represent the expected value of the signal and its variance. A great deal of effort has been made during the past decade to increase the SNR of single molecule biosensor in a variety of ways, as also discussed in many other chapters of this book, since improved SNR promises to allow the observation of new biological phenomena. In this section of this chapter we will examine the signal magnitude available to interferometric single molecule biosensors, and the variety of noise sources present in such measurements.

1.3.1 Signal Strength

In interferometric single molecule sensing, the molecule is illuminated with an optical probe field as shown in Figs. 1.3 and 1.4. In the typical case where it is smaller than the wavelength of the probe field, it will be polarised and form an electric dipole that oscillates at the same frequency as the probe field. This dipole emits optical radiations as a scattered field, which is the signal that biosensor aims to detect.

1.3.1.1 Photon Flux from Dipole Scattering

The mean photon number $\langle n_{scat} \rangle$ contained in the scattered field from the biomolecule is given by

$$\langle n_{scat} \rangle = \frac{\sigma}{4\pi w^2} \langle n_{in} \rangle \tag{1.2}$$

where σ is the scattering cross-section of the molecule, w is the waist of the probe field at the position of the particle, and $\langle n_{in} \rangle$ is the mean number of photons incident from the probe field in the measurement duration [17, 25]. In many cases it is reasonable to approximate the biomolecule as an isotropic sphere of radius r (see e.g. [17, 26, 27]), in which the scattering cross-section can be expressed as:

$$\sigma = \frac{8\pi k^4 r^6}{3} \left(\frac{m^2 - 1}{m^2 + 2} \right)^2, \tag{1.3}$$

where k is the norm of the probe field wave vector, and $m = n_p / n_m$ is the ratio of the refractive indices of the molecule and the surrounding medium [25]. For a protein molecule, the bulk refractive index is generally in the range off 1.35–1.45.

From Eqs. (1.2) and (1.3) one can calculate that with a beam waist of 1 μm and a probe wavelength of $\lambda = 2\pi / k = 780$ nm, a 10 nm silica nano-sphere ($n_p = 1.45$) in water ($n_m = 1.33$) will scatter one photon for every $\sim 10^{11}$ incident probe photons. For a BSA molecule approximated as a uniform sphere of refractive index 1.45 and radius equal to the Stokes radius of BSA (3.5 nm), only two photons will be scattered every $\sim 10^{14}$ incident probe photons. While this level of scattering is relatively low, it compares favourably to most inelastic scattering processes.

The magnitude of scattering ultimately determines the maximum rate at which a biomolecule can be measured, since at least one photon must be scattered in the measurement time. As an example, let us consider again a BSA molecule illuminated with a probe beam of 1 μm radius and wavelength $\lambda = 2\pi / k = 780$ nm. If 1 mW of probe power is used, corresponding to $\sim 10^{16}$ probe photons per second, $\sim$200 photons are scattered per second. Assuming that every one of these photons is detected and all noise sources (apart from the shot noise introduced by the very quantisation of light) are removed, the maximum possible measurement rate in this scenario is therefore 200 Hz. This is fast enough to capture the slow diffusion and conformational changes of biomolecules.

1.3.1.2 Atomistic Model of Molecular Polarizability

The approximations that biomolecules are uniform spheres as performed above, or that they are simple arrangements of uniform blocks with bulk properties, are commonly used in biosensing [17, 26–28]. However, for sufficiently small molecules it can be expected that their atomistic nature will play an important role in determining how they polarize and therefore their scattering cross-section. It is especially

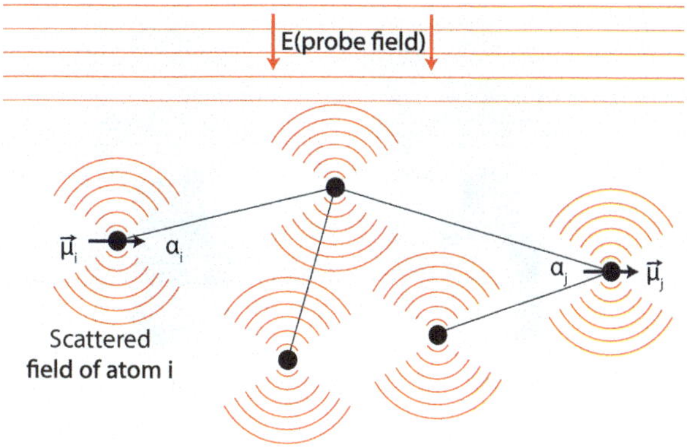

Fig. 1.6 Scattering of a five atoms (black spheres) molecule based on the atomistic model of molecular polarizability. A probe field E illuminates the atoms from above, polarising them. Each polarized atom emits a dipole field which induces its own polarization on the other atoms. In this way the total scattering depends on the arrangement of atoms

important to understand how the shifting coordinates of atoms in a molecule affect their scattering when the aim is to observe conformational changes and other motional dynamics.

Booth et al. [29] develop a method to perform atomistic modelling for scattering-based single-molecule biosensors. Following earlier proposals by [30, 31], the method considers that at any given time the molecule is a rigid ensemble of atoms. Each atom has its own atomic polarizability, therefore, when illuminated by the probe field it forms an induced dipole and create its own scattered field (as was the case in the previous section). The scattered field of each atom contributes to the total scattering of the molecule. However it also changes the field experienced by neighbour atoms (see Fig. 1.6). The total induced dipole moment μ_i of each atom i therefore depends both on its own atomic polarizability α_i and on the induced dipole moments of every other atom μ_j, which themselves depend on the atomic polarizability of atom i. Booth et al. [29] show that it is possible to account for these couplings in a computationally efficient way, allowing the excess polarizability of α_{ex} of molecules that are as large as 1600 kDa (110.000 atoms) to be calculated from their atomic structure. The excess polarizability can be related to the scattering cross-section by

$$\overline{\alpha_{ex}} = \left(\frac{6\pi n_m^4}{k^4} \sigma \right)^{1/2} . \tag{1.4}$$

The coordinates of the atoms in a biomolecule can be calculated as a function of time using molecular dynamics simulations. The approach of Booth et al. [29] then provides straightforward predictions of the magnitude of scattered light, and how it changes over time. For example, Booth et al. used molecular dynamic simulation of BSA to extract the variation of its polarizability due to thermal fluctuations.

Molecules that exhibit different conformations have been studied with single molecule biosensors [28]. Kim et al. observed, for the first time without label, the dynamics of DNA polymerase as it changes conformation using a plasmonically enhanced whispering gallery mode microcavity sensors (see Chap. 2). Prediction of

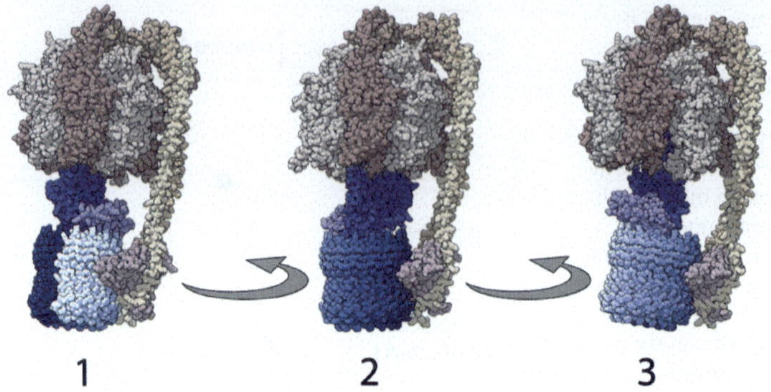

Fig. 1.7 Three conformational steps taken by a Chloroplast F1F0 ATPase during the decomposition of ATP into ADP. Structures were determined in Ref. [32] by cryo-EM and are generated from the PDB files 6FKF, 6FKH, and 6FKI using UCSF Chimera

the signal was computed using the bulk model as DNA polymerase executes large motion within the non-uniform electric field of the sensor. In this case, the dominant modulation of the signal does not originate from changes in polarizability but rather from the DNA polymerase moving from a high intensity probing field region to a low intensity region. However, many molecules undergo rotation and smaller-scale conformational rearrangements, for which such model would predict no signal. The approach of Booth et al. allows the signal due to these effects to be predicted. For instance, in the case of Chloroplast F1F0 ATPase shown in Fig. 1.7, rotation is predicted to result in a significant change in polarizability. To reach this conclusion, Booth et al. compare the optically-induced polarization of each conformation that a F1F0 ATPase molecule takes during the decomposition of adenosine triphosphate (ATP) into adenosine diphosphate (ADP). The atomic structures of these conformations have been determined by cryogenic electron microscopy [32] and are available from the Protein Data Bank (PDB). Using them, Booth et al. predict that the polarizability should change by as much as 350 Å^3 (4.3%) between conformations. This signal is within reach of interferometric single molecule sensors, as well as other single molecule sensors such as the optoplasmonic sensors discussed in Chap. 2.

1.3.1.3 Methods to Enhance the Signal Strength

As seen in Eq. 1.2, the scattered photon flux is proportional to the mean incident photon flux divided by the waist of the incident beam. It can therefore be increased simply by increasing the mean intensity of the incident field at the biomolecule. The benefit of this approach is constrained, however, by photodamage to the specimen which can occur at the high intensities using in single molecule biosensing experiments. This photodamage includes photochemical effects, local heating and physical damage and can affect growth, viability and function [33–36].

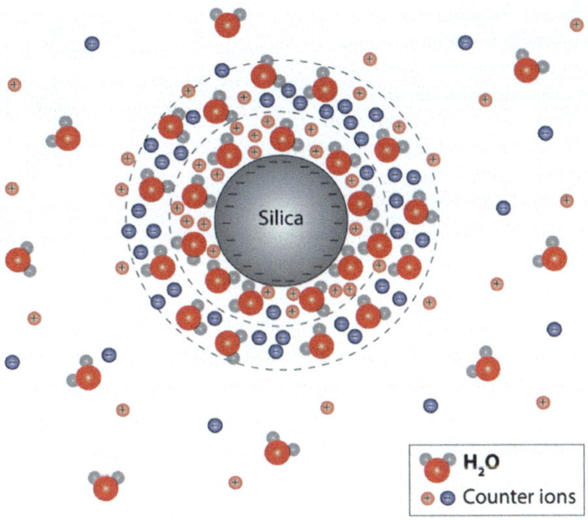

Fig. 1.8 Schematic of the electric double layer formation around a silica nanosphere

To further increase the scattering of biomolecules, their scattering cross-section can be enhanced. This can be done,for example, by attaching a label to the molecule [34], by using fluorescence or by changing the surrounding medium to increase the refractive index contrast. A disadvantage of this approach is that it negates the benefit of observing the molecule, and its behaviour, in its natural state [37–43].

An alternative form of scattering enhancement can be achieved directly from the interaction of the molecule with ions in the surrounding fluid. When particles are immersed in aqueous solution, surface-charges assemble at the liquid-solid interface of the molecule. Counter ions from the solution are attracted to screen these surface charges (see Fig. 1.8), creating an electric double layer around the particle [44]. The electric double layer alters the refractive index of a film of water close to the molecule. Since the area of this shell of altered refractive index scales with the radius of the molecule squared, while the volume of the molecule scales with radius cubed, for sufficiently small molecules the electric double layer will scatter more light than the molecule itself. In this way, the molecule interacts with light much more strongly than one would naively expect from dipole scattering [17, 45, 46]. The thickness of the electric double layer can be determined from the Debye length, which is inversely proportional to the ionic strength of the solution. Therefore, perhaps counter-intuitively, the less ions there is in the solution, the thicker the electric double layer will be and the greater the enhancement of scattering.

The increased interaction due to the electric double layer was first observed in unexpectedly large trapping forces for large particles such as 200 nm micelles in Ref. [45]. Trapping of micelles with optical tweezers was possible only at low ion concentration for which the electric double layer was large enough to contribute to the scattering cross-section of the micelles. The enhanced scattering due to the increased interaction was first confirmed using dark-field heterodyne sensor [17]. Detection of

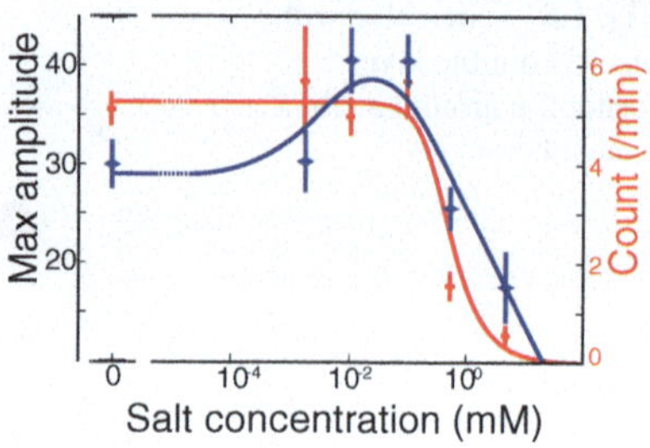

Fig. 1.9 Signal strength of nanoparticles detection and detection count as function of concentration of salt in the buffer solution observed with darkfield heterodyne sensor. Figure adapted from Ref. [17] with permissions

nanoparticles with radii as small as 5 nm and biomolecules with radii of 3.5 nm was found to only be possible when the salt concentration in the fluid was below a few millimoles, as shown in Fig. 1.9. Above this concentration, only particles larger than 25 nm were observable. The experiments demonstrated an absolute enhancement of photon flux by three orders of magnitude.

The signal strength can also be increased using an optical cavity. Single molecule measurements using shifts induced on the frequency of a cavity and using nonlinear effects in a cavity are discussed in Chaps. 2 and 3 respectively. The optical scattering from the biomolecule can also be increased via the Purcell effect that a cavity introduces [47–49]. In this case, the enhancement occurs because scattering from the molecule that is reflected by a cavity mirror and fed back onto the molecule stimulates further scattering.

The concept of Purcell enhancement can be understood from the basic one-dimensional scattering configurations shown in Fig. 1.10. In both configurations the field from the molecule scatters equally to the left and the right. In configuration (a), it is directly detected on both sides. The total photocurrent i_{dir} created is proportional to the norm of the field amplitude squared:

$$i_{dir} \propto |E_{scat}|^2 + |E_{scat}|^2 = 2|E_{scat}|^2 \tag{1.5}$$

with E_{scat} the amplitude of the scattered field. In case (b), a mirror is placed on the right side, reflecting the scattered field propagating in that direction back onto the molecule. This field then interferes with the left-going field. Constructive interference (as occurs on resonance in a cavity) increases the total scattered power, while destructive interference (as occurs off resonance in a cavity) decreases it. In the case of constructive interference we find that the photocurrent i_{mirror} is

$$i_{mirror} \propto |E_{scat} + E_{scat}|^2 = 4|E_{scat}|^2. \tag{1.6}$$

I.e. it is twice as large as the photocurrent in the direct detection configuration.

When using a cavity rather than a single mirror the enhancement can be much greater, due to the multiple reflections each constructively interfering. One can show that due to the presence of the cavity, the dipole scattering power P_{scat} of a particle can be enhanced by [50]:

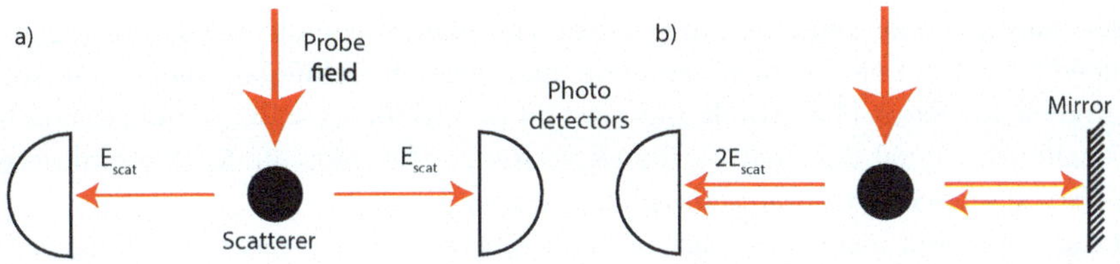

Fig. 1.10 Purcell enhancement from a single mirror in one dimension. **a** One dimension scheme directly detecting the total scattered field going leftward and rightward. **b** When a mirror is used to retroreflect the rightwards going field, constructive interference of the two fields increases the detected power

$$P_{cav} = \frac{3}{4\pi^2}\frac{Q\lambda^3}{V}P_{scat}. \tag{1.7}$$

where Q is the quality factor of the cavity and V is the cavity volume. As can be seen from this expression, the Purcell enhancement is proportional to the ratio of the quality factor and volume of the cavity. For typical high quality optical microcavities the quality factor can reach 10^8 with a $350\,\mu m^3$ mode volume. This leads to a Purcell enhancement of as much as 10^5 at 1550 nm.

1.3.2 *Noise Sources*

As can be seen from the discussion in the previous subsection, the photon flux from typical biomolecules and the change of photon flux due to molecular conformational changes are both typically very small. In order to detect single molecules with integration times of millisecond or less it is therefore crucial to minimise the noise of the sensor. Various forms of noise exist in interferometric sensors, including technical noise and quantum shot noise. We discuss the primary sources in this subsection.

1.3.2.1 Technical Noise

Technical noise in an interferometric single molecule sensor has several origins that can typically be separated in three categories: electronic noise, technical laser noise and optical background noise.

Electronic noise
Electronic noise arises from unwanted motion of electrons in the detection system used to measure the scattered field. Typically, it takes the form of amplified Johnson-Nyquist thermal noise. The most challenging source of electronic noise to minimise is the noise in the photodiode itself, since amplification within the photoreceiver or later electronics increases the signal level so that it can generally be arranged to be

larger than down stream electronic noise. The photodiode electronic noise can be reduced by careful choice of photodiode, and through techniques such a dark-field heterodyne detection [17] which uses a bright optical local oscillator field to amplify the scattering signal detected by the photodiode. The photodiode electronic noise can also be reduced by cooling the photoreceiver.

Line noise and electronic pick-up, can introduce noise tones at characteristic frequencies, though due to their near mono-frequency nature these noise sources are typically not problematic. Conversion of the analog photocurrent to a digital signal (for example using a data acquisition card or oscilloscope) can also introduce noise, both digitisation noise and aliasing of high frequency noise or signals to low frequencies. Digitisation noise can be challenging if a high dynamic range is desired, in which case a data acquisition system with a high bit depth is advisable. Aliasing can be avoided using an appropriate antialiasing filter.

Technical laser noise
Light emitted by lasers inherently contains intensity and frequency noise.

The intensity, or amplitude, noise is any unwanted fluctuations of the field intensity and has a variety of origins. Relaxation noise of the laser medium due to the lasing process is one important source of intensity noise that often occurs at frequencies relevant to biosensors. Intensity fluctuations can also be transferred from the laser pump, and polarization noise due to relative phase fluctuations between two orthogonal polarization components can be converted to amplitude noise by polarization selective elements such as waveguide modulators or polarization beam splitters.

Intensity noise can be directly be detected on a photodetector, and analysed via power spectral analysis, as shown for example in Fig. 1.11a. In this case, the primary amplitude noise arises from laser relaxation oscillation, and features a characteristic plateau at low analysis frequencies, followed by a resonance, and decay as the frequency increases. Considerable efforts have been made to reduce the intensity noise of laser system to increase the sensitivity of precision optical sensors (for example see Refs. [51, 52]). Intensity noise can be suppressed by using a noise eater where typically, fluctuations of the intensity are detected and fed back to an attenuator (see Fig. 1.12) that correct them (see e.g. traces B in Fig. 1.11). The feedback can also be directly applied to the laser itself. These feedback systems can reduce the intensity noise by hundreds of decibel at low frequencies (100 Hz) but their efficiency decreases with frequency. However, as shown in Fig. 1.11, reduction of the relaxation noise by 50 decibels can still be achieved at frequencies above 100 kHz.

The frequency noise comes from drifts in the frequency of the laser, while phase noise is introduced by fluctuations of the optical path length. Since frequency noise also causes phase fluctuations, after the field propagates over a length, frequency and phase noise are often treated together. They can arise due to the laser cavity vibrations, acoustic vibrations of optical elements along the beam path, changes in temperature, and coupling of intensity noise into frequency noise via nonlinearities [53]. The frequency noise is responsible for the laser linewidth broadening. Frequency and phase noise are a major concern for interferometric sensors since the interferometric signals to be detected are intrinsically sensitive to the phase of the

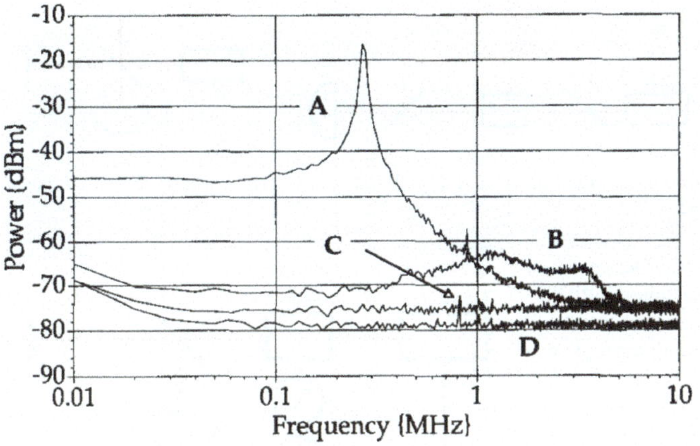

Fig. 1.11 power spectral density of the different noises in an optical system. **a** Sum of all the noises. **b** Intensity noise reduction using a feedback control loop. **c** Sum of the electronic noise and quantum noise. **d** Electronic noise. Figure adapted from Ref. [51] with permission

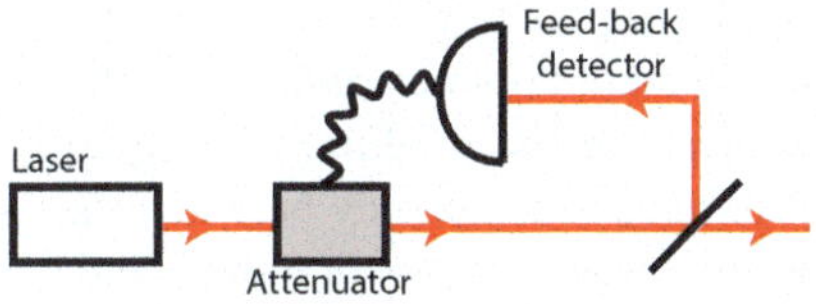

Fig. 1.12 Noise eater. Intensity fluctuations are detected and fed back to an attenuator

light. They often have a characteristic $1/f^\alpha$ scaling at low frequency where f is the frequency variable and $1 < \alpha < 2$ [54]. Most accessible biomolecular signals occur at frequencies ranging from a hertz to a kilohertz, a frequency range for which $1/f^\alpha$ noise is often highly problematic.

Unlike intensity noise, frequency and phase noise are not trivial to distinguish separately from the measured signal in interferometric single molecule biosensors. They can however be detected, if a reference low noise field is available to act as a phase reference, or using a highly stabilised optical cavity as a frequency reference. Phase noise that is introduced within the sensing device, as opposed to within the source laser, can also be detected using a delayed self-heterodyne system (see Fig. 1.13a) or an unbalanced interferometer (see Fig. 1.13b). In the case of the unbalanced interferometer, the laser light is split in two paths with lengths that are typically different by several tens of meters. This length difference is chosen, ideally, to be large compared to the coherence length of the light. In this case the phase noise on the two arms becomes uncorrelated, so that phase differences can be observed on interference. The detected interference signal can then be used directly to cancel the phase noise on the measurement, either in post-processing [55] or via feedback to the laser. However, this approach requires that the amplitude noise on the light is sufficiently low so that it does not contaminate the feedback signal. An alternative approach, immune to amplitude noise, is the Pound-Drever-Hall locking technique where frequency

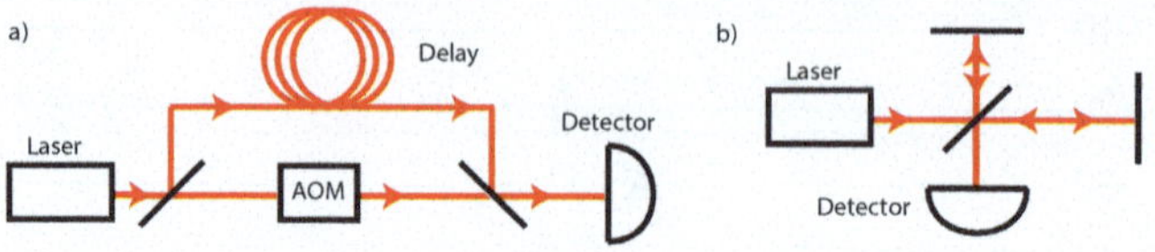

Fig. 1.13 Phase noise detection. **a** delayed self-heterodyne system. **b** Unbalanced Michaelson interferometer

Fig. 1.14 Pound-Drever-Hall locking technique. PBS = polarised beam splitter and $\lambda/4$ = quarter waveplate

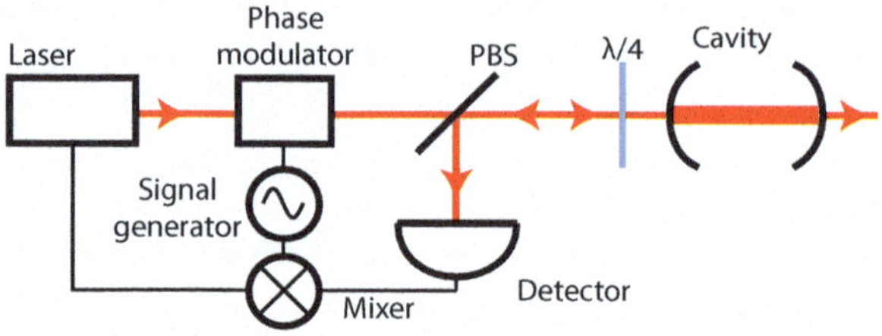

stabilisation of the laser is achieved with reference to an exterior optical cavity (see Fig. 1.14) [56, 57]. In this configuration, a phase modulated laser is reflected from a highly stable optical cavity, and the reflected field detected. Demodulating the photocurrent at the modulation frequency provides a signal that is proportional to the difference in frequency between the cavity and the laser. Frequency noise reduction of more than a factor of a thousand at low frequencies has been demonstrated using this technique [58].

Optical background noise
In addition to technical laser noise, any other external optical field can obscure the small scattering signal from a single biomolecule. Sunlight, light created by ambient illumination, and probe light scattered from other optical elements are all typically orders of magnitude stronger than the scattered field from a biomolecule and must be suppressed. Techniques used in single molecule biosensors to remove these spurious field are detailed in the next section (Sect. 1.3.2.2). Scattering from background particles and other contaminants in the solution with the biomolecule that cannot be distinguished from the scattered field of the nanoparticle can also introduce background noise, so that single molecule biosensors often need to be housed in clean environments.

1.3.2.2 Using Interference to Suppress Technical Noise in Biosensing

Interference of the scattered light with a much brighter local oscillator field can be used both to suppress background photons and to, as discussed earlier, amplify the electrical signal from the scattered light above the electronic noise floor of the detection device. It suppresses background photons since the frequency of the interference beat-note is given by the difference frequency between the local oscillator and the field it interferes with. Only background photons with frequencies very close to the

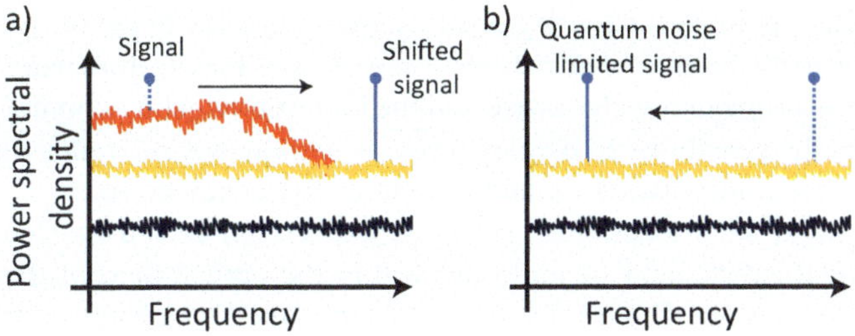

Fig. 1.15 **a** Scheme of the heterodyne frequency shift to move the signal into a frequency region where the local oscillator is quantum noise limited **b** Scheme of the quantum noise limited signal shifted back to its original frequency using a dual quadratures lock-in amplifier

frequency of the scattered field produce a beat-note that is indistinguishable from that of the scattered field. Indeed, so long as the frequency of the background light is different from that of the scattered light by at least the frequency resolution of the measurement, it can be fully suppressed. The frequency resolution is equal to the inverse of the measurement duration, and so can often be beneath a hertz. We can see, therefore, that an interferometer is equivalent to an exceptionally narrow band frequency filter—one that is much narrower than could be built conveniently through other means, and one that also amplifies the signal above electronic noise.

In iSCAT a homodyne configuration is used to interfere the scattered signal with a local oscillator that has the same frequency [59]. The challenge with this configuration is that, since there is no frequency difference between signal and local oscillator, slow phase fluctuations can masquerade as signals. The extreme stability of the iSCAT configuration, in which the reference beam and the scattered beam follow almost identical paths, allows common mode rejection of most sources of this phase noise.

In dark field heterodyne detection [17], a local oscillator is also used to remove background photons. However, the technical laser and background noise is avoided using a heterodyne configuration where the local oscillator and the scattered signal have different frequencies (see Fig. 1.15). This shifts the interference beat-note to high frequencies (typically tens of megahertz) where the local oscillator is only limited by the optical shot noise (see Fig. 1.15). After detection, a dual quadratures lock-in amplifier is then used to electronically remove the technical noise on the detected photocurrent and shift the shot noise limited signal back to low frequencies (see Fig. 1.15).

1.3.2.3 Removing Technical Noise Using Two Optical Cavity Modes

Biosensors that use optical ring cavities, such as like whispering gallery resonators (e.g. see [28, 48, 60–64]) typically monitor frequency shift of the cavity due to the presence of the molecule of interest (see Chap. 2). Therefore this types of measurement is intrinsically degraded by the frequency noise of the laser light. However, due to the bidirectional nature of ring cavities they feature pairs of degenerate modes, one propagating clockwise and the other propagating counter-clockwise. This provides the possibility to use one mode as a frequency reference for frequency shifts of the other, introducing immunity to laser frequency noise.

To understand such techniques we must understand how a molecule interacts with the optical cavity field. A molecule within the optical field of the two cavity modes introduces scattering from one mode to the other, which causes the modes to hybridise into a symmetric standing-wave mode and an asymmetric standing-wave mode. The symmetric mode as an intensity peak at the molecule position, while the asymmetric mode has zero field intensity at this position. The molecule then only interacts with the symmetric mode, shifting its frequency relative to the asymmetric mode. As such, the asymmetric mode can be used as a frequency reference (see Fig. 1.16). Since both modes are subject to the same technical and laser frequency noise, difference in their frequencies (often termed the "mode splitting") is therefore immune to these noise sources and provides a signal that is directly proportional to the scattering cross-section of the molecule.

This concept was first demonstrated in Ref. [60], where mode splitting between the two modes was measured by scanning the laser frequency across the optical resonance, and the effect of nanoparticles on this splitting characterised. In that work and following papers [65], impressive performance was achieved, with laser frequency noise essentially eliminated. One challenge, however, with this technique is that the scattering particle must be large enough to induce a resolvable splitting in the optical modes—that is, the splitting must be larger than the optical decay rate which is typically on the order of a megahertz. This has precluded its application with small proteins and other biomolecules in aqueous solution.

Several techniques have been proposed to circumvent this problem, although as yet they also have not been applied to biomolecules. One technique nests the whispering gallery mode resonator within an optical interferometer [66]. This allows the symmetric and antisymmetric modes to be separately accessed, one from each of the output ports of the interferometer, and in this way removes the requirement that the modes are resolvable. Another technique measures the intensity of light coupled from the clockwise (excited) whispering gallery mode into the counter-clockwise (backscattered) whispering gallery mode due to the presence of the molecule [67] (see Fig. 1.16). In Ref. [67] it was shown that this technique both allows sensitivity equal to that achievable from frequency shifts of a single optical mode and is free of frequency noise when the laser frequency matches the cavity frequency [67]. Other techniques introduce a gain medium into the optical cavity to reduce its decay rate [65] or take advantage of an exceptional point to increase the frequency shift [68].

1.3.2.4 Photon Shot Noise

The particle nature of the light introduces another type of noise, *shot noise*, which is inherent to every optical field and has origins fundamentally different from technical

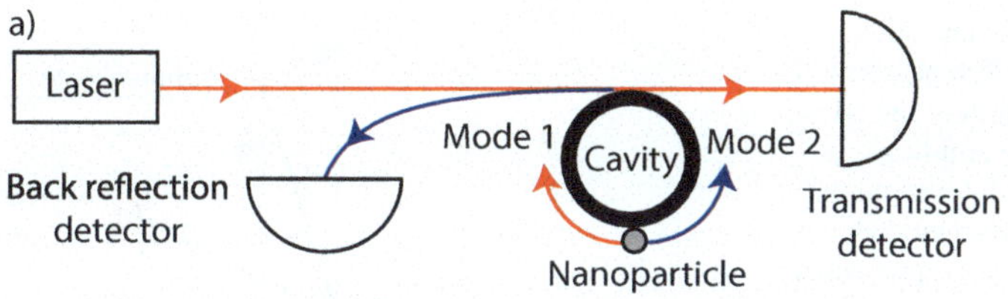

Fig. 1.16 Scheme of the self-referencing technique monitoring mode splitting via transmission of the cavity (right detector) or monitoring back reflection from the cavity (left detector). The two counter-propagating modes are represented by the red and blue arrow

noise. In quantum optics, optical fields are considered as a flux of photons particles, each with energy given by $\hbar\omega$, where $\hbar$ is the reduced Planck's constant and ω is the angular optical frequency. In the usual case that each photon is uncorrelated to the other photons in the optical field, the photons arrive at random times at the detector. This creates *photon shot noise*. The detection is governed by Poissonian statistics, and is therefore a random white process with a flat response in frequency (and time) as shown in Fig. 1.11c. Since the shot noise originates from the fundamental quantized nature of light, the techniques to suppress noise described in the previous section have no effect on it.

Suppression of the shot noise is not a trivial task and quantum engineered light need to be prepared, exhibiting a particular form of quantum correlations between photons (see Sect. 1.3.3.4). When the measurement of a biosensor is free of technical noise and the detected light does not exhibit quantum correlations, the sensor reaches a fundamental maximum in sensitivity, the *shot noise limit*. As we will see in the following section, this dictates the smallest particle that can be detected with the particular apparatus, without recourse to quantum correlations.

Photon statistics

Shot noise has been studied for several decades in the field of quantum optics. Using the formalism of quantum optics it can be shown that for a coherent state of light,[1] the probability to detect n photons during a given measurement time is [69]:

$$P(n) = \frac{e^{-\langle n\rangle}\langle n\rangle^n}{n!}, \tag{1.8}$$

where $\langle n\rangle$ is the mean number of photons detected. This correspond to a Poissonian distribution. A characteristic property of this distribution is that the variance of the distribution $Var[n]$ equals its mean. Coherent states are the ideal state emitted by a low noise laser, and are one of a set of "minimum uncertainty states" that carry the smallest amount of quantum noise allowed by the Heisenberg uncertainty of quantum mechanics. They can be represented in the phase space by a "ball and stick" diagram (see Fig. 1.17). In this representation, the length of the stick represent the amplitude

[1] This is the state of a laser free of technical noise and without any quantum correlations between photons.

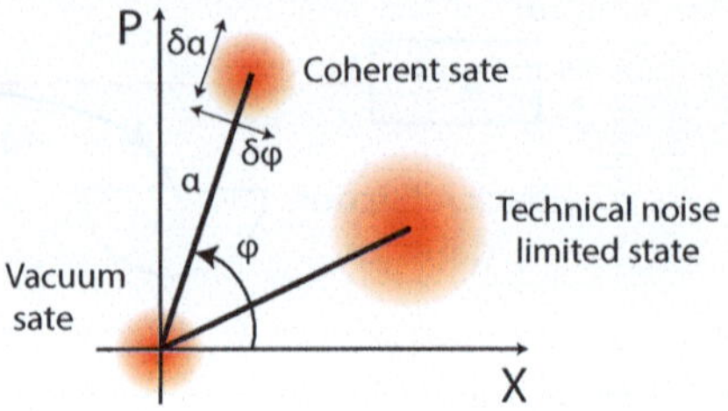

Fig. 1.17 Ball and stick diagram of different states of light. The length of the stick represents the amplitude of the field α and the angle the phase ϕ. A coherent and vacuum state are represented as well as a technical noise limited state

(α) of the optical field and the angle the phase (ϕ). The ball represent the fluctuation in phase $\delta\phi$ and amplitude $\delta\alpha$. Compared to a shot-noise-limited laser, the larger noise fluctuations of technical-noise-limited light are represented via a larger ball in the phase space diagram.

1.3.3 Shot Noise Limit to Signal-to-Noise

1.3.3.1 Shot Noise Limit of Direct Detection

The shot noise limit governs the smallest particle detectable by a biosensor for given apparatus and measurement time, and in the absence of quantum correlations between photons. As an example, let us consider a biosensor that directly detects the scattered field of a spherical nanoparticle with 100% efficiency as shown in Fig. 1.18. Assuming that the signal detected n_{scatt} is dominated by shot noise rather than technical noise, according to Eq. 1.1 the signal-to-noise is:

$$\text{SNR} = \langle n_{\text{scat}} \rangle, \tag{1.9}$$

where we have used the fact that the mean and variance of a Poissonian distribution are equal. This has an intuitive physical meaning: to be able to detect the scattering (SNR$>$ 1), at least one photon must be scattered on average during the measurement time.

Defining a particle as resolvable so long as the signal-to-noise is greater than one, it is possible to calculate the minimum scattering cross-section that is detectable above the quantum fluctuations σ_{min}. Using Eqs. 1.2 and 1.9 we find:

$$\sigma_{\text{min}} = \frac{4\pi w^2}{\langle n_{\text{in}} \rangle}. \tag{1.10}$$

For an input power $P_{\text{in}} = \hbar\omega\langle n_{\text{in}}\rangle/\tau$ with ω the angular frequency of the light and τ the measurement time, this shot noise limit can also be expressed as:

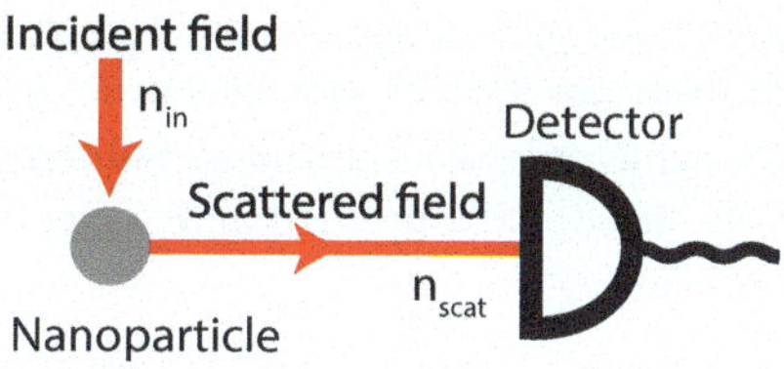

Fig. 1.18 Schematic of the direct detection of the scattered field of a nanoparticle

$$\sigma_{\text{min}} = \frac{4\pi w^2 \hbar\omega}{\tau P_{\text{in}}}. \tag{1.11}$$

For a particle in water with a refractive index of 1.45, an input power of 1 mW, a wavelength of 780 nm, a beam size of 1 μm, and a measurement time of 1 ms, we find that the quantum noise limit to the particle cross-section that can be resolved with direct measurement is $\sigma_{\text{min}} = 3.2\,\text{nm}^2$.

Assuming that the particle is spherical with a constant refractive index, one can also determine the radius r_{min} of the smallest particle detectable when at the shot noise limit using Eq. (1.3) which gives:

$$r_{\text{min}} = \frac{1}{2}\left(\frac{\lambda^2 w}{\pi^2}\right)^{1/3}\left(\frac{6\hbar\omega}{\tau P_{\text{in}}}\right)^{1/6}\left(\frac{m^2+2}{m^2-1}\right)^{1/3}. \tag{1.12}$$

For the same parameters as above, we find that the minimum resolvable particle radius is $r_{\text{min}} = 5.4\,\text{nm}$. Thus, we see that direct detection is in principle able to detect relatively large protein molecules such as antibodies, but with the scattering cross-section scaling as r^6 rapidly becomes ineffective at smaller sizes. We further see the importance of achieving shot noise limited performance. In particular, it becomes increasingly challenging to suppress background optical and electronic noise to levels below the scattering rate for increasingly small particles.

1.3.3.2 Shot Noise Limit of Interferometric Sensing

As discussed earlier, in interferometric sensing, the scattered light is interfered with a bright local oscillator beam. Since interference offers a means to strongly filter background noise, shot noise limited measurements can be achieved using these techniques. The presence of the local oscillator alters the shot noise limit somewhat. Detailed derivations for the cases of homodyne and heterodyne detection can be found in the supplementary of Ref. [17]. These show that for homodyne detection the SNR is equal to

$$\text{SNR}_{\text{Homodyne}} = 2\eta\langle n_{\text{scat}}\rangle \tag{1.13}$$

and for heterodyne,

$$SNR_{\text{Heterodyne}} = \eta\langle n_{\text{scat}}\rangle. \tag{1.14}$$

with η the percentage of scattered photon collected by the detector. These shot noise limit are similar to direct detection, showing the advantage of interferometric sensing: the same or better shot-noise limited sensitivity can be achieved as direct detection, with the signal amplified above electronic noise and background optical noise suppressed.

1.3.3.3 Optical Noise Scaling

To determine if a sensor operates at the shot noise limit, it is not necessary to be able to resolve single photon detection events. Instead, to test whether the sensor is limited by technical or shot noise, one can study how the noise scales with optical power.

Let us consider the experimental setup of Fig. 1.19. Here a laser field E_{in} containing some signal passes through a variable efficiency γ beam splitter used to change the optical power arriving at a sample and a photo-sensitive detector. When the optical field at the detector E_{out} is transformed into a photocurrent i via the photoelectric effect one can show that if E_{out} is dominated by technical noise,

$$i_{\text{tech}} \approx \gamma\langle n\rangle + 2\gamma\sqrt{\langle n\rangle}\delta_{\text{sig}} + 2\gamma\sqrt{\langle n\rangle}\delta_{\text{tech}} \tag{1.15}$$

where δ_{sig} is the signal amplitude, δ_{tech} is the fluctuation amplitude due to the technical noise, $\langle n\rangle$ is the mean incident photon number, and we have assumed that $\langle n\rangle \gg \{\delta_{\text{sig}}^2, \langle\delta_{\text{tech}}\rangle^2\}$. This assumption is generally valid. For a micro-watt optical field $\langle n\rangle \approx 10^{12}$ and $\langle\delta_{\text{tech}}\rangle^2$ is typically in the range of 100–10,000. From this relation, we can see that the signal and technical noise both scale with the detected optical power. This is expected since when attenuating or amplifying the laser power, by changing γ, the signal or fluctuations of the laser intensity will be reduced or amplified accordingly.

Equation 1.15 also shows that increasing the mean incident photon number ($\langle n\rangle$) amplifies both the signal and the technical noise contribution and therefore will not increase the signal to noise ratio when technical noise dominates. However, when increasing the intensity at the particle of interest, more photons interact with the particle (δ_{sig} increases) and if $\langle n\rangle$ is kept constant, δ_{tech} stays constant and the signal to noise ratio can be increased. This is the technique leveraged by the WGM and plasmonic sensors to detect label free single proteins when limited by technical noise.

The scaling of the quantum noise fluctuations with optical power is different. In the quantum picture, when optical power fluctuates, by changing γ in Fig. 1.19, this results in adding or removing quantum vacuum fluctuations to the optical field. Counter intuitively, even if it contains zero photons, vacuum noise is a coherent state of light and contains the same quantum fluctuations in phase and amplitude as a coherent state with many photons. Therefore, vacuum noise creates additional quantum noise and prevent the total quantum noise to be attenuated or amplified as fast as the technical noise. Semi-classically, the vacuum noise can be approximated as an additional noise created because the beam splitter randomly chooses which photons

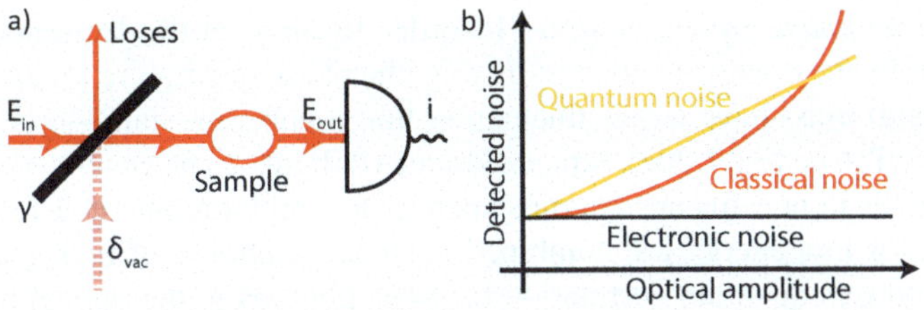

Fig. 1.19 **a** Scheme of light detection with a variable ratio $\gamma(t)$ beam splitter. Note that the beam splitter can be placed anywhere before the detector along the horizontal beam **b** Noise scaling as function of optical amplitude. The technical noise scales and quantum noise scale quadratically and linearly with optical amplitude while the electronic noise is constant

are transmitted and which photons are reflected creating additional quantum noise which does not affect the technical noise. Therefore, when detecting the photocurrent from an optical field containing technical and quantum noise, we have:

$$i_{\text{tot}} \approx \gamma \langle n \rangle + 2\gamma \sqrt{\langle n \rangle} \left(\delta_{\text{sig}} + \delta_{\text{tech}} + \delta_{\text{q}}\right) + 2\sqrt{\langle n \rangle \gamma (1-\gamma)} \delta_{\text{vac}} \tag{1.16}$$

with δ_q the amplitude of the quantum shot noise of the E_{in} and δ_{vac} the amplitude of the vacuum noise created by the beam splitter. Since the different noises are uncorrelated and knowing that the quantum shot noise and the vacuum noise are both coherent state therefore have equal variance, one can show that the power spectral density I_{tot} of the the photocurrent is equal to:

$$I_{\text{tot}} \approx 4\gamma^2 \langle n \rangle \left(\text{Var}[\Delta_{\text{sig}}] + \text{Var}[\Delta_{\text{tech}}]\right) + 4\gamma \langle n \rangle \text{Var}[\Delta_{\text{q}}] \tag{1.17}$$

where Δ_{sig}, Δ_{tech} and Δ_{q} are the Fourier transform of the signal, technical noise and quantum noise respectively and $Var[X]$ is the variance of X over the resolution bandwidth of the measurement.[2] Equation 1.17 shows that, if the dominant noise scales linearly with γ, that is to say linearly with the detected field amplitude, it has a quantum origin or if it scales quadratically, it has a technical origin [69] as shown in Fig. 1.19. From Eq. 1.16, we also see that the quantum noise will dominates at low power and therefore, biosensors detecting small amount of power can be assumed to be quantum noise limited.

1.3.3.4 Overcoming the Shot Noise Limit

It is possible to create states of light which have smaller uncertainty in some quadrature in phase space than a coherent state. This type of light is called squeezed light [70]. For instance, an optical field that has less noise than a coherent state in its amplitude, is called an *amplitude squeezed state*, and one that has less noise

[2] $Var[X](f) = \int_{f-\text{RBW}/2}^{f+\text{RBW}/2} |X(f')|^2 df'$ whith RBW the resolution bandwidth.

in its phase as *phase squeezed state*. In order to obey to the Heisenberg uncertainty principle, the conjugate quadrature (e.g. the phase quadrature for an amplitude squeezed state) must have larger fluctuation that a coherent state—it is said to be "anti-squeezed" (see Fig. 1.20a). Squeezed states of light can be generated using non-linear processes in non-linear materials such as second harmonic generation where two photons of low energy are combined to create a photon of high energy. This process introduces quantum correlation between photons in the optical field which can decrease the measured noise [71].

Optical biosensors can benefit from squeezing if they are already quantum noise limited. The sensitivity can be increased depending on how much squeezing is achieved. Noise variance reductions beneath the shot noise as large as a factor of thirty (−15 dB) have been achieved [72], and could potentially increase the signal-to-noise ratio of biosensors by the same amount. However, the use of squeezed light in biosensing is challenging because it is difficult to create squeezed state of light and they are fragile—any optical losses reduces the correlations between photons and therefore degrades the squeezing [70]. These losses can be expected to be significantly large in biological sensing apparatuses. Nevertheless, biological quantum enhanced measurement using squeezed states has been demonstrated in optical tweezers where the motion of particles within yeast cells has been probed with precision 46% beyond the quantum noise limit [73], the resolution of photonic-force microscopy have been enhanced by 14% [74], and nonlinear microscopy has been performed with signal-to-noise improved by 35% compared to the shot-noise limit [75]. Furthermore, similar techniques have been developed to increase the sensitivity of gravitational wave detectors and improve the efficiency quantum information storage, and showed that the SNR can be increased by a factor 4.5 (6.5 dB) in a frequency band relevant to biophysics experiments [76].

The sensitivity of interferometric single molecule sensing could be enhanced by squeezed light in a similar way to the enhancement of optical tweezers-based particle tracking in Ref. [73]. For example, vacuum squeezed states can be focused to the sample and collected With the scattered field to decrease the amount of vacuum shot noise as seen in Fig. 1.20.

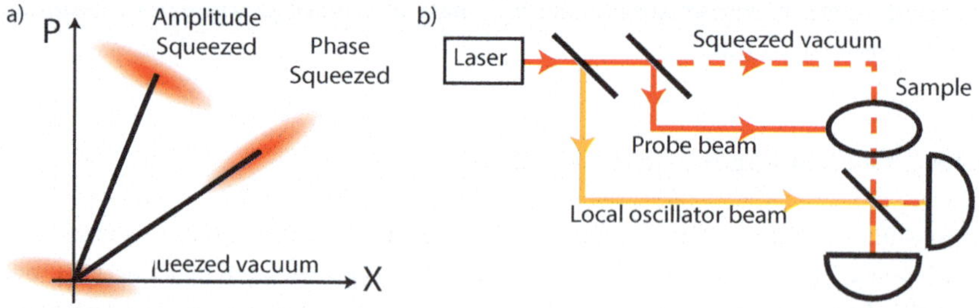

Fig. 1.20 **a** Quantum squeezed states in phase space. **b** Scheme of a quantum enhanced interferometric sensor

1.4 Applications

The main applications of interferometric sensing so far have been in single-particle tracking and molecule size detection.

Single particle tracking in biology is commonly performed via fluorescence measurements. However, this type of measurement suffers from relatively slow speed due fluorescence saturation limiting the emitted photon flux, as well as limited acquisition times over which the particle can be tracked due to photobleaching of label molecules. Although interferometric sensing does not suffer from these issues, fluorescence is the primary approach to single-particle tracking used so far in biology. The reason for this is that in fluorescence imaging it is easy to filter out scattering that does not come from molecules of interest (see Sect. 1.3.2). This is not so trivial in case of interferometric sensing, and as a consequence interferometric techniques have been primarily used in an environment without scatterers other than the particle of interest. Most interferometric sensing applications demonstrated to date have used iSCAT microscopy, as it provides a high degree of spatial information.

At a scale of nanometers, gold nanoparticles were among the first to be detected and tracked by an interferometric technique [79]. Because of their strong scattering and biocompatibility, they are often used as labels in image scanning microscopy. Gold nanoparticles as small as 2 nm have been imaged with iSCAT [80]. Larger particles in the 20 nm to micrometer range scatter more light (see Sect. 1.3.1.1) enabling faster and more precise localisation of particles [81]. However, this comes with the compromise of greater invasion into the natural dynamics of the biosystem, and lower sensitivity to environmental forces due to the increased mass. Rotational diffusion of gold nanorods has also been investigated using iSCAT [82]. The speed of iSCAT measurements is limited by the cameras used for detection, so other techniques like dark-field heterodyne detection [17] are more suitable for achieving measurements at speeds in the hundreds of kilohertz and above.

Tracking of viruses was demonstrated in the early days of interferometric sensing, thanks to their large size between tens and hundreds of nanometers. Compared to these earlier experiments, the two dimensional spatial resolution of iSCAT can provide additional information. For instance, it can be used to determine whether the virus contains DNA or not [77] (Fig. 1.21), since the presence of DNA affected the polarizability.

Even though much smaller than viruses, single proteins have been detected, starting with the detection of myosin VA (500 kDa) [20] (Fig. 1.22). By detecting 74 nm steps of myosin along actin filaments at speeds unattainable for fluorescence microscopy, it was possible to unravel additional dynamics of molecular motion. Later experiments were able to detect molecules down to about 50 kDa in size, both with dark-field heterodyne sensing [17] and iSCAT microscopy [26]. Heterodyne sensing has demonstrated quantum limited detection and can be naturally integrated with microfluidics using hollow core fibers [83].

Tracing viruses and proteins on the surface of a cell is of great importance for biology. Unfortunately, it is not straight forward to maintain the same tracking

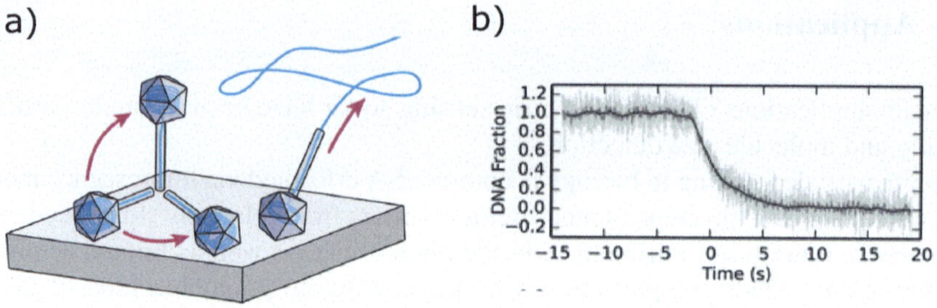

Fig. 1.21 **a** Schematic of a bacteriophage virus attached to a cover slip. Motion of the head of the bacteriophage can be followed with iSCAT, as can the release of the DNA [77]. **b** Fraction of DNA contained in the virus over time after the release process is triggered. Figure adapted from [77]. Used with permission

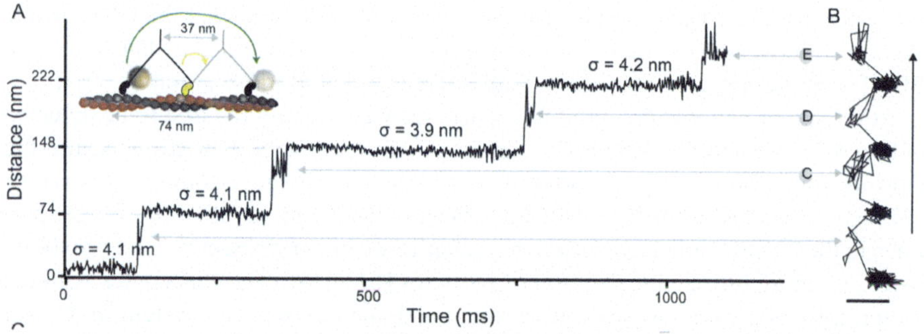

Fig. 1.22 **a** iSCAT microscopy of 20 nm gold nanoparticle attached to myosin, Fast tracking of myosin shows dynamics of its 74 nm steps. **b** Two dimensional tracking. Scale bar: 50 nm. Figure adapted from [78]. Creative Commons Attribution license

performance as has been achieved in clean solutions in the presence of cells, since they scatter a lot of light creating a large non-stationary background. Still, it has been shown that with advanced data processing one can perform three-dimensional tracing of viruses and proteins labelled with gold nanoparticles. Spatial and temporal resolutions of 5 nm and 50 μs have been achieved [84].

The intensity of light scattering from a particle depends on its polarizability which in turn depends on its refractive index and volume, as discussed in Sect. 1.3.1.1. In the case of biomolecules, the refractive index has been found to not vary greatly across different molecules. The density is also found to be relatively constant across different biomolecules. As a consequence, one can use dipole scattering to distinguish between molecules of different mass. This has been demonstrated in the case of iSCAT microscopy [16], and the developed technique termed *mass photometry*. The standard deviation of refractive index and specific volume has been calculated for $>10^5$ protein sequences to be 0.3 and 1.2% respectively [59]. In order to measure molecular mass, the technique uses 'calibration molecules' with known mass and compares their light scattering to that from molecules of interest. In this way,

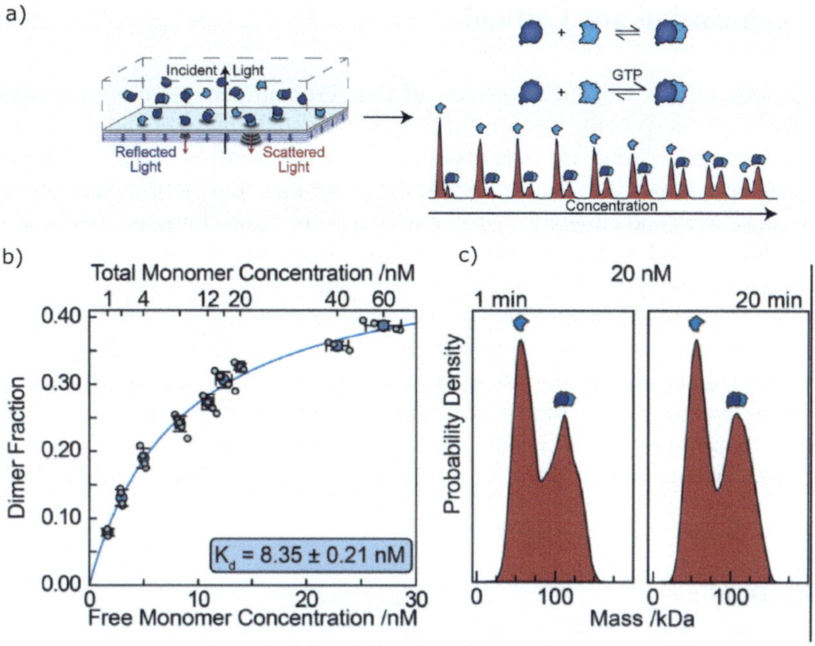

Fig. 1.23 **a** Mass photometry of tubulin using iSCAT. The number of bound and free monomers can be counted, as these exhibit different interference contrast. This provides information on interactions between protein monomers. **b** The fraction of tubulin dimers as a function of free monomer concentration. **c** Mass distributions in sample over time. Figure adapted from [23]. Creative Commons CC-BY license

mass photometry has been found to be capable of detecting the mass distribution of solutions of different proteins or different oligomeric states, providing information on interactions between proteins [23] (Fig. 1.23). It also allows the heterogeneity of a sample to be determined [85], which is of importance for characterizing protein purification. This shows the potential of mass photometry as a replacement for more complex characterization techniques such as negative stain electron microscopy [85].

As iSCAT can discern molecules of different sizes, it has been also used to monitor the assembly and disassembly of biomolecules. For example, the length of microtubules can be monitored [21], even down to individual tubulin dimers [86]. Formation of supported lipid bilayers has been investigated in [87]. This offers the prospect to study biomolecular interactions with cellular membrane bilayer models, such as their interactions with drugs and small molecules, and transport processes [84].

1.5 Conclusion and Outlook

This chapter has provided an overview of interferometric single-molecule biosensors. Such sensors provide a powerful approach to study the size and dynamics of single molecules, offering the benefits of label-free operation and relative simplicity. The chapter outlines the basic principles of how they function, details the various noise sources and challenges, discusses experimental techniques, and describes recent advances and applications. It is clear from the breadth of applications that have already been demonstrated that interferometric single-molecule biosensors will have a significant role to play in building our understanding of proteins and how they interact. A significant outstanding question is whether these techniques can be extended to provide detailed information about the structure of biomolecules and to track how this structure changes over time. Recent theory suggests that interferometric single-molecule biosensors may be capable of such measurements [29]. This would open an exciting avenue to understand biomolecular processes such as protein folding.

References

1. Toru Yoshizawa. Handbook of optical metrology principles and applications, 2017.
2. Jeremy B Sanderson. *Phase Contrast Microscopy*. American Cancer Society, 2001.
3. Jianling Chen, Yan Xu, Xiaohua Lv, Xiaomin Lai, and Shaoqun Zeng. Super-resolution differential interference contrast microscopy by structured illumination. *Opt. Express*, 21(1):112–121, 2013.
4. Dan P Popescu, Costel Flueraru, Youxin Mao, Shoude Chang, John Disano, Sherif Sherif, Michael G Sowa, et al. Optical coherence tomography: fundamental principles, instrumental designs and biomedical applications. *Biophysical reviews*, 3(3):155–169, 2011.
5. Yongkang Gao, Qiaoqiang Gan, Zheming Xin, Xuanhong Cheng, and Filbert J Bartoli. Plasmonic mach–zehnder interferometer for ultrasensitive on-chip biosensing. *ACS nano*, 5(12):9836–9844, 2011.
6. PI Nikitin, AN Grigorenko, AA Beloglazov, MV Valeiko, AI Savchuk, OA Savchuk, G Steiner, C Kuhne, A Huebner, and R Salzer. Surface plasmon resonance interferometry for micro-array biosensing. *Sensors and Actuators A: Physical*, 85(1-3):189–193, 2000.
7. Tyler S Ralston, Daniel L Marks, P Scott Carney, and Stephen A Boppart. Interferometric synthetic aperture microscopy. *Nature Physics*, 3(2):129–134, 2007.
8. Ivano Alessandri and John R. Lombardi. Enhanced raman scattering with dielectrics. *Chemical Reviews*, 116(24):14921–14981, 2016.
9. Joshua A Riback, Micayla A Bowman, Adam M Zmyslowski, Kevin W Plaxco, Patricia L Clark, and Tobin R Sosnick. Commonly used fret fluorophores promote collapse of an otherwise disordered protein. *Proceedings of the National Academy of Sciences*, 116(18):8889–8894, 2019.
10. Michael S Woody, Donald A Winkelmann, Marco Capitanio, E Michael Ostap, and Yale E Goldman. Single molecule mechanics resolves the earliest events in force generation by cardiac myosin. *Elife*, 8:e49266, 2019.
11. Mayumi Nakanishi-Matsui, Sachiko Kashiwagi, Hiroyuki Hosokawa, Daniel J. Cipriano, Stanley D. Dunn, Yoh Wada, and Masamitsu Futai. Stochastic high-speed rotation of escherichia coli atp synthase f1 sector. *The Journal of Biological Chemistry*, 281:4126–4131, Feb 2006.
12. Mika Hilvo, Lina Baranauskiene, Anna Maria Salzano, Andrea Scaloni, Daumantas Matulis, Alessio Innocenti, Andrea Scozzafava, Simona Maria Monti, Anna Di Fiore, Giuseppina De

Simone, Mikaela Lindfors, Janne Jänis, Jarkko Valjakka, Silvia Pastoreková, Jaromir Pastorek, Markku S. Kulomaa, Henri R. Nordlund, Claudiu T. Supuran, and Seppo Parkkila. Biochemical characterization of ca ix, one of the most active carbonic anhydrase isozymes*. *Journal of Biological Chemistry*, 283(41):27799–27809, 2008.

13. Haruto Ishikawa, Kyungwon Kwak, Jean K. Chung, Seongheun Kim, and Michael D. Fayer. Direct observation of fast protein conformational switching. *Proceedings of the National Academy of Sciences*, 105(25):8619–8624, 2008.
14. Joshua C Vaughan, Shu Jia, and Xiaowei Zhuang. Ultrabright photoactivatable fluorophores created by reductive caging. *Nature methods*, 9(12):1181–1184, 2012.
15. Jonathan B Grimm, Brian P English, Jiji Chen, Joel P Slaughter, Zhengjian Zhang, Andrey Revyakin, Ronak Patel, John J Macklin, Davide Normanno, Robert H Singer, et al. A general method to improve fluorophores for live-cell and single-molecule microscopy. *Nature methods*, 12(3):244–250, 2015.
16. Gavin Young, Nikolas Hundt, Daniel Cole, Adam Fineberg, Joanna Andrecka, Andrew Tyler, Anna Olerinyova, Ayla Ansari, Erik G Marklund, Miranda P Collier, et al. Quantitative mass imaging of single biological macromolecules. *Science*, 360(6387):423–427, 2018.
17. NP Mauranyapin, LS Madsen, MA Taylor, M Waleed, and WP Bowen. Evanescent single-molecule biosensing with quantum-limited precision. *Nature Photonics*, 11(8):477–481, 2017.
18. Joanna Andrecka, Jaime Ortega Arroyo, Katie Lewis, Robert A. Cross, and Philipp Kukura. Label-free imaging of microtubules with sub-nm precision using interferometric scattering microscopy. *Biophysical Journal*, 110(1):214–217, 2016.
19. Lewis S Mosby, Nikolas Hundt, Gavin Young, Adam Fineberg, Marco Polin, Satyajit Mayor, Philipp Kukura, and Darius V Köster. Myosin ii filament dynamics in actin networks revealed with interferometric scattering microscopy. *Biophysical Journal*, 2020.
20. J Ortega Arroyo, J Andrecka, KM Spillane, N Billington, Y Takagi, JR Sellers, and P Kukura. Label-free, all-optical detection, imaging, and tracking of a single protein. *Nano letters*, 14(4):2065–2070, 2014.
21. Joanna Andrecka, Jaime Ortega Arroyo, Katie Lewis, Robert A Cross, and Philipp Kukura. Label-free imaging of microtubules with sub-nm precision using interferometric scattering microscopy. *Biophysical journal*, 110(1):214–217, 2016.
22. M. Noto, M. Khoshsima, D. Keng, I. Teraoka, V. Kolchenko, and S. Arnold. Molecular weight dependence of a whispering gallery mode biosensor. *Applied Physics Letters*, 87(22):223901, 2005.
23. Adam Fineberg, Thomas Surrey, and Philipp Kukura. Quantifying the monomer–dimer equilibrium of tubulin with mass photometry. *Journal of molecular biology*, 432(23):6168–6172, 2020.
24. Anirban Mitra, Filipp Ignatovich, and Lukas Novotny. Real-time optical detection of single human and bacterial viruses based on dark-field interferometry. *Biosensors and Bioelectronics*, 31(1):499–504, 2012.
25. John David Jackson. Classical electrodynamics, 1999.
26. Marek Piliarik and Vahid Sandoghdar. Direct optical sensing of single unlabelled proteins and super-resolution imaging of their binding sites. *Nature communications*, 5(1):1–8, 2014.
27. Fea Vollmer, Daniela Braun, Albert Libchaber, Masoud Khoshsima, Iwao Teraoka, and Susan Arnold. Protein detection by optical shift of a resonant microcavity. *Applied physics letters*, 80(21):4057–4059, 2002.
28. Eugene Kim, Martin D Baaske, Isabel Schuldes, Peter S Wilsch, and Frank Vollmer. Label-free optical detection of single enzyme-reactant reactions and associated conformational changes. *Science advances*, 3(3):e1603044, 2017.
29. Larnii S. Booth, Eloise V. Browne, Nicolas P. Mauranyapin, Lars S. Madsen, Shelley Barfoot, Alan Mark, and Warwick P. Bowen. Dynamic polarizability of macromolecules for single-molecule optical biosensing. *arXiv*, 2021.
30. Jon Applequist, James R Carl, and Kwok-Kueng Fung. Atom dipole interaction model for molecular polarizability. application to polyatomic molecules and determination of atom polarizabilities. *Journal of the American Chemical Society*, 94(9):2952–2960, 1972.

31. B Th Thole. Molecular polarizabilities calculated with a modified dipole interaction. *Chemical Physics*, 59(3):341–350, 1981.
32. Alexander Hahn, Janet Vonck, Deryck J Mills, Thomas Meier, and Werner Kühlbrandt. Structure, mechanism, and regulation of the chloroplast atp synthase. *Science*, 360(6389), 2018.
33. Utkur Mirsaidov, Winston Timp, Kaethe Timp, Mustafa Mir, Paul Matsudaira, and Gregory Timp. Optimal optical trap for bacterial viability. *Physical Review E*, 78(2):021910, 2008.
34. Yoshiyuki Sowa, Alexander D Rowe, Mark C Leake, Toshiharu Yakushi, Michio Homma, Akihiko Ishijima, and Richard M Berry. Direct observation of steps in rotation of the bacterial flagellar motor. *Nature*, 437(7060):916–919, 2005.
35. Markita P Landry, Patrick M McCall, Zhi Qi, and Yann R Chemla. Characterization of photoactivated singlet oxygen damage in single-molecule optical trap experiments. *Biophysical journal*, 97(8):2128–2136, 2009.
36. Sina Wäldchen, Julian Lehmann, Teresa Klein, Sebastian Van De Linde, and Markus Sauer. Light-induced cell damage in live-cell super-resolution microscopy. *Scientific reports*, 5(1):1–12, 2015.
37. Marina S Dietz, S Sophia Wehrheim, Marie-Lena IE Harwardt, Hartmut H Niemann, and Mike Heilemann. Competitive binding study revealing the influence of fluorophore labels on biomolecular interactions. *Nano letters*, 19(11):8245–8249, 2019.
38. Sofie Fogh Hedegaard, Mohammed Sobhi Derbas, Tania Kjellerup Lind, Marina Robertnova Kasimova, Malene Vinther Christensen, Maria Høtoft Michaelsen, Richard A Campbell, Lene Jorgensen, Henrik Franzyk, Marité Cárdenas, et al. Fluorophore labeling of a cell-penetrating peptide significantly alters the mode and degree of biomembrane interaction. *Scientific reports*, 8(1):1–14, 2018.
39. Ágnes Szabó, Tímea Szendi-Szatmári, László Ujlaky-Nagy, Ildikó Rádi, György Vereb, János Szöllősi, and Peter Nagy. The effect of fluorophore conjugation on antibody affinity and the photophysical properties of dyes. *Biophysical journal*, 114(3):688–700, 2018.
40. Jutta Pauli, Marieke Pochstein, Andrea Haase, Joanna Napp, Andreas Luch, and Ute Resch-Genger. Influence of label and charge density on the association of the therapeutic monoclonal antibodies trastuzumab and cetuximab conjugated to anionic fluorophores. *ChemBioChem*, 18(1):101–110, 2017.
41. Carolina Sanchez-Rico, Lena Voith von Voithenberg, Lisa Warner, Don C Lamb, and Michael Sattler. Effects of fluorophore attachment on protein conformation and dynamics studied by spfret and nmr. *Chemistry (Weinheim an der Bergstrasse, Germany)*, 23(57):14267, 2017.
42. Kapil Dave, Hannah Gelman, Chu Thi Hien Thu, Drishti Guin, and Martin Gruebele. The effect of fluorescent protein tags on phosphoglycerate kinase stability is nonadditive. *The Journal of Physical Chemistry B*, 120(11):2878–2885, 2016.
43. Linliang Yin, Wei Wang, Shaopeng Wang, Fenni Zhang, Shengtao Zhang, and Nongjian Tao. How does fluorescent labeling affect the binding kinetics of proteins with intact cells? *Biosensors and Bioelectronics*, 66:412–416, 2015.
44. Amy E Larsen and David G Grier. Like-charge attractions in metastable colloidal crystallites. *Nature*, 385(6613):230–233, 1997.
45. Seongmin Park, Siyoung Q Choi, Chaeyeon Song, Mahn Won Kim, and Myung Chul Choi. Surface charge effects on optical trapping of nanometer-sized lipid vesicles. *Soft matter*, 10(42):8406–8412, 2014.
46. Kevin Namink, Xuanhui Meng, Marc TM Koper, Philipp Kukura, and Sanli Faez. Electric-double-layer-modulation microscopy. *Physical Review Applied*, 13(4):044065, 2020.
47. Matthew R Foreman, Jon D Swaim, and Frank Vollmer. Whispering gallery mode sensors. *Advances in optics and photonics*, 7(2):168–240, 2015.
48. Eugene Kim, Martin D Baaske, and Frank Vollmer. Towards next-generation label-free biosensors: recent advances in whispering gallery mode sensors. *Lab on a Chip*, 17(7):1190–1205, 2017.
49. Tobias J Kippenberg, AL Tchebotareva, J Kalkman, A Polman, and Kerry J Vahala. Purcell-factor-enhanced scattering from si nanocrystals in an optical microcavity. *Physical review letters*, 103(2):027406, 2009.

50. Michael Motsch, Martin Zeppenfeld, Pepijn WH Pinkse, and Gerhard Rempe. Cavity-enhanced rayleigh scattering. *New Journal of Physics*, 12(6):063022, 2010.
51. CC Harb, MB Gray, H-A Bachor, R Schilling, P Rottengatter, I Freitag, and H Welling. Suppression of the intensity noise in a diode-pumped neodymium: Yag nonplanar ring laser. *IEEE journal of quantum electronics*, 30(12):2907–2913, 1994.
52. Michèle Heurs, Volker M Quetschke, Benno Willke, Karsten Danzmann, and Ingo Freitag. Simultaneously suppressing frequency and intensity noise in a nd: Yag nonplanar ring oscillator by means of the current-lock technique. *Optics letters*, 29(18):2148–2150, 2004.
53. B Daino, P Spano, M Tamburrini, and S Piazzolla. Phase noise and spectral line shape in semiconductor lasers. *IEEE Journal of Quantum Electronics*, 19(3):266–270, 1983.
54. Aditya Kakkar, Jaime Rodrigo Navarro, Richard Schatz, Xiaodan Pang, Oskars Ozolins, Aleksejs Udalcovs, Hadrien Louchet, Sergei Popov, and Gunnar Jacobsen. Laser frequency noise in coherent optical systems: spectral regimes and impairments. *Scientific reports*, 7(1):1–10, 2017.
55. Tao Lu, Hansuek Lee, Tong Chen, Steven Herchak, Ji-Hun Kim, Scott E Fraser, Richard C Flagan, and Kerry Vahala. High sensitivity nanoparticle detection using optical microcavities. *Proceedings of the National Academy of Sciences*, 108(15):5976–5979, 2011.
56. Eric D Black. An introduction to pound–drever–hall laser frequency stabilization. *American journal of physics*, 69(1):79–87, 2001.
57. Jon D Swaim, Joachim Knittel, and Warwick P Bowen. Detection of nanoparticles with a frequency locked whispering gallery mode microresonator. *Applied Physics Letters*, 102(18):183106, 2013.
58. Daryl T Spencer, Michael L Davenport, Tin Komljenovic, Sudharsanan Srinivasan, and John E Bowers. Stabilization of heterogeneous silicon lasers using pound-drever-hall locking to si 3 n 4 ring resonators. *Optics express*, 24(12):13511–13517, 2016.
59. Gavin Young and Philipp Kukura. Interferometric scattering microscopy. *Annual review of physical chemistry*, 70:301–322, 2019.
60. Jiangang Zhu, Sahin Kaya Ozdemir, Yun-Feng Xiao, Lin Li, Lina He, Da-Ren Chen, and Lan Yang. On-chip single nanoparticle detection and sizing by mode splitting in an ultrahigh-q microresonator. *Nature photonics*, 4(1):46, 2010.
61. Jon D Swaim, Joachim Knittel, and Warwick P Bowen. Detection limits in whispering gallery biosensors with plasmonic enhancement. *Applied Physics Letters*, 99(24):243109, 2011.
62. Andrea M Armani, Rajan P Kulkarni, Scott E Fraser, Richard C Flagan, and Kerry J Vahala. Label-free, single-molecule detection with optical microcavities. *science*, 317(5839):783–787, 2007.
63. F Vollmer, S Arnold, and D Keng. Single virus detection from the reactive shift of a whispering-gallery mode. *Proceedings of the National Academy of Sciences*, 105(52):20701–20704, 2008.
64. Martin D Baaske, Matthew R Foreman, and Frank Vollmer. Single-molecule nucleic acid interactions monitored on a label-free microcavity biosensor platform. *Nature nanotechnology*, 9(11):933, 2014.
65. Lina He, Şahin Kaya Özdemir, Jiangang Zhu, Woosung Kim, and Lan Yang. Detecting single viruses and nanoparticles using whispering gallery microlasers. *Nature nanotechnology*, 6(7):428–432, 2011.
66. Joachim Knittel, Terry G McRae, Kwan H Lee, and Warwick P Bowen. Interferometric detection of mode splitting for whispering gallery mode biosensors. *Applied Physics Letters*, 97(12):123704, 2010.
67. Joachim Knittel, Jon D Swaim, David L McAuslan, George A Brawley, and Warwick P Bowen. Back-scatter based whispering gallery mode sensing. *Scientific reports*, 3(1):1–5, 2013.
68. Weijian Chen, Şahin Kaya Özdemir, Guangming Zhao, Jan Wiersig, and Lan Yang. Exceptional points enhance sensing in an optical microcavity. *Nature*, 548(7666):192–196, 2017.
69. Hans-Albert Bachor, Timothy C Ralph, St Lucia, and Timothy C Ralph. *A guide to experiments in quantum optics*, volume 1. Wiley Online Library, 2004.
70. Michael A Taylor and Warwick P Bowen. Quantum metrology and its application in biology. *Physics Reports*, 615:1–59, 2016.

71. SF Pereira, Min Xiao, HJ Kimble, and JL Hall. Generation of squeezed light by intracavity frequency doubling. *Physical review A*, 38(9):4931, 1988.
72. Henning Vahlbruch, Moritz Mehmet, Karsten Danzmann, and Roman Schnabel. Detection of 15 db squeezed states of light and their application for the absolute calibration of photoelectric quantum efficiency. *Physical review letters*, 117(11):110801, 2016.
73. Michael A Taylor, Jiri Janousek, Vincent Daria, Joachim Knittel, Boris Hage, Hans-A Bachor, and Warwick P Bowen. Biological measurement beyond the quantum limit. *Nature Photonics*, 7(3):229–233, 2013.
74. Michael A Taylor, Jiri Janousek, Vincent Daria, Joachim Knittel, Boris Hage, Hans-A Bachor, and Warwick P Bowen. Subdiffraction-limited quantum imaging within a living cell. *Physical Review X*, 4(1):011017, 2014.
75. Catxere A Casacio, Lars S Madsen, Alex Terrasson, Muhammad Waleed, Kai Barnscheidt, Boris Hage, Michael A Taylor, and Warwick P Bowen. Quantum correlations overcome the photodamage limits of light microscopy. *arXiv preprint* arXiv:2004.00178, 2020.
76. Henning Vahlbruch, Simon Chelkowski, Karsten Danzmann, and Roman Schnabel. Quantum engineering of squeezed states for quantum communication and metrology. *New Journal of Physics*, 9(10):371, 2007.
77. Aaron M Goldfain, Rees F Garmann, Yan Jin, Yoav Lahini, and Vinothan N Manoharan. Dynamic measurements of the position, orientation, and dna content of individual unlabeled bacteriophages. *The Journal of Physical Chemistry B*, 120(26):6130–6138, 2016.
78. Joanna Andrecka, Jaime Ortega Arroyo, Yasuharu Takagi, Gabrielle de Wit, Adam Fineberg, Lachlan MacKinnon, Gavin Young, James R Sellers, and Philipp Kukura. Structural dynamics of myosin 5 during processive motion revealed by interferometric scattering microscopy. *Elife*, 4:e05413, 2015.
79. Kalkbrenner Lindfors, Thomas Kalkbrenner, Patrick Stoller, and Vahid Sandoghdar. Detection and spectroscopy of gold nanoparticles using supercontinuum white light confocal microscopy. *Physical review letters*, 93(3):037401, 2004.
80. Kristỳna Holanová, Milan Vala, and Marek Piliarik. Optical imaging and localization of prospective scattering labels smaller than a single protein. *Optics & Laser Technology*, 109:323–327, 2019.
81. Ying-Hsiu Lin, Wei-Lin Chang, and Chia-Lung Hsieh. Shot-noise limited localization of single 20 nm gold particles with nanometer spatial precision within microseconds. *Optics Express*, 22(8):9159–9170, 2014.
82. Mahdi Mazaheri, Jens Ehrig, Alexey Shkarin, Vasily Zaburdaev, and Vahid Sandoghdar. Ultrahigh-speed imaging of rotational diffusion on a lipid bilayer. *Nano letters*, 20(10):7213–7219, 2020.
83. Nicolas P Mauranyapin, Lars S Madsen, Larnii Booth, Lu Peng, Stephen C Warren-Smith, Erik P Schartner, Heike Ebendorff-Heidepriem, and Warwick P Bowen. Quantum noise limited nanoparticle detection with exposed-core fiber. *Optics express*, 27(13):18601–18611, 2019.
84. Richard W Taylor and Vahid Sandoghdar. Interferometric scattering microscopy: seeing single nanoparticles and molecules via rayleigh scattering. *Nano letters*, 19(8):4827–4835, 2019.
85. Adar Sonn-Segev, Katarina Belacic, Tatyana Bodrug, Gavin Young, Ryan T VanderLinden, Brenda A Schulman, Johannes Schimpf, Thorsten Friedrich, Phat Vinh Dip, Thomas U Schwartz, et al. Quantifying the heterogeneity of macromolecular machines by mass photometry. *Nature communications*, 11(1):1–10, 2020.
86. Keith J Mickolajczyk, Elisabeth A Geyer, Tae Kim, Luke M Rice, and William O Hancock. Direct observation of individual tubulin dimers binding to growing microtubules. *Proceedings of the National Academy of Sciences*, 116(15):7314–7322, 2019.
87. Joanna Andrecka, Katelyn M Spillane, Jaime Ortega-Arroyo, and Philipp Kukura. Direct observation and control of supported lipid bilayer formation with interferometric scattering microscopy. *ACS nano*, 7(12):10662–10670, 2013.

Chapter 2
Optoplasmonic Whispering Gallery Mode Sensors for Single Molecule Characterization: A Practical Guide

Sivaraman Subramanian, Kulathunga Mudalige Kalani Perera, Srikanth Pedireddy, and Frank Vollmer

Abstract Optoplasmonic sensors are a new class of single-molecule sensors that leverage the exceptional quality factor of dielectric micro cavities and the strong localization of electromagnetic fields by metal nanoparticles. These sensors have even enabled the detection of the smallest entities in solution, that is, single atomic ions. This chapter describes the working principle of these sensors with a focus on the experimental aspects of the sensor. An overview of various dielectric micro cavities and their fabrication techniques is provided. The synthesis and fabrication of metal nanoparticles and the techniques for combining them with micro cavities are also discussed. Various aspects of the sensor such as the sensitivity, limits of detection, time resolution, methods for functionalization, and robustness are discussed. Finally, solutions and approaches to overcome many of the limitations are presented and an outlook on the future applications of these sensors is presented.

2.1 Introduction

Optoplasmonic sensors are a class of sensors developed in the past decade that combine the high quality factors of dielectric microcavities and the small nanometer-scale localization of electric fields by metal nanoparticles to achieve exceptional sensitivity for detecting down to single molecules in solution. These sensors have enabled the detection of even the smallest chemical species in solution, that is, single ions [1]. This chapter describes the working principle of these sensors with a focus on the

S. Subramanian · K. M. Kalani Perera · S. Pedireddy · F. Vollmer (✉)
Living Systems Institute, University of Exeter, Stocker Road, Exeter EX4 4QD, UK
e-mail: f.vollmer@exeter.ac.uk

S. Subramanian
e-mail: s.subramanian2@exeter.ac.uk

K. M. Kalani Perera
e-mail: kp478@exeter.ac.uk

S. Pedireddy
e-mail: s.pedireddy@exeter.ac.uk

W. Bowen et al. (eds.), *Single Molecule Sensing Beyond Fluorescence*,
Nanostructure Science and Technology,
https://doi.org/10.1007/978-3-030-90339-8_2

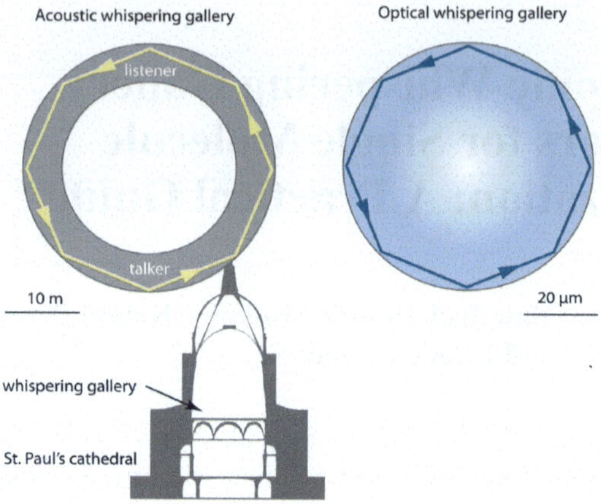

Fig. 2.1 Representations of whispering gallery modes due to total internal reflection of an acoustic wave (left) or an optical wave (right). Image adapted with permission from Foreman et al. [2]. Copyright (2015) OSA

experimental aspects of the sensor. Although multiple dielectric microcavities exist, optical whispering gallery modes in spherical micro cavities (WGMs, see Fig. 2.1) have been predominantly explored for use in these sensors. Hence, the chapter will describe optical WGMs as the system of choice. Nonetheless, the general description of the sensing principle can be applied to any cavity geometry. Any formulations explicit for optical WGMs are highlighted. The plasmonic part of the sensor typically uses rod-shaped gold nanoparticles for sensing. Hence, the experimental details will describe methods mainly considering gold nanoparticles as the system of choice. The theoretical descriptions are however applicable to any metal nanoparticle material and geometry unless stated otherwise.

Optical microcavities and plasmonic nanoparticles fall in the general category of resonance-based sensors. Specific boundary conditions in these systems result in the formation of optical or plasmonic resonances. Similar to a tuning fork, the frequency of the resonances is morphology dependent, that is, dependent on the geometry of the structures, the properties of the material, and the surrounding medium. The interaction of a biomolecule with these resonances results in a change in the resonance spectral properties which can be used to infer details about the interacting species. Figure 2.2 illustrates the optoplasmonic sensing technique. Figure 2.2A shows a sketch of the sensor, in this case, a spherical WGM resonator with a plasmonic gold nanorod attached on its surface. The zoom-in shows a small biomolecule (an enzyme) attached to the gold surface that further interacts with other molecules (substrates in this case). Figure 2.2B shows the transmission spectrum of WGM before (blue) and after (orange) the binding of the enzyme to the gold nanorod surface. The zoom-in shows that the binding of the enzyme causes a redshift of the WGM resonance. The WGM spectrum can be obtained continuously in time by scanning the laser wavelength around the WGM transmission dip and extract the resonance position into time traces as shown in Fig. 2.2C. The binding of a molecule to the

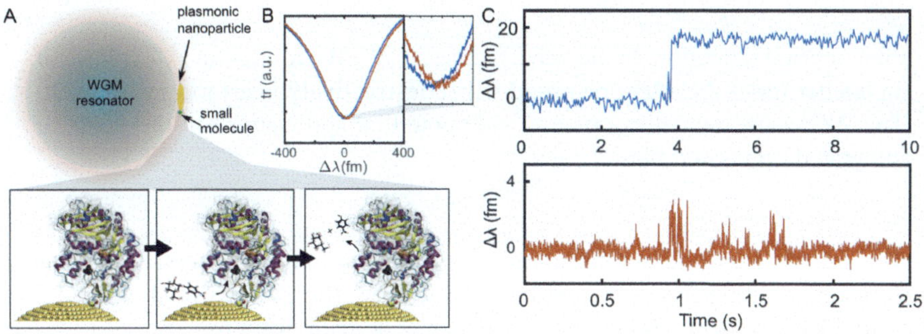

Fig. 2.2 Illustration of single molecule sensing using optoplasmonic sensors. A Schematic of an optoplasmonic sensor made of a WGM resonator and a plasmonic gold nanoparticle. The zoom-in shows a single molecule (an enzyme) attached on the gold surface and its interaction with its ligands (a substrate). **B** Plot shows the WGM resonance spectra before (blue) and after (orange) the binding of a single enzyme on the gold nanoparticle. The binding of the enzyme to the gold nanoparticle translates into a red-shift of the WGM resonance as shown in the inset. **C** Time traces of the WGM resonance. A step-like transition (top) in the time trace indicates binding of a molecule to the gold nanorod and the spike-like transitions (bottom) show transient interactions of molecules in the near-field of the plasmonic gold nanoparticle. Resonance wavelength shifts $\Delta\lambda$ are in units of femtometer (fm). Image adapted with permission from Subramanian et al. [3]. Copyright (2021) American Chemical Society

optoplasmonic sensor results in a step response as shown in the plot on top. Transient molecular interactions within the near-field of the plasmonic nanoparticle for example during substrate turnover by the enzyme result in a spike-like signal as seen in the plot on the bottom.

2.2 Background Information

Understanding the theoretical underpinnings of the optoplasmonic sensor is required for good sensor design and development. In this section, we will describe some useful parameters that enable a physical understanding of the sensor system.

Whispering Gallery Modes

The whispering gallery modes (WGMs) were first introduced by Lord Rayleigh in 19th century to describe the curvilinear propagation of the sound waves under the dome of St. Paul's Cathedral in London. The relevant concept has been generalized to the light waves tangentially reflected at the closed concave surface of an optical cavity such as a 100 um glass microsphere, see Figure 2.1. The confinement of the light on optical WGMs originates from the continuous total internal reflection. In the wave picture, a WGM acquires an extra phase of an integer times 2π after one round-trip, constructively interfering with itself. The WGM microcavities possess wide potential applications in the fields of integrated optics and biosensing [4].

The signals from the optoplasmonic sensor are primarily observed as changes in the spectral properties of the optical microcavity. The power stored in the cavity exhibits a Lorentzian lineshape in the frequency domain given by,

$$P(\omega) = P\frac{(\kappa/2)^2}{(\omega-\omega_0)^2 + (\kappa/2)^2}, \tag{2.1}$$

where P is the power, κ is the full-width-at-half-maximum (FWHM) of the cavity resonance spectrum, and ω_0 the resonance frequency [141]. Although direct observation of the cavity power spectrum is usually not feasible, it can be inferred indirectly. For example, the transmitted/reflected light from the cavity can be used to obtain the cavity spectrum. Figure 2.3 shows an example transmission spectrum of an optical microcavity.

A dimensionless parameter Q termed the quality factor captures the temporal confinement of light in the cavity as,

$$Q = \frac{\omega_0}{\kappa} = \omega_0\tau \tag{2.2}$$

where τ is the cavity lifetime [142]. The quality factor is influenced by many loss mechanisms in the cavity such as radiation, material absorption, and surface and bulk scattering. The intrinsic quality factor of the cavity is can be written in the form,

$$Q_{int}^{-1} = Q_{mat}^{-1} + Q_{surf}^{-1} + Q_{scat}^{-1} + Q_{rad}^{-1}. \tag{2.3}$$

Here, Q_{int} is the intrinsic quality factor of the resonator, Q_{mat} describes the intrinsic material absorption, Q_{surf} surface absorption losses, Q_{scat} describes the scattering losses (such as due to imperfections in the form of surface roughness) and Q_{bend} describes bending loss (or radiation loss or tunnel loss) [142].

The transmitted power spectrum shown in Fig. 2.3 is typically obtained indirectly using the light transmitted or reflected from an evanescent coupler such as tapered optical fiber or a prism. Hence, the observed $Q_{spectrum}$ is a combination of the intrinsic and coupling Q given as,

$$Q_{spectrum}^{-1} = Q_{int}^{-1} + Q_{coupling}^{-1} \tag{2.4}$$

where additional losses occur when using the coupler. Typical loaded $Q_{spectrum}$ are in the range of 10^7–10^8 in aqueous medium for glass microsphere resonators of diameter ~80 μm used in optoplasmonic sensing.

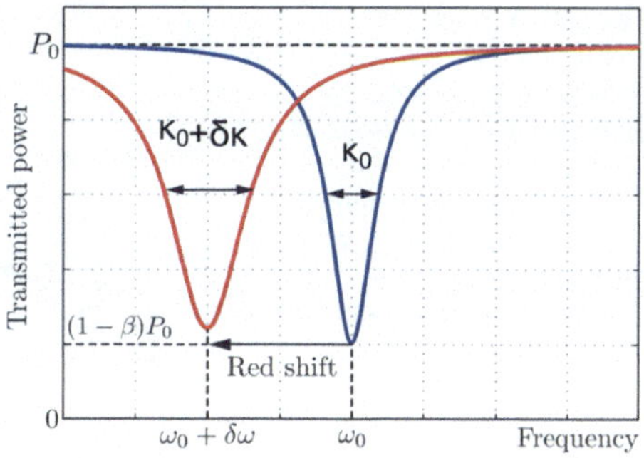

Fig. 2.3 Example transmission spectrum of an optical microcavity, before (blue) and after (red) the binding of a spherical dielectric particle. Image adapted with permission from Foreman et al. [2]. Copyright (2015) OSA

The finesse F of the microsphere cavity is determined from $F \sim Q(\lambda/2\pi R)$, i.e. for $R = 40\,\mu$m, $\lambda = 760$ nm and $Q = 2 \times 10^6$ one calculates $F = 6048$ which corresponds to the number of roundtrips the light takes around the circumference of the microsphere before the intensity decays to $1/e$ of the starting value. The more than 6000 roundtrips of light taken during steady-state measurements enhance the sensitivity of the micro cavity sensor.

The WGM spectrum and Q factor of glass microspheres

WGM transmission spectra are most easily accessed with micrometer glass spheres [5], see Fig. 2.4A. An approximately $R = 200\,\mu$m radius microsphere is easily fabricated by melting a standard (SMF-28, Corning) optical fiber using a micro torch flame. Light is coupled to the microsphere using an optical fiber that is tapered, see Fig. 2.4B. Making the tapered optical fibers requires a fabrication rig to pull and heat the fiber. The glass microsphere is positioned right next to the taper-fiber region using mechanical stages. A distributed feedback (DFB) laser diode with a nominal wavelength around 1340 nm and a near-infrared photodetector are connected to the input and output of the optical fiber. The WGM spectra Fig. 2.4C is acquired by ramping the DFB laser diode current which, above the lasing threshold, tunes the emission wavelength continuously over an approx. 0.2 nm range with a tuning coefficient of approx. 0.01 nm/mA. By tuning the narrow linewidth (1–10 MHz) DFB laser emission it is possible to resolve WGM spectra and their full-width-half-max (FWHM) up to Q factors of more than 10^8.

Another parameter describing the spatial confinement of light in the resonator is the mode volume $\mathcal{V}$. Many definitions are found in the literature to quantify the mode volume of a resonator. For high Q resonators considered in this work, the mode volume can be defined as the energy density integrated over the whole space normalized to the maximum within the optical. Hence, the mode volume is given as,

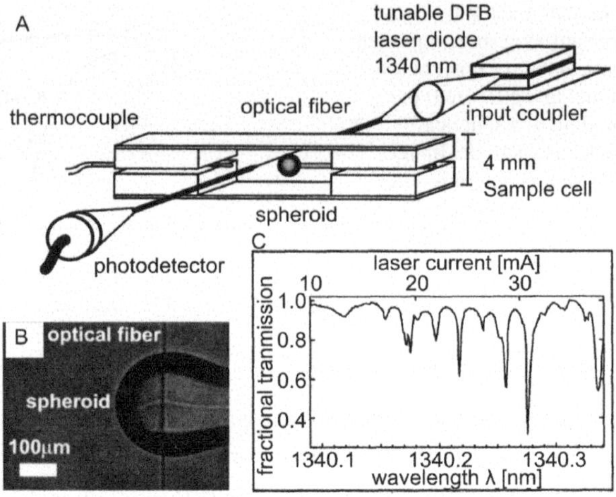

Fig. 2.4 **A** Setup for measuring the WGM transmission spectrum by tuning a DFB laser diode. **B** Glass microsphere made by heating the tip of an optical fiber. **C** Example for a WGM transmission spectrum acquired by ramping the laser diode current. Image adapted with permission from Vollmer et al. [5]. Copyright (2002) American Institute of Physics

$$\mathcal{V} = \frac{\int_v \epsilon(\mathbf{r})|\mathbf{E}(\mathbf{r})|^2 dV}{\max[\epsilon(\mathbf{r})|\mathbf{E}(\mathbf{r})|^2]}. \tag{2.5}$$

where, $\epsilon(r)$ is the permittivity, $E(r)$ is the electric field, and V is the volume [140].

The mode volume of glass microsphere resonators typically used for sensing applications ranges from 10^2–10^3 μm^3.

The ratio of the parameters Q and the mode volume $\mathcal{V}$ (alternatively finesse F) are the important factors determining the sensitivity of a cavity-based sensor. As we will see, maximizing the Finesse and improving the overlap of the circulating light with the target analyte molecules (by using the plasmonic near-field enhancement of gold nanorods) i.e. lowering the mode-volume of the sensor enables the highly sensitive detection of single-molecules.

Trading Q for small V—the optimal microsphere size

The sensitivity of any cavity-based sensor is proportional to the ratio of the parameters $Q/\mathcal{V}$. Hence, it is important to obtain high Q and low $\mathcal{V}$ microcavities. However, a trade-off exists between the maximum Q and the minimum $\mathcal{V}$ achievable. This is because a lower mode-volume $\mathcal{V}$ requires a smaller cavity size, while the cavity Q reduces with cavity size. For example, in the case of spherical WGM cavities, as the cavity size decreases, the radiation losses increases due to the increased curvature of the sphere, decreasing the overall Q. The value of Q also depends on the wavelength of excitation due to the wavelength dependence of absorption-induced losses. There is an optimal microsphere radius for the most sensitive nanoparticle detection. The optimal radius depends on laser wavelength, the cold cavity Q-factor of the fabricated microsphere, the absorption of the light in the surrounding medium (water), and the coupling method (tapered fiber or prism), see Foreman et al. [6]. For a typical optoplasmonic WGM biosensing experiment with excitation of WGM in the VIS—NIR wavelength range, the optimal microsphere diameter is about 80–90 μm. Practically, it is not necessary to control the diameter of the microsphere to better than 5 μm, which can be achieved by real-time imaging during the CO_2 laser-based microsphere fabrication; see also Sect. 2.5. When more of the optical fiber is pushed into the CO_2 laser beam more glass material is melted and a larger sphere is formed, up to a certain limit.

2.3 Why Optoplasmonic Sensing?

Biosensing with cavities initially utilized either microcavities or plasmonic nanoparticles separately. A diverse range of biomolecules could be detected using each of these platforms separately. Plasmonics has a longer history in biodetection as compared to WGM and hence a large volume of literature can be found on the use of plasmonic nanoparticles for biosensing. Although, sensing of small concentrations (nM) of molecules was routinely achieved using plasmonic nanoparticles, attaining the ultimate goal of single molecules evaded researchers until recently [7–9]. Although plasmonic nanoparticles possess a very small mode volume, the extremely poor Q results in a low signal-to-noise ratio (SNR) when using plasmonic nanoparticles as single-molecule sensors.

On the other hand, optical microcavities can possess exceptionally high Q. Nonetheless, the larger mode volume of these resonators results in small enhancements of the single-molecule signals. Shattering a 10 nm-diameter nanoparticle detection limit to enter the arena of single-molecule sensing proved to be difficult just based on dielectric microcavities (mainly using WGMs), see Fig. 2.5. An 80 μm-diameter microsphere cavity with Q of 2×10^6 enabled single Influenza A virus detection with a virus particle diameter of ~100 nm by monitoring the MHz frequency shifts of the WGM resonances [10]. The detection of single BSA protein molecules with the same resonator would require observing a few kHz in WGM frequency shifts. The BSA protein with approximately 3.2 nm Stokes radius has a polarizability that is more than 3500 times smaller than that of the 50 nm radius Influenza A virus. Even with an ultimate Q of 10^9 that one can achieve in principle in WGM silica glass cavities, the noise sources such as the laser noise, thermorefractive noise, etc. make it difficult to reach single-molecule sensing, see Fig. 2.5.

The detection limits for sensing single viruses for a realistic WGM microsphere sensor were considered by Arnold et al. [11]. A more rigorous estimate of the detection limit for small biomolecules based on the Cramer-Rao Lower Bound was estimated by Foreman et al. [6]. The minimum number N of biomolecules (which are considered to be spherical with the polarizability α, required to produce a detectable frequency shift for a WGM sensor where spectra are acquired by laser scanning are calculated as [6],

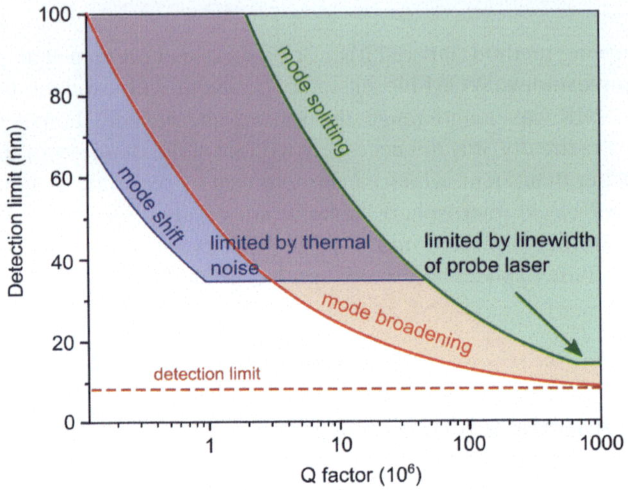

Fig. 2.5 Limits for detecting spherical polystyrene nanoparticles in air through the resonance shift, mode splitting and mode broadening. Reprinted from [12] with permission from John Wiley & Sons, Inc

$$N = \frac{(n_s^2 - n_m^2)}{\Re[\alpha]} \frac{R^3}{|Y_{ll}(\pi/2)|^2} \frac{F}{Q_0} \frac{(1 + Q_c/Q_0)^3}{4Q_c^2/Q_o^2} \tag{2.6}$$

where, n_s is the refractive index of the cavity material, n_m is the refractive index of surrounding medium, Q_0 is the intrinsic quality factor of the WGM, Q_c is the coupling quality factor, Y_{ll} the usual spherical harmonic function and F is a sensitivity factor that depends on experimental configuration. The minimum number N of BSA molecules (polarizability $\alpha = 4\pi\epsilon_0 \times 3.85 \times 10^{-21}$ cm^3, see Eq. 12 for how the molecular polarizability of BSA was determined) that causes a detectable (MHz) shift to a WGM excited in spherical resonator with $R = 40\,\mu$m and $\lambda \approx 780$ nm is approximately 10^4.

Hence, the idea of combining plasmonics with optical microcavities came about to boost the WGM shift signal above the noise, leveraging a reasonably high Q of the optical microcavities and the small-mode volume of the plasmonic nanoparticles. This form of sensing, termed optoplasmonic sensing, detects single molecules as changes in the spectral properties of the optical microcavity when the molecule interacts with the near-field of the plasmonic nanoparticle [13]. The plasmonic nanoparticle provides a signal boost by more than a factor of 1000 (practically often more than 10000). Adding plasmonic nanoparticles does reduce the Q of the optical microcavities. However, the signal enhancement provided by the nanoparticles compensates for this loss of Q by bringing the frequency (wavelength) shifts of the optical microcavity (WGM) above 1 MHz (1 femtometer, fm) even for very small less than 70 Da molecules/ions, see Table 2.1 [14]. Single-molecule detection takes place in the tiny detection volume of the plasmonic near-field (the plasmonic hotspot) thus isolating for single-molecule signals even in the presence of large backgrounds.

Additionally, the strong non-linear decay of the plasmonic near-field enables studying not only the presence of relatively large molecules such as proteins, enzymes, or other nanomachines but also their conformational dynamics due to the different enhancements experienced by different parts of the protein [3, 15].

Adding the plasmonic nanoparticle in optoplasmonic sensing does not fundamentally change the noise floor in practice. Typically, a 1.3–1.5 times increase in WGM linewidth is observed and thus the noise floor of the estimated WGM resonance frequency does not change significantly. Hence, leveraging the near-field enhancements of plasmonic nanorods and taking into account the only mild suppression of the Q factor due to the excitation of the localized plasmon resonances can explain the foundations for the large gains in sensitivity that enable the detection of single molecules on the optoplasmonic sensors. Table 2.1 gives an overview of some of the small molecules and ions which have been detected on plasmonic nanorods coupled to WGM glass microspheres.

2.3.1 Optoplasmonic Sensing Theory

As introduced previously, the optoplasmonic sensor is made up of two components, an optical microcavity and one or more plasmonic nanoparticles attached to the cavity. The principle of optoplasmonic sensing is based on the changes to the spectral properties of the optical microcavity upon the interaction of analytes with the plasmonic nanoparticle excited by the optical microcavity. There exist three main spectral changes that can be monitored to infer about the analytes. These are the shift of the cavity resonance frequency, the change in the cavity linewidth, or the splitting of the cavity resonance.

The case of small particles such as micro/nanoparticles of dielectrics as well as biomolecules such as proteins can be considered using a perturbation theory approach [5, 19]. In Fig. 2.6A, consider the small dielectric particle (in red) diffusing into the evanescent field of the WGM. Assuming the interaction of the particle with WGM is in the weak coupling regime (as opposed to strong coupling which can occur in the case of two coupled resonances), the fractional shift in WGM wavelength (frequency) can be given as [2, 20, 141],

$$\frac{\Delta\lambda}{\lambda} = -\frac{\Delta\omega}{\omega} \approx \frac{\int_{V_p} (\epsilon_p(\mathbf{r}) - \epsilon_m)\mathbf{E}(\mathbf{r})^* \acute{\mathbf{E}}(\mathbf{r}) d\mathbf{r}}{2\int_V \epsilon(\mathbf{r})\mathbf{E}(\mathbf{r})^* . \acute{\mathbf{E}}(\mathbf{r}) d\mathbf{r}} \tag{2.7}$$

where $\mathbf{E}^*(\mathbf{r})$ is the Hermitian adjoint of $\mathbf{E}(\mathbf{r})$, $\epsilon_p(\mathbf{r})$ is the permittivity of the dielectric particle, V is volume of all space, V_p is the volume of the particle, $\mathbf{E}(\mathbf{r})$ is the electric field before the perturbation, $\acute{\mathbf{E}}(\mathbf{r})$ is the perturbed electric field, $\epsilon(\mathbf{r}) = \epsilon_\mathbf{s}$ (microcavity permittivity) for $r < R$ and $\epsilon(\mathbf{r}) = \epsilon_\mathbf{m}$ (host medium permittivity) otherwise. Equation 2.7 can be rewritten in terms of the difference in polarizability of the particle and surrounding medium α_{ex} as,

Table 2.1 Optoplasmonic single molecule/particle detection with/without gold nanorods

Molecule/ nanoparticle	Molecule/particle weight	Sensor system (wavelength) and AuNR plasmon enhancer	Resonance wave-length&linewidth shifts, $\Delta\lambda$&$\Delta\kappa$ if applicable	Noise floor
DNA oligonucleotides [16]	~10 kDa	~80 μm sphere + 40 nm × 12 nm AuNR (780 nm)	~10 fm	~1 fm
DNA polymerase (nm conformational change) [15]	~90 kDa	~90 μm sphere + 35 nm × 10 nm AuNR (642 and 780 nm)	~15 fm	~1 fm
MalL mutant (Ao conformational change) [3]	~66 kDa	~80 μm sphere + 35 nm × 10 nm AuNR (790 nm)	~20 fm binding ~5 fm turnover	~1 fm
Amino acids (Gly, Cys) [14]	Gly ~ 75 Da, Cys ~ 121 Da	~80 μm sphere + 24 nm × 10 nm AuNR (640 nm)	$\Delta\lambda$ ~ 4–30 fm (on gold), $\Delta\kappa$ ~ 4 fm (via disulfide)	~1 fm
Attomolar cysteamine (amine-gold interaction) [14]	~77 Da	~80 μm sphere + 24 nm × 10 nm AuNR (640 nm)	$\Delta\lambda$ ~ 7 fm (max), $\Delta\kappa$ ~ 4 fm (max)	~1 fm
Attomolar dopamine (amine-gold interaction, initial results)	~153 Da	~88 μm sphere + 48 nm × 12 nm AuNR (780 nm)	$\Delta\lambda$ ~ 10 fm (max), $\Delta\kappa$ ~ 4 fm (max)	~ <1fm
Ions (zinc, mercury) [1]	Zn: 65 Da, Hg: 201 Da	~80 μm sphere + 42 nm × 10 nm AuNR (780 nm)	~10 fm	~1 fm
DNA (double-stranded)-XPA protein interaction [17]	~26 kDa	PhC nanocavity + 50 nm gold sphere (1500 nm)	dsDNA ~ 1 pm; gold nanosphere: 440 pm	~0.3 pm
Influenza A virus (110 nm) [18]	~10 MDa, ~5×10^{-16}g	~80 μm sphere, no plasmon enhancer (763 nm)	~11 fm (max)	~4 fm
Adenovirus (100 nm, initial results)	~4 MDa	~90 μm sphere, no plasmon enhancer (640 nm)	~7 fm	~ <1 fm
Polystyrene particle (200 nm, initial results)	1.05 g/cm^3	~90 μm sphere, no plasmon enhancer (640 nm)	~100 fm (max)	~ <1 fm
AuNR (42 nm × 12 nm; 756 nm plasmon resonance) [16]	19.32 g/cm^3	~70 μm sphere, no plasmon enhancer (780 nm)	$\Delta\lambda$ ~ 40 fm (max), $\Delta\kappa$ ~ 90 fm (max)	~1 fm fm: femto-meter

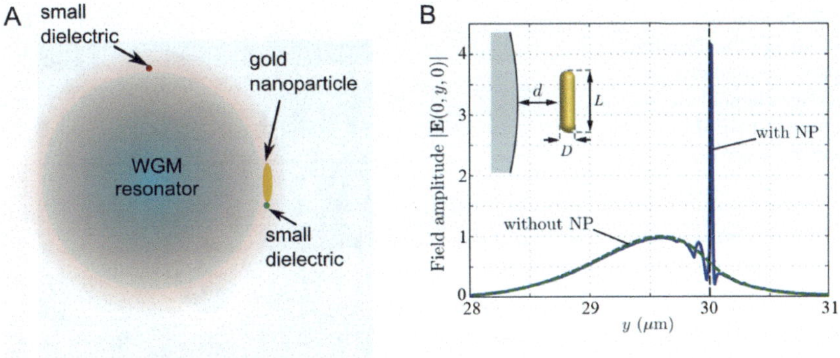

Fig. 2.6 **A** Illustration of a small spherical dielectric nanoparticle interaction with a WGM microcavity surface (red) and another dielectric (green) interacting within the enhanced near-field of the plasmonic nanoparticle attached to the microcavity. **B** Simulated electric field amplitude with and without the presence of the plasmonic enhancement. The field experienced by the green dielectric is at least an order of magnitude higher than that of the red dielectric. Image reprinted with permission from [16] Copyright 2014, Rights Managed by Nature Publishing Group@

$$\frac{\Delta\lambda}{\lambda} = -\frac{\Delta\omega}{\omega} \approx \frac{\Re[\alpha_{ex}]}{2} \frac{f_d |\mathbf{E}(\mathbf{r_p})|^2}{\int_V \epsilon(\mathbf{r})|\mathbf{E}(\mathbf{r})|^2 d\mathbf{r}} \tag{2.8}$$

where r_p is the position of the nanoparticle, f_d is a correction term to account for the exponential decay of the evanescent field [2]. For very small nanoparticles, $f_d \approx 1$. Thus, optical microcavities can be used for detecting small dielectrics (particle in red in Fig. 2.6A) by the shift of the resonance frequency due to the energy stored in polarizing the particle.

The smallest size of a particle detected by a bare microcavity is typically in the range of $d > 10$ nm, where d is the particle diameter. Various sources of noise prevents the detection of smaller entities, especially biological molecules whose refractive index is close to that of water. We will discuss more about noise and sensitivity limits of WGM sensors in Sect. 2.8. For now, it is sufficient to say that additional signal enhancement is required to detect single biomolecular entities in solution. This sensitivity enhancement is provided by the plasmonic aspect of the optoplasmonic sensors. Figure 2.6A also shows a small dielectric particle (in green) interacting within the near-field of the plasmonic nanoparticle attached to the optical microcavity. Now, the signal obtained as the shift in the WGM frequency is orders of magnitude higher (as compared to the particle in red interacting directly with the microcavity) due to the plasmonic enhancement.

A intuitive understanding of this enhanced frequency shift may be obtained by rewriting Eq. 2.8 as,

$$\frac{\Delta\omega}{\omega_0} = \frac{-\Delta\lambda}{\lambda_0} \approx -\frac{\Re[\alpha]}{2} \frac{\epsilon_m f^2(\mathbf{r_p})}{\mathcal{V}} \tag{2.9}$$

where, $f^2(\mathbf{r_p}) = |\mathbf{E}(\mathbf{r_p})|^2 / \max[\epsilon(\mathbf{r})|\mathbf{E}(\mathbf{r})|^2]$ is the normalized mode distribution, $\mathcal{V} = \int_V \epsilon(\mathbf{r})|\mathbf{E}(\mathbf{r})|^2 d\mathbf{r} / \max[\epsilon(\mathbf{r})|\mathbf{E}(\mathbf{r})|^2]$ is the mode volume and $\epsilon(\mathbf{r})$ is the per-

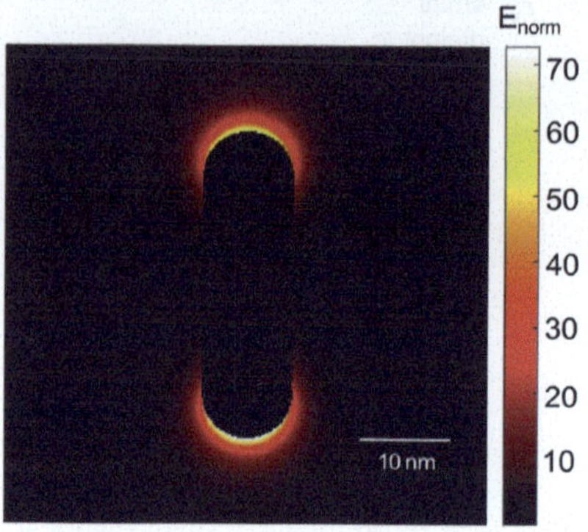

Fig. 2.7 Near-field enhancement of a plasmonic nanoparticle simulated with a boundary-element method using the MNPBEM software package [21]

mittivity of glass inside the resonator and of the surrounding medium outside. The fractional shift in the wavelength (or frequency) is then proportional to the factor $f^2(\mathbf{r_p})/\mathcal{V}$. For a bare WGM cavity, the value of $f(\mathbf{r_p}) << 1$ as the magnitude of the electric field at the particle location (WGM cavity surface) is much smaller than the maximum of the field within the cavity.

Figure 2.6B shows the normalized field amplitude of a WGM cavity with and without plasmonic enhancement; this figure is adapted from the Supplementary Information of Baaske et al. [16]. The boundary (surface) of the WGM cavity is at $y = 30\,\mu\text{m}$. When a plasmonic nanoparticle is added to the WGM cavity surface as shown in Fig. 2.6A, the maximum of the electric field is shifted to the near-field of the plasmonic nanoparticle due to the plasmonic enhancement. The local field enhancement of the plasmonic nanoparticle can now be considered in Eq. 2.9 without including any effects of the hybridization of the WGM and plasmonic resonances as the coupling between the two resonances is weak. The weak coupling assumption is valid in this case as the two resonators have orders of magnitude difference in their decay rates. When a small molecule now binds to the tips of the nanoparticle, the shifts induced in the WGM resonance are higher than when the molecule binds directly to the microsphere surface. The value of $f^2(\mathbf{r_p})$ is now equal to 1 as the maximum field intensity is at the molecule location. Further, the mode volume of the combined system $\mathcal{V}$ is much smaller than the bare WGM as the maximum field intensity is enhanced by the plasmonic nanoparticle. Thus the combination of an enhanced field experienced by the molecule and a reduction in system mode volume results in the observed signal enhancement.

Optoplasmonic sensing within the near-field of gold nanorods

The value of $f^2(\mathbf{r_p})$ within the near-field of a plasmonic nanoparticle such as a gold nanorod can be estimated from various methods such as finite-difference-time-domain (FDTD), Finite Element Method (FEM) and Boundary-Element

Method (BEM) simulations and is typically several orders of magnitude larger as compared to anywhere else at or near the optoplasmonic sensor surface. By adding the plasmon enhancer to the microcavity, ultra-sensitive nanoscale detection volumes are created in the plasmonic sensing hotspots, which extend about 5 nm from the tips of gold nanorods, see Fig. 2.7. In these sensing hotspots, single molecules are detected when they enter the plasmon-enhanced sensing volume where the molecule is polarised. According to Eq. 9, the energy associated with the polarization of the molecule (the induced dipole moment) results in the detectable WGM frequency (wavelength) shift signals, the magnitude of which are proportional to $f^2(\mathbf{r_p})$ and typically on the order of several MHz (several femtometers, fm); see Fig. 2.2C for some example single-molecule signal traces. Detection of molecules results in distinct signal patterns that appear typically either as steps or spikes in the signal traces; these distinct single-molecule signals can be clearly discerned from any background signals that may occur i.e. due to a slow variation of temperature of the sample cell or due to any changes in the composition/refractive index of the sample cell solution, see also Sect. 11. Very often in biosensing, a receptor molecule is immobilized on the gold nanorod in order to detect the specific ligands as they bind to or transiently interact with the receptor molecule. When receptor molecules are immobilized on the nanorod, only those molecules that are immobilized at the tip produce the detectable signals because only at the tip of the nanorod the signal boost by $f^2(\mathbf{r_p})$ is large enough to bring the single-molecule WGM frequency (wavelength) shifts above the noise which is typically about 1 MHz (1 fm).

Freely-diffusing molecules are typically not detected as they pass a 10 nm wide detection volume within less than a microsecond which is the time resolution of a standard optoplasmonic sensor, see Sect. 2.9.

The size and aspect ratio of the nanorod is usually chosen such that the WGM excites the localized surface plasmon near/at resonance, resulting in a large $f^2(\mathbf{r_p})$. Nanorods with plasmon resonances in the VIS to NIR spectral range are commercially available or they can be synthesized in the laboratory, see Sect. 6. The frequency/wavelength shift signal of the optoplasmonic sensor does typically not increase with the application of higher optical power, see Eq. 9; this is different from other plasmon-based sensing techniques such as those based on surface-enhanced Raman scattering, see Chap. 5. In optoplasmonic sensing, the coupling of a large power to the WGM can be used to exert optical forces on nanoparticles [22] and possibly molecules [23], see also Chaps. 7 and 8. The thermo-optical and non-linear optical effects have to be taken into account when large powers are coupled to the WGMs, see Chap. 2.12.

Adding plasmonic nanoparticles can result in observable WGM mode-split

The microsphere supports clockwise (CW) and counter-clockwise (CCW) traveling WGM modes that have equal (degenerate) resonance frequency. When a strong scatterer such as a plasmonic nanoparticle is added to the microcavity surface, the CW and CCW WGM couple with each other through the backscattering and Rayleigh scattering. This, as a result, gives rise to the spectral splitting of the previously degenerate WGM modes. If the Q-factor of the WGM cavity is high enough, typically $>10^7$, a separation (splitting) of the WGM modes can be observed by the appearance of two closely-spaced peaks in the transmission spectra. It is worth noting that the small splitting distance between the two spectral peaks may either increase or decrease as additional plasmonic nanoparticles are added because of the phase differences of the scattered light. Very often during the optoplasmonic sensor assembly, when the plasmonic nanoparticles are added, the double-peaks are not resolved and therefore the mode-split is usually observed as part of an overall increase of the FWHM linewidth which occurs each time another plasmonic nanoparticle is added. The linewidth changes together with the wavelength shifts are used for the real-time monitoring of the number of nanoparticles that are attached to the microcavity during assembly, see Sect. 7.

In the case that the mode-split is resolved, each one of the split WGM resonances interacts with plasmonic nanoparticles differently; $f^2(\mathbf{r_p})$ in Eq. 9 is now different for each one of the split modes. For the example of just one plasmonic nanoparticle attached to the WGM microcavity and excited red-detuned from plasmon resonance, the lower-frequency WGM split-mode is associated with a larger $f^2(\mathbf{r_p})$ because the plasmon resonance is excited more efficiently and resulting in the shift of this split mode to the lower frequency as the particle is added. At the same time, the linewidth of the lower-frequency split-mode increases. The higher-frequency split mode, on the other hand, excites the plasmon resonance less effectively and $f^2(\mathbf{r_p})$ for this mode is smaller. The lower-frequency split-mode sets up a CW/CCW interference pattern at the microsphere surface such that the plasmonic nanoparticle is at/closer to an anti-node (maximum), whereas the higher-frequency WGM split-mode sets up a CW/CCW interference pattern such that the plasmonic nanoparticle is at/closer to a node (minimum).

The maximum nanoparticle mode-split (shift) of 2 g can be measured on cavities. The 2 g mode split is twice as large as the corresponding WGM shift $\Delta\lambda$ measured at lower Q when the WGM mode-split is not resolved: $\Delta\lambda = g$.

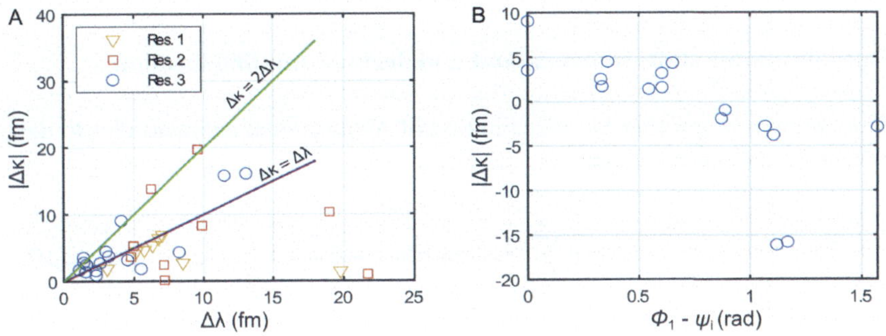

Fig. 2.8 **A** WGM linewidth shifts ($|\Delta\kappa|$) magnitude versus WGM resonance wavelength shift ($\Delta\lambda$) for binding of small sub-kDa molecules measured using three different resonators. Solid lines of $\Delta\kappa = \Delta\lambda$ and $\Delta\kappa = 2\Delta\lambda$ are also plotted for reference. **B** Relative angular position ($\psi_1 - \psi_i$) of the molecules binding on resonator 3 estimated from the ratio of $\Delta\kappa$ and $\Delta\lambda$. Image adapted with permission from Subramanian et al. [25]. Copyright (2021) American Institute of Physics

Single-molecule linewidth shifts can be larger than the wavelength shifts

For the case that the plasmonic-nanoparticle induced mode-split is not resolved, single-molecule signals can be observed not only from the WGM frequency/wavelengths shifts but also from the changes of the effective FWHM linewidths of the unresolved mode-split. The perturbation by the molecule (which is proportional to $f^2(\mathbf{r_p})$) is different for each mode [14]. Figure 2.8 shows that the single-molecule FWHM linewidth shifts can provide higher amplitude signals (points between the green and purple line) than previously predicted using a first-order perturbation theory that did not consider the mode splitting introduced by the plasmonic nanoparticles [25]. The optoplasmonic single-molecule linewidth shift $\Delta\kappa$ can vary from $-2g$ to $2g$ with $g = \Delta\lambda$ being the predicted wavelength shift magnitude from the perturbation theory Eq. 9. Recently, Vincent et al. [14] showed linewidth shifts $\Delta\kappa$ for the interaction of single-L-cysteine and DTNB (5,5-dithio-bis-(2-nitrobenzoic acid)) with the thiol group of cysteamine molecules using a similar sensor. They showed that $|\Delta\kappa| > 2\Delta\lambda$, that is, the points would fall above the green line in Fig. 2.8A. The shifts in $\Delta\kappa$ on the order of 8 fm with no significant shifts in $\Delta\lambda$ above the noise are attributed to an alternate sensing mechanism that requires additional investigation and is not additional discussed here. Note that the change in the sign of the linewidth shifts can be either positive or negative. This shows that the observed single-molecule linewidth shifts do not arise from additional losses induced by the molecules which are non-absorbing at the WGM wavelength. The shifts rather arise from the change in the frequency splitting of the standing wave modes. Since the small frequency splitting is unresolved, this is observed as the change in resonance linewidth.

Optoplasmonic shifts inform about a molecules' binding location

The frequency splitting S_N after binding of N plasmonic nanoparticles to the WGM can be given as [26]

$$S_N = \sum_{i=1}^{N} 2g_i \cos(2\psi_1 - 2\psi_i), \tag{2.10}$$

where, $g_i = \Delta\omega_i \equiv -\Delta\lambda_i$ is the splitting caused by the ith particle, ψ_1 is the spatial phase difference between the first particle and the anti-node of one standing wave mode, and ψ_i is the angular distance between the ith particle and the first particle. When a molecule interacts with the near-field of the plasmonic nanoparticles, it can be seen as introducing a shift of g' according to 2.8. A corresponding change in the splitting S_N is introduced depending on the position of the gold nanoparticle where the molecule binds. This change in splitting translates into a change in the linewidth as the splitting is unresolved. A value of $\Delta S = S'_N - S_N < 0$ provides a negative shift (decrease) in the linewidth, and $\Delta S > 0$ provides a positive shift (increase) in the linewidth. The relative azimuthal position of the binding molecules can be estimated using Eq. 2.10 as plotted in Fig. 2.8B. It is seen from the plots that the binding events cluster around six different values of $\psi_1 - \psi_i$. The clustering of the events indicates the relative azimuthal position of the gold nanoparticles attached to the WGM resonator. The differences in amplitude of $\Delta\kappa$ within each cluster arise from the different enhancements experienced by the molecules attached to different locations on the same nanoparticle. Moreover, the polar position of the nanoparticles can be estimated using multiple WGMs of different polar orders [27].

2.4 Sensing Single-Molecule Reactions and Interactions

Numerous single-molecule systems, from interaction of small sub-kDa molecules to the conformational dynamics of large polymerase proteins have been explored using optoplasmonic sensors. A few examples of the systems studied in literature and a description of the experiments are presented in this section.

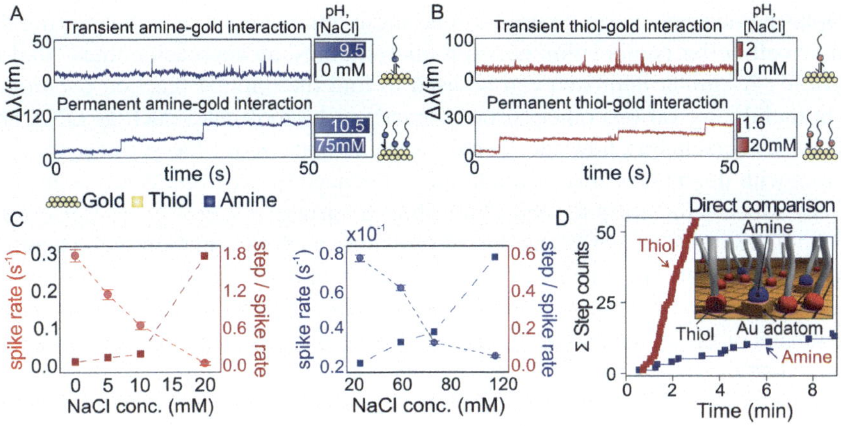

Fig. 2.9 **A** WGM wavelength shift signals $\Delta\lambda$ in units of femtometer measured for the amine head group of a single stranded DNA oligonucleotide (26 bases long) at different pH and salt (NaCl) concentrations. **B** WGM wavelength shift signals measured for the thiol head group of a single stranded DNA oligonucleotide (26 bases long). **C** Optimizing the yield of the surface reactions (step/spike rate) by varying NaCl concentration. **D** Number of cumulative binding steps for thiol in comparison to amine DNA molecules on the same gold nanorods. Reprinted from [28] by permission of John Wiley & Sons, Inc

2.4.1 Sensing Ligand Surface Reactions

Ligand surface reactions from low to high affinities were studied for the thiol and amine gold reactions. The interaction of a primary amine head group of a modified DNA oligonucleotide (26 bases) with CTAB covered gold nanorods in the low to neutral pH range is transient. In the low pH range, the amine remains in its nonreactive, charged form (NH_3^+). Gold nanoparticles on optoplasmonic sensors can monitor these transient interactions as they occur with off-rate constants ($1/k_{off}$) on the order of milliseconds, see Fig. 2.9A. At a high pH, most of the amine groups are deprotonated (NH_2) and the lone electron pair of the nitrogen atom can consequently interact with gold atoms. Covalent bonds are formed which are observed from steps in the optoplasmonic sensor traces.

An opposite trend with pH is observed for the thiol-gold interaction of DNA oligonucleotides (26 bases) with a thiol head group, see Fig. 2.9B. Only below a pH of approximately 2.5, spikes are observed with comparably high rates. This trend continues with the reduction of the pH to 2. At around pH 1.8 step events are observed, occurring simultaneously with a drop in the spike rate as the total number of gold surface atoms available for the thiol-gold bond formation are depleted. Starting from the optimal pH values, equal to 2 for thiol and 10.5 for amine, the affinity between both head groups and the surface gold atoms can be further increased by the introduction of sodium chloride (NaCl) into the solution. Plotting the ratio of the step to spike rate identifies the optimal reaction conditions, see Fig. 2.9C. The optoplasmonic sensors can be used to find these optimal conditions by varying the sample cell buffer composition i.e. with microfluidics, an approach coined "dial-a-reaction". A similar approach can be used to find the optimal reaction conditions for immobilizing various other surface ligands or biomolecules such as DNA and protein. Understanding how the ligand molecule approaches the surface, starts to interact with it, and how this interaction is influenced by environmental parameters are key aspects for grasping and controlling a surface reaction to its full extent. The optoplasmonic sensor is capable of providing such information on a chemical reaction by discerning transient from permanent interactions and thus discerning ligand's access to the surface from a bond formation. For example, the cumulative step counts of the amine and thiol interactions on the gold nanorods reveal that the thiol and amine head groups bind to different gold atoms, see Fig. 2.9D. It is known that amines bind to low coordinated surface gold atoms (adatoms) whereas thiols bind to (100) and (111) gold surfaces.

Attomolar single-molecule detection

Optoplasmonic sensors have recorded single-molecule signals at less than fM (femtomolar) concentrations. *Vincent* et al. [14] reported the real-time detection of molecules binding to gold nanoparticles at concentrations of approx. 100 aM (attomolar). The electrostatic trapping of molecules near/at the nanosensor surface may aid the detection at ultra-low concentration levels. Note that the success of such ultra-sensitive experiments is highly dependent on the type of molecule, here 77 Da cysteamine, the surface interactions (here an amine-gold interaction), the ionic strength, and other buffer conditions. Three other key points should be considered: (1) lowering of the Debye length (1 M NaCl), (2) pH-based enforcement of the molecular charge (thiolate/amine deprotonation above both pKa), and (3) choosing a small molecule such as cysteamine with high diffusivity and low molecular mass. The additional issue to consider is directing the molecules to the sensing site by passivating the glass with a non-fouling coating such as polyethyleneglycol (PEG) as not to sequester analyte molecules and limit their capture by any surface other than the gold nanorod surface. The attomolar detection limit could be optimized further if more nanorods can be bound to the microsphere to increase the number of plasmonic sensing sites.

2.4.2 DNA Sensing by Hybridization

Thiol-modified oligonucleotides 5'-thiol-TTTT-GAGATAAACGAGAAGGAT TGAT were immobilized on CTAB-modified gold nanorods as the receptor strands

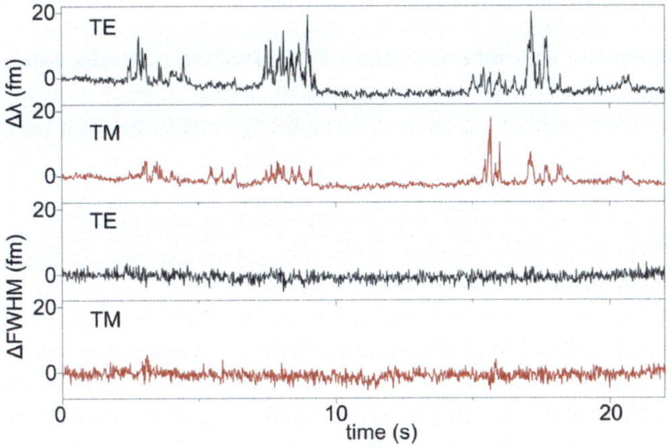

Fig. 2.10 WGM wavelength shift signals recorded the for hybridization of a 22-mer oligonucleotide with its three base-pair-mismatched complementary strand immobilized on a gold nanorod of the optoplasmonic sensor. WGM wavelength shifts were recorded for a TE and a TM mode. No shifts were recorded for the WGM linewidth in this example. Adapted from [16]. Copyright (2014), Rights Managed by Nature Publishing Group

for the specific detection of complementary DNA strands by hybridization [16]. The typical medium condition used for immobilization of the thiol-modified DNA oligonucleotide receptor strands onto the gold nanorods was a pH of 1.8 and increasing concentrations of NaCl in steps of 5 mM up to 50 mM. The number of immobilized oligonucleotides was tracked in real-time. The protocol results in an average of about 10 oligonucleotides per nanorod contributing to the sensing signals in this experiment. The affinity of DNA hybridization to the complementary strand depends on many factors that include the number of complementary base pairs, the specific DNA sequence, pH, salt, and temperature. In this experiment, base mismatches were introduced so that the hybridization kinetics was transient.

Once the nucleotide probe strand was immobilized on the CTAB gold nanorod, the interaction with the (partially) complementary oligonucleotide strand was investigated. Figure 2.10 shows the WGM wavelength shift signals that were recorded for the interaction of the gold-nanorod-bound docking strand with a 22mer analyte strand that contained 3 mismatched base pairs. The melting temperature for this DNA interaction was below room temperature and therefore the transient interaction kinetics between the complementary DNA strands were observed. For the perfectly matching strands, on the other hand, the melting temperature is above room temperature and the single-stranded DNA hybridization to the receptor strands is permanent and recorded as step signals. No WGM linewidths shift signals were recorded in this experiment [16].

How to determine the polarizability of the analyte particle/molecule?

A dielectric nanoparticle's excess polarisability in solution can be expressed as

$$\alpha_{ex} = 4\pi a^3 \frac{\epsilon_p - \epsilon_m}{\epsilon_p + 2\epsilon_m} \tag{2.11}$$

where ϵ_p is the relative permittivity of the particle and ϵ_m is the relative permittivity of the surrounding medium i.e. aqueous solution, $\epsilon = n^2$ with n the refractive index [142]. Using this equation can be a convenient way of estimating the polarizabily of bioparticles such as virus particles with $a = 20$–1000 nm in radius and a typical refractive index of about n = 1.45–1.5. The excess polarizability α_{ex} for protein or DNA is determined from the measurements of the refractive index increment dn/dc of the pure protein/DNA aqueous solution. The excess polarizability is estimated from the refractive index increment by

$$\alpha_{ex} = 2\epsilon_0 n_m \times (\mathrm{dn/dc})_{\mathrm{protein/DNA}} \times \mathrm{MW}_{\mathrm{protein/DNA}} \tag{2.12}$$

where ϵ_0 is the free space permittivity, MW is the molecular weight of the protein/DNA molecule and dn/dc of protein and DNA are approx. 0.183 cm^3/g and 0.166 cm^3/g, respectively [2, 145]. The dn/dc value can be determined from the refractive index measurements of the pure protein/DNA solutions on a standard Abbe refractometer.

Optoplasmonic sensing compared with other single-molecule techniques

Single-molecule techniques influence a molecule system in different ways. For example, fluorescence labels may affect the kinetics and dynamics of biomolecular interactions, while the optoplasmonic approaches use local field enhancements that are not uniform across a plasmonic nanostructure, both of which can play a significant role in the observed statistics. Before one can combine the information obtained from fluorescence and optoplasmonic techniques, single-molecule experiments must be compared and cross-validated.

The detection of DNA hybridization on fluorescence-based DNA-PAINT and optoplasmonic sensors has been cross-validated [29]. The study investigated the impact of (i) the presence of labels, and (ii) the potential influence of the plasmonic nanoparticle surface, and (iii) plasmonic heating effects. The measurements reveal that the dissociation rates of hybridized DNA strands are the same for both techniques within the experimental error. The compatible k_{off} values suggest that there is no significant local temperature increase due to the near-field enhancement of the gold nanorods at the light intensities used with both techniques. On the other hand, the rates of association of the complementary strands show large discrepancies which originate from differences in the surface heterogeneities of the plasmonic nanoparticles, as well as from the fraction of DNA strands that are contributing to the signals.

2.4.3 Sensing Enzyme Conformational Dynamics

The plasmonic near-field of optoplasmonic sensors defines a detection length scale that, fortuitously, is on the order of the diameter of a protein such as BSA with a protein diameter estimated from x-ray crystallography of about 3–4 nm. Any changes in the chemical structure or physical shape (conformation) of a protein such as an enzyme that remains immobilized within the nanometer-scale plasmonic sensing hotspot can be observed in real-time [15] (for a discussion of the time resolution of optoplasmonic sensors see Sect. 2.9). Optoplasmonic sensors are thus sensitive to structural changes in proteins because the detection signal, the resonance frequency or wavelength shift, changes as the overlap of the protein with the highly localized near field changes. Likewise, any chemical changes in the composition of the immobilized protein or the binding of a substrate to an enzyme or a ligand to a receptor would all contribute to the shifts. The variations of the wavelength shift signal $\Delta\lambda$ caused by e.g. conformational (shape) changes of a protein can be estimated from:

$$\Delta\lambda \propto \alpha'_{ex}\Big(\int_{v_m(\alpha_{ex}t_2)} |E(r)|^2 dV - \int_{v_m(\alpha_{ex}t_1)} |E(r)|^2 dV\Big) \tag{2.13}$$

The magnitude and sign of these $\Delta\lambda$ wavelength shifts are proportional to the changes in the electric field intensity integrated over the volume occupied by the molecule $v_m(t)$ at the times t_1 and t_2, where α'_{ex} is the excess volume polarizability of the protein, and $v_m(t)$ is the volume taken up by the protein which depends on the conformation (shape) of the protein adopted at time points t_1 and t_2 [143].

Optoplasmonic sensing of conformational changes has been used to study substrate turnover of active enzymes. Many enzymes undergo conformational changes to find the reaction-ready conformation of the enzyme-substrate complex. A study with the glucosidase enzyme MalL [3] shows that the optoplasmonic sensor traces likely comprise information about chemically distinct processes such as (i) rapid binding of substrate to give ES; (ii) conformational fluctuation/rearrangement to form the catalytically competent binary complex, ES* (a transition state-like conformation); (iii) decay of this complex and chemical turnover to form product EP; (iv) product release and new substrate binding. Figure 2.11A shows the concept for sensing active protein dynamics and the conformational changes associated with substrate turnover. Figure 2.11B shows the specific signals that were obtained for the active enzyme MalL, a glucosidase enzyme. This particular signal pattern has been

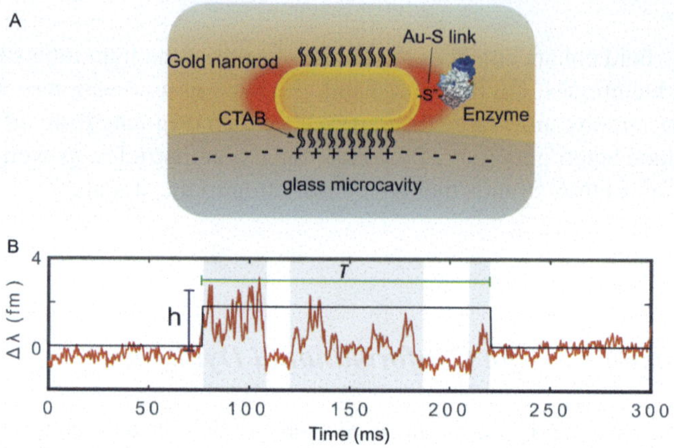

Fig. 2.11 **A** Concept for optoplasmonic sensing of enzyme dynamics, the conformational changes associated with substrate turnover. **B** WGM wavelength shift signals recorded for enzyme Mall turning over substrate pNPG. The gray shaded areas can be found in 10–20% of the WGM signals and may indicate different substates of Mall finding the reaction ready conformation within time period T. The k_{cat} (number of substrate turnovers per second) of a mutant of Mall at the surface was found to be in the same order of magnitude as for the enzyme in solution. Adapted with permission from Subramanian et al. [3]. Copyright (2021) American Chemical Society

observed repeatedly during substrate turnover and may indicate conformational substates of the enzyme searching the reaction ready conformation within time interval τ. It is worth noting that the approximately 4–5 fm WGM wavelength shift signals recorded for MalL correspond to very small conformational/structural changes of MalL that molecular dynamics simulations reveal to be on the order of Angstroms [30]. In another optoplasmonic single-molecule study with the active polymerase enzyme pfu, much larger 15–30 fm WGM shift signals were recorded for what are in that case nm-scale conformational motions of the pyrococcus furiosus polymerase which resemble the opening and closing motion of a hand [15]. It is worth noting that the MalL study required microseconds of time resolution to reveal the enzyme's dynamics whereas a millisecond time resolution was sufficient in the polymerase study.

Temperature dependence of enzymatic rates reveal thermodynamics

The temperature dependence of enzyme turnover is commonly used to study the fundamental thermodynamics and processes involved in catalysis. The thermodynamic parameters are often extracted by fitting kinetic data to the Eyring equation, giving an activation enthalpy and entropy. Recently, however, some enzyme kinetics do not conform to the Eyring model. Hobbs et al. proposed a new model for interpreting these data called macromolecular rate theory (MMRT). This model postulates an activation heat capacity ($\Delta C_P^\ddagger$) associated with the change in conformational dynamics (along the reaction coordinate) between the enzyme-substrate complex and the enzyme-transition state complex,

$$\ln k = \ln \frac{k_B T}{h} - \frac{\Delta H_{T_0}^\ddagger + \Delta C_P^\ddagger (T - T_0)}{RT} + \frac{\Delta S_{T_0}^\ddagger + \Delta C_P^\ddagger (\ln T - \ln T_0)}{R} \tag{2.14}$$

where $\Delta H_{T_0}^\ddagger$ is the change in enthalpy and $\Delta S_{T_0}^\ddagger$ is the change in entropy between the ground and transition state of the reaction at an arbitrary reference temperature (T_0) [144]. Here, $\Delta G^\ddagger$ is the Gibb's free energy, k_B is the Boltzman constant, h is the Planck's constant and R is the universal gas constant.

Optoplasmonic sensing of the conformational dynamics of active enzyme MalL was used to test the MMRT model [3]. WGM wavelength shift signals for the active enzyme were recorded in presence of the substrate pNPG. The signals are grouped to identify the time interval Δt between single substrate turnover events.

For MalL, the kinetic data from the distribution of the interval times Δt, can be fit by more than one exponential $A(t) = \sum_{i=1}^{n \leq 2} A_i \exp(-k_i t)$ [144] revealing that more than one enzyme-substrate conformational state is contributing to the kinetic data. For both of these substates we find that the temperature dependence of their rates follows the MMRT theory confirming a negative $\Delta C_P^\ddagger$ in single-molecule measurements.

2.5 Fabrication of Optical Microcavities

Numerous resonator types have been realized in literature. The optical resonators can be generally categorized according to the principle of operation into whispering gallery mode (WGM) cavities, Fabry-Perot cavities, and photonic crystal (PhC) cavities. Each of these types of cavities is distinguished by the mechanism of trapping light in the cavity. WGM cavities are structures with some circular symmetry where light is trapped due to multiple total internal reflections and the consequent constructive interference of light of specific frequencies. On the other hand, Fabry-Perot cavities are made of two highly reflective surfaces where light is trapped due to reflection at the surface. Finally, PhCs are cavities where light is trapped due to the formation of the so-called photonic band-gaps in dielectric (typically silicon) substrates with structured refractive index profile at the nanoscale (typically by patterning nanoholes on the substrate). Figure 2.12 lists some of the different types of cavities, including waveguide coupled microsphere, fiber-coupled microtoroid, waveguide coupled ring resonator, fiber-coupled capillary (LOORR), bottleneck resonator, micro-bubble, Fabry Perot resonator, fiber-based resonator, and PhC

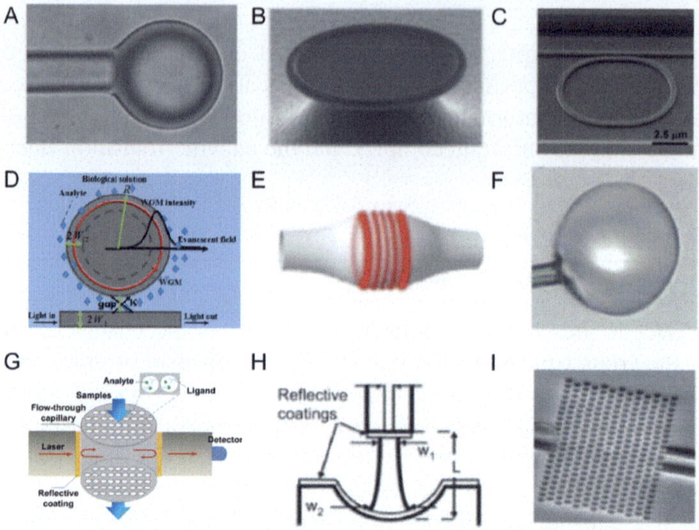

Fig. 2.12 Microcavity resonator sensor geometries. Some of the more common design include **A** waveguide coupled microsphere (reprinted with permission from [31] Copyright 2012 De Gruyter), **B** fiber coupled microtoroid (reprinted with permission from [31] Copyright 2012 De Gruyter), **C** waveguide coupled ring resonator (reprinted with permission from [32] Copyright 2013 Elsevier), **D** fiber coupled capillary (LCORR), **E** bottleneck resonator (reprinted with permission from [33] Copyright 2009 American Physical Society), **F** micro-bubble (reproduced with permission from [34], Copyright 2011 Optical Society of America), **G** Fabry Perot resonator (reprinted with permission from [35], Copyright 2011 AIP Publishing), **H** fiber based resonator (reprinted with permission from [36], Copyright 2005 AIP Publishing) and **I** photonic crystal resonator (adapted with permission from [37] Copyright 2007 Optical Society of America)

resonator, respectively. Table 2.2 summarizes some optical microcavity resonators, their size, fabrication procedure, and quality factor Q which have been utilized for biosensing applications [31].

Amongst the structures presented in Table 2.2, microsphere resonators are extremely easy to fabricate with a high Q factor (10^{10}) in low-loss dielectrics [19, 38]. Glass microspheres can be fabricated by two main techniques; based on the melting process and the sol-gel approach. The melting approach is based on temperature, surface tension, and subsequent cooling processes. In general, the melting process will deteriorate the crystallinity, stoichiometry, and purity of the material. One can initiate the fabrication by melting the raw glass components and then pour the viscous glass onto a mold (spinning plate) or transfer it to a proper stream of liquid nitrogen [39, 40]. On the other hand, one can start the fabrication from the mechanical step of grinding the glass into millimeter or micron-size particles and then sending them into a microwave plasma torch or into a vertical tube furnace. The plasma torch is useful only for melting silica glass, whereas the furnace is designed for melting glasses which have comparatively low glass transition temperatures such as the heavy metal fluoride glasses, the chalcogenide glasses, and the phosphate glasses. The rotating electrical arc was developed to alter the powder of arbitrary glass shapes into micro-

Table 2.2 A summary of some common resonator geometries, Q factor, size, and fabrication methods.

Resonator type	Q factor in water	Typical size	Fabrication methods
Spherical [5, 10, 11, 20, 43–49]	10^8	50–500 μm	CO_2 laser melting, butane/oxygen flame melting of optical fiber tips
Microtoroid [50–53]	10^8	30–200 μm	CO_2 reflow of an under-cut silica micro-disk on silicon wafer
Ring-resonator [54–59]	10^5	20–200 μm	Lithography techniques; substrates include silicon nitride (visible operation), silicon (near-IR) and polymer thin-films
Capillary [60, 61]	10^{7-8}	150 μm	Softening by CO_2 laser and stretching a fused silica capillary
Disk resonator [62, 63]	10^4	10–100 μm	Fabricated from siliconoxy- nitride film on a silicon wafer by lithography technology
Bottleneck resonator [33]	10^8	40 μm	Fabricated from standard optical glass using a two-step heat and pull process
Microtube ring resonator [64, 65]	10^8	10 μm	Roll-up of a strained SiO/SiO_2 Nano membrane to form a micro-tube with a thin wall
Microbubble [34, 66]	10^{3-7}	30–500 μm	CO_2 heating and pulling of glass capillary
Micro-coil [67, 68]	10^6	500 μm	Wind a microfiber coil on a cylindrical rod with a lower refractive index
Fiber-based [36, 69]	10^6	20–200 μm	Resonator is formed between the plane tip of a fiber and a concave micro-mirror fabricated by standard silicon etching and optical coating techniques
Fabry-Perot [70]	10^{4-6}	40 μm	The resonator is formed between two reflectors composed of two period Si/SiO_2 Bragg structures
Photonic crystal cavity [37, 71]	10^6	<10 μm	e-beam lithography and reactive ion etching

spheres [41, 42]. With this technique, a large number of spheres can be made at once, and with high throughput. The size distribution of the spheres is seen as drifting to a large size and hence controlling the fabrication of very small spheres <100 μm can become difficult. Furthermore, fabricating spheres of an exact size repeatedly is challenging; an optical microscope is required to sort out the microspheres of the desired size. The microspheres can be picked up with the help of a glass capillary which is connected to a vacuum pump. Once picked up, the microspheres can be glued to the end of an optical fiber facet [38].

Alternatively, spherical glass resonators can be formed by melting a glass rod or a standard telecom fiber to the desired diameter. Upon heating, the end of the glass fiber forms a spherical volume under the surface tension. This method is reproducible and inexpensive, which gives high-quality spheres. In addition to that, this technique produces a highly spherical, uniform, and very low intrinsic roughness microsphere due to the glass's high viscosity and hence helps to reduce the surface scattering energy losses. Different heating sources, including high-power (e.g., a CO_2) laser, an oxygen/butane or nitrous oxide/butane micro torch, or an electric arc available in a commercial fiber splicer, have been used to initiate the reflow process [43, 72, 73]. The major drawback of these techniques is that they only create one microsphere cavity at a time, and the micro flame process produces asymmetrical spheres [38, 74]. Figure 2.13A illustrates the fabrication of silica microspheres using a CO_2 laser.

The interesting fact of the sol-gel method is that this process starts from the precursor level and ends up with a spherical microsphere and the same process can be used to coat a spherical cavity. For example, silica spheres have been synthesized via the sol-gel method using base-catalyzed hydrolysis of tetraethyl orthosilicate. If the earth metals are incorporated with the synthesis, it is necessary to use an acid catalyst to overcome the hydroxide formation [38]. Furthermore, the sol-gel technique is considered the best method to coat a uniform layer on top of a spherical microresonator [2]. The Q factor of a microsphere is generally reduced by the coated material due to the absorption of the light by the coating and due to the scattering at the interfaces. The microspheres have been coated with many materials including the conjugated polymers, PMMA, nanocrystalline silicon, silica-hafnia glass, and many more. Recently, other methods such as low-pressure chemical vapor deposition (LPCVD) and thermal annealing have been introduced as alternative methods for coating a thin film onto a microsphere [75–77].

The micro bottle resonators are fabricated from the silica fiber by changing the fiber morphology into a bump, which looks like a bottle. Previously, micro bottle resonators were fabricated using the heat and pull method. The most basic method consists of thinning the fiber at two positions such that the thicker area in between these segments forms the bottle shape, as shown in Fig. 2.13B. As per the first demonstration of the micro bottle resonator fabrication, a small section of a fiber capillary is exposed to the melting process using the CO_2 laser beam by pulling the optical fiber [80]. The computer control of the heating and pulling process has been enabled the fabrication of bottle resonators with a Q factor up to 10^8. [33, 81, 82]. In an alternate process, Ganapathy et al. used short sections of optical fiber in a standard fusion splicer to form micro bottles, which is a thermo-mechanical process. The cleaved ends of two optical fibers are pushed towards each other while heated to a particular temperature at which it softens and fuses. The heating process was done using the arc discharge with an arc duration of one second, where it was exploited splicer actions on a piece of fiber to soften a small region with simultaneous compression. The combined splicer efforts resulted in a significant bulge along with the fiber. In addition to that, multiple short arcs were used to soften the glass controllably [81].

Toroid microcavities are made from a 2–3 μm layer of silicon dioxide on the surface of the silicon wafer. The fabrication process consists of four main steps. The first step is photolithography is used to define a disk-shaped photoresist pad on a silicon substrate that undergoes wet thermal oxidation in a horizontal tube furnace. Secondly, this pad pattern acts as a etch mask under hydrofluoric acid (HF). The third step is a selective etching process where the edges of the SiO_2 circular pad are equally undercut with xenon difluoride (XeF_2) just leaving circular silicon pillars to support

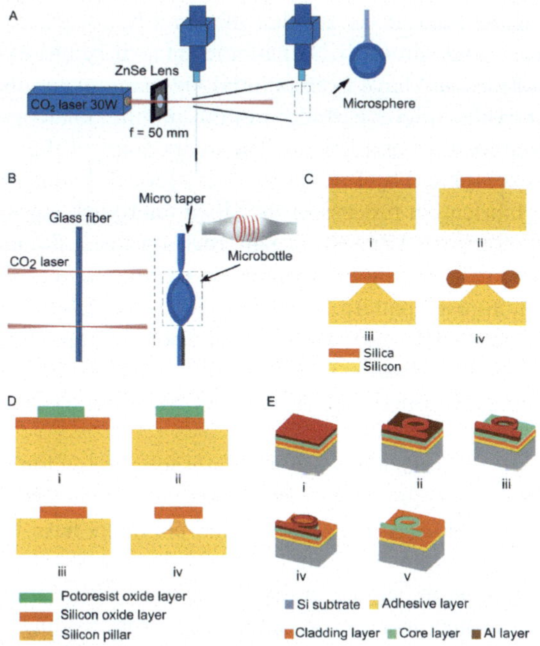

Fig. 2.13 The fabrication process of a **A** microsphere resonator; **B** microbottle resonator; **C** microtoroid resonator: (i) SiO_2 is deposited on a silicon wafer, (ii) HCl etching is applied to create the disk structure on top of the wafer, (iii) XeF_2 etching is used to create a post structure, (iv) CO_2 laser irradiation for the structure to smoothen to form the toroid structure [78]; **D** microdisk resonator: (i) photolithography to pattern the photoresist onto a silicon oxide layer, (ii) reactive ion etching to transfer the pattern of the photoresist into the silicon oxide layer, (iii) remove the residual photoresist using photoresist remover and H_2SO_4/H_2O_2 solution, (iv) undercut the silica microdisk resonator using XeF_2dry etching to form a silicon pillar [78]; and **E** polymer-based waveguide ring resonator: (i) the core layer was spin-coated on the adhesive layer and Al layer deposition and photoresist layer spin-coated on the core layer, respectively, (ii) photoresist layer patterning, (iii) double layers patterning, (iv) core layer patterning, (v) removing the mask, cladding and annealing [79]

the larger SiO_2 disks. Lastly, silica edge of the undercut disk is selectively melt and reflow under a CO_2 laser without disturbing the underlying silicon support pillar. The silica disk shrinks and etches at the border to form the toroidal morphology mainly due to the surface tension. The main fabrication stages of the microtoroid resonators are depicted in Fig. 2.13C [83, 84]. Recently, electron beam lithography has been utilized to pattern a silicon layer via reactive ion etched. This step was followed by non-selective etching of the native oxide and the main etching using Cl_2, HBr, and O_2 to obtain torroidal resonators [85]. The on-chip fabrication of microtoroids via lithography provides a more precise method to tailor the desired geometry [84].

Microdisk fabrication is quite similar to microtoroids and fabricated via lithography method (wet and dry chemical etching). The reflow process is absent in the microdisk fabrication process, and this tends to result in a lower Q factor due to the higher surface roughness of the disk resonator as compared to the microtoroid. The main advantages of the microdisks are that they can be integrated with micromechanical devices (e.g. high-stress Si_3N_4 nanomechanical beam) to detect the force and mechanical motions and have been tailored via many materials such as silicon nitride, titanium dioxide, silicon carbide, lithium niobite, chalcogenide glass, and polymers in addition to silica and silicon. The major steps of the fabrication of silica microdisk are exemplified by Fig. 2.13D (i–iv) [86, 87] which was reported by Guanyu et al. The fabrication process of the silica microdisk consists of four main steps. Firstly, a photoresist (AZ6130) is patterned on top a thermally grown silicon oxide layer. The resultant mold is subjected to bake to get rid of the sidewall roughness of the photoresist pattern. Then a reactive ion etching step is performed with $SF_6/CHF_3/HeSF_6/CHF_3/He$ to transfer the photoresist pattern onto the silicon oxide coating. The residual photoresist is removed by the solution of H_2SO_4/H_2O_2. As the final step, the silica microdisk is undercut XeF_2 by dry etching to form the silicon pillar [78].

Microrings have been fabricated via photolithography and etching methods. Moreover, these resonators are coupled via a bus waveguide , which is fabricated on the same chip. Silicon insulators, silicon nitride, and polymers have been used as fabrication materials [82]. As an illustration, the fabrication steps of polymer-based waveguide ring resonator are shown in Fig. 2.13E (i–v) [79]. Firstly, a thin film of the adhesive layer is coated on a silicon wafer, and lower cladding is spin-coated on the top of the adhesive layer. The core layer of the same thickness is placed on top of the cladding layer. The aluminum (Al) film is coated on the surface of the core by thermal evaporation deposition. The photoresist layer is spin-coated on the layer of Al, patterned and developed. A mixture of hydrogen peroxide and phosphoric acid is used to remove the exposed part of the Al layer. The double-layer pattern is used as the mask for deep reactive ion etching of the core layer. The solution of hydrogen peroxide and phosphoric acid helps to get rid of the mask. Once the core fabrication is done, it is exposed to spin coating to thicken the upper cladding layer and is cured by ultraviolet light. Typically the Q factor of the ring resonators falls within the range 10^{4-5}. Silicon microrings have been multiplexed and integrated with microfluidics. Additonally, space couplers have also developed [82].

2.6 Plasmonic Gold Nanoparticles: Synthesis and Functionalization

In general, metal nanoparticles are fabricated using two approaches: top-down and bottom-up approaches. The former approach uses traditional lithographic techniques to pattern nanostructures onto a substrate. Though this technique has good control over structures such as clean surfaces, control of composition and to achieve periodic arrays, it involves complex and multistep procedures which may also result in undesirable polycrystalline nanostructures. The cost and time invested increase with the decrease in the size of the nanoparticles. These disadvantages are overcome by the latter approach which uses chemical synthesis routes to directly synthesize the metal nanostructures on a large scale with tailored morphologies. Here we focus on only

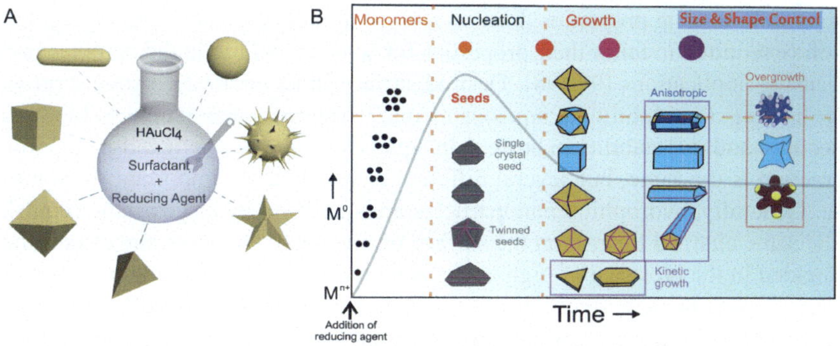

Fig. 2.14 **A** Schematic for chemical synthesis of anisotropic gold nanoparticles ranging from spherical to nanostar morphology. **B** An illustration of synthetic strategies for controlling size and shape of metal nanoparticles along with LaMer diagram

noble metal nanoparticles because of their enhanced optical and catalytic properties. A variety of anisotropic noble metal nanoparticles such as nanorods [88], nanostars [89], nanocubes [90], nanowires [91], nanoprisms [92], octahedron [93] and so forth were prepared using various chemical approaches (Fig. 2.14A).

The synthesis of larger nanoparticles from small seeds is hindered by the formation of a wide distribution of sizes without uniformity. This problem could be overcome by seed-mediated synthesis using step by step growth on seeds and preventing secondary nucleation to control the size of the nanoparticles. In this method nucleation and growth, steps are separated to get better control over the morphology of the nanoparticles. Typically, seeds are synthesized by the reduction of $HAuCl_4$ in a boiling aqueous sodium citrate solution. When the sodium citrate is added, this will reduce the Au^{3+} ions to gold atoms (Fig. 2.14B) [94, 95]. As time proceeds the concentration of gold atoms rises rapidly until it exceeds saturation, resulting in the formation of nuclei. The free gold ions in the solution will then bind to the nuclei surface and growth continues. In this method, the citrate ions act as both reducing agents and surface stabilizers. The size of the nanoparticles can be tuned by choosing the proper ratios of $HAuCl_4$ and sodium citrate. The synthesis typically proceeds by the injection of seed (small cluster of metal) solution to the growth solution which is prepared by the reduction of metal precursor; and by controlling the parameters such as concentration of seed solution, the morphology of seeds, temperature, etc. The seed concentration greatly affects the kinetics of the metal reduction [96].

The metal nanostructures are composed of a metal core which is stabilized by a ligand shell. Ligands, in general, are bifunctional molecules; one end binds to the nanoparticle core-forming compact monolayers and the other end (functional groups such as $-OH$, $-NH_2$, $-SO_3H$ and $-COOH$) faces outwards from the core nanoparticle and dictates the molecular interactions with the surrounding environment and with the biomolecules. As prepared, metal nanoparticles cannot be readily applied in optoplasmonic sensing due to incompatibility of the ligand(s) with the desired nanoparticle sensing properties. Further surface modification of the nanoparticles is therefore required to tailor their properties for sensing and use in the optoplasmonic biosensing applications [97, 98]. Desired surface functionalization should provide the following: (i) solubility in aqueous media, (ii) specific interactions to biological molecules, and (iii) stability in buffers. In aqueous solutions, the ligand-nanoparticle interaction is the same, but several different effects that are important for stability arise. Generally, hydrophilic nanoparticles are stabilized by electrostatic repulsion by the same charged ligands on the surface of the metal. However, the electric field is shielded in the presence of high salt concentrations, and hence, the nanoparticles can come close to each other until the attractive Van der Waals (VdW) forces take over, eventually causing agglomeration. Depending on the isoelectric point and the pH of the solution, nanoparticles can also lose or change the sign of their charge (Fig. 2.15).

Ligand exchange can be used to incorporate a wide variety of functional ligands into the ligand shell of the metal nanoparticle after its synthesis. Ligand exchange can be carried out via the following approaches:

- single-phase ligand exchange,
- ligand exchange via the intermediate ligand and,
- bi-phase ligand exchange with conditions to exchange from hydrophobic to hydrophilic and vice versa.

2.6.1 Single Phase Ligand Exchange

This approach is straightforward and can be used to introduce multiple functionalities on the metal nanoparticles. Typically, the weak binding ligands such as citrate, PVP, and CTAB can be readily exchanged by adding an excess of stronger binding thiolated ligands and stirring overnight at room temperature. The ligand exchange occurs because of the high affinity of thiolate groups for gold surfaces. Because of this high affinity of thiol groups for gold, the thiolate molecules replace the other ligands such as amines. The ligand exchange reaction relies on the equilibrium between the thiolate ligands on the surface of nanoparticles and those in the solution. Hence, using an excess of thiolate ligands in the solution drives the ligand-exchange reaction to completion. For example, complete ligand exchange of CTAB to methoxy poly(ethylene glycol) thiol (mPEG-SH) on gold nanorods requires 12–24 h. This reaction time and kinetics for the complete ligand exchange can be significantly reduced to typically 30 min by carrying out the ligand exchange in Tris buffer with a pH value of 3.0 [99]. Furthermore, the surface modification of gold nanorods (AuNRs) can be preferentially carried out after the synthesis of CTAB coated AuNRs. It has been proposed that CTAB preferentially binds to the 100 and 110 facets of the AuNRs, thus affecting the anisotropic growth. The loosely bound CTAB molecules at the AuNRs tips can be displaced by molecules with higher affinity to the gold

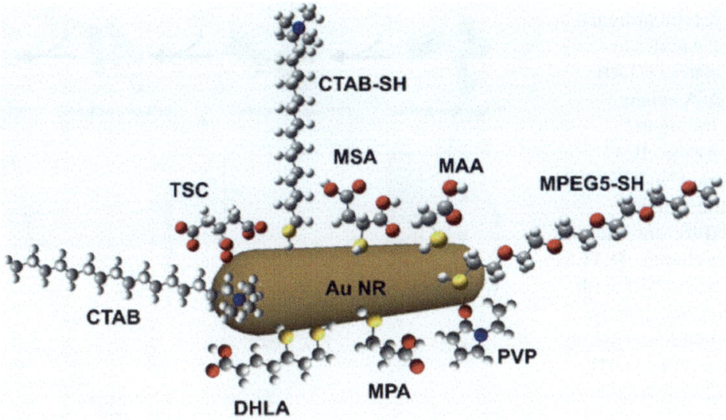

Fig. 2.15 A gold nanorod with different hydrophilic ligand molecules. The molecules are drawn to scale in the above schematics. cetyltrimethylammonium bromide (CTAB), trisodium citrate (TSC), thiolated cetyltrimethylammonium bromide (CTAB-SH), mercaptosuccinic acid (MSA), mercaptoacetic acid (MAA), mPEG5-SH, Polyvinylpyrrolidone (PVP), mercaptopropionic acid (MPA), and dihydrolipid acid (DHLA)

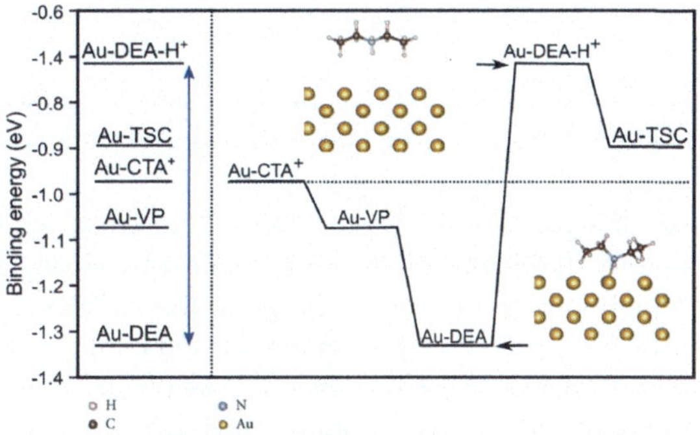

Fig. 2.16 DFT calculations. Left: calculated binding energies ($E_{binding}$) of different capping ligands adsorbed on the Au (110) surface. Right: a typical ligand exchange process based on the pH-tunable binding energy of DEA. Inset: optimized structures of DEA and DEA-H$^+$ on the Au (110) surface. Reproduced under terms of the CC-BY licence [103]. Copyright (2020) AAAS

surface. It has been reported that molecules such as cysteine, cysteamine, PEG-SH, etc., can selectively displace the CTAB molecules at the tips [100–102].

We give two examples of the ligand exchange protocols. Figure 2.16 shows the calculated binding energies of diethylamine (DEA), cetyltrimethylammonium (CTA), trisodium citrate (TSC), and N-vinylpyrrolidone (VP) on Au (110) surface [103]. DEA demonstrated the highest binding energy which is attributed to the formation of strong Au-N bond and increasing order of binding strengths are Au–DEA-H$^+$ (-0.72 eV) $<$ Au-TSC (-0.90 eV) $<$ Au-CTA (-0.97 eV) $<$ Au–VP (-1.07 eV) $<$ Au–DEA (-1.33 eV). Based on the binding strengths the ligand exchange can proceed only via TSC $- >$ CTAB $- >$ PVP. Hence, DEA facilitates the exchange of

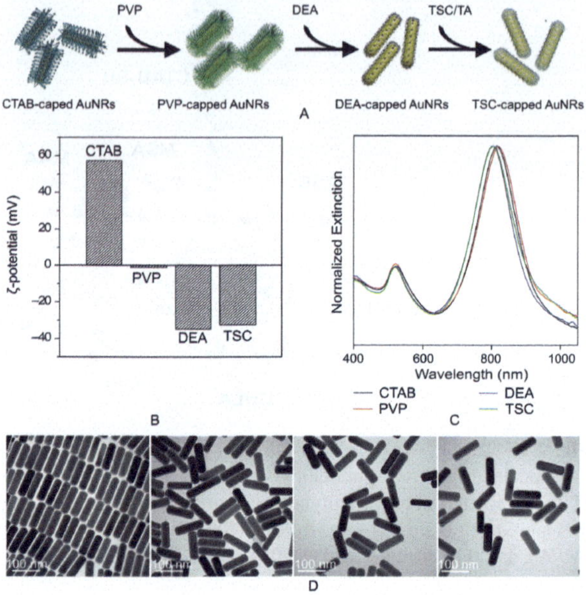

Fig. 2.17 A typical ligand exchange of AuNRs to replace the native CTAB with TSC. **A** A scheme illustrating the ligand exchange process. **B, C** Potentials and UV-VIS-NIR spectra of the AuNRs obtained at different stages of the ligand exchange. **D** TEM images of the AuNRs with different capping ligands. Reproduced under terms of the CC-BY licence [103]. Copyright (2020) AAAS

many commonly used ligands from noble metal nanoparticle surfaces. Interestingly, protonated form of DEA (DEA-H^+) becomes the weakest ligand favoring subsequent ligand exchange with many common ligands. Firstly, the CTAB molecules on the AuNRs are exchanged with PVP molecules in an ethanolic solution by stirring overnight. Secondly, PVP molecules are displaced by strong DEA molecules by stirring for 2 h which impart a negative charge to the AuNRs. Finally, the DEA molecules on the surface are protonated using tannic acid in the presence of TSC ligand for the completed exchange of DEA with TSC. As result, the CTAB on AuNRs was exchanged to weak ligand TSC by exploiting pH switchable binding affinity of DEA on the metal surface. No changes in the shape and size of the AuNRs were discernible by the transmission electron microscopy (TEM) imaging (Fig. 2.17), which could be attributed to the mild conditions that have been involved in the whole ligand exchange process.

In another example, Wei et al., demonstrated the exchange of cetyltrimethylammonium bromide (CTAB)-stabilized Au NRs to citrate stabilized Au NRs using polystyrenesulfonate (PSS) as an intermediate ligand. Yin et al., demonstrated the ligand exchange on CTAB coated Au NRs coating and etching of Cu_2O [104]. The CTAB-coated Au NRs are exchanged with poloxamer ligand F127 via coating and exchange of Cu_2O as a sacrificial layer on the Au NRs.

2.6.2 *Bi-Phasic Ligand Exchange*

Some of the metal nanoparticles are synthesized in organic solvents with hydrophobic ligands such as tetraalkylammonium salts and phosphine ligands as the stabilizing agent. These nanoparticles must be transferred to the aqueous phase and remain in water with no loss of the physical or chemical properties over extended periods. Caruso et al., successfully demonstrated the phase transfer of tetra alkyl ammonium supported Au nanoparticles from toluene to water using 0.1 M 4-dimethylaminopyridine (DMAP) solution [105]. Phase transfer of gold nanoparticles was initiated instantaneously upon the addition of the DMAP solution, with direct transfer across the phase boundary completed within one hour without any agitation of the solution. In another study, Kang et al. reported phase transfer-mediated ligand exchange method is developed for highly selective and rapid synthesis of colloidal phospholipid bilayer-coated gold nanocrystals. The complete replacement of strongly bound surface ligands such as CTAB and citrate by phospholipid bilayer can be quickly achieved by water-chloroform phase transfer [106].

2.7 Optoplasmonic Sensor Assembly: Combining Plasmonic Nanoparticles with Optical Micro Cavities

The sensor assembly is the process of combining the plasmonic nanoparticles with the optical microcavity so that the cavity excites the localised surface plasmon resonances (LSPRs) in the nanoparticles. Depending on the nanoparticle fabrication approach, they are either directly patterned on the cavity surface in case of top-down fabrication or they are assembled using various chemical techniques for a bottom-up approach. The top-down assembly process relies on lithography techniques and other cleanroom processes, while the bottom-up approach can be performed under normal laboratory conditions.

2.7.1 *Bottom-Up Approaches*

Figure 2.18 illustrates the bottom-up approaches for sensor assembly. In the bottom-up approaches, the optical cavity and the plasmonic nanostructures are fabricated separately. They are then assembled in solution by chemical methods by modifying the surface of the optical cavity and/or the surface of the plasmonic particle. Typically, the optical cavity is fabricated from glass surfaces that are negatively charged at a neutral solution pH. Hence, one approach is to use the positively charged nanoparticles such as gold nanorods coated with cetrimonium bromide (CTAB) as the surfactant to electrostatically deposit the nanoparticles on the cavity surface. The exact mechanism of binding of CTAB coated nanoparticles to the glass surface is not known. However, the following procedure has been reported to attach CTAB nanoparticles to glass microspheres. A $\sim$1 pM solution of CTAB coated gold nanoparticles are added

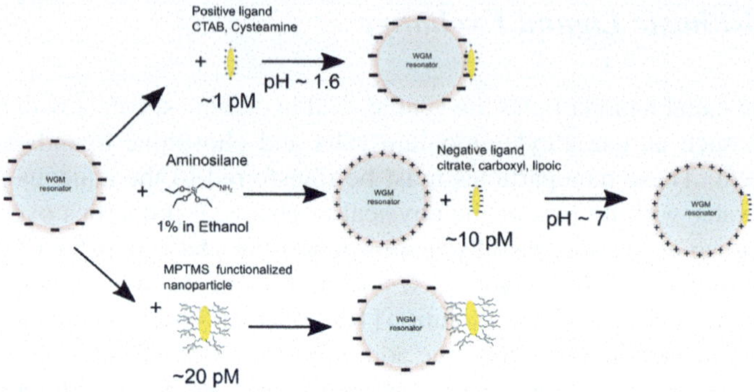

Fig. 2.18 Bottom-up assembly of an optoplasmonic sensor. In the bottom-up approaches, the sensor is assembled in solution via chemical methods

to a chamber containing the microsphere at a solution pH of 1.6. CTAB coated nanoparticles attach to the cavity surface as can be observed via binding step in the WGM wavelength/frequency/FWHM traces [16].

Alternatively, the surface charge of the glass microsphere is transformed by amino silanization of the microcavity. Molecules such as 3-aminopropyltriethoxysilane (APTES) or N-(2-aminoethyl)-3-aminopropyltrimethoxysilane (AEAPTMS) contain silanol (Si-OH) groups that can hydrolyze on the surface of the glass where they form the siloxane (Si-O-Si) linkages [107, 108]. The solvent-exposed amino groups are then typically positively charged at neutral pH, allowing electrostatic attachment of the negatively charged nanoparticles (e.g. citrate, carboxyl, lipoic capped) as shown in Fig. 2.18. The aminosilanization further allows for amide linkages between the primary amines on the glass and carboxyl groups on the gold nanoparticles by the well-known carbodiimide cross-linking reaction [109]. This heterobifunctional cross-linking reaction can be carried out in solution or the vapor phase. First, the surface of the microcavity is activated by immersing the cavity in a piranha solution ($7xH_2SO_4$:$3xH_2O_2$) for 10–20 mins. For solution-phase deposition, a 1–2% solution of the aminosilane is prepared in water, ethanol, or anhydrous toluene. Although using water as a solvent makes the solution preparation simple, the formation of polymer chains of the silanes results in uneven multilayered deposition and a loss of Q. Hence, anhydrous toluene or ethanol is preferred. For a complete deposition, a reaction time of over 24 h may be required. Since partial coverage is sufficient, reaction times as short as 45 s can be used. However, the variation in the layer thickness could result in loss of cavity Q. Aminosilanes deposited from solution have been found to form unstable layers that degrade upon exposure to water over a long time [108]. Alternatively, the vapor phase deposition provides better (thinner) layer thickness with small standard deviations. The vapor phase deposition can be performed by placing the cavity in a chamber with a small amount (~0.5 ml) of the aminosilane. The chamber can then be heated to 90 °C or evacuated to deposit the aminosilane on the microcavity surface. For both solution or vapor phase, the cavity is washed in toluene (2x), ethanol (2x), and water (2x) to displace any weakly bound silanes. The cavity is then typically dried and cured at 110 °C for 15 mins to promote the formation of siloxane bonds.

Finally the surface of the nanoparticles can be coated with an mercaptosilane such as (3-Mercaptopropyl)trimethoxysilane (MPTMS). The thiol (SH) group in MPTMS binds to the gold nanoparticle whereas the silanol groups can bind to the glass following the above procedure for APTES silanization of glass. Once the ligand exchange (from CTAB or citrate to MPTMS) is performed, the gold nanoparticles can be attached to the glass surface in solution at neutral pH conditions.

Table 2.3 lists the common nanorod functionalizations used for immobilizing the nanorods on the resonator. The pros and cons of each functionalization are also listed. Typically, CTAB (cetrimonium bromide) or citrate capped nanorods are easily available via commercial sources. MPTMS ((3-mercaptopropyl)trimethoxysilane) functionalized nanorods have to be synthesized in the lab starting from citrate capped nanorods and exchanging the ligand. CTAB capped nanorods have been predominantly used in the past as they can be attached via a one-step immobilization procedure. This reduces measurement time significantly while increasing the likelihood of obtaining a high Q resonator before nanorod attachment. Citrate capped resonators require the resonator surface to be pre-functionalized with aminosilanes. This process can result in the reduction of the Q due to the non-uniform coating of the aminosilanes. Additional issues with aminosilanization of the resonators arise due to the high positive charge on the surface. Biomolecules having negatively charged regions attach to the resonator surface in bulk further reducing the Q. Hence, anti-fouling agents have to be added to the resonator [14]. However, perfect coverage of the surface with an anti-fouling agent is very difficult to achieve.

2.7.2 *Top-Down Approaches*

Unlike the bottom-up approaches, the top-down approaches require clean-room lithography processes. Figure 2.19 illustrates some top-down approaches for assembly of optoplasmonic sensors. The microcavities such as toroids, disks or PhCs can be fabricated using lithography techniques as described in Sect. 2.5. Then, a photo resist is spin coated on top of the cavity and patterned into the form of nanostructures using E-beam lithography techniques. Then gold layers are fabricated i.e. by sputtering. The resist layer is then lifted-off to obtain gold nanostructures on the microcavity surface (Fig. 2.19A). Alternatively, the gold nanostructures can be patterned on a glass substrate and the microcavity can be placed on top of the nanostructures as shown in (Fig. 2.19B). Since lithography techniques are mainly applicable to planar substrates, these approaches are useful for planar cavities such as PhCs. Due to the requirement of expensive equipment and the long parameter optimization time required, these approaches have not been explored sufficiently in the literature.

Table 2.3 Immobilization of nanorods with different surface functionalizations, and their pros and cons.

Nanorod functionalization	Immobilization method	Pros	Cons
CTAB	Van der Waal's/Electrostatic	Simple one-step process	Not stable above pH 8
		Relatively stable over pH 3–8	CTAB may prevent easy binding of molecules
		Well tested protocols for single-molecule studies in lab	Charged surface may lead to electrostatic interactions with molecules
MPTMS	Covalent	High stability over the entire pH range	Multi-step process
		NPs can be pre-functionalized with receptor/analyte	Nanorods functionalization to MPTMS required
			MPTMS surface density tuning is tricky
Citrate	Electrostatic/covalent	Very stable over large pH range	Silanization required. Multi step process
		Citrate allows easy binding of molecules	Silanization often ruins Q
		Most protocols in the literature use citrate capped nanorods	Positive charge on resonator. Debye screening is difficult

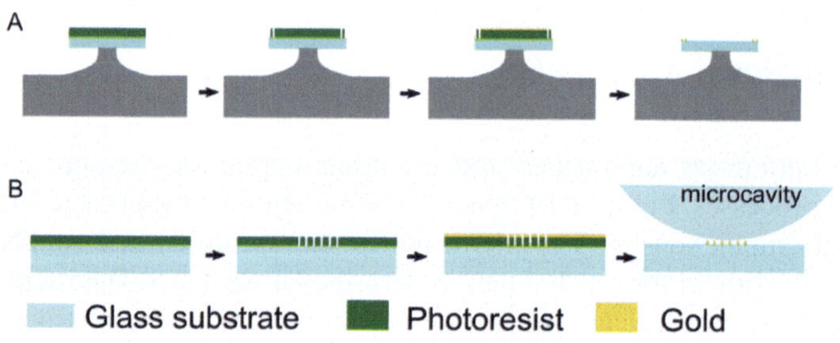

Fig. 2.19 Illustration of the top-down assembly approaches of an optoplasmonic sensor. **A** Gold nanostructures fabricated on the microcavity disk surface by sputtering. **B** Gold nanostructures patterned on a glass substrate that is then brought near the glass microsphere microcavity

2.8 Sensitivity, Signal-to-Noise and Detection Limits

Although optoplasmonic sensors have exceptional sensitivity for detecting single molecules, an important question to address is the ultimate detection limits of such sensors. The fundamental limit of the detection in the classical regime is the shot-noise limit that arises from the quantum nature of light. Although it is possible to circumvent this limit using the quantum properties of light [110], applications of such methods for sensing are challenging and have not yet been shown. For a sensor based on the resonance shift of the WGM, the shot noise detection limit in steady-state is given as [111],

$$\frac{\delta\omega_{min}}{\omega} \geq \frac{1}{Q}\sqrt{\frac{\hbar\omega}{P_0\beta\eta\tau}} \tag{2.15}$$

where $\delta\omega_{min}$ is the minimum resolvable shift, Q is the quality factor of the cavity, $\hbar$ is the Planck's constant, ω is the laser frequency, P_0 is the input power, β is the coupling efficiency, η is the photodetector quantum efficiency and τ is the integration time. For a typical WGM with $Q = 1 \times 10^7, \omega/2\pi = 3.84 \times 10^4$ Hz, $P_0 = 100\,\mu$W, $\eta = 0.9$, $\beta = 0.1$ and $\tau = 20$ ms, the shot noise limited minimum detectable shift is $\delta\omega_{min} =$ 5.75 Hz. The expected resonance shift from single protein of diameter 10 nm and refractive index $n_p = 1.54$ is much higher and is approximately 1.8 kHz. Hence, for a shot-noise limited system using high-Q WGMs, sensing of single proteins should be easily achievable. However, in reality, each measurement system is affected by other noise sources depending on the timescales of interaction. Biological sensors typically have to work in the μs timescale, where noise sources other than the shot-noise limit the detection. Figure 2.20 plots the Allan variance measured for two microtoroid resonators at various integration times [112]. The theoretically estimated detection limit based on thermorefractive noise [113] and shot noise from Eq. 2.15 for a resonator of radius $R = 40\,\mu$m excited at $\lambda \approx 780$ nm are also plotted for reference. The figure is reprinted from Foreman et al. [2]. The figure shows that, at the μ s-ms timescale, the so-called thermorefractive [2, 113] noise limits the sensitivity. Another factor that affects the noise at these timescales is the laser frequency jitter. This puts the minimum detectable shift to much higher values in the range of 10–100 kHz. At longer time scales in the range of 1s or longer, other noise sources such as thermal and laser drifts increase the noise further putting the minimum detectable shift to >1 MHz.

Laser linewidth

The laser linewidth is an important factor for the noise in the sensor. Acquiring the spectra of high-Q cavities benefits from narrow-linewidth lasers. Initial demonstrations of biosensing with high-Q microcavities [5] utilized micro-spherical WGM microcavities and DFB lasers with diode current tuning. The

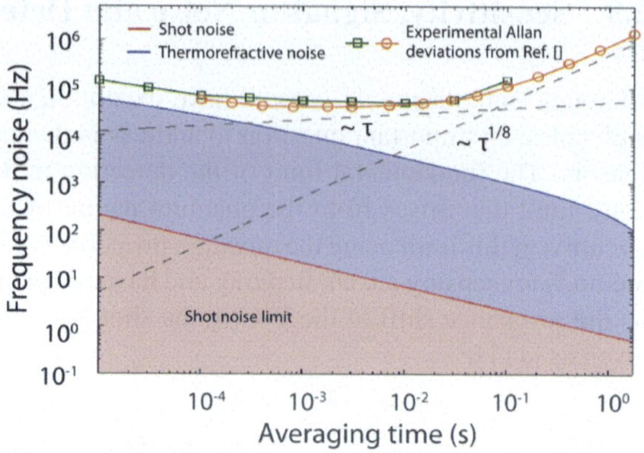

Fig. 2.20 Allan variance of the a WGM resonance frequency for two microtoroid resonators measured by Dobrindt [112]. The theoretical detection limits estimated based on the thermorefractive noise [113] and shot noise are plotted for reference. Image reprinted with permission from Foreman et al. [2]. Copyright (2015) OSA

light was coupled into the microsphere from another single-mode optical fiber that was tapered in the coupling region. Both, WGM microsphere and tapered fiber were immersed in a liquid sample cell. The DFB laser linewidth was (1–10 MHz) at a wavelength of 1340 nm and tuning was performed over a 0.2 nm range with a current tuning coefficient of approximately 0.01 nm/mA. Subsequent demonstrations, especially for single-molecule sensing [16, 114] utilized ECDLs with linewidths <1 MHz reducing the system noise considerably [115].

2.9 Time Resolution

The time resolution of a cavity-based sensor is ultimately limited by the cavity lifetime, that is the time required for the light to decay from the cavity. The quality factor Q of a cavity is related to the cavity lifetime as $Q = \omega_0 \tau$, where ω_0 is the angular resonance frequency in radians and τ is the cavity lifetime in seconds. Typical cavity lifetimes for the resonators described in previous sections are of the order of 0.1–10 ns where a lower Q resonator provides a lower cavity decay time τ. Hence, the signals can be acquired with nanosecond time resolution. The trade-off here is one of the signal bandwidth to sensitivity. As the quality factor of the resonator is inversely proportional to the time resolution, faster acquisition of the signal results in a lower sensitivity. One can imagine this as a higher Q resonator confining light longer and hence accumulates more of the molecular information more before reaching the detector. Hence, the longer confinement leads to a lower time resolution and higher sensitivity, and vice-versa.

Bandwidth and noise trade-off

The implications of the noise as the acquisition bandwidth increases has to be considered. An increase in acquisition bandwidth also increases the integrated frequency noise at each time point.

Typical quality factors for resonators that enable single-molecule sensing are in the range of 10^6 and hence a time resolution of approx. 5 ns can be theoretically achieved. In practice, however, the signals are typically acquired by scanning a tunable laser over a bandwidth B around the cavity resonance to obtain the complete cavity spectrum. This limits the scan time to be slow enough to cover the bandwidth of interest,

$$s > \frac{B}{\kappa}\tau, \tag{2.16}$$

where s is the scan time, κ is the cavity linewidth in Hz, τ is the cavity lifetime in seconds and B is the bandwidth of interest in Hz. The two main methods of tuning a laser are by tuning the temperature of the laser module or by tuning a grating using a piezo. The temperature methods allow for a larger scan bandwidth B but are limited to a time resolution of approx. 100 ms. The piezo tuning has a smaller bandwidth but allows a higher time resolution in the order of 100 kHz. Typically, tuning frequencies are however <1 kHz.

Some alternative methods have been proposed to overcome the technical limitations of laser scanning and acquire the single-molecule signals with a nanosecond time resolution. A frequency lock-in method based on the Pound-Drever-Hall (PDH) lock-in method was used by Swaim et al. [116] to detect single gold nanoparticles in solution with a ms time resolution. This method was extended by Subramanian et al. to detect conformational dynamics of single enzymes using the optoplasmonic sensor with a time resolution of 10 μs. The idea is to lock the laser frequency to the cavity resonance peak using a feedback controller. Fluctuations in the cavity frequency can then be read out continuously by direct measurement of the feedback error signal. Rosenblum et al. [117] proposed another method they termed cavity ring-up spectroscopy. They used laser pulses with sharp rise times slightly detuned from the cavity resonance to obtain interferograms from which the cavity resonance can be obtained by a simple Fourier transform. Although this method has not been applied in the context of bio-sensing so far, it shows promise for acquiring the complete cavity spectrum with a nanosecond resolution.

Locking to the cavity resonance is challenging as the resonance spectrum cannot be directly used as it is symmetric on both sides of the resonance as shown in Fig. 2.21A (top). Nonetheless, the derivative of the cavity resonance as shown in Fig. 2.21A (bottom) is antisymmetric and hence can be used to lock the laser to the cavity. Since there are no methods to directly find the derivative of the cavity spectrum, an indirect method based on frequency modulation is used.

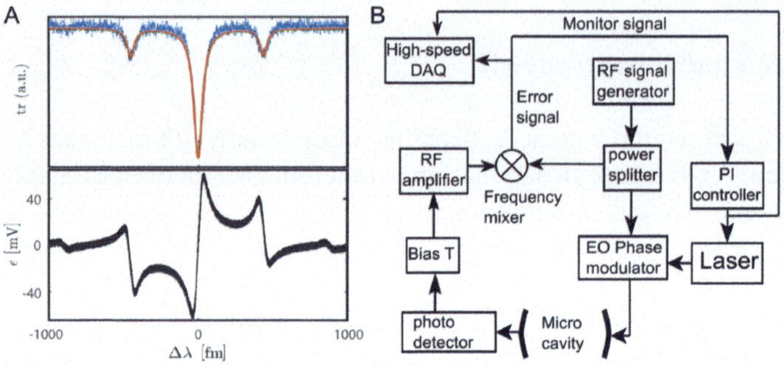

Fig. 2.21 **A** Measured cavity transmission spectrum with sidebands modulated at $w_0 \pm \Omega$. **B** Error signal (derivative if cavity spectrum) obtained after analog demodulation and lowpass filtering

A fast electro-optic modulator is used to modulate the phase of the cavity's incident field at a frequency Ω, such that the incident field takes the form $E_{inc} = E_0 \exp(\omega_0 t + \beta \sin \Omega t)$, where $E_0 \exp(\omega_0 t)$ is the field before the phase modulation and β is the modulation depth. As a result, sidebands are produced at $\omega \pm \Omega$, $\omega \pm 2\Omega$ and so on as shown in Fig. 2.21A. If the modulation frequency is much higher than cavity linewidth, then the sidebands do not accumulate any phase information of the cavity. However, the beat note between the carrier and sidebands contains the relative phase information between the incident and cavity fields.

This modulated signal is then mixed down using an analog mixer and low pass filtered to extract the derivative signal ϵ (also known as the error signal) as shown in the bottom plot of Fig. 2.21A. This error signal is then proportional to the detuning of the laser frequency from the cavity resonance and is given by [118],

$$\epsilon \approx -8\sqrt{P_c P_s}\frac{\delta f}{\kappa} \tag{2.17}$$

where P_c is the power in the carrier, P_s is power in the sidebands and f is the laser frequency and κ is the cavity linewidth. As long as the cavity linewidth is constant, a simple calibration can be used to convert the measured feedback voltages to the frequency shifts.

Figure 2.21B shows a schematic of the circuit diagram to achieve a PDH lock-in of the laser to the microcavity resonance. The PDH lock requires the modulation frequency to be much larger than the cavity linewidth. Hence, the external electro-optic modulator capable of modulations >200 MHz is required. Additionally, RF frequency generators capable of driving high-frequency modulators are also required. The phase-modulated laser is input to the cavity the transmitted/reflected light is collected via a high-speed photodetector (the detector bandwidth should be greater than the modulation frequency). The DC and RF components are split using a Bias-T and the RF component is mixed with the modulation frequency to derive the error signal. The error signal is then fed to a PID controller to provide feedback control to the high-speed modulation input of the laser. The DC component can be used for monitoring purposes and to initially set up the lock.

2.9.1 Cavity Ring-Up Spectroscopy

Although the frequency lock-in provides a fast technique for detecting shifts in the cavity resonance, the complete cavity spectrum cannot be obtained. An alternative approach, cavity ring-up spectroscopy (CRUS) as recently published by Rosenblum et al. [117] uses far detuned pulses to perform a heterodyne measurement to acquire the complete cavity spectrum in the nanosecond time scale. The short rise time of the laser pulses corresponds to a large bandwidth given by

$$B \sim 1/t_r \tag{2.18}$$

where t_r is the rise time of the pulse [117]. If the detuning of the laser from the cavity resonance δ is less than B, then light couples into the cavity and interferes with the specularly reflected light-producing interferograms as shown in Fig. 2.22.

Consider the input field $x(t)$ producing the output field $y(t)$, see Fig. 2.22B, the ring-up signal is then defined $s(t) = (|y(t)|^2 - |x(t)|^2)/x(t)$. Since, the output field is obtained by the interference of the input and cavity fields, the output field in frequency domain can be written as $Y(\omega) = X(\omega) + X(\omega)H(\omega)$ where $H(\omega)$ is the cavity transfer function. Fourier transforming $s(t)$ and noting that $H(\omega)$ only contains positive frequencies due to the large detuning, we get $H(\omega) = S(\omega)/X(\omega)$ [117]. Then, the transmission spectrum $T(\omega) = |Y(\omega)/X(\omega)|^2$ can be rewritten as

$$T(\omega) = |1 + S(\omega)/X(\omega)|^2 \tag{2.19}$$

The resolution of the spectrum obtained is limited only by the Fourier transformation which in turn depends on the signal-to-noise of $s(t)$. Hence, a high coupling efficiency of light to the cavity resonance is required which can be challenging to achieve in practice.

2.10 Optoplasmonic Sensor Instrumentation

The sensor instrumentation for exciting the microcavities and monitoring them can be fairly simple. Typically, a tunable laser source such as a distributed feedback (DFB) laser diode or an external cavity diode laser (ECDL) is coupled into the microcavity using a coupling setup. The laser is then scanned around the optical modes with a scan rate in the range of 0.1–100 Hz using temperature, diode current, or piezo control. Figure 2.23A shows a simple schematic of the instrumentation required for exciting the microcavities and acquiring the spectra. A triangular scanning waveform

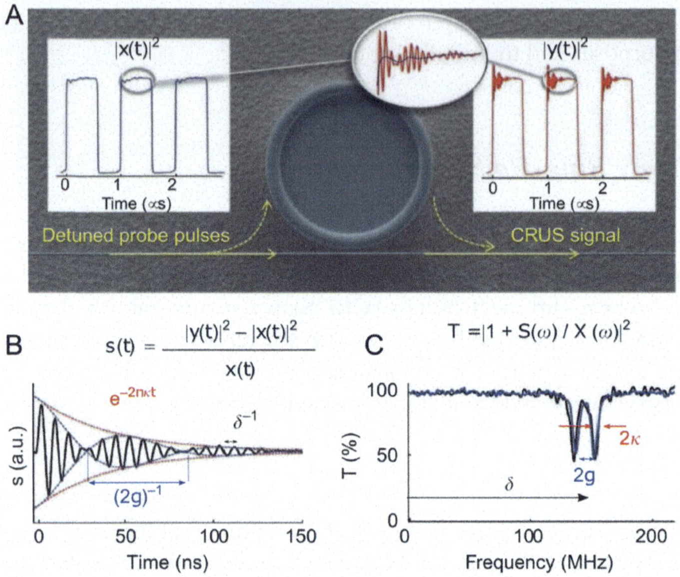

Fig. 2.22 Figure shows the implementation of CRUS by Rosenblum et al. reprinted from [117]. **A** SEM of a fiber coupled microtoroid. Sharp probe pulse (left) and beating signal (right). **B** Shows the signal $s(t)$. The detuning δ is exhibited by the fast oscillations, the linewidth 2κ is given by the exponential decay and beat note gives the g the mode splitting. **C** Fourier transforming $s(t)$ allows the complete retrieval of of the transmission spectrum (black). A comparison (blue) is provided with the spectrum obtained by a scan four orders of magnitude slower. Image reprinted with permission from Rosenblum et al. [117]. Copyright 2015, Rights Managed by Nature Publishing Group

is typically used to tune the laser wavelength over a certain bandwidth around the cavity resonance. The scanning rate depends on the type of tuning performed. In DFB lasers, the temperature of the diode is typically tuned limiting the scan rate to 1 Hz (due to the slow rate of temperature stabilization). Alternatively, the laser diode can be ramped at a rate 100 Hz, hysteresis in laser output typically prevent the faster scan rates. In ECDLs scan rates up to 1 kHz can be achieved as the external cavity is tuned using a fast piezo that houses a grating. The light from a moderate bandwidth photodetector and digitized using an oscilloscope or a data acquisition card (DAQ). For much higher time resolution, the laser is locked to the cavity with a PDH feedback loop that can have a microsecond time constant, see Sect. 2.9. The laser wavelengths can be calibrated using i.e. a high-resolution Fizeau wavemeter.

The coupling of light into the microcavity is straightforward for Fabry-Perot cavities where free-space beams can be coupled via the input mirror or through an optical fiber in the case of a fiber-based Fabry-Perot type resonator. The coupling of light into more exotic cavities such as WGMs and PhCs is slightly more complex due to the high-Q of these cavities. Free-space beams do not easily couple into the optical modes due to the low radiation loss channels in these cavities. Hence, evanescent coupling methods are typically employed. Figure 2.23B shows various schemes for

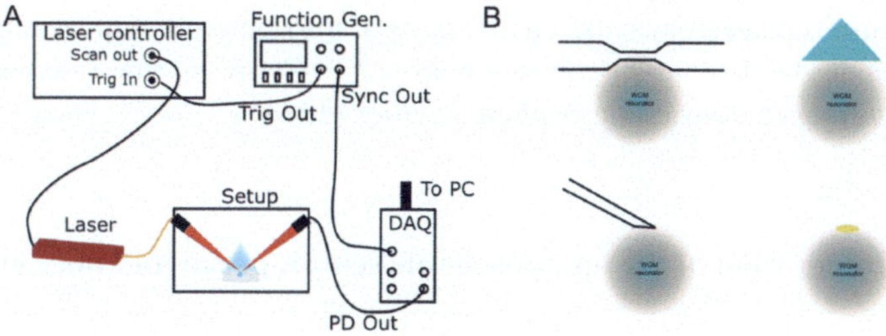

Fig. 2.23 **A** A typical setup to acquire the cavity spectrum. **B** Different methods of coupling light into the optical microcavity

evanescent coupling of light into a WGM microcavity. Many of these methods can also be employed to couple light into other microcavities such as PhCs.

Among these methods, tapered fiber coupling is the most popular in literature [119]. Tapered fibers are fabricated on pulling rigs. The fibers are clamped onto two mechanical stages that slowly move apart as the midsection of the fiber is heated with a brush of a propane/oxygen flame. It is possible to optimize the movement of the stages and that of the flame (flame-brushing technique) in order to fabricate adiabatic tapers with very small insertion loss and that provide the critical coupling to the WGM resonators [120].

Alternatively, a prism based coupling method has been employed for exciting WGMs in spherical and disk resonators. Although the coupling efficiencies using a prism coupler is lower than tapered fibers, robustness of the coupling is improved. Although initial alignment is time consuming, coupling efficiencies of up to 80% can be achieved [121] with robustness against mechanical, thermal and pressure noise sources. Coupling using a prism is based on frustrated total internal reflection (FTIR). Light tunnels from a prism to a WGM resonator when the evanescent field of the latter is overlapped with the evanescent field due to total internal reflection at a prism surface and the phase matching condition is met. Using a coupled mode approach, Gorodetsky et al. derive the conditions for coupling as [121],

$$\sin\Phi = \frac{l}{n_p k R} \tag{2.20a}$$

$$\Delta\Phi^2 = \frac{\sqrt{n_s^2 - 1}}{n_p^2 k R \cos^2\Phi} \tag{2.20b}$$

$$\Delta\Theta^2 = \frac{n_s + \sqrt{n_s^2 - 1}}{n_p^2 k R} \tag{2.20c}$$

where n_p is the refractive index of the prism material, n_s is the refractive index of the sphere, $k = 2\pi/\lambda$ is the wavenumber, R is the sphere radius, l is the mode number, Φ is the angle of incidence at the prim surface, $\Delta\Phi$ and $\Delta\Theta$ are the divergence angles of the Gaussian input beam in the Φ and Θ directions.

Another method of coupling light into microcavities is using scattering from nanoscatterers. A nanoscale Rayleigh structure scatters light in all directions. When such a structure is placed close to a cavity, most of the scattered light is collected into the cavity mode due to Purcell enhancement [122]. Coupling to a free-space beam is possible with asymmetric microsphere cavities [123].

Fluidic integration of the optoplasmonic sensor, prism and fiber coupler

The integration of the optoplasmonic WGM sensor with fluids can be based on the prism coupling and the tapered fiber coupling. Figure 2.24A shows the example for the design of a prism-coupled setup that integrates the sphere melted at the end of the single-mode fiber within a 400 μl U-shaped sample cell made of PDMS. The PDMS gasket is pressed against the prism to make the seal. For temperature control, a thermoelectric element is mounted behind the (thick) glass plate that holds the PDMS gasket. Fluid is exchanged by pipetting. The alignment of the sphere to the prism requires the inclusion of a small glass plate, here triangularly shaped and made of the same glass as the prism. This small glass platelet is glued onto the large prism using index-matching UV curable adhesive. The small glass plate serves as the spacer to be able to align the microsphere to the glass surface where the light is reflected by near total internal reflection. For alignment and during sensing, the fiber with the microsphere is held in a ceramic ferrule mounted on an XYZ stage. The

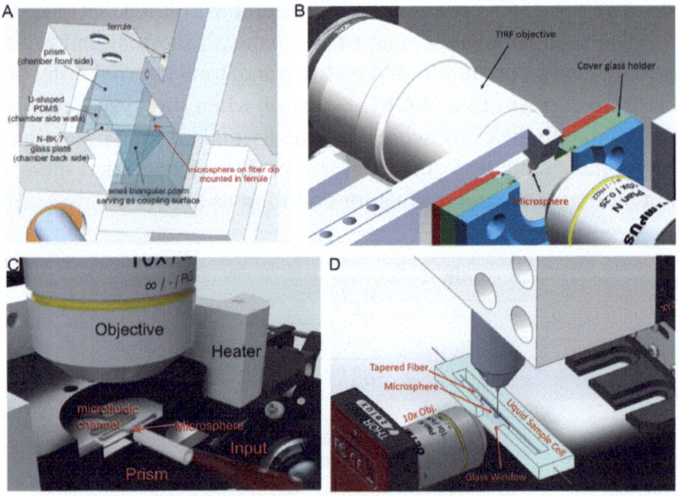

Fig. 2.24 **A** Glass microsphere integrated with PDMS sample cell (trough). Image adapted from [1]. **B** Fluidic cell integraed with glass microsphere and TIR objective. **C** Glass microsphere integrated in a horizontal PDMS fluidic channel. **D** Fluidic sample cell integrated with tapered optical fiber and that is open to the top from where microsphere is positioned

precise positioning of the microsphere is required to control the gap between the microsphere and the prism surface with about 100 nm precision. The optimal gap size is a bit less than 1 μm in a typical WGM sensing experiment in water [6]. For mechanical stability, the mounts for the prism and the XYZ stage that holds the microsphere are both bolted onto a thick base plate. The camera images the reflecting prism surface and the image provides a visual aide for the positioning of the microsphere with respect to the light spot that is indicating where the light is reflected. The spot size is controlled i.e. with a beam expander and lenses to allow for good coupling efficiency which practically can reach anywhere 30–80% [124] for optoplasmonics sensing. Alternatively, the WGM sensing setup uses a prism cut out with a semi-hole of a few mm in diameter. The hole is introduced to position the microsphere at the prism surface despite being held in the ferrule. In prism-coupling setups, the refractive index of the prisms is always higher than that of the microsphere; very often NSF11 or SF9 prims have been used because they provide for a good coupling efficiency and convenient angles for the laser beam alignment.

Figure 2.24B shows the excitation of WGM with a total-internal-reflection (TIR) objective. This design for the optoplasmonic sensor allows single-molecule fluorescence imaging in parallel to the label-free optoplasmonic sensing. For example, DNA-PAINT and other super-resolution microscopy techniques such as PALM and STORM that are based on the use of fluorescence-labeled analyte molecules can be implemented together with the single-molecule optoplasmonic sensing on gold nanorods [29].

Figure 2.24C shows a microsphere integrated with a horizontal PDMS channel on a prism. The fluid enters the channel via standard microfluidic tubing and exits the channel after the microsphere into a waste bin. This design allows the buffer and sample solution to be exchanged with minimal mechanical perturbation and with laminar flow profiles [125]. The microfluidic sample delivery via the channel enables step-changes in buffer composition and higher throughput in sensing. Furthermore, the consumption of sample volume can be minimal.

Figure 2.24D shows a tapered fiber integrated with a horizontally arranged PDMS fluidic chamber that is open to the top from where the microsphere is inserted. This tapered-fiber optoplasmonic setup has the advantage that the light can be critically coupled if one fabricated a sufficiently thin fiber and positions the sphere at the location where the critical (almost 100%) coupling is achieved. Then, almost all of the light transmitted through the fiber is coupled to the microsphere and one can obtain the maximum circulating power in the WGM which can be particularly important when using the WGM microcavity for non-linear sensing, see Chap. 2.12. A drawback of the tapered fiber is that the taper often collects and binds the nanoparticles during sensor assembly such that transmission is lost. Different from the prism, the taper cannot be cleaned after use.

2.11 Signal Acquisition and Analysis

The full spectrum of the WGM is obtained using laser scanning. The recorded spectra are first processed to obtain a WGM resonance position (λ) and linewidth (κ) using a centroid fitting algorithm [1]. Figure 2.25A shows the photodetector voltage obtained using the setups described in the previous section. The x-axis in Fig. 2.25A is represented in points that can be easily transformed to the laser frequency using a simple scaling factor obtained previously using a wavemeter for calibration. Typically, a triangular waveform is used to tune the laser wavelength. Either the upscan (as shown in Fig. 2.25A) or the down scan of the triangular waveform can be used to track the WGM transmission dip (black trace). As seen in the plot, the output power of the laser varies with the triangular waveform. Hence, a background trace before coupling of the WGM is recorded.

First, the position of the WGM transmission dip is found by aligning the resonator with the coupler and the coupling is optimized. Then, the resonator is moved away from the coupler using the micrometer stage, and the background (red trace) is recorded with no coupling. The resonator is moved back to the position of maximum coupling. The WGM spectra at each sweep are divided by the background to normalize the transmission spectrum as shown in Fig. 2.25B. In experiment, the laser power output fluctuates over time and small differences in voltage can be observed over time (the black trace is recorded 10 mins after recording the background red trace). Hence, after dividing the WGM spectrum by the background, the trace is re-normalized to unity as maximum as shown in Fig. 2.25B (blue trace). Then, the

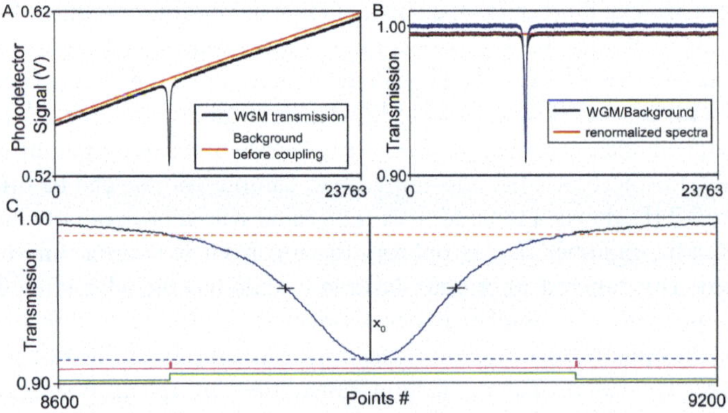

Fig. 2.25 Illustration of the extraction of the WGM resonance wavelength and FWHM. **A** Photodetector voltage of the WGM transmission (black) measured ≈10 min after start of measurement and the background (red) measured before start of measurement. **B** Renormalization of the WGM spectra using the background to compensate for intensity drifts. **C** A zoom of the normalized spectrum and extraction of the mode position and FWHM. Image adapted with permission from Baaske et al. [1]. Copyright 2016, Rights Managed by Nature Publishing Group

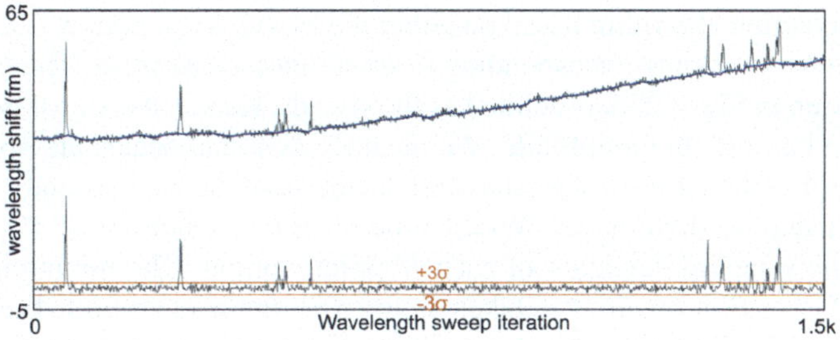

Fig. 2.26 WGM wavelength in fm plotted over the wavelength sweep iteration. Each iteration point corresponds to a 20 ms time interval. The plot on top has been offset for clarity. The WGM wavelength over time is composed of a slow background drift (due to temperature and pressure), a high frequency noise (corresponding to laser linewidth as indicated by 3σ, bottom trace) and the single molecule signals (spike-like transitions). The slow background drift is removed by subtracting the trace (top) with a filtered version of the signal to obtain the detrended trace (bottom). Image adapted with permission from Baaske et al. [1]. Copyright 2016, Rights Managed by Nature Publishing Group

WGM resonance peak position and linewidth are extracted using a centroid algorithm [1] using a threshold (red, dashed) as shown in the zoom-in of Fig. 2.25C.

A custom Labview (National Instruments Inc., USA) program is used to record and process the raw WGM spectra and used to track the WGM resonance position in the PDH lock-in scheme. Once the WGM λ and κ time traces are obtained, the data is analyzed for peaks using the MATLAB GUI. After extraction of the WGM wavelength and FWHM, time traces as shown in Fig. 2.26 (top) are obtained. The WGM wavelength is plotted as a function of the sweep iteration. Each iteration point corresponds to a 20 ms time interval. The plot on top has been offset for clarity. The WGM wavelength over time is composed of a slow background drift (due to ambient temperature and pressure fluctuation [126]), a high-frequency noise (corresponding to laser linewidth as indicated by σ, bottom trace), and the single-molecule signals (spike-like transitions). Hence, drift correction is applied to remove the slow variation of the resonance position ($\Delta\lambda$). The drift of $\Delta\lambda$ occurs due to the slow variation of temperature which is typically on the order 0.2 K over the course of the experiment if the sample-cell temperature is not actively controlled. In the case of the PDH signal, most of the drift is automatically accounted for by the PID feedback to the laser. However, large amplitude changes in pressure and temperature are still measured by the error signal.

A first-order Savitzky-Golay filter with a window length (typically 101 points) depending on the sampling rate is applied to the signal. A low pass filtered version of the time trace (Fig. 2.26 blue trace) is obtained upon filtering. Then, the low-pass filtered trace (blue) is subtracted from the original time trace to obtain the detrended trace as shown in Fig. 2.26 (bottom). Now, only the random noise remains which can be quantified as the minimum standard deviation σ of the time trace. Any transitions in the detrended trace with magnitude above a level of 3σ as marked by the orange bounds (bottom trace) are considered as the signal for further peak analysis.

A peak detection algorithm based on either the MATLAB function *findpeaks* is used to find the $\Delta\lambda$ values corresponding to single-molecule events. The 'spike' like transitions seen in Fig. 2.27A is defined as the signals. Hence, the signal is similar to the noise but has a higher amplitude. We quantify as signal peaks, all 'spikes' with amplitude higher than $3 - 5\sigma$ (the standard deviation of the background). The value of σ is evaluated by dividing the WGM time trace into windows of N points and evaluating the standard deviation of each N-point window. The minimum value of the standard deviation obtained is taken as the σ of the background. Typically, the value is 0.4–0.5 fm. Then, the threshold of detecting peaks using *findpeaks* is chosen manually between $3 - 5\sigma$. Figure 2.27B shows the peaks detected in a segment of the WGM time trace. The triangles show all the individual peaks detected above the threshold. The detected peaks are then classified into single-molecule 'events'. A single molecule 'event' defines the set of signal peaks that belong to one set of interactions (one enzyme-substrate interaction in our case). The detected peaks are coalesced into 'events' using the nearest neighbor search based on a grouping threshold. Essentially, any consecutive peaks within a time separation of τ are coalesced into one single-molecule 'event'. The grouping threshold τ used in this work is between 50–150 ms. Figure 2.27C shows the peaks detected from Fig. 2.27B grouped into single-molecule events using $\tau = 50$ ms. Further, single-molecule events can contain sub-domains. These are extracted by using a smaller grouping threshold of 10–30 ms (Fig. 2.27D). The signal peaks likely indicate the conformational fluctuations of the enzyme after substrate binding as described in the manuscript. The source code of the GUI can be obtained from https://github.com/ssubram905/WGM_DataAnalysis.git.

2.12 Outlook

For optoplasmonic sensors to find more widespread use, a more versatile sensor setup is needed; one that allows for the multiplexed detection of biomarkers in parallel. This platform could use microspheres fabricated by one of the glass melting techniques or by the sol-gel synthesis approaches, see Sect. 2.5. Fabricated in bulk, the free-floating glass microspheres could then be aligned in a microfluidic channel integrated on a suitable prism coupling device. The assembly and disassembly of the array of microspheres could be controlled by actuating flow, or by pressing/releasing microspheres via suitable air channels that are integrated as the control channels together with the fluidic channels in the same PDMS structure [127]. The microspheres could be fabricated in batches; they could be batch-functionalized with different biorecognition elements such as antibodies, aptamers, or proteins; for the specific detection of important biomarkers such as viral biomarkers [128]. Imaging the light scattered by each one of the microspheres as the laser wavelength is tuned across the WGM resonances could enable a parallel readout. Detecting the WGM spectra from the scattered light is furthermore a suitable, nearly background-free approach when the coupling efficiency is low. To couple many microspheres in parallel, the prism coupler could be replaced by a slab waveguide coupler.

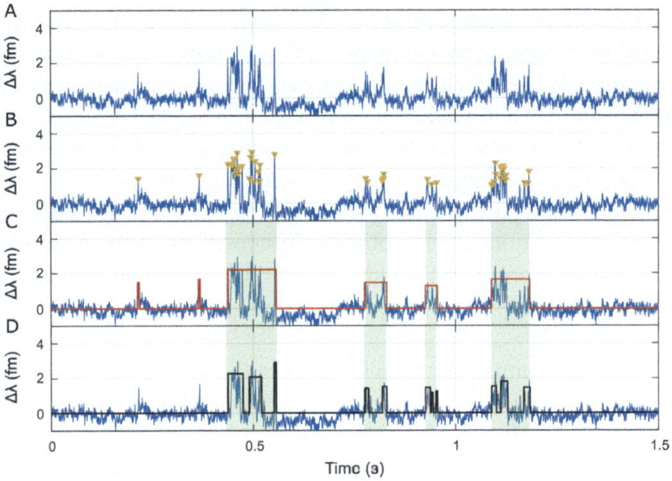

Fig. 2.27 **A** WGM time traces after drift correction. **B** Individual peaks detected using the peak-detection algorithm. **C** Grouped peaks (solid line) with a time threshold of 50 ms. **D** Re-grouping peaks to identify sub-domains with a smaller grouping threshold of 20 ms. Image adapted with permission from Subramanian et al. [3]. Copyright (2021) American Chemical Society

The optoplasmonic sensors sensitivity depends on the orientation of the gold nanorods with respect to the WGM electric field polarization. Optoplasmonic sensing can advance further if the orientation of the gold nanorods is controlled. Here, bottom-up assembly methods are needed that can achieve the assembly of the optoplasmonic sensors with nanorods of defined orientation. For example, a gold nanorod array could be created by assembling the nanorods on a substrate where they are aligned with the help of lithographically (electron beam) defined nanopatterns fabricated in silicon nitride films/wafers [129]. The nanorods can be trapped in the nanopatterns with the help of a receding meniscus as the drop of the nanoparticle solution is drawn across the hydrophobic silicon nitride substrate. Once assembled in the nanopatterns, the nanorods could be transferred onto the surface of a glass microsphere using a PDMS stamp.

The assembly of gold nanorod dimers can provide further gains in sensitivity. For example, a gold nanorod dimer can enhance optical signals by more than a factor of 4 [130]. The DNA Origami technique might be another useful approach for assembling nanorod dimers with a sub-nm precision [131]. Furthermore, the use of plasmonic nanoparticles such as the nanostars can potentially enhance optoplasmonic sensitivity even further; and together with advancing the interferometric measurement schemes, see Chap. 1, this may allow the optoplasmonic sensors to study properties of molecules previously inaccessible such as the single-molecule chirality [132], see also Chap. 9. The automation of the microfluidic flow can enable the rapid screening of the molecules of many different samples. With high-throughput, one could probe drugs for their interactions with target receptors. For example, one could screen a library of drugs for those drugs that trigger a desired conformational change in the target receptor. Here, the target receptor could be one of the G-protein coupled receptors (GPCR). Approximately 35% of approved drugs target GPCRs [133]. Screening for allosteric drugs [134] can help the pharmaceutical industry with developing drugs that have higher efficacy and fewer side effects.

Further, an optoplasmonic sensor with automated microfluidics and data acquisition, real-time signal analysis and sensor feedback could greatly facilitate the fundamental single-molecule studies. Programming of the microfluidic flow would enable the automated bottom-up optoplasmonic sensor assembly. Automated single-molecule measurement protocols could facilitate the repetitive measurement cycles and the implementation of more complicated flow profiles including concentration gradients.

The magnitude and temporal characteristics of the sensor signals contain important information about microscopic dynamics and the identity of molecular species present in the sensing volume. More sophisticated signal-analysis methods will be needed as more complex biomolecular systems are probed. Analysis methods are needed that can scrutinized the WGM sensor signals for all of the information that they carry about the molecular process. The signal analysis might be customized for the specific signal shapes and patterns associated with the biomolecular process. For example, some of our initial results indicate that the turnover of substrate molecules by hinge-bending enzymes such as the phosphoglycerate kinase enzyme PGK produces a characteristic double-peak sensor signal. The double-peak signals may be fit by two Gaussian line profiles to scrutinize the fitting parameters for the information they contain about the molecular aspects of the enzymatic reaction. With improved control over the orientation of an enzyme on the gold nanorod, [3], reproducible sensor signals (with reproducible amplitude levels and/or overall signal patterns) can be observed and this could be used to identify some of the (conformational) sub-states as they are accessed by enzymes or molecular motors [135]. The predictive capabilities of molecular dynamics (MD) simulations could be combined with electromagnetic simulations of the optoplasmonic sensor response. A coarse-grained MD model of a protein combined with the simulation of the corresponding optoplasmonic sensor signal might allow the full reconstruction of the 3D movements; a first step towards reconstructing molecular movies from the optoplasmonic sensor measurements. Reconstructing the full 3D movements of enzymes and nanomachines could be aided by the use of the near-field of an optoplasmonic sensor that could be used to probe different parts of the enzyme/nanomachine. By recording sensor signals with lasers operating at different wavelengths one could probe different volume sections of an enzyme/nanomachine for its movements. This is possible since the decay length of the plasmonic near-field is proportional to the wavelength of the probing light. By exciting WGM in the VIS to near-IR, one could 'scan' the detection volume on a nanometer length scale to obtain sensor signals that contain complementary spatial information about the dynamics of a biomolecular process. Importantly in this context, it has been shown that the optoplasmonic single-molecule signals can be obtained even far-detuned from the plasmon resonance [17].

More advanced signal analysis of the signal patterns may be needed, one that identifies the principal components that may vary slightly because of the internal degrees of freedom of the enzyme/nanomachine. A wavelet-based algorithm, suitable for denoising aperiodic streaming data, could be applied. Combined with the high sampling rate (up to MHz) of the optoplasmonic sensor, the wavelet-based signal processing methodology could enable the real-time detection of the principal components. Furthermore, the real-time (microsecond) signal analysis on field-programmable gate arrays could be used for the real-time control of an enzyme's activity. A feedback loop that controls the enzyme activity could be implemented for example by the rapid heating/cooling of the gold nanorod the enzyme is attached to. In this way, it may become possible to control synthetic biomolecular processes where one steers the enzyme such as a polymerase along its reaction pathway to achieve specific biochemical outcomes. Controlling the synthesis of complex enzymes such as polymerases could become useful for example for the on-demand de novo synthesis of DNA strands.

In the future, the optoplasmonic sensor with its high detection sensitivity and speed could be used to probe biomolecular dynamics and properties that have been difficult to access with other single-molecule techniques. For example, more complex biomaterial systems could be probed such as those that are based on membranes. Lipid membranes could be coated around the plasmonic nanoparticle [136] or they could be deposited by fusing lipid vesicles onto the microsphere sensor at the location of the plasmon enhancer. Binding entire synthetic cells to the sensors may enable us to study the transport of small molecules across passive and active membrane channels. Single-molecule sensing of membranes might also provide the opportunity for combining or complementing the optoplasmonic sensing capabilities with those of the nanopore sensors, see Chaps. 10–13. Combining optoplasmonic sensing with other single-molecule techniques can broaden the applications of the technique. One idea is to combine the optoplasmonic sensor with the optical tweezer technique, see also Chaps. 7 and 8. One could use the plasmonic hotspot of the nanoparticle to exert the optical trapping forces on a protein, i.e. plasmonic tweezer. If successful, the optoplasmonic tweezer could apply very small, femtonewton (fN) forces to make some conformational states of an enzyme more likely than others, and this could be used to control enzymatic activity. The optoplasmonic force sensor could then detect fN forces of enzymes through the application of the minute counteracting optoplasmonic tweezer forces which affect the enzyme movement/function. Furthermore, the combination of the optoplasmonic sensing with the DNA-Paint technique [29] and with other super-resolved microscopies may provide a powerful platform for combining high-resolution sensing and imaging of biomolecular process such as those of biomolecular motors. Separating the molecular signals that originate from the optical force from the signals due to any temperature changes will be one of the challenges.

Spectroscopy could be added to optoplasmonic sensing. An amplitude-modulated pump beam could be used to cause vibrations of the protein due to electrostrictive forces caused by the near-field of the plasmonic nanostructure. Alternatively, the interference of two laser beams could be used to apply the electrostrictive forces at the frequency of the beat note. This could enable a technique for protein fingerprinting by observing the movements of low frequency (GHz) extraordinary acoustic Raman (EAR) modes [137], see also Chap. 8. Optoplasmonic sensors will be capable of detecting GHz motions of protein domains because, from a time-averaging perspective, the spatial overlap of a vibrating protein domain with the probing near field is slightly different as compared to the overlap of a static domain. This change in overlap causes the sensor signal to change. A sweep of the amplitude modulation frequency from 10–100 GHz could provide EAR signals in optoplasmonic spectroscopy. This may enable a novel approach for the identification of a protein from its EAR spectrum.

Extracting most information per photon resource will be ideally suited to study biological samples with minimal perturbation, ultimately probing single molecules with single photons. This would allow optoplasmonic sensors to probe a photosensitive biomolecule for longer time periods than otherwise possible. This approach could also make use of quantum-optical measurement techniques. Applied to optoplasmonic single-molecule sensing, one could achieve the highest possible sensitivity per photon budget [138].

Ensemble measurements of nanoparticles do not provide a detailed understanding of how nanoparticle's size and composition affect the activity and selectivity of a reaction occurring at their catalytic surface. Also, it is important to track the catalyst surface changes and dynamics of reactions occurring at the surface. These surface changes are asynchronous, which are difficult to characterize in ensemble measurements. The choice of nanoparticles on optoplasmonic sensors that catalyze inorganic/organic reactions can provide a window into observing those catalytic processes that have evaded a more detailed analysis, such as the catalytic reactions of single enzymes, nanozymes, metal nanoclusters, and the analysis of low-yield reaction pathways that usually go undetected in bulk measurements. The optoplasmonic sensor system could make contributions to the analysis of these nano chemical events, events that occur at the location of single atoms and molecules and that occur unsynchronized and sporadically in time. Preparing the suitable nanoparticles on the sensor will be one of the challenges to overcome.

To date, the optoplasmonic single-molecule sensing capability has not been commercialized. One way to approach the development of a commercial optoplasmonic sensor platform could be to design throw-away PDMS sample cells that include the microspheres in a fluidic channel on a glass substrate and that are mounted on a stationary WGM reader i.e. with a prism coupler. The plasmonic nanostructure/particle-coupled microspheres could be fabricated to scale in bulk, with customized plasmonic nanoparticles and surface functionalities. The microsphere sensors could even contain a laser-barcode to identify each one of the different sensing spheres and its surface functionalities [139]. Another approach to commercializing the optoplasmonic sensor platform could make use of the fabrication method shown in Fig. 18. Plasmonic nanostructure arrays could be fabricated on a planar substrate that is then combined with the microspheres. With a potential for attomolar single-molecule detection capabilities, the possibility for multiplexed sensing, and automated sample analysis, optoplasmonic sensors could provide for extremely sensitive nanosensors that quickly and accurately detect novel viruses, identify critical health-related biomarkers, and uncover harmful toxins in your drinking water.

In closing, we hope that this manual on optoplasmonic single-molecule sensing will facilitate the uptake of the technique as a research tool. The authors hope this book chapter will inspire many new single-molecule optoplasmonic experiments, together with new optoplasmonic sensor instrumentations and the development of new optoplasmonic detection modalities. We hope that this book chapter is a useful guide for implementing the optoplasmonic sensor and sensing technique in the laboratory. For a more general discussion of the physics of the optical whispering gallery modes and their applications in sensing, we refer the reader to [4].

Acknowledgements We thank Dr Hsin-Yu Wu and Dr Samir Vartabi Kashanian for contributing to Fig. 2.24. FV acknowledges funding from EPSRC EP/R031428/1 and EP/T002875/1 and from the Royal Society (Wolfson Research Merit Award).

References

1. Martin D. Baaske and Frank Vollmer. Optical observation of single atomic ions interacting with plasmonic nanorods in aqueous solution. *Nature Photonics*, 10(11):733–739, 2016.
2. Matthew R. Foreman, Jon D. Swaim, and Frank Vollmer. Whispering gallery mode sensors. *Advances in Optics and Photonics*, 7(2):168, jun 2015.
3. Sivaraman Subramanian, Hannah B.L. Jones, Simona Frustaci, Samuel Winter, Marc W. van der Kamp, Vickery L. Arcus, Christopher R. Pudney, and Frank Vollmer. Sensing enzyme activation heat capacity at the single-molecule level using gold-nanorod-based optical whispering gallery modes. *ACS Applied Nano Materials*, 2021.
4. Frank Vollmer and Deshui Yu. *Optical Whispering Gallery Modes for Biosensing: From Physical Principles to Applications*. Springer International Publishing, Cham, 2020.
5. F. Vollmer, D. Braun, A. Libchaber, M. Khoshsima, I. Teraoka, and S. Arnold. Protein detection by optical shift of a resonant microcavity. *Applied Physics Letters*, 80(21):4057–4059, may 2002.
6. Matthew R Foreman, Wei-Liang Jin, and Frank Vollmer. Optimizing detection limits in whispering gallery mode biosensing. *Optics express*, 22(5):5491–511, 2014.
7. Irene Ament, Janak Prasad, Andreas Henkel, Sebastian Schmachtel, and Carsten Sönnichsen. Single Unlabeled Protein Detection on Individual Plasmonic Nanoparticles. *Nano Letters*, 12(2):1092–1095, feb 2012.
8. Peter Zijlstra, Pedro M.R. Paulo, and Michel Orrit. Optical detection of single non-absorbing molecules using the surface plasmon resonance of a gold nanorod. *Nature Nanotechnology*, 7(6):379–382, 2012.
9. Adam B. Taylor and Peter Zijlstra. Single-Molecule Plasmon Sensing: Current Status and Future Prospects. *ACS Sensors*, 2(8):1103–1122, aug 2017.
10. F Vollmer, S Arnold, and D Keng. Single virus detection from the reactive shift of a whispering-gallery mode. *Proceedings of the National Academy of Sciences of the United States of America*, 105(52):20701–20704, 2008.
11. S. Arnold, R. Ramjit, D. Keng, V. Kolchenko, and I. Teraoka. MicroParticle photophysics illuminates viral bio-sensing. *Faraday Discuss.*, 137:65–83, 2008.
12. Linbo Shao, Xue-Feng Jiang, Xiao-Chong Yu, Bei-Bei Li, William R. Clements, Frank Vollmer, Wei Wang, Yun-Feng Xiao, and Qihuang Gong. Detection of Single Nanoparticles and Lentiviruses Using Microcavity Resonance Broadening. *Advanced Materials*, 25(39):5616–5620, oct 2013.
13. Miguel A. Santiago-Cordoba, Svetlana V. Boriskina, Frank Vollmer, and Melik C. Demirel. Nanoparticle-based protein detection by optical shift of a resonant microcavity. *Applied Physics Letters*, 99(7):073701, 2011.
14. Serge Vincent, Sivaraman Subramanian, and Frank Vollmer. Optoplasmonic characterisation of reversible disulfide interactions at single thiol sites in the attomolar regime. *Nature Communications*, 11(1):2043, dec 2020.

15. Eugene Kim, Martin D Baaske, Isabel Schuldes, Peter S Wilsch, and Frank Vollmer. Label-free optical detection of single enzyme-reactant reactions and associated conformational changes. *Science Advances*, 3(3):e1603044, mar 2017.
16. Martin D Baaske, Matthew R Foreman, and Frank Vollmer. Single-molecule nucleic acid interactions monitored on a label-free microcavity biosensor platform. *Nature Nanotechnology*, 9(August):1–7, 2014.
17. Feng Liang, Yuzheng Guo, Shaocong Hou, and Qimin Quan. Photonic-plasmonic hybrid single-molecule nanosensor measures the effect of fluorescent labels on dna-protein dynamics. *Science Advances*, 3(5), 2017.
18. Frank Vollmer, Stephen Arnold, and David Keng. Single virus detection from the reactive shift of a whispering-gallery mode. *Proceedings of the National Academy of Sciences*, 105(52):20701–20704, 2008.
19. S Arnold, M Khoshsima, I Teraoka, S Holler, and F Vollmer. Shift of whispering-gallery modes in microspheres by protein adsorption. *Optics letters*, 28(4):272–274, 2003.
20. Iwao Teraoka and Stephen Arnold. Theory of resonance shifts in TE and TM whispering gallery modes by nonradial perturbations for sensing applications. *Journal of the Optical Society of America B*, 23(7):1381, jul 2006.
21. Ulrich Hohenester and Andreas Trügler. MNPBEM - A Matlab toolbox for the simulation of plasmonic nanoparticles. *Computer Physics Communications*, 183(2):370–381, feb 2012.
22. S. Arnold, D. Keng, S. I. Shopova, S. Holler, W. Zurawsky, and F. Vollmer. Whispering gallery mode carousel – a photonic mechanism for enhanced nanoparticle detection in biosensing. *Opt. Express*, 17(8):6230–6238, Apr 2009.
23. Miguel A. Santiago-Cordoba, Murat Cetinkaya, Svetlana V. Boriskina, Frank Vollmer, and Melik C. Demirel. Ultrasensitive detection of a protein by optical trapping in a photonic-plasmonic microcavity. *Journal of Biophotonics*, 5(8-9):629–638, 2012.
24. Deshui Yu and Frank Vollmer. Allan deviation tells the binding properties in single-molecule sensing with whispering-gallery-mode optical microcavities. *Phys. Rev. Research*, 3:023087, May 2021.
25. Sivaraman Subramanian, Serge Vincent, and Frank Vollmer. Effective linewidth shifts in single-molecule detection using optical whispering gallery modes. *Applied Physics Letters*, 117(15):151106, oct 2020.
26. Lina He, Şahin Kaya Özdemir, Jiangang Zhu, Faraz Monifi, Huzeyfe Yılmaz, and Lan Yang. Statistics of multiple-scatterer-induced frequency splitting in whispering gallery microresonators and microlasers. *New Journal of Physics*, 15(7):073030, jul 2013.
27. D. Keng, X. Tan, and S. Arnold. Whispering gallery micro-global positioning system for nanoparticle sizing in real time. *Applied Physics Letters*, 105(7), 2014.
28. Eugene Kim, Martin D. Baaske, and Frank Vollmer. In Situ Observation of Single-Molecule Surface Reactions from Low to High Affinities. *Advanced Materials*, 28(45):9941–9948, 2016.
29. Narima Eerqing, Sivaraman Subramanian, Jesús Rubio, Tobias Lutz, Hsin-Yu Wu, Janet Anders, Christian Soeller, and Frank Vollmer. Comparing transient oligonucleotide hybridization kinetics using dna-paint and optoplasmonic single-molecule sensing on gold nanorods, 2021.
30. Marc W. van der Kamp, Erica J. Prentice, Kirsty L. Kraakman, Michael Connolly, Adrian J. Mulholland, and Vickery L. Arcus. Dynamical origins of heat capacity changes in enzyme-catalysed reactions. *Nature Communications*, 9(1):1177, dec 2018.
31. Frank Vollmer and Lan Yang. Review Label-free detection with high-Q microcavities: a review of biosensing mechanisms for integrated devices. *Nanophotonics*, 1(3-4):267–291, 2012.
32. Mi Kyoung Park, Jack Sheng Kee, Jessie Yiying Quah, Vivian Netto, Junfeng Song, Qing Fang, Eric Mouchel La Fosse, and Guo-Qiang Lo. Label-free aptamer sensor based on silicon microring resonators. *Sensors and Actuators B: Chemical*, 176:552–559, 2013.
33. M. Pöllinger, D. O'Shea, F. Warken, and A. Rauschenbeutel. Ultrahigh-Q Tunable Whispering-Gallery-Mode Microresonator. *Physical Review Letters*, 103(5):053901, jul 2009.

34. Amy Watkins, Jonathan Ward, Yuqiang Wu, and Síle Nic Chormaic. Single-input spherical microbubble resonator. *Optics letters*, 36(11):2113–2115, 2011.
35. Yunbo Guo, Hao Li, Karthik Reddy, Hrishikesh S. Shelar, Vasuki R. Nittoor, and Xudong Fan. Optofluidic Fabry-Pérot cavity biosensor with integrated flow-through micro-/nanochannels. *Applied Physics Letters*, 98(4):041104, jan 2011.
36. Michael Trupke, Edward A Hinds, S Eriksson, EA Curtis, Zakaria Moktadir, Elena Kukharenka, and Michael Kraft. Microfabricated high-finesse optical cavity with open access and small volume. *Applied Physics Letters*, 87(21):211106, 2005.
37. Mindy Lee and Philippe M. Fauchet. Two-dimensional silicon photonic crystal based biosensing platform for protein detection. *Opt. Express*, 15(8):4530–4535, Apr 2007.
38. Giancarlo C Righini and Silvia Soria. Biosensing by WGM microspherical resonators. *Sensors*, 16(6):905, 2016.
39. K J Vahala. Optical microcavities. *Nature*, 424(6950):839–846, 2003.
40. Kiyotaka Miura, Koichiro Tanaka, and Kazuyuki Hirao. CW laser oscillation on both the 3/2-11/2 and 3/2-13/2 transitions of Nd3+ ions using a fluoride glass microsphere. *Journal of Non-Crystalline Solids*, 213-214:276–280, may 1997.
41. Ioan Bica. Formation of glass microspheres with rotating electrical arc. *Materials Science and Engineering: B*, 77(2):210 212, 2000.
42. Jibo Yu, Elfed Lewis, Gerald Farrell, and Pengfei Wang. Compound glass microsphere resonator devices. *Micromachines*, 9(7), 2018.
43. Kerry A Wilson, Craig A Finch, Phillip Anderson, Frank Vollmer, and James J Hickman. Whispering gallery mode biosensor quantification of fibronectin adsorption kinetics onto alkylsilane monolayers and interpretation of resultant cellular response. *Biomaterials*, 33(1):225–236, 2012.
44. Carol E Soteropulos, Heather K Hunt, and Andrea M Armani. Determination of binding kinetics using whispering gallery mode microcavities. *Applied physics letters*, 99(10):103703, 2011.
45. Niranjan M. Hanumegowda, Caleb J. Stica, Bijal C. Patel, Ian White, and Xudong Fan. Refractometric sensors based on microsphere resonators. *Applied Physics Letters*, 87(20):201107, nov 2005.
46. Julie Lutti, Wolfgang Langbein, and Paola Borri. A monolithic optical sensor based on whispering-gallery modes in polystyrene microspheres. *Applied Physics Letters*, 93(15):151103, 2008.
47. Julie Lutti, Wolfgang Langbein, and Paola Borri. High Q optical resonances of polystyrene microspheres in water controlled by optical tweezers. *Applied Physics Letters*, 91(14):141116, oct 2007.
48. Vladimir S Ilchenko, X Steve Yao, and Lute Maleki. Pigtailing the high-q microsphere cavity:? a simple fiber coupler for optical whispering-gallery modes. *Optics letters*, 24(11):723–725, 1999.
49. Vladimir S Ilchenko and Andrey B Matsko. Optical resonators with whispering-gallery modes-part ii: applications. *IEEE Journal of selected topics in quantum electronics*, 12(1):15–32, 2006.
50. Heather K Hunt, Carol Soteropulos, and Andrea M Armani. Bioconjugation strategies for microtoroidal optical resonators. *Sensors*, 10(10):9317–9336, 2010.
51. Lina He, Şahin Kaya Özdemir, Jiangang Zhu, Woosung Kim, and Lan Yang. Detecting single viruses and nanoparticles using whispering gallery microlasers. *Nature Nanotechnology*, 6(7):428–432, 2011.
52. Tao Lu, Hansuek Lee, Tong Chen, Steven Herchak, Ji-Hun Kim, Scott E Fraser, Richard C Flagan, and Kerry Vahala. High sensitivity nanoparticle detection using optical microcavities. *Proceedings of the National Academy of Sciences*, 108(15):5976–5979, 2011.
53. Lina He, Şahin Kaya Özdemir, Jiangang Zhu, and Lan Yang. Ultrasensitive detection of mode splitting in active optical microcavities. *Physical Review A*, 82(5):053810, 2010.
54. Muzammil Iqbal, Martin A Gleeson, Bradley Spaugh, Frank Tybor, William G Gunn, Michael Hochberg, Tom Baehr-Jones, Ryan C Bailey, and L Cary Gunn. Label-free biosensor arrays based on silicon ring resonators and high-speed optical scanning instrumentation. *IEEE Journal of selected topics in quantum electronics*, 16(3):654–661, 2010.

55. Camille Delezoide, Marion Salsac, Joseph Lautru, Hervé Leh, Claude Nogues, Joseph Zyss, Malcolm Buckle, Isabelle Ledoux-Rak, and Chi Thanh Nguyen. Vertically coupled polymer microracetrack resonators for label-free biochemical sensors. *IEEE Photonics Technology Letters*, 24(4):270–272, 2011.
56. A. Ksendzov and Y. Lin. Integrated optics ring-resonator sensors for protein detection. *Optics Letters*, 30(24):3344, 2005.
57. Ayca Yalcin, Ketul C Popat, John C Aldridge, Tejal A Desai, John Hryniewicz, Nabil Chbouki, Brent E Little, Oliver King, Vien Van, Sai Chu, et al. Optical sensing of biomolecules using microring resonators. *IEEE Journal of Selected Topics in Quantum Electronics*, 12(1):148–155, 2006.
58. Tom Claes, Jordi Girones Molera, Katrien De Vos, Etienne Schacht, Roel Baets, and Peter Bienstman. Label-free biosensing with a slot-waveguide-based ring resonator in silicon on insulator. *IEEE Photonics journal*, 1(3):197–204, 2009.
59. Arthur Nitkowski, Antje Baeumner, and Michal Lipson. On-chip spectrophotometry for bioanalysis using microring resonators. *Biomedical optics express*, 2(2):271–277, 2011.
60. John T Gohring, Paul S Dale, and Xudong Fan. Detection of her2 breast cancer biomarker using the opto-fluidic ring resonator biosensor. *Sensors and Actuators B: Chemical*, 146(1):226–230, 2010.
61. Ian M White, Hesam Oveys, and Xudong Fan. Liquid-core optical ring-resonator sensors. *Optics letters*, 31(9):1319–1321, 2006.
62. Robert W Boyd and John E Heebner. Sensitive disk resonator photonic biosensor. *Applied optics*, 40(31):5742–5747, 2001.
63. Aaron Schweinsberg, Sandrine Hocdé, Nick N Lepeshkin, Robert W Boyd, Christopher Chase, and Julie E Fajardo. An environmental sensor based on an integrated optical whispering gallery mode disk resonator. *Sensors and Actuators B: Chemical*, 123(2):727–732, 2007.
64. Elliot J Smith, Sabine Schulze, Suwit Kiravittaya, Yongfeng Mei, Samuel Sanchez, and Oliver G Schmidt. Lab-in-a-tube: detection of individual mouse cells for analysis in flexible split-wall microtube resonator sensors. *Nano letters*, 11(10):4037–4042, 2011.
65. Stefan M Harazim, Vladimir A Bolaños Quiñones, Suwit Kiravittaya, Samuel Sanchez, and Oliver G Schmidt. Lab-in-a-tube: on-chip integration of glass optofluidic ring resonators for label-free sensing applications. *Lab on a Chip*, 12(15):2649–2655, 2012.
66. Rico Henze, Tom Seifert, Jonathan Ward, and Oliver Benson. Tuning whispering gallery modes using internal aerostatic pressure. *Optics letters*, 36(23):4536–4538, 2011.
67. Jacob Scheuer and Mikhail Sumetsky. Optical-fiber microcoil waveguides and resonators and their applications for interferometry and sensing. *Laser & Photonics Reviews*, 5(4):465–478, 2011.
68. Fei Xu, Peter Horak, and Gilberto Brambilla. Optical microfiber coil resonator refractometric sensor. *Optics express*, 15(12):7888–7893, 2007.
69. Frank Vollmer and Peer Fischer. Ring-resonator-based frequency-domain optical activity measurements of a chiral liquid. *Optics letters*, 31(4):453–455, 2006.
70. David A Bergstein, Emre Ozkumur, Arthur C Wu, Ayca Yalcin, Jeremy R Colson, James W Needham, Rostem J Irani, Jonathan M Gershoni, Bennett B Goldberg, Charles DeLisi, et al. Resonant cavity imaging: A means toward high-throughput label-free protein detection. *IEEE Journal of Selected Topics in Quantum Electronics*, 14(1):131–139, 2008.
71. Sudeep Mandal, Julie M Goddard, and David Erickson. A multiplexed optofluidic biomolecular sensor for low mass detection. *Lab on a Chip*, 9(20):2924–2932, 2009.
72. Lefevreseguin Collot, V Lefevre-Seguin, M Brune, JM Raimond, and S Haroche. Very high-q whispering-gallery mode resonances observed on fused silica microspheres. *EPL (Europhysics Letters)*, 23(5):327, 1993.
73. Pham Van Hoi, Chu Thi Thu Ha, and Hoang Quang Hung. Long-band emission of microsphere lasers based on erbium-doped sol-gel silica-alumina glasses. *Applied Physics Letters*, 87(16):161110, 2005.

74. Lu Cai, Junyao Pan, Yong Zhao, Jin Wang, and Shiyuan Xiao. Whispering Gallery Mode Optical Microresonators: Structures and Sensing Applications. *physica status solidi (a)*, 217(6):1900825, mar 2020.
75. D. Ristić, V. Holý, M. Ivanda, M. Marciuš, M. Buljan, O. Gamulin, K. Furić, M. Ristić, S. Musić, M. Mazzola, A. Chiasera, M. Ferrari, and G.C. Righini. Surface characterization of thin silicon-rich oxide films. *Journal of Molecular Structure*, 993(1-3):214–218, may 2011.
76. Shancheng Yang, Yue Wang, and Handong Sun. Advances and prospects for whispering gallery mode microcavities. *Advanced Optical Materials*, 3(9):1136–1162, 2015.
77. Davor Ristić, Mile Ivanda, Giorgio Speranza, Zdravko Siketić, Ivančica Bogdanović-Radović, Marijan Marciuš, Mira Ristić, Ozren Gamulin, Svetozar Musić, Krešimir Furić, Giancarlo C Righini, and Maurizio Ferrari. Local Site Distribution of Oxygen in Silicon-Rich Oxide Thin Films: A Tool to Investigate Phase Separation. *The Journal of Physical Chemistry C*, 116(18):10039–10047, may 2012.
78. Guanyu Li, Pei Liu, Xiaoshun Jiang, Chao Yang, Jiyang Ma, Hongya Wu, and Min Xiao. High-Q silica microdisk optical resonators with large wedge angles on a silicon chip. *Photonics Research*, 3(5):279, oct 2015.
79. Zuo Feng, Yuming He, Wei Yan, Fuhua Yang, Weihua Han, and Zhaofeng Li. Progress of waveguide ring resonators used in micro-optical gyroscopes. *Photonics*, 7(4):1–16, 2020.
80. G. Kakarantzas, T. E. Dimmick, T. A. Birks, R. Le Roux, and P. St. J. Russell. Miniature all-fiber devices based on CO_2 laser microstructuring of tapered fibers. *Optics Letters*, 26(15):1137, aug 2001.
81. Ganapathy Senthil Murugan, James S. Wilkinson, and Michalis N. Zervas. Selective excitation of whispering gallery modes in a novel bottle microresonator. *Optics Express*, 17(14):11916, jul 2009.
82. G. C. Righini, Y. Dumeige, P. Féron, M. Ferrari, G. Nunzi Conti, D. Ristic, and S. Soria. Whispering Gallery Mode microresonators: Fundamentals and applications. *Rivista del Nuovo Cimento*, 34(7):435–488, 2011.
83. D K Armani, T J Kippenberg, S M Spillane, and K J Vahala. Ultra-high-Q toroid microcavity on a chip. *Nature*, 421(6926):925–928, 2003.
84. Xuefeng Jiang, Abraham J. Qavi, Steven H. Huang, and Lan Yang. Whispering-Gallery Sensors. *Matter*, 3(2):371–392, aug 2020.
85. Mark Lawrence, David R. Barton, Jefferson Dixon, Jung-Hwan Song, Jorik van de Groep, Mark L. Brongersma, and Jennifer A. Dionne. High quality factor phase gradient metasurfaces. *Nature Nanotechnology*, 15(11):956–961, nov 2020.
86. E. Gavartin, P. Verlot, and T. J. Kippenberg. A hybrid on-chip optomechanical transducer for ultrasensitive force measurements. *Nature Nanotechnology*, 7(8):509–514, 2012.
87. T. Wienhold, S. Kraemmer, S. F. Wondimu, T. Siegle, U. Bog, U. Weinzierl, S. Schmidt, H. Becker, H. Kalt, T. Mappes, S. Koeber, and C. Koos. All-polymer photonic sensing platform based on whispering-gallery mode microgoblet lasers. *Lab on a Chip*, 15(18):3800–3806, 2015.
88. Xiaolong Xu, Yuanyuan Zhao, Xiangdong Xue, Shuaidong Huo, Fei Chen, Guozhang Zou, and Xing Jie Liang. Seedless synthesis of high aspect ratio gold nanorods with high yield. *Journal of Materials Chemistry A*, 2(10):3528–3535, 2014.
89. Andrés Guerrero-Martínez, Silvia Barbosa, Isabel Pastoriza-Santos, and Luis M. Liz-Marzán. Nanostars shine bright for you. Colloidal synthesis, properties and applications of branched metallic nanoparticles. *Current Opinion in Colloid and Interface Science*, 16(2):118–127, 2011.
90. Sara E. Skrabalak, Leslie Au, Xingde Li, and Younan Xia. Facile synthesis of Ag nanocubes and Au nanocages. *Nature Protocols*, 2(9):2182–2190, 2007.
91. Elcin Dertli, Sahin Coskun, and Emren Nalbant Esenturk. Gold nanowires with high aspect ratio and morphological purity: Synthesis, characterization, and evaluation of parameters. *Journal of Materials Research*, 28(2):250–260, 2013.

92. Leonardo Scarabelli, Marc Coronado-Puchau, Juan J. Giner-Casares, Judith Langer, and Luis M. Liz-Marzán. Monodisperse gold nanotriangles: Size control, large-scale self-assembly, and performance in surface-enhanced raman scattering. *ACS Nano*, 8(6):5833–5842, 2014.
93. Yonggang Lu, Haibin Zhang, Fan Wu, Hong Liu, and Jingzhong Fang. Size-tunable uniform gold octahedra: fast synthesis, characterization, and plasmonic properties. *RSC Advances*, 7(30):18601–18608, 2017.
94. G. Frens. Particle size and sol stability in metal colloids. *Kolloid-Zeitschrift & Zeitschrift für Polymere*, 250(7):736–741, 1972.
95. John Turkevich, Peter Cooper Stevenson, and James Hillier. A study of the nucleation and growth processes in the synthesis of colloidal gold. *Discussions of the Faraday Society*, 11(c):55–75, 1951.
96. Sandeep Ghosh and Liberato Manna. The Many "facets" of Halide Ions in the Chemistry of Colloidal Inorganic Nanocrystals. *Chemical Reviews*, 118(16):7804–7864, 2018.
97. Luca Guerrini, Ramon A. Alvarez-Puebla, and Nicolas Pazos-Perez. Surface modifications of nanoparticles for stability in biological fluids. *Materials*, 11(7):1–28, 2018.
98. Meghana Matur, Harishkumar Madhyastha, T. S. Shruthi, Radha Madhyastha, S. P. Srinivas, P. N. Navya, and Hemant Kumar Daima. Engineering bioactive surfaces on nanoparticles and their biological interactions. *Scientific Reports*, 10(1):1–14, 2020.
99. Zhong Zhang and Mengshi Lin. Fast loading of PEG-SH on CTAB-protected gold nanorods. *RSC Advances*, 4(34):17760–17767, 2014.
100. Brian Szychowski, Haixu Leng, Matthew Pelton, and Marie Christine Daniel. Controlled etching and tapering of Au nanorods using cysteamine. *Nanoscale*, 10(35):16830–16838, 2018.
101. Xin Shi, Hannah L. Perry, and James D. E. T. Wilton-Ely. Strategies for the functionalisation of gold nanorods to reduce toxicity and aid clinical translation. *Nanotheranostics*, 5(2):155–165, 2021.
102. Matthew R. Hauwiller, Xingchen Ye, Matthew R. Jones, Cindy M. Chan, Jason J. Calvin, Michelle F. Crook, Haimei Zheng, and A. Paul Alivisatos. Tracking the Effects of Ligands on Oxidative Etching of Gold Nanorods in Graphene Liquid Cell Electron Microscopy. *ACS Nano*, 14(8):10239–10250, 2020.
103. Qikui Fan, Hao Yang, Juan Ge, Shumeng Zhang, Zhaojun Liu, Bo Lei, Tao Cheng, Youyong Li, Yadong Yin, and Chuanbo Gao. Customizable Ligand Exchange for Tailored Surface Property of Noble Metal Nanocrystals. *Research*, 2020:1–12, 2020.
104. Chunyu Zhou, Yaocai Bai, Fan Yang, Tao Sun, Liang Zhang, Yuanqing Cai, Tao Gu, Yun Liu, Mingfu Gong, Dong Zhang, and Yadong Yin. Ligand exchange on noble metal nanocrystals assisted by coating and etching of cuprous oxide. *Materials Chemistry Frontiers*, 4(6):1614–1622, 2020.
105. David I. Gittins and Frank Caruso. Spontaneous phase transfer of nanoparticulate metals from organic to aqueous media. *Angewandte Chemie - International Edition*, 40(16):3001–3004, 2001.
106. Youngjae Lee, Jungwoo Jang, Jinho Yoon, Jeong Woo Choi, Inhee Choi, and Taewook Kang. Phase transfer-driven rapid and complete ligand exchange for molecular assembly of phospholipid bilayers on aqueous gold nanocrystals. *Chemical Communications*, 55(22):3195–3198, 2019.
107. E. Metwalli, D. Haines, O. Becker, S. Conzone, and C. G. Pantano. Surface characterizations of mono-, di-, and tri-aminosilane treated glass substrates. *Journal of Colloid and Interface Science*, 298(2):825–831, 2006.
108. Mojun Zhu, Maria Z. Lerum, and Wei Chen. How To Prepare Reproducible, Homogeneous, and Hydrolytically Stable Aminosilane-Derived Layers on Silica. *Langmuir*, 28(1):416–423, jan 2012.
109. Thermo Fisher. Chemical Reactivity of Crosslinkers and Modification Reagents. Technical report, Thermo Fisher Scientific, 2012.

110. Changhyoup Lee, Frederik Dieleman, Jinhyoung Lee, Carsten Rockstuhl, Stefan A. Maier, and Mark Tame. Quantum Plasmonic Sensing: Beyond the Shot-Noise and Diffraction Limit. *ACS Photonics*, 3(6):992–999, jun 2016.
111. R. W P Drever, J. L. Hall, F. V. Kowalski, J. Hough, G. M. Ford, a. J. Munley, and H. Ward. Laser phase and frequency stabilization using an optical resonator. *Applied Physics B Photophysics and Laser Chemistry*, 31(2):97–105, 1983.
112. Jens M Dobrindt. *Bio-sensing using toroidal microresonators & Theoretical cavity optomechanics*. PhD thesis, Ludwig-Maximilians-University of Munich, 2012.
113. Jon D. Swaim, Joachim Knittel, and Warwick P. Bowen. Detection limits in whispering gallery biosensors with plasmonic enhancement. *Applied Physics Letters*, 99(24):243109, dec 2011.
114. Jiangang Zhu, Sahin Kaya Ozdemir, Yun-Feng Xiao, Lin Li, Lina He, Da-Ren Chen, and Lan Yang. On-chip single nanoparticle detection and sizing by mode splitting in an ultrahigh-Q microresonator. *Nature Photonics*, 4(1):46–49, 2010.
115. Marina S. Dietz, S. Sophia Wehrheim, Marie Lena I.E. Harwardt, Hartmut H. Niemann, and Mike Heilemann. Competitive Binding Study Revealing the Influence of Fluorophore Labels on Biomolecular Interactions. *Nano Letters*, 19(11):8245–8249, 2019.
116. Jon D. Swaim, Joachim Knittel, and Warwick P. Bowen. Detection of nanoparticles with a frequency locked whispering gallery mode microresonator. *Applied Physics Letters*, 102(18):2011–2015, 2013.
117. Serge Rosenblum, Yulia Lovsky, Lior Arazi, Frank Vollmer, and Barak Dayan. Cavity ring-up spectroscopy for ultrafast sensing with optical microresonators. *Nature Communications*, 6:6788, 2015.
118. Eric D. Black. An introduction to Pound-Drever-Hall laser frequency stabilization. *American Journal of Physics*, 69(1):79, 2001.
119. J. C. Knight, G. Cheung, F. Jacques, and T. A. Birks. Phase-matched excitation of whispering-gallery-mode resonances by a fiber taper. *Opt. Lett.*, 22(15):1129–1131, Aug 1997.
120. R Garcia-Fernandez, W Alt, F Bruse, C Dan, K Karapetyan, O Rehband, A Stiebeiner, U Wiedemann, D Meschede, and A Rauschenbeutel. Optical nanofibers and spectroscopy. *Applied Physics B*, 105(1):3, 2011.
121. Michael L. Gorodetsky and Vladimir S. Ilchenko. Optical microsphere resonators: optimal coupling to high-Q whispering gallery modes. *J. Opt. Soc. Am.*, 16(1):14, 1998.
122. Jiangang Zhu, Şahin K. Özdemir, Huzeyfe Yilmaz, Bo Peng, Mark Dong, Matthew Tomes, Tal Carmon, and Lan Yang. Interfacing whispering-gallery microresonators and free space light with cavity enhanced Rayleigh scattering. *Scientific Reports*, 4(1):6396, may 2015.
123. Zachary Ballard, Martin D. Baaske, and Frank Vollmer. Stand-off biodetection with free-space coupled asymmetric microsphere cavities. *Sensors*, 15(4):8968–8980, 2015.
124. Yu A Demchenko, I A Bilenko, and M L Gorodetsky. Optimisation of the prism coupling of optical whispering-gallery-mode microcavities. *Quantum Electronics*, 47(8):743–747, aug 2017.
125. Ying-Jen Chen, Ulrike Schoeler, Chung-Hsuan (Benjamin) Huang, and Frank Vollmer. Combining whispering-gallery mode optical biosensors with microfluidics for real-time detection of protein secretion from living cells in complex media. *Small*, 14(22):1703705, 2018.
126. Arturo Bianchetti, Alejandro Federico, Serge Vincent, Sivaraman Subramanian, and Frank Vollmer. Refractometry-based air pressure sensing using glass microspheres as high-q whispering-gallery mode microresonators. *Optics Communications*, 394:152–156, 2017.
127. Jessica Melin and Stephen R. Quake. Microfluidic large-scale integration: The evolution of design rules for biological automation. *Annual Review of Biophysics and Biomolecular Structure*, 36(1):213–231, 2007. PMID: 17269901.
128. Nikita Toropov, Eleanor Osborne, Tina Loveleen Joshi, James Davidson, Caitlin Morgan, Joseph Page, Justin Pepperell, and Frank Vollmer. Sars-cov-2 tests: Bridging the gap between laboratory sensors and clinical applications. *ACS Sensors* 6(8), 2815–2837, 2021.
129. Valentin Flauraud, Massimo Mastrangeli, Gabriel D Bernasconi, Jeremy Butet, Duncan T L Alexander, Elmira Shahrabi, Olivier J F Martin, and Juergen Brugger. Nanoscale topographical control of capillary assembly of nanoparticles. *Nature Nanotechnology*, 12(1):73–80, 2017.

130. Maxim L Nesterov, Xinghui Yin, Martin Schäferling, Harald Giessen, and Thomas Weiss. The Role of Plasmon-Generated Near Fields for Enhanced Circular Dichroism Spectroscopy. *ACS Photonics*, 3(4):578–583, apr 2016.
131. Meng Wang, Jinyi Dong, Chao Zhou, Hao Xie, Weihai Ni, Shun Wang, Huile Jin, and Qiangbin Wang. Reconfigurable plasmonic diastereomers assembled by dna origami. *ACS Nano*, 13(12):13702–13708, 2019. PMID: 31550129.
132. Aneeth Kakkanattu, Narima Eerqing, Shahin Ghamari, and Frank Vollmer. Review of optical sensing and manipulation of chiral molecules and nanostructures with the focus on plasmonic enhancements. *Opt. Express*, 29(8):12543–12579, Apr 2021.
133. Krishna Sriram and Paul A. Insel. G protein-coupled receptors as targets for approved drugs: How many targets and how many drugs? *Molecular Pharmacology*, 93(4):251–258, 2018.
134. Cody J. Wenthur, Patrick R. Gentry, Thomas P. Mathews, and Craig W. Lindsley. Drugs for allosteric sites on receptors. *Annual Review of Pharmacology and Toxicology*, 54(1):165–184, 2014. PMID: 24111540.
135. Salma Kassem, Thomas van Leeuwen, Anouk S. Lubbe, Miriam R. Wilson, Ben L. Feringa, and David A. Leigh. Artificial molecular motors. *Chem. Soc. Rev.*, 46:2592–2621, 2017.
136. Joanna L. Richens, Jonathan P. Bramble, Jennifer Bain, and Paul O'Shea. Plasmon-based determination of macromolecular interactions with membrane-encapsulated nanoparticles. In David L. Andrews, Angus J. Bain, Jean-Michel Nunzi, and Andreas Ostendorf, editors, *Nanophotonics VII*, volume 10672, pages 20 – 35. International Society for Optics and Photonics, SPIE, 2018.
137. Timothy DeWolf and Reuven Gordon. Theory of acoustic raman modes in proteins. *Phys. Rev. Lett.*, 117:138101, Sep 2016.
138. Jolly Xavier, Deshui Yu, Callum Jones, Ekaterina Zossimova, and Frank Vollmer. Quantum nanophotonic and nanoplasmonic sensing: towards quantum optical bioscience laboratories on chip. *Nanophotonics*, 10(5):1387–1435, 2021.
139. Nikita Toropov, Gema Cabello, Mariana P Serrano, Rithvik R Gutha, Matías Rafti, and Frank Vollmer. Review of biosensing with whispering-gallery mode lasers. *Light: Science & Applications*, 10(1):42, 2021.
140. J. Xavier, S. Vincent, F. Meder and F. Vollmer (2018) Advances in optoplasmonic sensors–combining optical nano/microcavities and photonic crystals with plasmonic nanostructures and nanoparticles. *Nanophotonics, 7*(1), 1–38. De Gruyter.
141. Baaske, Martin D and Foreman, Matthew R and Vollmer, Frank (2014) Single-molecule nucleic acid interactions monitored on a label-free microcavity biosensor platform. *Nature nanotechnology, 9*(11), 933–939. Nature Publishing Group.
142. Vollmer, Frank and Yang, Lan (2012) Review label-free detection with high-Q microcavities: a review of biosensing mechanisms for integrated devices. *Nanophotonics, 1*(3–4), 267–291. Walter de Gruyter.
143. Kim, Eugene and Baaske, Martin D and Schuldes, Isabel and Wilsch, Peter S and Vollmer, Frank (2017) Label-free optical detection of single enzyme-reactant reactions and associated conformational changes. *Science advances, 3*, e1603044. American Association for the Advancement of Science.
144. Subramanian, Sivaraman, Jones, Hannah BL, Frustaci, Simona, Winter, Samuel, van der Kamp, Marc W, Arcus, Vickery L, Pudney, Christopher R and Vollmer, Frank (2021) Sensing Enzyme Activation Heat Capacity at the Single-Molecule Level Using Gold-Nanorod-Based Optical Whispering Gallery Modes. *ACS Applied Nano Materials*. ACS Publications.
145. Tathyana Tumolo, Lucio Angnes and Mauricio S. Baptista (2004) Determination of the refractive index increment (dn/dc) of molecule and macromolecule solutions by surface plasmon resonance. *Analytical Biochemistry, 333*(2), 273–279. 0003-2697. https://doi.org/10.1016/j.ab.2004.06.010.

Chapter 3
Nonlinear Optical Microcavities Towards Single-Molecule Sensing

Wenjing Liu, Shui-Jing Tang, and Yun-Feng Xiao

Abstract Optical sensing techniques are of importance for detecting and understanding nanoscale objects. Optical microcavities can significantly enhance light-matter interactions via strong field confinement, which enable optical detection with high temporal and spatial resolution down to the single-molecule level. Besides the extensively demonstrated linear optical approach, nonlinear optical processes are also strongly amplified in microcavities, and exploring them in sensing applications may not only allow label-free detection with further increased sensitivity and precision, but also reveal unique spectral fingerprints that are hidden in the linear interaction regime. This chapter will review molecular sensing mechanisms, modalities, and recent advances based on nonlinear optical effects.

3.1 Introduction

Optical microcavities confine light to small volumes by resonant recirculation, giving rise to an enhancement of the light-matter interaction. They are employed extensively for a wide range of applications and studies, such as nonlinear optics, cavity quantum quantum electrodynamics, and sensing. An ideal cavity would confine light indefinitely (that is, without loss) and would have resonant frequencies at precise values. Deviation from this ideal condition is described by the cavity quality (Q) factor, which is proportional to cavity lifetime τ of photons, that is $Q = \omega\tau$ with ω being the resonant angular frequency. On the other hand, the spatial confinement of photons in the cavity is typically described by the mode volume (V). According to

W. Liu
State Key Laboratory for Mesoscopic Physics and Frontiers Science Center for Nano-optoelectronics, School of Physics, Peking University, Beijing, China

S.-J. Tang
Collaborative Innovation Center of Extreme Optics, Taiyuan, Shanxi, China

Y.-F. Xiao (✉)
Peking University Yangtze Delta Institute of Optoelectronics, Nantong, Jiangsu, China
e-mail: yfxiao@pku.edu.cn

W. Bowen et al. (eds.), *Single Molecule Sensing Beyond Fluorescence*, Nanostructure Science and Technology,
https://doi.org/10.1007/978-3-030-90339-8_3

light confinement mechanisms, there are four kinds of representative optical microcavites, including Fabry–Pérot cavities, plasmonic cavities, photonic crystal cavities and whispering-gallery-mode (WGM) cavities. Fabry–Pérot cavities are composed of two parallel reflectors with light bouncing back and forth between them, and resonant modes are formed by constructive interference between multiple beams [1]. Surface plasmon resonances arise from the collective oscillations of electrons at the surface of metal films or nanoparticles [2]. Photonic crystal cavities are formed by introducing a point defect in periodic photonic structures with a photonic band gap [3, 4]. The propagation of light within the frequency range of the photonic band gap is forbidden and the photons are thus trapped inside the defect. Whispering-gallery-mode (WGM) cavities confine light by total internal reflections near the cavity perimeter, and such modes have therefore been called "whispering-gallery" modes. A summary of typical Q and V values for different types of cavities is presented in Fig. 3.1. WGM microcavities, benefiting from ultrahigh Q factors and small V, can significantly enhance light-matter interactions, providing an ideal platform for exploring a broad range of nonlinear optical effects, and ultrahigh-sensitivity detection. Over the past decades, researchers have developed ultra-high Q WGM microresonators using different structures (i.e. microsphere [5], microtoroid [6], and microbubble [7]) and materials (e.g. silica, $LiNbO_3$ [8], Si_3N_4 [9]) as shown in Fig. 3.2.

At the same time, optical WGM devices have attracted considerable attention in label-free detection of nano-objects, such as nanoparticles and biomolecules [10]. Microcavity sensors operated in the linear light-matter interaction regime generally rely on the refractive index change, the elastic scattering, or the absorption introduced by nanoparticles entering their evanescent field. Such interactions correspond to mode shift [11, 12], mode splitting [13, 14], and mode broadening [15], respectively, which can be detected in real-time with an optical fiber taper or other waveguides. The concentration of nanoparticles or molecules and their approximate sizes can be determined by analyzing these signals [16, 17]. Numerous experiments have been performed that have successfully detected various particles in different environments. For example, single Influenza A and Lentivirus nanoparticles with 110–120 nm diameters have been detected directly in both air and liquid [11, 13, 15]; Single ribosome nanoparticles with diamters down to 30 nm have been detected in aqueous

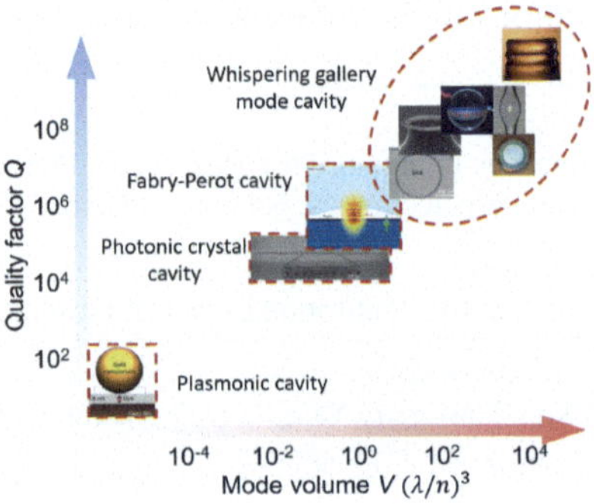

Fig. 3.1 Quality factor and mode volume of representative plasmonic cavity, photonic crystal cavity, Fabry–Pérot cavity, and whispering-gallery mode

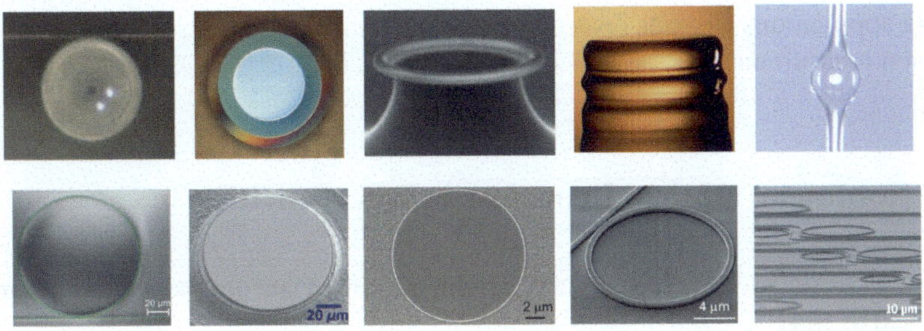

Fig. 3.2 Nonlinear WGM microcavities with different materials and structures. From left to right: SiO_2 microsphere [5], SiO_2 wedge microcavity [24], SiO_2 microtoroid [25], CaF_2 rod [26], SiO_2 microbubble [27] (top); Si_3N_4microring [9], $LiNbO_3$ microdisk [28], organometallic halide perovskite microdisk [29], AlGaAs microring [30], diamond microring [31] (bottom)

solution [18]. Furthermore, combining with plasmonic resonances and mode-locking techniques, the sensitivity of WGM microresonaters has been improved to single biomolecules [19] and even single atomic ions [20] level. The rapid progress in microcavity sensing based on linear mechanisms has been summarized in several recent reviews [21–23].

On the other hand, owing to their high Q factors, WGM microcavities serve as ideal platforms for studying nonlinear optical effects. Efficient nonlinear interaction allows the observation of rich phenomena, including low threshold Raman and Brillouin lasing [32–34], harmonic generation [35, 36], spontaneous symmetry breaking [37], nonreciprocal light transmission [38, 39], and optical frequency comb generation [40, 41]. The nonlinear processes can produce new optical signals at Different from evanescent sensors operating at the linear sensing regime, such as microcavity and nanowaveguide sensors [16, 42, 43], the nonlinear processes may produce new optical frequencies away from the excitation beam, which can significantly enhance the signal-to-noise ratio. Moreover, nonlinear effects introduced by the molecules themselves are often associated with unique spectral fingerprints, such as the identification of molecular vibrational modes in Raman spectroscopy (Fig. 3.3), which is useful for specific identification of nanoparticles and molecules in complex media. In this chapter, we start with a brief introduction of the nonlinear processes within optical microcavities that are actively involved in nanoparticle and biochemical sensing, followed by a detailed discussion of the emerging nonlinear

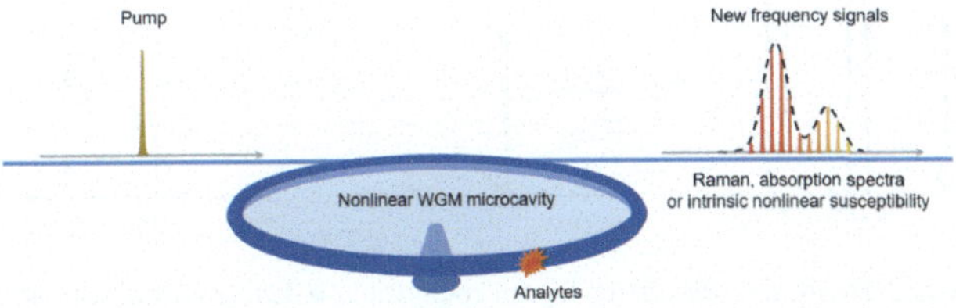

Fig. 3.3 Schematic of nonlinear molecular sensing with WGM microcavity

sensing applications towards few- and single-molecule detection, identification, and precision measurements.

3.2 Nonlinear Optical Processes in Microcavities

Under electromagnetic excitation, the polarization of an optical material can be expressed as [44]

$$P = \epsilon_0(\chi^{(1)}E + \chi^{(2)}EE + \chi^{(3)}EEE + \cdots) \tag{3.1}$$

where ϵ_0 is the vacuum permittivity, E is the electric field, and $\chi^{(n)}$ represents the nth order susceptibility. With a strong excitation, the material itself is changed by the electromagnetic field, leading to optical nonlinearity. Nonlinearity can either modify the optical response at the same wavelength such as Kerr effect or generate photons at different frequencies from the excitation such as harmonic generation. Efficient nonlinear conversion of the optic signal requires the conservation of both the energies and momentum of all quasiparticles involved, which is termed as the phase matching condition [45–47]. Therefore, careful dispersion engineering may be necessary to achieve high nonlinear efficiency, especially for processes spanning a large frequency range.

3.2.1 *Raman and Brillouin Scattering*

Raman and Brillouin scattering processes are amongst the most extensively studied nonlinear optical effects because they can be observed at any excitation wavelengths, at modest power, and with broad types of analytes. As third order nonlinear effects, they arise from inelastic scattering processes between photons and the vibrations of the medium. During the scattering process, a photon loses or gains energy from the vibrational modes, creating Stokes or anti-Stokes side-bands separated from the excitation by the phonon frequencies (Fig. 3.4a).

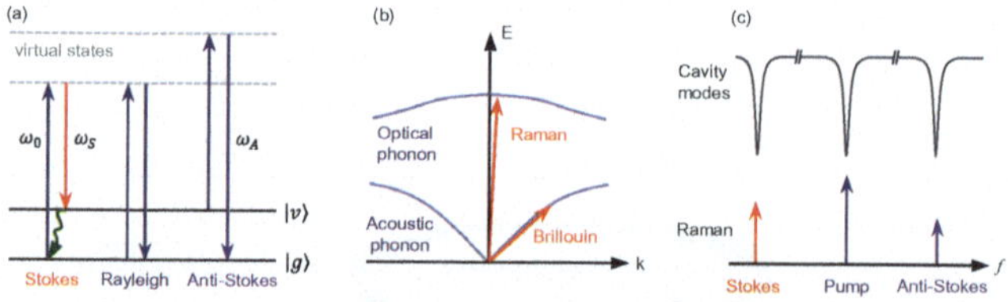

Fig. 3.4 Raman and Brillouin scattering in microcavities. **a** Energy diagram of the elastic and inelastic scattering processes. **b** Different phonon branches involved in the Raman and Brillouin scattering. (c) Cavity enhancement of the Raman scattering of the initial and final states

Raman scattering resulted from the scattering between photons and optical phonons, which have relatively high phonon energy and near-zero momentum. On the other hand, Brillouin scattering corresponds to the scattering with acoustic phonons, or equivalently, sound wave of the medium, hence the scattering-induced energy shift is much smaller (Fig. 3.4b). As mentioned earlier, the phase-matching condition requires both energy and momentum conservation during the scattering process. The optical phonon branch is quite flat, hence the phase-matching condition can be readily satisfied by phonons with momentum spanning a broad range. In contrast, the acoustic phonons have nearly linear dispersion with a constant momentum, thus careful engineering of the phase-matching condition is critical for efficient Brillouin scattering to take place. In WGM microcavities, specifically, each (unperturbed) mode is two-fold degenerate with clock-wise and counterclock-wise propagation direction, respectively. Brillouin scattering occurs in two configurations, i.e., the forward scattering that couples the two co-propagating modes, and the backward scattering that couples the two counter-propagating modes. Each processes, fulfilling different phase matching conditions, occurs at a particular frequency with a narrow bandwidth (Fig. 3.4b).

The Stokes scattering rate can be expressed as $\rho = G_s N_p(N_s + 1)$, in which G_s is the Raman/Brillouin gain coefficient that is proportional to the corresponding $\chi^{(3)}$ and cavity configuration, N_p and N_s denotes the mean photon numbers of the pump and the Stokes field, respectively [44]. Analogous to the Einstein coefficients in the linear regime, elastic scattering manifests spontaneous and stimulated processes. With $N_s \ll 1$, spontaneous scattering dominates, where the Stokes scattering rate is proportional to the pump field. This is the case for most free-space Raman spectroscopy. Stimulated scattering starts to become significant when $N_s{\sim}1$, leading to Raman/Brillouin gain and superlinearly increased intensity. Lasing action is initiated when the gain fully compensates the system loss.

In optical cavities and resonators, Raman/Brillouin scattering processes can be greatly enhanced by both enhancing the pump beam and increasing the final density of states of the scattered states. (Fig. 3.4c). In plasmonic nanoresonators (see also Chaps. 2 and 5), the overall enhancement can reach 10^8–10^{10} [48, 49]. In ultrahigh-Q optical microcavities, comparable enhancement has been theoretically predicted under double-resonance conditions [50, 51], while up to 10^4 enhancement has been reported in experiment [52]. On the other hand, ultrahigh-Q microcavities are ideal platforms for Raman and Brillouin lasers and applications. A silica WGM microcavity with 10^7 quality factor can support Raman lasers with sub-miliwatts threshold under continuous wave (CW) operation [53].

Useful spectroscopy technique have also been developed at the anti-Stokes side of the scattering spectrum. Coherent anti-Stokes Raman scattering (CARS) is a coherent four wave mixing processes (discussed in more details in Sect. 3.2.3) that is widely applied in biochemical sensing. In this process, a pump laser at ω_0 and a Stokes laser at ω_s interact in the nonlinear analyte, generating an anti-Stokes signal at $(2\omega_0 - \omega_s)$. When the difference between the pump and the Stokes beam matches the vibrational modes of the analyte, the CARS process is resonantly enhanced. Compared with the relatively weak spontaneous Raman scattering, coherent CARS is much more efficient, while the analytes are virtually unperturbed during the process. These advantages make it suitable in biochemical sensing and imaging. However, the trade-off is that two lasers fulfilling the phase matching condition are required in the measurement. On the other hand, stimulated anti-Stokes Raman scattering (ASRS) can be generated with one pump beam. This pump beam excite Stokes photons in the medium, which then interact with the pump to produce a coherent anti-Stokes signal.

As a brief summary, arising from the photon-phonon interaction, Raman/Brillouin spectroscopy is an important tool to identify vibrational modes of the analyte molecules. In sensing, it can reveal spectral fingerprinting of the analyte molecules interacting with the evanescent field of the cavity, which is hidden in linear sensing methods. Moreover, Raman and Brillouin lasers exhibits narrow linewidths compared to passive microcavity modes, which can be utilized in sensing devices with enhanced sensitivity and lower detection limit. We will discuss these two sensing schemes in Sects. 3.3.1–3.3.3, respectively.

3.2.2 *Sum Frequency and Harmonic Generation*

Sum frequency generation (SFG) refers to the process in which several photons interact with the nonlinear media and convert into one photon. Specifically, in a nth order nonlinear process, $\hbar\omega_1 + \hbar\omega_2 + \cdots + \hbar\omega_{n-1} \rightarrow \hbar\omega_n$. Particularly, when identical photons converts into one photon with multiplied frequency, the process is termed as harmonic generation. Second and third harmonic generations (SHG and THG) are the most widely observed harmonic generation processes, belonging to the second and third order nonlinear effect, respectively. These two processes can be expressed as [44]

$$P^{(2)}(2\omega) = 2\epsilon_0 \chi^{(2)}(2\omega; \omega; \omega) E(\omega) E(\omega), \tag{3.2}$$

$$P^{(3)}(3\omega) = 6\epsilon_0 \chi^{(3)}(3\omega; \omega; \omega; \omega) E(\omega) E(\omega) E(\omega), \tag{3.3}$$

respectively. The process is enabled via interaction between photons and the virtual energy states of the nonlinear medium, as seen in Fig. 3.5a. When the fundamental or

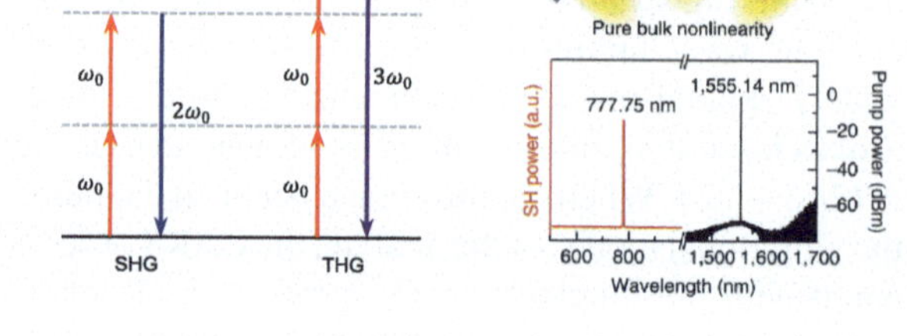

Fig. 3.5 Sum frequency and Harmonic generation. **a** Energy diagram of SHG and THG. **b** Surface SHG in WGM microcavities via symmetry-breaking-induced nonlinearity.Adapted with permission from Ref. [35] Copyright 2019 Springer Nature

harmonic photon matches a transition between real states, the nonlinear susceptibility is resonantly enhanced.

Sum frequency and harmonic generation processes serve as invaluable tools in sensing applications due to their sensitivity to structural symmetry. Especially, even order susceptibilities vanishes within materials possessing inversion symmetry, even order processes can thus be exploited to probe surface properties, detect analytes with broken inversion symmetry, and to characterize the structure of the analytes. Recently, in silica WGM microcavities, strong surface SHG has been observed due to structural symmetry breaking and cavity enhancement on both the fundamental and SHG fields (Fig. 3.5b) [35], demonstrating the feasibility of efficiently detecting harmonic generation in these systems. Efforts have been devoted in WGM SHG or THG sensors targeting thin layers of organic molecules coated on the cavity surface [36, 54]. Organic molecules can possess large nonlinear susceptibilities, making sum frequency and harmonic generation a efficacious tool that is able to sense small quantities of molecules. We will discuss these recent advances in Sect. 3.4.

3.2.3 Four-Wave Mixing (FWM) and Optical Frequency Combs

Four-wave mixing (FWM) is a third order nonlinear effect involving four photons. Frequently observed FMW processes induces the conversion of two photons into two new photons at shifted frequency, or three photons converted to one. The CARS and THG introduced in Sects. 3.2.1 and 3.2.2 are among typical FWM processes observed in optical microcavities. In this section, we focus on degenerate and nondegenerate FWM and their applications in optical frequency comb generation. In these processes, two photons interact with the nonlinear medium and subsequently produce two new photons, defined as the signal and idler photons, respectively. The process is termed as degenerate FWM if the two pump photons are identical, otherwise it is termed as nondegenerate FWM, as presented in Fig. 3.6a. As a third order nonlinear process, its occurrence is not constrained by the structural inversion symmetry, hence it is allowed in most optical materials including SiO_2, SiN_x, and Si. It serves as one of the most commonly observed mechanisms for generating light with new frequencies in optical microcavities, and in particular, it leads to the generation of optical frequency combs.

Optical frequency combs, and especially, dissipative Kerr solitons (DKSs) are cascaded nonlinear effects occurring in high Q WGM microcavities generally promoted by FWM [40, 41]. DKSs are broadband, coherent waveforms with precisely equidistant frequency lines, which emerge as a promising tool for ultrafast ranging [55, 56], optical atomic clock [57], and molecular spectroscopy [58–60]. The formation of DKS is enabled by a delicate balance between nonlinearity and dispersion, as well as loss and gain. DKS formation is generally enabled by scanning a laser across a cavity mode from the blue detuned side to the red detuned side. It starts with incoherent and continuous frequency comb (modulational instability comb, MI comb) generation through degenerate FWM of the pump beam, followed by nondegenerate

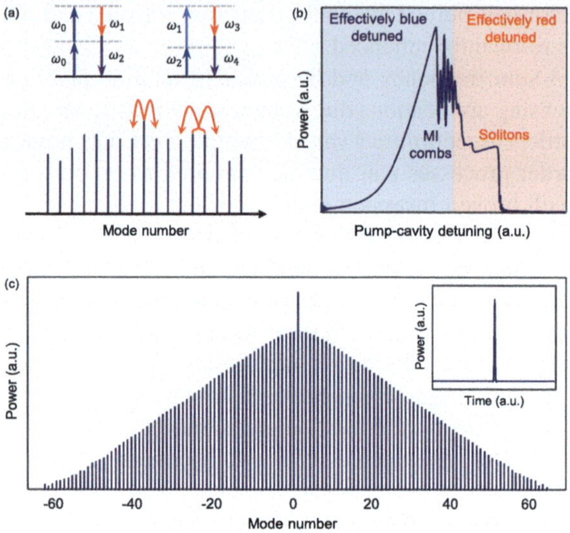

Fig. 3.6 Dissipative Kerr solitons in WGM microcavities. **a** Frequency comb generation through degenerate and nondegenerate four wave mixing. **b** The MI comb and soliton formation during the pump scanning across a cavity mode. **c** The single DKS spectrum in the frequency and time domain

FWM from the comb teeth, as shown in Fig. 3.6a. MI combs are not phase-locked, exhibiting noisy and chaotic temporal oscillations as observed in their energy spectrum (Fig. 3.6b). Around the point at which the effective detuning changes from blue to red and with a proper pump intensity, the output signal experiences a sudden drop of its noise, characterizing the entering of the phase-locked, pulsed DKS states. A serials of steps with equal heights may be observed in this state, resulting from different numbers of DKSs formed in the system. The single DKS state manifests an envelop of the sech^2 function, with equidistant comb teeth corresponding to the FSR of the microcavity. More detailed description of DKSs formation in WGM can be found in Ref. [41].

Here, we focus on the application of microcavity DKSs for molecular spectroscopy. This novel light source has enabled fast, on-chip spectrometers with ultrahigh precision. Different measurement schemes and their figures of merit will be discussed in Sect. 3.5.

3.3 Molecular Sensing Based on WGM Raman Spectroscopy

3.3.1 Surface Enhanced Raman Spectroscopy

As a distinguished platform for nanoparticle detection with single-molecule sensitivity, WGM sensors generally operate in the linear light-matter interaction regime, which measures the linear polarizability of nanoparticles entering the evanescent field of WGM cavity. In order to specifically identify molecules, extra chemical functionalization on cavity surface is required to bind analyte molecules with receptor molecules with a high specification and selectivity. Therefore, the specificity of WGM microsensor mainly relies on the chemical composition and functionalization of the device. On the contrary, Raman spectroscopy provides a powerful way for specificity identification without receptor molecules via the molecular fingerprints, i.e., their vibrational modes. Hence, surface enhanced Raman spectroscopy (SERS), and cavity enhanced Raman spectroscopy (CERS) in WGM microcavities have attracted considerable interests in the past decades.

Conventional plasmonic SERS techniques are based on noble metal nanoparticles, while Raman sensors could be more biocompatible relying on dielectric WGM microcavities. For examples, the heating effects are negligible because the low absorption of dielectric materials and the low excitation powers benefiting from high Q factors. Moreover, all-dielectric sensors are more robust against harsh electrochemical conditions, in which metals would be attacked to shift plasmonic resonances. Distinct from plasmonic SERS techniques that enhance both the pump and the Raman fields simultaneously in a broadband plasmonic resonance, WGM microcavities exhibit many optical modes with ultra-narrow bandwidths determined by ultrahigh Q factors. Therefore, Raman enhancement in WGM microresonators relies on the resonance conditions and field overlapping at both the excitation and the Stokes wavelengths.

In 2008, Ausman and Schatz [50] firstly built the formulas of WGM enhanced Raman scattering based on the Lorentz-Mie theory. In their theory, a microsphere with radii of 5–20 μm is illuminated by a plane wave light field to calculate the field enhancement at the pump frequency in the visible wavelength range. The Raman emission is investigated by calculating the far-field radiation intensity of a dipolar emitter located at the cavity surface. As compared to a free-space excitation of the same point dipole, an enhancement factor of 10^3–10^4 is obtained for an analyte located on the WGM hotspot when the pump light is enhanced by the cavity resonance. When both the pump and Stokes-shifted Raman peaks are in resonance, the Raman enhancement increases to 10^8. Then, the theory was soon been exploited in several experimental works. In 2010, Anderson [61] experimentally explored SERS in silica spheres with diameters of 5–10 μm with molecular films, and an enhancement factor as high as 10^3 was obtained for two touching microspheres. Later, Raman sensing of solution and gas was also demonstrated by using SiO_2-TiO_2 core-shell microspheres. Since the TiO_2 shell has a higher refractive index, the WGM was excited in a TiO_2 shell so that the electric field is tightly confined [62, 63]. With self-assembled and multi-stacked core-shell microspheres allowing for large area detection, an overall enhancement up to 140-fold was experimentally observed (Fig. 3.7).

Efforts towards few- and single- particle sensing have been devoted to enhancing the signal-to-noise ratio, via optimizing the sensing scheme and parameters. Limited by fabrication techniques, the early progresses in WGM Raman sensing mainly utilize microspheres with moderate Q, and conventional Raman spectroscopy system equipped with free-space illumination and collection geometry. More recently, with the rapid development of microcavity fabrication and coupling techniques, SERS could be further enhanced by evanescently-coupled ultrahigh-Q WGM microresonators. While ultrahigh-Q microcavities could enhance intracavity power by sev-

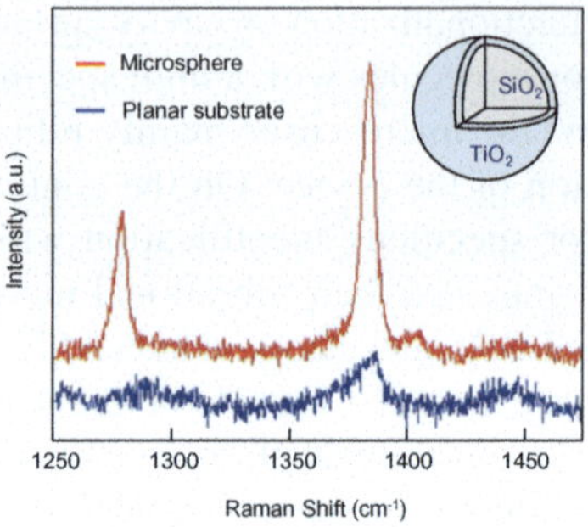

Fig. 3.7 Raman spectra of CO_2 adsorbed onto SiO_2-TiO_2 core-shell microspheres and TiO_2 substrate. Adapted with permission from Ref. [62]. Copyright The Royal Society of Chemistry 2016

eral orders of magnitude, the most efficient methods to excite these WGM modes are evanescent coupling schemes utilizing optical fibers taper, on-chip waveguide, and prism, due to the momentum mismatch between the WGMs and photons in vacuum. These near-field coupling approach can reach up to near 100% coupling efficiency, hence greatly improve the signal-to-noise ratio. Therefore, it opens an avenue to Raman detection towards single-nanoparticle/molecule level with optical WGM microcavity. In 2015, Liu et al. [51] theoretically investigated the Raman enhancement from a single nanoparticle on an ultrahigh-Q WGM microcavity coupled with a tapered fiber.

In that work, both the pump and Stokes fields are assumed to be in resonance with the cavity modes. Through the Maxwell nonlinear equation and under the slowly varying envelop approximation, the rate equations of the spontaneous Raman scattering processes in a WGM cavity can be written as [44]

$$\frac{dN_s}{dt} = -\kappa_s N_s + g_s N_p \tag{3.4}$$

and the Raman yield

$$Y = \frac{P_s}{P_{in}} = \frac{2\kappa_{s,1}\kappa_{p,1}g_s}{\kappa_s\kappa_p^2} \tag{3.5}$$

Here N_s and N_p are the mean photon numbers of the Stokes and pump fields, κ_s and κ_p are their dissipation rates, respectively. g_s is the Raman gain coefficient of the nanoparticle proportional to $\mathrm{Im}(\chi^{(3)})/V_R$ with V_r representing the mode volume. P_s and P_{in} are the collected Raman power and input pump power, and $\kappa_{s,1}$ and $\kappa_{p,1}$ are the taper-cavity coupling rate of the cavity mode corresponding to the Stokes and pump fields (Fig. 3.8a).

The total dissipation rate of each mode is composed of three parts; the intrinsic cavity loss, the taper-cavity coupling rate, and the loss induced by the Rayleigh scattering of the nanoparticle. The Rayleigh scattering of the nanoparticle lifts the degeneracy of the counterpropagating clockwise and anti-clockwise WGM modes, forming two standing waves, i.e., the symmetric mode with the nanoparticle at the anti-node and the anti-symmetric mode with the nanoparticle at the node, respectively. Therefore, only the symmetric mode interacts with the nanoparticle, and its loss rate is modified to $\kappa_i = \kappa_{i,r} + \kappa_{i,0} + \kappa_{i,1}$. Here $\kappa_{i,0}$ and $\kappa_{i,1}$ are the intrinsic cavity loss and the taper-

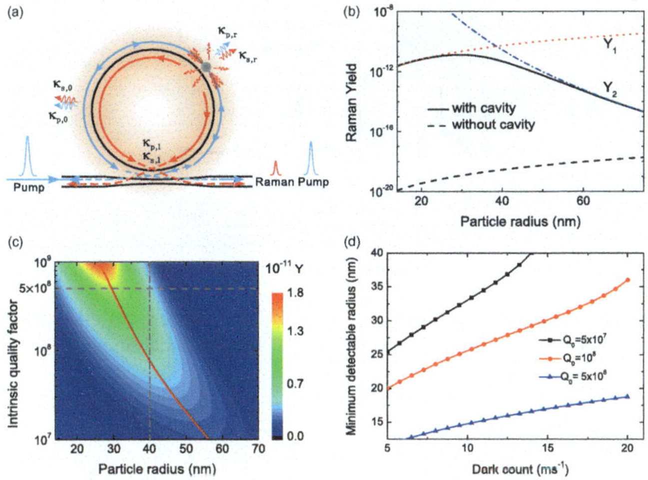

Fig. 3.8 Single particle Raman spectroscopy in WGM microsensors. **a** The illustration of a nanoparticle interacting with the WGM Raman sensor. **b** The Raman yield as a function of the particle radius. Y_1: the Raman yield when the scattering loss is neglected. Y_2 the Raman yield when the scattering loss is dominated. **c** the Raman yield as a function of cavity intrinsic quality factor and particle radius. **d** The detection limit as a function of detector dark count with different microcavity quality factors. Adapted with permission from Ref. [51]. Copyright 2015 American Physical Society

cavity coupling rate, respectively, while $\kappa_{i,r}$ denotes the Rayleigh scattering induced loss. The Raman yield exhibits a nonmonotonic dependence on particle size, and reaches the maximum when the Rayleigh scattering loss is comparable to the intrinsic cavity loss (Fig. 3.8b). The Raman signal also strongly depends on the intrinsic quality factor of the microcavity. Specifically, when the strongest Raman signal is generated, the particle radius is proportional to $Q^{-1/6}$, resulting from the reduced Rayleigh scattering cross section of smaller particle sizes, as shown in Fig. 3.8c. The detection limit is then derived as a function of the intrinsic quality factor of the microcavity. Unlike optical sensors based on linear response, Raman sensors detect Raman scattering away from the pump frequency, and hence are insensitive to the noise of the pump beam and thermorefractive fluctuations. Therefore, the detection limit of a Raman microcavity sensor depends mainly on the dark count rate of the photodetector. As demonstrated in previous discussions, a higher cavity quality factor results in a higher Raman signal and hence a lower detection limit, which is quantitatively estimated in Fig. 3.8d. For example, a single nanoparticle with a diameter down to 30 nm may be detectable with Q of 5×10^8. This value is comparable to linear microcavity sensors based on mode splitting.

In 2018, Huang et al. experimentally observed enhanced Raman scattering from rhodamine 6G molecules coated on silica microspheres with an ultrahigh-Q of 2×10^7 by a fiber-coupling scheme and mode-locking technique, as shown in Fig. 3.9a [52]. In contrast to earlier approaches relying on far-field excitation, in this work, more than 99% of the pump light could be coupled into a single WGM by a tapered fiber to excite Raman signal of molecules with high efficiency. It allows for a thorough experimental study of the Raman enhancement by the WGM mechanism. A total

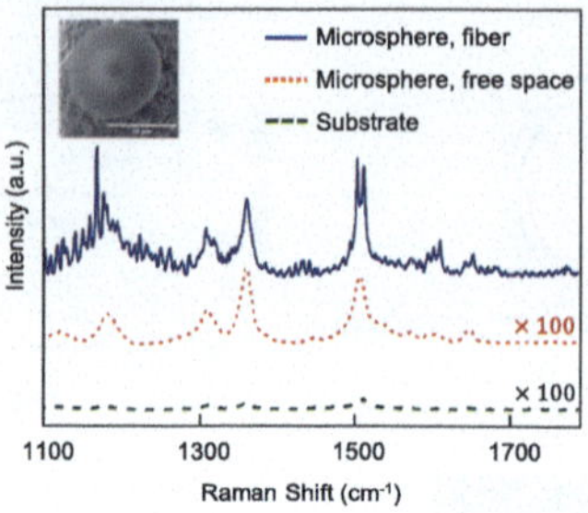

Fig. 3.9 Cavity enhanced Raman spectroscopy in a WGM microsphere coated with R6G molecules. Adapted with permission from Ref. [52]. Copyright 2018 The Optical Society

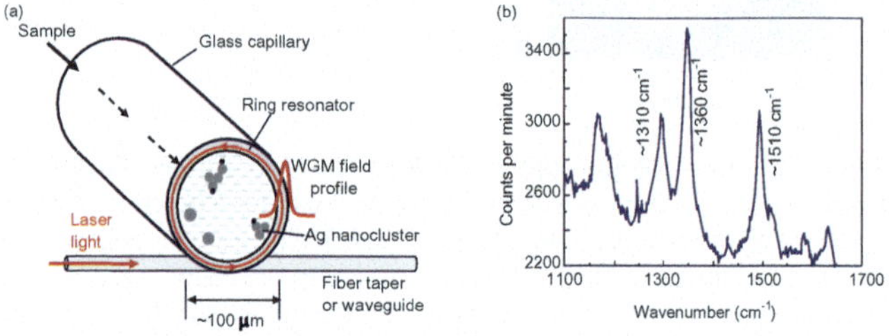

Fig. 3.10 SERS in liquid core ring resonators decorated by Ag nanoclusters. **a** Schematic of the experimental setup. **b** Raman spectrum for 33 nM R6G in Ag colloid. Adapted with permission from Ref. [64]. Copyright 2007 The Optical Society

Raman enhancement factor of 1.4×10^4 was observed experimentally compared with free-space excitation. Within this, a minor factor of 1.134 can be attributed to Purcell enhancement at stokes wavelength matching WGMS, while the remaining 1.2×10^4 attributed to the pump enhancement at the WGM resonance.

Analogous to microcavity sensors in the linear regime, the combination of dielectric WGM microcavities with plasmonic nanoparticles could enhance the Raman signals further for molecular specificity identification. For example, in 2007, White et al. developed a Raman-based sensor in a optofluidic ring resonator (Fig. 3.10) [64]. The device is composed of a glass capillary to move the sample past the optical ring resonator, which is present in the circumference of the glass capillary wall. Ag nanoclusters and molecules of interest that sent through the microfluidic channel can be excited by the WGM mode, enabling Raman detection of a few hundreds of R6G molecules.

3.3.2 Stimulated Raman and Stimulated Anti-Stokes Raman spectroscopy

In the previous section, we mainly discussed Raman spectroscopy based on surface enhanced spontaneous scattering. As spontaneous scattering is intrinsically weak,

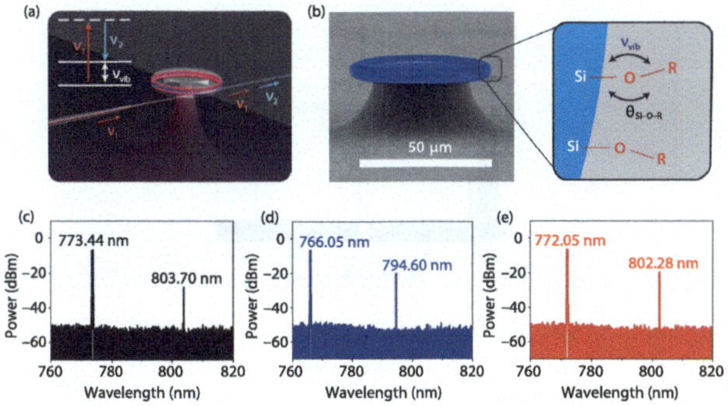

Fig. 3.11 Raman lasing from WGM microcavity coated with monolayer molecule. **a** Illustration of experimental scheme. **b** Scanning electron microscope image of the WGM microcavity. **c–e** Representative Raman emission spectral from three different molecular coatings **c** OH, **d** MS, and **e** DMS. Adapted with permission from Ref. [54]. Copyright 2017 Springer Nature

it requires careful optimization to enhance the signal-to-noise ratio. In this section, we explore coherent Raman processes and their potential applications in molecular spectroscopy. In these coherent nonlinear processes, the photons scattered from the analyte add up coherently and their flux increases superlinearly with the pump power, resulting in strong spectral lines with narrow linewidth. State-of-the-art experiments have reported Raman lasing with a monolayer of molecules, demonstrating the potential to detect a small number of, or single, molecules in experiment.

Recently, Shen et al. successfully realized stimulated Raman scattering and Raman lasing with monolayer organic molecules bounded on the silica toroidal microcavity, as presented in Fig. 3.11 [54]. Careful surface modification was implemented to enhance the Raman signal and achieve the lasing threshold, in which the polarization of molecules was aligned to the WGM resonance. Three different monolayer functionalized groups were studied; the hydroxyl (OH) layer intrinsic to the silica surface, as well as the organic methylsiloxane (MS) and dimethylsiloxane (DMS) molecular monolayers. Raman lasing was achieved with low threshold of 200 μW. A relevant but different experiment was carried out in similar molecule-coated microspheres [65], in which lower threshold parametric oscillation was observed, demonstrating an alternative approach for nonlinear molecular sensing.

From the anti-Stokes side of the spectrum, Soltani et al. reported SARS from a hybrid system composed of WGM microcavities coated with Au nanorods that are functionalized by Polyethylene Glycol (PEG) [66]. The hybrid system produces efficient SARS scattering when it is pumped by a laser beam (Fig. 3.12). The strong SARS emission is attributed to the high $\chi^{(3)}$ of the PEGylated Au nanorod. Although more thorough investigation may be needed to further understand and optimize the device for sensing applications, the current observation indicates that it may be possible to exploit the coherent anti-Stokes spectrum for few- and single molecule sensing and spectroscopy.

In the two sections above, we discussed Raman molecular spectroscopy in WGM microcavity sensors with both spontaneous and stimulated scattering. The experi-

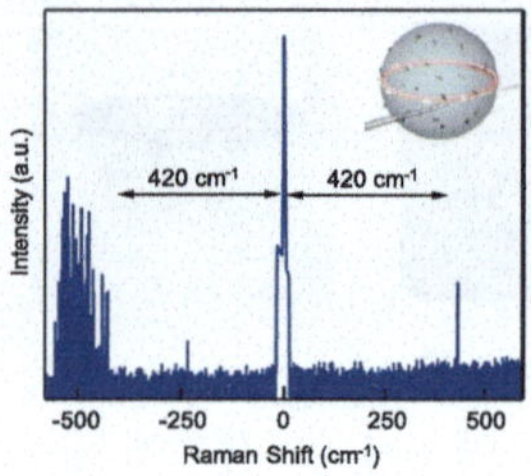

Fig. 3.12 Stimulated anti-Stokes Raman emission spectrum from PEGylated Au nanorod coated WGM microsphere. Inset: Schematic of the device. Adapted with permission from Ref. [66]. Copyright 2018 American Chemical Society

Table 3.1 Selected experimental results of Raman spectroscopy in WGM microcavities

References	Microcavity	Analytes	Q factor	Enhancement factor
Shen et al. [54]	Taper coupled silica microtoroid	Monolayer MS and DMS	$>10^7$	–
Huang et al. [52]	Taper coupled silica microsphere	Nanometer-thick R6G molecule film	2×10^7	1.4×10^4
Bontempi et al. [62]	SiO_2-TiO_2 core-shell microspheres	Environmental CO_2	–	140
Alessandri [63]	SiO_2-TiO_2 core-shell microspheres	Methylene blue	–	10–100
Anderson [61]	Silica microspheres	Molecular films	–	10^3
White et al. [64]	Optical microfluidic ring resonator	400 pM R6G with Ag colloid	$>10^6$	–

mental parameters of the state-of-the-art WGM Raman sensors are summarized in Table 3.1.

Despite the rapid experimental progress, several challenges need to be overcome to further enhance the Raman efficiency towards single molecule detection. Specifically, double resonance at both the pump and Raman wavelength is required to push the detection limit to the single-molecular level, as predicted by the theoretical model. In particular, for coherent anti-Stokes processes, the phase matching condition needs to be fulfilled. Moreover, although a high enhancement factor can be achieved due to the high Q-factor, the overall Raman enhancement may be limited if the linewidth of the WGM mode is much narrower than that of the Raman peak. Hence, to overcome these limitations and further improve sensitivity, the microcavity material, geometry, and measurement schemes should be carefully optimized. Optoplasmonic microcavities may be exploited, analogous to single molecule detectors in the linear optical regime, as discussed in Chap. 2, to enhance the light-matter interactions. With further improvement, combining microcavity based single-nanoparticle/molecule Raman spectroscopy and ultrasensitive linear WGM sensing scheme is promising to achieve real-time, label-free nanoparticle/molecule counting and identification.

3.3.3 Stimulated Scattering Based Sensitivity Enhancement

The sensitivity of microcavity based sensors is highly dependent on the spectral linewidth of WGM mode in the linear regime. Stimulated nonlinear processes can be introduced to act as an intrinsic gain mechanism to compensate the loss of microcavities, and thus reduce the mode linewidth and further lower the detection limit. In 2014, Özdemir et al. [67] and Li et al. [53] used mode splitting of Raman laser to detect single particle with the detection limit of 10 nm.

The initially degenerate clockwise and counterclockwise propagating modes are coupled via backscattering derived from the nanoparticle at the WGM cavity surface. Once the coupling exceeds the loss rate in the system, the mode splitting can be observed. When the pumping power is higher than the Raman lasing threshold (Fig. 3.13a), the two split modes generate a beat oscillation in the time domain that the beatnote frequency corresponds to the mode splitting, as shown in Fig. 3.13b. By monitoring the beatnote frequency change, the number of single particles attached onto the microcavity can be detected in real-time (Fig. 3.13c, d). In these studies, the Raman gain significantly compensates the intrinsic microcavity loss, allowing

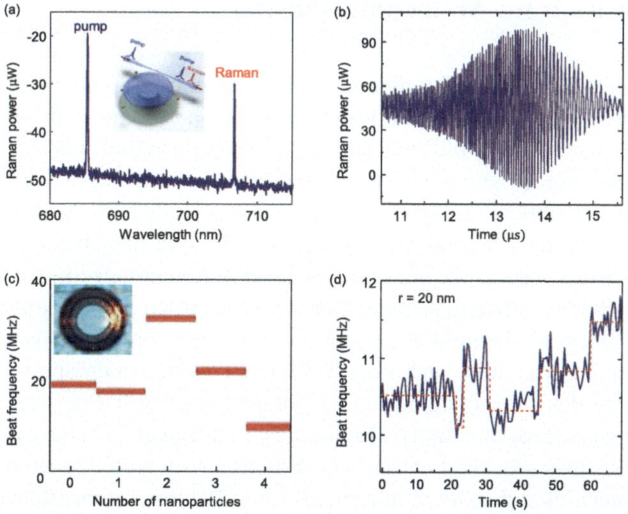

Fig. 3.13 Split-mode Raman laser based single particle detection. **a** Schematic of the experiment and Raman lasing spectrum. **b** Beatnote generated by the interference between the split mode lasers. **c** Beat frequency as a function of the number of nanoparticles attached onto the microcavity sensor. **d** Beat frequency as a function of time. Adapted with permission from Ref. [53]. Copyright 2014 National Academy of Sciences

the emerging of mode splitting that is otherwise absent in the linear regime, hence effectively lowered the detection limit. On the other hand, the beatnote between the two supermodes provides an observable that allows real-time monitoring of the minor changes in the mode splitting. Compared to conventional stimulated emission and lasing processes, the intrinsic gain mechanism of stimulated Raman scattering does not need to introduce external gain such as rare-earth ion [68], and also operate at arbitrary wavelengths. An Analogous sensing scheme has been demonstrated with stimulated Brillouin scattering. Yao et al. reported gas detection via graphene enhanced stimulated Brillouin scattering in a microfluidic resonator [69]. A layer of nm-thick reduced graphene oxide (rGO) film was deposited onto the inner surface of the microfluidic resonator, whose Brillouin Stokes shifts changes by the molecular adsorption induced surface elastic modulation. Such a change is monitored by the beatnote frequency change between the pump and the Stokes beam, in which the sensitivity reached 200 kHz/ppm in ammonia gas detection.

Besides molecular sensing, the enhancement scheme can be applied in detecting and measuring other physical quantities. For example, stimulated Brillouin scattering has been exploited to develop sensitive optical sensors such as microcavity gyroscopes. In these structures, when the sensor rotates, the counter-propagating Brillouin lasers exhibit different frequency shifts arising from the Sagnac effect, and thus the rotation rate can be detected by the heterodyne beatnotes [70, 71].

3.4 Surface Enhanced Harmonic and Sum Frequency Generation for Molecular Sensing

Harmonic generation can be significantly enhanced in high-Q microcavities by resonantly enhancing the fundamental pump light, the harmonic light, or both, given that the phase-matching condition is satisfied. In microcavities made of low nonlinearity materials such as silica, sensing based on second or higher order harmonic signals can be advantageous for detecting analytes with high nonlinear susceptibilities. Especially, as a second order process, SHG can only be significant at the microcavity surface due to the presence of inversion symmetry in the microcavity bulk [35], which is sensitive in probing molecules adsorbed onto the microcavity surface. In 2008, Xu et al. theoretically proposed efficient parametric down conversion, i.e., the inverse process of second order SFG, from a silica microsphere coated with nonlinear molecules [72], opening up the possibility of utilizing such a scheme in molecular sensing. Experimentally, Dominguez-Juarez et al. [73] have reported SHG sensing of molecules adsorbed onto a WGM microcavity with detection limit as low as 50–100 molecules, a density requirement that is 10^4 times lower than on a flat surface (Fig. 3.14). Due to the dispersion of the material and the geometrical structure, the effective refractive index n_{eff} at the fundamental and harmonic wavelengths are different, leading to phase mismatch. The periodic grating patterned onto the surface of the microcavity introduces an additional wave vector $k_{grating}$ that compensates such mismatch, which is generally termed as quasi-phase matching. Quantitatively,

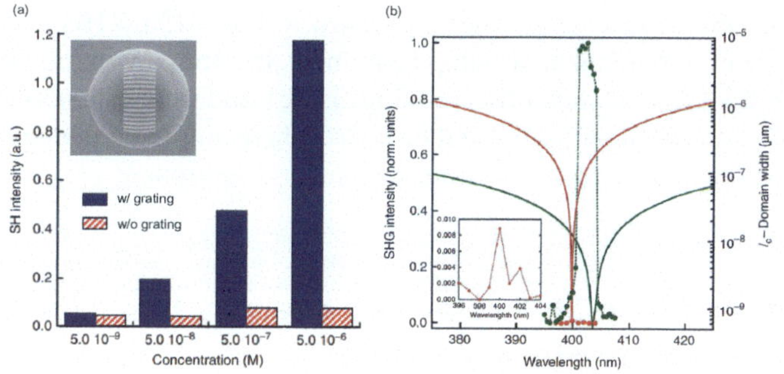

Fig. 3.14 Molecular sensing by SHG in WGM microcavity sensor. **a** SH intensity as a function of molecule concentration with and without grating. Inset: SEM image of the microcavity patterned with periodic grating. **b** Second harmonic generation as a function of half the wavelength of the fundamental wave when the coating solution was 2.5×10^{-4} M (green dots) and when the coating solution was 5×10^{-9} M (red dots), and their corresponding phase matching condition. Phase matching is evaluated by Experimental domain width minus the calculated coherence length at the equator as a function of half the wavelength of the fundamental wave. The inset shows an enlarged view of the red dotted curve. Adapted with permission from Ref. [73]. Copyright 2011 Springer Nature, licensed under a Creative Commons Attribution (CC BY) license: https://creativecommons.org/licenses/

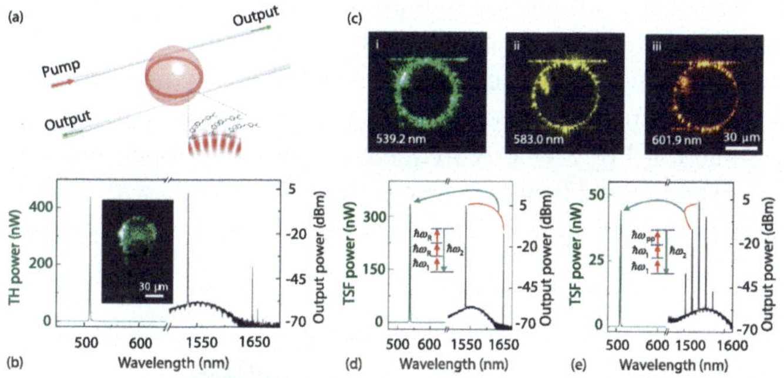

Fig. 3.15 THG and SFG in microspheres coated with organic molecules. **a** Schematic of the experimental setup. **b** Measured TH spectrum (green) and infrared pump spectrum (black). Inset: Scattering CCD image of the TH signal. **c** Real-space image of SFG at 539.2, 583.0, and 601.9 nm. **d** Measured spectra of Raman-scattering-assisted SFG. One pump photon (ω_1) and two Raman Stokes photons (ω_R) are annihilated to create a visible photon (ω_2). **e** Measured spectra of parametric-oscillation-assisted SFG. Two pump photons (ω_1) and one parametric photon (ω_{pp}) create a visible photon (ω_2). Adapted with permission from Ref. [36]. Copyright 2019 American Physical Society

at a particular fundamental wavelength where the quasi-phase-matching condition is reached (dip of the solid curves in Fig. 3.14b), the SH intensity exhibits a sharp enhancement.

The potential of molecular sensing with THG has also been demonstrated by Chen et al. [36], exploiting the large third order susceptibility of organic molecules, as presented in Fig. 3.15. In that work, a microsphere resonator was coated with a thin layer of DSP organic molecules, as shown in Fig. 3.15a. THG and SFG were investigated with a fundamental pump beam near 1550 nm. With the DSP coating, bright TH and SF signals were observed (Fig. 3.15c), and were enhanced by 4-orders of magnitude compared with a bare silica microsphere, implying that the TH and SF signal may serve as an effective tool for molecular detection.

3.5 High Precision Molecular Spectroscopy Based on Microcavity Frequency Combs

Dissipative Kerr solitons (DKSs), consisted of equidistant laser lines, have revolutionized precision measurements of time and frequency, leading to applications such as optical clocks, metrology and spectroscopy. Especially, microcavity based frequency combs provide a compact and chip-scale platform to real-time identification of molecular fingerprints with high spectral resolution. In general, DKSs, serving as broadband coherent light sources, can be used to detect the absorption spectrum, Raman spectrum and other spectral responses of the analyte with comb lines of extremely accurate frequencies. Among various spectroscopy configurations with optical frequency combs (reviewed in detail in Ref. [60]), microcavity-based dual-comb spectroscopy attracts the highest interest in the past five years due to the miniature size and fast measurement speed over a broad spectral ranges. A typical microcavity-based dual-comb spectrometer consists of two microresonators that can generate two optical frequency combs with slightly different repetition frequency Δf_{rep}, as presented in Fig. 3.16a. A frequency comb with repetition frequency f_{rep} interrogates the sample and beats on a single fast photodiode with a reference comb with repetition frequency $f_{\text{rep}} + \Delta f_{\text{rep}}$. The interference signal between the two

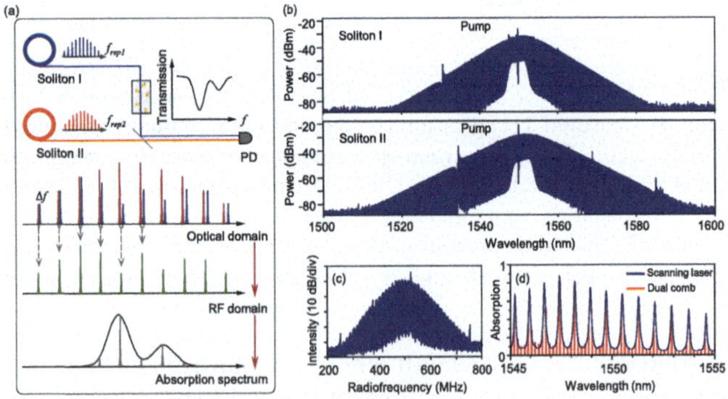

Fig. 3.16 Precision molecular measurement based on Dual-comb spectroscopy. **a** Schematic of dual-comb spectroscopy. **b** Experimentally obtained soliton spectra from two WGM cavities with slightly different repetition frequencies. **c** The RF comb resulted from the interference between the two optical combs in (**b**). **d** Measured absorption spectrum of $H^{13}CN$ in the $2\nu_3$ band. **b–d** are adapted with permission from Ref. [58]. Copyright 2016 AAAS

combs is then recorded as a function of time and Fourier transformed into a radio-frequency (RF) comb with the repetition frequency Δf_{rep}. Therefore, the optical spectral features are down-converted to the RF domain, thus allowing direct measurement of the analyte (i.e. absorption) spectrum using mature electrical detectors with high sensitivity and fast acquisition rate.

Microcavity-based dual-comb spectroscopy was first reported by Suh et al. in 2016, to reveal absorption spectrum of the $H^{13}CN$ gas (Fig. 3.16b–d) [58] at the telecom wavelength range. In that work, two soliton combs with a small Δf_{rep} of 2.6 MHz were generated and stabilized in two silica microdisk resonators precisely controlled by the fabrication process. Upon the interference between the two combs, the absorption spectrum of the $H^{13}CN$ in the $2\nu_3$ band was directly read out in the electrical domain with a center frequency of 500 MHz over 30 dB signal-to-noise-ratio near central lines and with sub-millisecond acquisition time. Later, in 2018, Yu et al. demonstrated a microcavity-based dual-comb spectrometer in the mid-infrared domain [74], which is of vital importance to molecular fingerprinting, but challenging using conventional spectrometers owing to the lack of sensitive, fast mid-infrared detector arrays. Using two mutually coherent mode-locked frequency combs spanning from 2.6 to 4.1 in two silicon microresonators, the absorption spectrum of acetone spanning from 2900 to 3100 nm is measured at 127-GHz resolution. This work extends microcomb-based sensing applications to liquid/condensed matter phase studies.

Besides the dual-comb spectrometers, several methods have been explored to simplify the device configurations and/or enhance their capability. For example, in 2019, Yang et al. reported a vernier spectrometer in a single WGM microresonator (Fig. 3.17) [75]. In this scheme, two counter-propagating, mutually phase-locked soliton combs with slightly different repetition frequency in a single high-Q silica

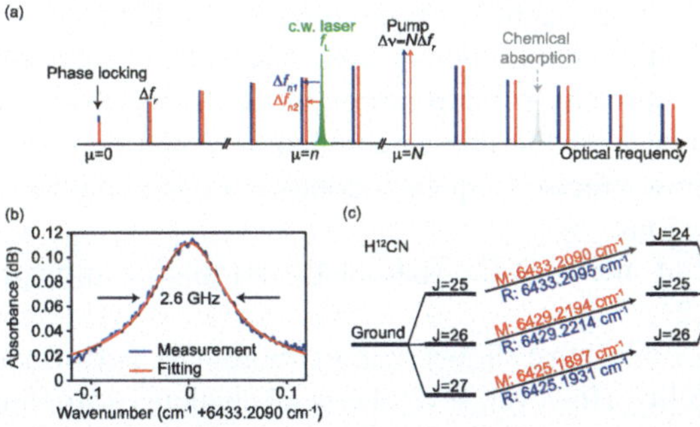

Fig. 3.17 Vernier spectrometer using DKS in microresonators. **a** The concept of the measurement. Counterpropagating soliton frequency combs (red and blue) feature repetition rates that differ by Δf_r, phase-locking at the comb line with index $m = 0$ and effective locking at $m = N$, thereby setting up the vernier spectrometer. **b** Spectroscopy of $H^{12}CN$ gas. **c** Energy level diagram showing transitions between ground state and $2\nu_1$ levels. The measured and reference transition wave numbers are noted in red and blue, respectively. Adapted with permission from Ref. [75]. Copyright 2019 AAAS

microresonator were utilized to implement a frequency vernier scale, which enables rapid and broadband measurement of optical frequencies. The absorption spectrum of $H^{12}CN$ was measured with exceptionally high precision by this vernier spectrometer. Also in a single high-Q WGM microcavity, direct frequency comb [76] for performing high-precision atomic spectroscopy and stabilization, and frequency comb spectroscopy combined with a low-resolution Fourier transform spectrometer [77] have been demonstrated experimentally.

Compared with typical mode-locked lasers, the microcavity frequency combs with high repetition frequencies up to 10^1–10^3 GHz are desirable in many aspects, including the high detection speed scaling with the repetition frequency, the feasibility of chip-scale integration, and ability to resolve adjacent comb teeth via conventional spectral analyzers. However, the relatively large linespacing of a fixed microcomb limits the spectral resolution of the spectrometer, making it challenging to resolve narrow spectral features of the molecules of interest. Therefore, microcombs with precise frequency and repetition rate control are highly demanded in practical spectroscopy applications. In order to overcome such challenge, several efforts have been devoted, mainly by tuning the WGM resonances via the temperature control. Temperature tuning of microcomb frequency was first reported by Yu et al., and 12.5% of comb line spacing has been experimentally demonstrated with a bulk thermal electric cooler [59]. Later, they develop an integrated microheaters on top of the microresonator that can tune the comb tooth by over a full free spectral range for both single- and dual-comb spectrometers [77, 78]. Besides comb frequency tuning, temperature controlling of the microcavity resonance have shown to be able to help the DKS initiation and stabilization for high resolution spectroscopy. Later in 2020, Liu et al. [79] reported fast electrical tuning of the comb frequency up to MHz bandwidth via applying piezoelectric components. Very recently, a microcomb spectrometer with independent and simultaneous control of the comb frequency and repetition rate was demonstrated by Stern et al. [76], and it was employed to accurately resolve the hyperfine manifold of rubidium atoms. In that experiment, via the Pound-Drever-Hall technique, the pump laser was locked to the WGM resonance and fine tuned by the intracavity pump power, while the repetition rate was aligned with a seeding frequency of the phase modulation of the pump laser. Therefore, this configuration allow precise frequency control of each comb tooth suitable for spectroscopy applications.

The major figure of merits of the state-of-the-art microcomb spectrometers are presented in Table 3.2.

Chip-scale microcombs have shown their extraordinary performances with great robustness and tunability. However, several crucial challenges still hinder their practical spectroscopy applications, such as the low microcomb efficiency and the limitation of their working wavelength range. Owing to the small temporal overlap between the pulsed soliton state and the continuous pump light, the microcomb efficiency is generally low. The efficiency could be enhanced by applying gain materials in microcavities or exploiting dark solitons [41]. Moreover, current microcomb spectrometers are mainly limited to near- or mid-infrared bands, however, it would be important to extend their operation to the visible or even ultraviolet wavelength ranges, which is

Table 3.2 Selected experimental demonstrations of microcomb based molecular spectroscopy

References	Mechanism	Frequency span (THz)	Repetition frequency (GHz)	Minimum acquisition time
Stern et al. [76]	Direct comb spectroscopy	192–198	22	–
Yang et al. [75]	Vernier spectrometer	192–194	22	0.1 ps for laser chirping rate monitor
Yu et al. [78]	Direct comb spectroscopy	176–207	195	<2 s with 200 GHz FSR-spanning tuning
Yu et al. [74]	Dual-comb spectroscopy	73–115	127	78 ns
Dutt et al. [59]	Dual-comb spectroscopy	170–220	452	1.79 ns
Yu et al. [77]	Direct comb spectroscopy	76–124	127	–
Suh et al. [58]	Dual-comb spectroscopy	191–195	22	386 ns

crucial for the characterization of the electronic transitions of molecules and atoms. Lately, Chen et al. demonstrated microcomb of more than two-octave span ranging from 450 to 2000 nm through both second- and third-order nonlinearities in a deformed silica microcavity [80], holding great potential for unlocking the applications of microcomb spectroscopy in the visible band.

3.6 Emerging Sensing Methods via Other Nonlinear Processes

In the past decade, sensing techniques have been rapidly developed by incorporating rich nonlinear effects including photothermal effect and optomechanical interactions. Measurements of effects enable comprehensive evaluation of various physical quantities of the analytes, including temperature, optical susceptibility, mass, and vibrational modes, opening up opportunities for multidimensional sensing and precision measurements.

Thermal-optical effect describes the phenomenon that the refractive index of material changes with the temperature change induced by light. Specifically, in microcavity sensors, the heat due to absorption of the analytes can be dissipate into the cavity, leading to power dependent resonance frequency shifts. Measurements of such shifts have experimentally enabled detection of individual nanoparticles [81–83]. Moreover, this method may not only provide an alternative approach for molecular sensing, but also be utilized for studying the thermodynamic and dissipative properties of molecules. Reported by Heylman et al. [81] in 2016, a WGM microtoroid was able to be utilized as an ultrasensitive single nanoparticle absorption spectrometer. In that work, a gold nanorod as the analyte was excited by a pump beam, while a second off-resonance probe beam was employed to measure the local refractive index change by monitoring the shift of the cavity resonance. Ultrahigh sensitivity down to $\Delta \mathbf{T} \sim 100$ nK was enabled owing to the ultra-high Q factor of the cavity resonance combined with the Pound-Drever-Hall locking technique. As a result, a minimum detectable frequency shift as small 84 Hz was successfully realized. An analogous measurement scheme has later been demonstrated in an inverted excitation geometry, with the pump beam delivered from the substrate of a microtoroid cavity [82]. This alternation in the exprimental setup may be beneficial for optical integration and sensing in microfluidics. Furthermore, in a same gold nanorod-microtoroid coupled system, Pan et al. demonstrated simultaneous identification of scattering, absorption and transmission cross sections of the gold nanorod, by further combining the aforementioned absorption measurements with the two-side transmission measurement [83]. This experiment thus allowed the quantitative elucidation of all the energy dissipation pathways of the coupled system.

Optomechanical oscillators (OMO) have found applications in high sensitivity nanoparticle detection and precision measurements of force, mass, and accelerations by utilizing the coupling between the mechanical vibration and optical modes. In microcavities with simultaneous high-Q optical and mechanical modes, a tiny nanoparticle adsorption is enough to perturb the mechanical oscillation, which can be read-out by the modulation of the optical mode. Specifically, for single molecules with mass much smaller than the OMO sensor, the adsorption of the analytes can be viewed as adding a mass to the sensor, which leads to a redshift of the mechanical mode. Through this effect, sub-pg single-molecule sensing has been realized in microtoroid and microsphere OMA sensors [84, 85]. A detailed discussion on optomechanical sensing can be found in Chap. 5.

3.7 Conclusion and Outlook

To conclude, we have introduced commonly observed nonlinear effects in WGM microcavities and their applications in nonlinear optical sensing focusing on single- and few-molecule detection and precision measurement of atomic and molecular spectra. Compared with evanescent sensors [16, 17, 42, 43, 86, 87] operated in the linear optical regime, incorporating the rich nonlinear optical interactions greatly extended the sensing functionalities, i.e. enabling spectral fingerprinting and specific identification of nanoparticles and molecules, as well as precision measurements of thermal and mechanical biomaterial properties. While this chapter mainly focus on nanoparticle and molecule detection, nonlinear sensing can find widespread applications in diverse sensing fields such as optical ranging [55, 56] and quantum non-demolition measurements [88]. While theoretical estimations prove their promise, experimentally, the detection limit of nonlinear sensors should be further pushed for single-molecule sensing, identification, and measurement. The accomplishment of such target may require the cooperative efforts from different aspects, including the development of high-quality factor microcavities made of high nonlinearity materials, sophisticated dispersion engineering, and advanced measurement and signal-to-noise ratio techniques [89].

References

1. Daniel Najer, Immo Söllner, Pavel Sekatski, Vincent Dolique, Matthias C Löbl, Daniel Riedel, Rüdiger Schott, Sebastian Starosielec, Sascha R Valentin, Andreas D Wieck, et al. A gated quantum dot strongly coupled to an optical microcavity. *Nature*, pages 1–1, 2019.
2. Rohit Chikkaraddy, Bart De Nijs, Felix Benz, Steven J Barrow, Oren A Scherman, Edina Rosta, Angela Demetriadou, Peter Fox, Ortwin Hess, and Jeremy J Baumberg. Single-molecule strong coupling at room temperature in plasmonic nanocavities. *Nature*, 535(7610):127–130, 2016.
3. Yoshihiro Akahane, Takashi Asano, Bong-Shik Song, and Susumu Noda. High-q photonic nanocavity in a two-dimensional photonic crystal. *nature*, 425(6961):944–947, 2003.
4. Parag B Deotare, Murray W McCutcheon, Ian W Frank, Mughees Khan, and Marko Lončar. High quality factor photonic crystal nanobeam cavities. *Applied Physics Letters*, 94(12):121106, 2009.
5. Alessandro Chiasera, Yannick Dumeige, Patrice Feron, Maurizio Ferrari, Yoann Jestin, Gualtiero Nunzi Conti, Stefano Pelli, Silvia Soria, and Giancarlo C Righini. Spherical whispering-gallery-mode microresonators. *Laser & Photonics Reviews*, 4(3):457–482, 2010.
6. DK Armani, TJ Kippenberg, SM Spillane, and KJ Vahala. Ultra-high-q toroid microcavity on a chip. *Nature*, 421(6926):925–928, 2003.
7. Shui-Jing Tang, Zhihe Liu, Yan-Jun Qian, Kebin Shi, Yujie Sun, Changfeng Wu, Qihuang Gong, and Yun-Feng Xiao. A tunable optofluidic microlaser in a photostable conjugated polymer. *Advanced Materials*, 30(50):1804556, 2018.
8. Jintian Lin, Ni Yao, Zhenzhong Hao, Jianhao Zhang, Wenbo Mao, Min Wang, Wei Chu, Rongbo Wu, Zhiwei Fang, Lingling Qiao, et al. Broadband quasi-phase-matched harmonic generation in an on-chip monocrystalline lithium niobate microdisk resonator. *Physical review letters*, 122(17):173903, 2019.
9. Jacob S. Levy, Mark A. Foster, Alexander L. Gaeta, and Michal Lipson. Harmonic generation in silicon nitride ring resonators. *Optics Express*, 19(12):11415-11421, 2011.
10. Matthew R Foreman, Jon D Swaim, and Frank Vollmer. Whispering gallery mode sensors. *Advances in optics and photonics*, 7(2):168–240, 2015.
11. F Vollmer, S Arnold, and D Keng. Single virus detection from the reactive shift of a whispering-gallery mode. *Proceedings of the National Academy of Sciences*, 105(52):20701–20704, 2008.
12. Siyao Zhang, Shui-Jing Tang, Shengfei Feng, Yun-Feng Xiao, Wanyin Cui, Xinke Wang, Wenfeng Sun, Jiasheng Ye, Peng Han, Xinping Zhang, et al. High-q polymer microcavities integrated on a multicore fiber facet for vapor sensing. *Advanced Optical Materials*, 7(20):1900602, 2019.
13. Jiangang Zhu, Sahin Kaya Ozdemir, Yun-Feng Xiao, Lin Li, Lina He, Da-Ren Chen, and Lan Yang. On-chip single nanoparticle detection and sizing by mode splitting in an ultrahigh-q microresonator. *Nature photonics*, 4(1):46, 2010.
14. Yinglun Xu, Shui-Jing Tang, Xiao-Chong Yu, You-Ling Chen, Daquan Yang, Qihuang Gong, and Yun-Feng Xiao. Mode splitting induced by an arbitrarily shaped rayleigh scatterer in a whispering-gallery microcavity. *Physical Review A*, 97(6):063828, 2018.
15. Linbo Shao, Xue-Feng Jiang, Xiao-Chong Yu, Bei-Bei Li, William R Clements, Frank Vollmer, Wei Wang, Yun-Feng Xiao, and Qihuang Gong. Detection of single nanoparticles and lentiviruses using microcavity resonance broadening. *Advanced Materials*, 25(39):5616–5620, 2013.

16. Shui-Jing Tang, Shuai Liu, Xiao-Chong Yu, Qinghai Song, Qihuang Gong, and Yun-Feng Xiao. On-chip spiral waveguides for ultrasensitive and rapid detection of nanoscale objects. *Advanced Materials*, 30(25):1800262, 2018.
17. Xiao-Chong Yu, Yanyan Zhi, Shui-Jing Tang, Bei-Bei Li, Qihuang Gong, Cheng-Wei Qiu, and Yun-Feng Xiao. Optically sizing single atmospheric particulates with a 10-nm resolution using a strong evanescent field. *Light: Science & Applications*, 7(4):18003–18003, 2018.
18. Judith Su, Alexander F. G. Goldberg, and Brian M. Stoltz. Label-free detection of single nanoparticles and biological molecules using microtoroid optical resonators. *Light: Science & Applications*, 5(1):e16001–e16001, 2016.
19. Martin D Baaske, Matthew R Foreman, and Frank Vollmer. Single-molecule nucleic acid interactions monitored on a label-free microcavity biosensor platform. *Nature nanotechnology*, 9(11):933, 2014.
20. Martin D Baaske and Frank Vollmer. Optical observation of single atomic ions interacting with plasmonic nanorods in aqueous solution. *Nature Photonics*, 10(11):733–739, 2016.
21. Xuefeng Jiang, Abraham J Qavi, Steven H Huang, and Lan Yang. Whispering-gallery sensors. *Matter*, 3(2):371–392, 2020.
22. Yanyan Zhi, Xiao-Chong Yu, Qihuang Gong, Lan Yang, and Yun-Feng Xiao. Single nanoparticle detection using optical microcavities. *Advanced Materials*, 29(12):1604920, 2017.
23. Frank Vollmer and Stephen Arnold. Whispering-gallery-mode biosensing: label-free detection down to single molecules. *Nature methods*, 5(7):591, 2008.
24. Hansuek Lee, Tong Chen, Jiang Li, Ki Youl Yang, Seokmin Jeon, Oskar Painter, and Kerry J Vahala. Chemically etched ultrahigh-q wedge-resonator on a silicon chip. *Nature Photonics*, 6(6):369–373, 2012.
25. Tobias J Kippenberg, AL Tchebotareva, J Kalkman, A Polman, and Kerry J Vahala. Purcell-factor-enhanced scattering from Si nanocrystals in an optical microcavity. *Physical review letters*, 103(2):027406, 2009.
26. Tobias J Kippenberg, AL Tchebotareva, J Kalkman, A Polman, and Kerry J Vahala. Ultrahigh optical Q factors of crystalline resonators in the linear regime. *Physical Review A*, 74(6):063806, 2006.
27. Yong Yang, Jonathan Ward, and Síle Nic Chormaic. Quasi-droplet microbubbles for high resolution sensing applications. *Optics express*, 22(6):6881–6898, 2014.
28. Jintian Lin, Yingxin Xu, Jielei Ni, Min Wang, Zhiwei Fang, Lingling Qiao, Wei Fang, and Ya Chang. Phase-Matched Second-Harmonic Generation in an On-Chip $LiNbO_3$ Microresonator *Physical Review Applied*, 6(1):014002, 2016.
29. Nan Zhang, Wenzhao Sun, Sean P Rodrigues, Kaiyang Wang, Zhiyuan Gu, Shuai Wang, Wenshan Cai, Shumin Xiao, and Qinghai Song. Highly reproducible organometallic halide perovskite microdevices based on top-down lithography. *Advanced Materials*, 29(15):1606205, 2017.
30. Lin Chang, Weiqiang Xie, Haowen Shu, Qi-Fan Yang, Boqiang Shen, Andreas Boes, Jon D Peters, Warren Jin, Chao Xiang, Songtao Liu, et al. Ultra-efficient frequency comb generation in AlGaAs-on-insulator microresonators. *Nature communications*, 11(1):1–8, 2020.
31. BJM Hausmann, I Bulu, V Venkataraman, P Deotare, and Marko Lončar. Diamond nonlinear photonics. *Nature Photonics*, 8(5), 369–374, 2014.
32. TJ Kippenberg, SM Spillane, DK Armani, and KJ Vahala. Ultralow-threshold microcavity raman laser on a microelectronic chip. *Optics letters*, 29(11):1224–1226, 2004.
33. Bei-Bei Li, Yun-Feng Xiao, Meng-Yuan Yan, William R Clements, and Qihuang Gong. Low-threshold raman laser from an on-chip, high-q, polymer-coated microcavity. *Optics letters*, 38(11):1802–1804, 2013.
34. SM Spillane, TJ Kippenberg, and KJ Vahala. Ultralow-threshold raman laser using a spherical dielectric microcavity. *Nature*, 415(6872):621–623, 2002.
35. Xueyue Zhang, Qi-Tao Cao, Zhuo Wang, Yu-xi Liu, Cheng-Wei Qiu, Lan Yang, Qihuang Gong, and Yun-Feng Xiao. Symmetry-breaking-induced nonlinear optics at a microcavity surface. *Nature Photonics*, 13(1):21–24, 2019.

36. Jin-hui Chen, Xiaoqin Shen, Shui-Jing Tang, Qi-Tao Cao, Qihuang Gong, and Yun-Feng Xiao. Microcavity nonlinear optics with an organically functionalized surface. *Physical Review Letters*, 123(17):173902, 2019.
37. Qi-Tao Cao, Heming Wang, Chun-Hua Dong, Hui Jing, Rui-Shan Liu, Xi Chen, Li Ge, Qihuang Gong, and Yun-Feng Xiao. Experimental demonstration of spontaneous chirality in a nonlinear microresonator. *Physical Review Letters*, 118(3):033901, 2017.
38. JunHwan Kim, Mark C Kuzyk, Kewen Han, Hailin Wang, and Gaurav Bahl. Non-reciprocal brillouin scattering induced transparency. *Nature Physics*, 11(3):275–280, 2015.
39. Zhen Shen, Yan-Lei Zhang, Yuan Chen, Chang-Ling Zou, Yun-Feng Xiao, Xu-Bo Zou, Fang-Wen Sun, Guang-Can Guo, and Chun-Hua Dong. Experimental realization of optomechanically induced non-reciprocity. *Nature Photonics*, 10(10):657–661, 2016.
40. Pascal Del'Haye, Albert Schliesser, Olivier Arcizet, Tom Wilken, Ronald Holzwarth, and Tobias J Kippenberg. Optical frequency comb generation from a monolithic microresonator. *Nature*, 450(7173):1214–1217, 2007.
41. Tobias J. Kippenberg, Alexander L. Gaeta, Michal Lipson, and Michael L. Gorodetsky. Dissipative kerr solitons in optical microresonators. *Science*, 361(6402):eaan8083, 2018.
42. NP Mauranyapin, LS Madsen, MA Taylor, M Waleed, and WP Bowen. Evanescent single-molecule biosensing with quantum-limited precision. *Nature Photonics*, 11(8):477–481, 2017.
43. Hongtao Li, Yunyun Huang, Guanhua Hou, Aoxiang Xiao, Pengwei Chen, He Liang, Yugang Huang, Xiaotian Zhao, Lili Liang, Xinhuan Feng, et al. Single-molecule detection of biomarker and localized cellular photothermal therapy using an optical microfiber with nanointerface. *Science advances*, 5(12):eaax4659, 2019.
44. Robert W Boyd. *Nonlinear optics*. Academic press, 2019.
45. QingHai Song. Emerging opportunities for ultra-high q whispering gallery mode microcavities. *Science China Physics, Mechanics & Astronomy*, 62(7):74231, 2019.
46. Guoping Lin, Aurélien Coillet, and Yanne K Chembo. Nonlinear photonics with high-q whispering-gallery-mode resonators. *Advances in Optics and Photonics*, 9(4):828–890, 2017.
47. Yihang Li, Xuefeng Jiang, Guangming Zhao, and Lan Yang. Whispering gallery mode microresonator for nonlinear optics. arXiv preprint arXiv:1809.04878, 2018.
48. Katrin Kneipp, Yang Wang, Harald Kneipp, Lev T Perelman, Irving Itzkan, Ramachandra R Dasari, and Michael S Feld. Single molecule detection using surface-enhanced raman scattering (sers). *Physical review letters*, 78(9):1667, 1997.
49. Alan Campion and Patanjali Kambhampati. Surface-enhanced raman scattering. *Chemical society reviews*, 27(4):241–250, 1998.
50. Logan K. Ausman and George C. Schatz. Whispering-gallery mode resonators: Surface enhanced raman scattering without plasmons. *The Journal of Chemical Physics*, 129(5):054704, 2008.
51. Rui-Shan Liu, Wei-Liang Jin, Xiao-Chong Yu, Yong-Chun Liu, and Yun-Feng Xiao. Enhanced raman scattering of single nanoparticles in a high-q whispering-gallery microresonator. *Physical Review A*, 91(4):043836, 2015.
52. Steven H. Huang, Xuefeng Jiang, Bo Peng, Corey Janisch, Alexander Cocking, Şahin Kaya Özdemir, Zhiwen Liu, and Lan Yang. Surface-enhanced raman scattering on dielectric microspheres with whispering gallery mode resonance. *Photonics Research*, 6(5):346–356, 2018.
53. Bei-Bei Li, William R. Clements, Xiao-Chong Yu, Kebin Shi, Qihuang Gong, and Yun-Feng Xiao. Single nanoparticle detection using split-mode microcavity raman lasers. *Proceedings of the National Academy of Sciences*, 111(41):14657–14662, 2014.
54. Xiaoqin Shen, Hyungwoo Choi, Dongyu Chen, Wei Zhao, and Andrea M. Armani. Raman laser from an optical resonator with a grafted single-molecule monolayer. *Nature Photonics*, 14(2):95–101, 2020.
55. Myoung-Gyun Suh and Kerry J Vahala. Soliton microcomb range measurement. *Science*, 359(6378):884–887, 2018.
56. Philipp Trocha, M Karpov, D Ganin, Martin HP Pfeiffer, Arne Kordts, S Wolf, J Krockenberger, Pablo Marin-Palomo, Claudius Weimann, and Sebastian Randel. Ultrafast optical ranging using microresonator soliton frequency combs. *Science*, 359(6378):887–891, 2018.

57. Th Udem, Ronald Holzwarth, and Theodor W Hänsch. Optical frequency metrology. *Nature*, 416(6877):233–237, 2002.
58. Myoung-Gyun Suh, Qi-Fan Yang, Ki Youl Yang, Xu Yi, and Kerry J. Vahala. Microresonator soliton dual-comb spectroscopy. *Science*, 354(6312):600–603, 2016.
59. Avik Dutt, Chaitanya Joshi, Xingchen Ji, Jaime Cardenas, Yoshitomo Okawachi, Kevin Luke, Alexander L. Gaeta, and Michal Lipson. On-chip dual-comb source for spectroscopy. *Science Advances*, 4(3):e1701858, 2018.
60. Nathalie Picqué and Theodor W. Hänsch. Frequency comb spectroscopy. *Nature Photonics*, 13(3):146–157, 2019.
61. Mark S. Anderson. Nonplasmonic surface enhanced raman spectroscopy using silica microspheres. *Applied Physics Letters*, 97(13):131116, 2010.
62. Nicolò Bontempi, Luca Carletti, Costantino De Angelis, and Ivano Alessandri. Plasmon-free sers detection of environmental CO_2 on TiO_2 surfaces. *Nanoscale*, 8(6):3226–3231, 2016.
63. Ivano Alessandri. Enhancing raman scattering without plasmons: Unprecedented sensitivity achieved by tio2 shell-based resonators. *Journal of the American Chemical Society*, 135(15):5541–5544, 2013.
64. Ian M. White, John Gohring, and Xudong Fan. Sers-based detection in an optofluidic ring resonator platform. *Optics Express*, 15(25):17433–17442, 2007.
65. Xiaoqin Shen, Rigoberto Castro Beltran, Vinh M Diep, Soheil Soltani, and Andrea M Armani. Low-threshold parametric oscillation in organically modified microcavities. *Science advances*, 4(1):eaao4507, 2018.
66. Soheil Soltani, Vinh M. Diep, Rene Zeto, and Andrea M. Armani. Stimulated anti-stokes raman emission generated by gold nanorod coated optical resonators. *ACS Photonics*, 5(9):3550–3556, 2018.
67. Şahin Kaya Özdemir, Jiangang Zhu, Xu Yang, Bo Peng, Huzeyfe Yilmaz, Lina He, Faraz Monifi, Steven He Huang, Gui Lu Long, and Lan Yang. Highly sensitive detection of nanoparticles with a self-referenced and self-heterodyned whispering-gallery raman microlaser. *Proceedings of the National Academy of Sciences*, 111(37):E3836–E3844, 2014.
68. Xue-Feng Jiang, Yun-Feng Xiao, Chang-Ling Zou, Lina He, Chun-Hua Dong, Bei-Bei Li, Yan Li, Fang-Wen Sun, Lan Yang, and Qihuang Gong. Highly unidirectional emission and ultralow-threshold lasing from on-chip ultrahigh-q microcavities. *Advanced Materials*, 24(35):OP260–OP264, 2012.
69. Baicheng Yao, Caibin Yu, Yu Wu, Shu-Wei Huang, Han Wu, Yuan Gong, Yuanfu Chen, Yanrong Li, Chee Wei Wong, Xudong Fan, et al. Graphene-enhanced brillouin optomechanical microresonator for ultrasensitive gas detection. *Nano letters*, 17(8):4996–5002, 2017.
70. Jiang Li, Myoung-Gyun Suh, and Kerry Vahala. Microresonator brillouin gyroscope. *Optica*, 4(3):346–348, 2017.
71. Yu-Hung Lai, Myoung-Gyun Suh, Yu-Kun Lu, Boqiang Shen, Qi-Fan Yang, Heming Wang, Jiang Li, Seung Hoon Lee, Ki Youl Yang, and Kerry Vahala. Earth rotation measured by a chip-scale ring laser gyroscope. *Nature Photonics*, pages 1–5, 2020.
72. Yong Xu, Ming Han, Anbo Wang, Zhiwen Liu, and James R Heflin. Second order parametric processes in nonlinear silica microspheres. *Physical review letters*, 100(16):163905, 2008.
73. J. L. Dominguez-Juarez, G. Kozyreff, and Jordi Martorell. Whispering gallery microresonators for second harmonic light generation from a low number of small molecules. *Nature Communications*, 2(1):254, 2011.
74. Mengjie Yu, Yoshitomo Okawachi, Austin G. Griffith, Nathalie Picqué, Michal Lipson, and Alexander L. Gaeta. Silicon-chip-based mid-infrared dual-comb spectroscopy. *Nature Communications*, 9(1):1869, 2018.
75. Qi-Fan Yang, Boqiang Shen, Heming Wang, Minh Tran, Zhewei Zhang, Ki Youl Yang, Lue Wu, Chengying Bao, John Bowers, Amnon Yariv, et al. Vernier spectrometer using counter-propagating soliton microcombs. *Science*, 363(6430):965–968, 2019.
76. Liron Stern, Jordan R. Stone, Songbai Kang, Daniel C. Cole, Myoung-Gyun Suh, Connor Fredrick, Zachary Newman, Kerry Vahala, John Kitching, Scott A. Diddams, and Scott B. Papp. Direct kerr frequency comb atomic spectroscopy and stabilization. *Science Advances*, 6(9):eaax6230, 2020.

77. Mengjie Yu, Yoshitomo Okawachi, Austin G. Griffith, Michal Lipson, and Alexander L. Gaeta. Microresonator-based high-resolution gas spectroscopy. *Optics Letters*, 42(21):4442–4445, 2017.
78. Mengjie Yu, Yoshitomo Okawachi, Chaitanya Joshi, Xingchen Ji, Michal Lipson, and Alexander L. Gaeta. Gas-phase microresonator-based comb spectroscopy without an external pump laser. *ACS Photonics*, 5(7):2780–2785, 2018.
79. Junqiu Liu, Hao Tian, Erwan Lucas, Arslan S Raja, Grigory Lihachev, Rui Ning Wang, Jijun He, Tianyi Liu, Miles H Anderson, Wenle Weng, et al. Monolithic piezoelectric control of soliton microcombs. *Nature*, 583(7816):385–390, 2020.
80. Hao-Jing Chen, Qing-Xin Ji, Heming Wang, Qi-Fan Yang, Qi-Tao Cao, Qihuang Gong, Xu Yi, and Yun-Feng Xiao. Chaos-assisted two-octave-spanning microcombs. *Nature Communications*, 11(1):2336, 2020.
81. Kevin D. Heylman, Niket Thakkar, Erik H. Horak, Steven C. Quillin, Charles Cherqui, Kassandra A. Knapper, David J. Masiello, and Randall H. Goldsmith. Optical microresonators as single-particle absorption spectrometers. *Nature Photonics*, 10(12):788–795, 2016.
82. Kassandra A Knapper, Feng Pan, Morgan T Rea, Erik H Horak, Jeremy D Rogers, and Randall H Goldsmith. Single-particle photothermal imaging via inverted excitation through high-q all-glass toroidal microresonators. *Optics express*, 26(19):25020–25030, 2018.
83. Feng Pan, Kevin C. Smith, Hoang L. Nguyen, Kassandra A. Knapper, David J. Masiello, and Randall H. Goldsmith. Elucidating energy pathways through simultaneous measurement of absorption and transmission in a coupled plasmonic–photonic cavity. *Nano Letters*, 20(1):50–58, 2020.
84. Fenfei Liu, Seyedhamidreza Alaie, Zayd C Leseman, and Mani Hossein-Zadeh. Sub-pg mass sensing and measurement with an optomechanical oscillator. *Optics express*, 21(17):19555–19567, 2013.
85. Wenyan Yu, Wei C Jiang, Qiang Lin, and Tao Lu. Cavity optomechanical spring sensing of single molecules. *Nature communications*, 7(1):1–9, 2016.
86. Sanli Faez, Yoav Lahini, Stefan Weidlich, Rees F Garmann, Katrin Wondraczek, Matthias Zeisberger, Markus A Schmidt, Michel Orrit, and Vinothan N Manoharan. Fast, label-free tracking of single viruses and weakly scattering nanoparticles in a nanofluidic optical fiber. *ACS nano*, 9(12):12349–12357, 2015.
87. Shui-Jing Tang, Fei Gao, Da Xu, Yan Li, Qihuang Gong, and Yun-Feng Xiao. Rayleigh scattering in an emitter-nanofiber-coupling system. *Physical Review A*, 95(4):043801, 2017.
88. A Pontin, M Bonaldi, Antonio Borrielli, L Marconi, F Marino, Gregory Pandraud, GA Prodi, PM Sarro, Enrico Serra, and F Marin. Quantum nondemolition measurement of optical field fluctuations by optomechanical interaction. *Physical Review A*, 97(3):033833, 2018.
89. Wenjing Liu, You-Ling Chen, Shui-Jing Tang, Frank Vollmer, and Yun-Feng Xiao. Nonlinear sensing with whispering-gallery mode microcavities: From label-free detection to spectral fingerprinting. *Nano Letters*, 2020.

Part II
Optomechanical Sensing

Raman spectroscopy techniques are well developed to probe the mechanical vibrations of biological media. Combined with plasmonic nanoparticles, they allow vibrations sensing down to the single-molecule sensing. Surface enhanced Raman scattering (SERS) in particular is powerful tool that allows both fundamental studies of biomolecular interactions and the unlabeled detection of low-abundance biomarkers associated with specific diseases. However, Raman spectroscopy is well treated in a number of existing textbooks[1]. In this book we instead focus on two new areas of optomechanical sensing; the first using a mechanical resonator that is coupled to an optical cavity to detect the presence of single biomolecules, and the second using quantum optics theories to build a new understanding of SERS and of the new area of molecular optomechanics, where an optical nanocavity is used to both enhance and control the light-molecule interaction.

In Chap. 4 the authors show that the mechanical resonances of an optical whispering gallery mode cavity can be driven into coherent oscillation by radiation pressure, even in the presence of a surrounding fluid. They then show that by monitoring the frequency of the oscillation it is possible to resolve frequency shifts due to small molecules that bond to the surface of the cavity. In Chap. 5 the authors develop quantum optics models for the interactions of the vibrations of molecules with light confined within an optical cavity. They use these models to identify limits to the usual electric-field-to-the-fourth-power scaling of SERS signals with light and to predict that at high powers the scaling may experience a dramatic change to electric-field-to-the-eighth-power. They develop master equation approaches to model SERS within an optical cavity and explore the possibility of using the molecular optomechanics in the good cavity regime—where the cavity decay rate is slower than the molecular vibration frequency—to control the vibrations of the molecule.

---

[1] E.g. see Infrared and Raman Spectroscopy: Principles and Spectral Interpretation, P. J. Larkin, Elsevier, 2018.

Chapter 4
Optomechanical Sensing

Wenyan Yu, Wei C. Jiang, Qiang Lin, and Tao Lu

Abstract In this chapter, we present a cavity optomechanical sensor. Utilizing the rigidity of the optical spring in a quivering whispering gallery microcavity, single protein molecules can be detected at high signal-to-noise ratio. In addition to its high sensitivity, such a sensor also offers an effective detection area that is many orders of magnitude larger than plasmonic counterparts.

4.1 Introduction

Detecting individual nanoparticles, molecules and messenger ribonucleic acid (mRNA), etc., imaging them, finding their chemical composition and observing their mutual interactions are of constant interests to scientists and engineers for their broad medical, biological and chemical applications, e.g., early disease diagnosis and understanding how a drug interacts with its target molecules. In the past decades, a variety of tools have been developed to observe single particles down to molecular scale [1, 2], among which whispering gallery microcavities (WGMs) become one of the most attractive candidates [3–9] (also discussed in Chap. 2).

W. Yu · T. Lu (✉)
Department of ECE, University of Victoria, EOW 448, 3800 Finnerty Rd., Victoria V8P 5C2, Canada
e-mail: taolu@uvic.ca

W. C. Jiang
Science and Technology Division, Corning Research and Development Corporation, Corning, NY 14831, USA
e-mail: jiangwc@corning.com

Q. Lin
Institute of Optics, University of Rochester, Rochester, NY 14627, USA
e-mail: qiang.lin@rochester.edu

Department of Electrical and Computer Engineering, University of Rochester, Rochester, NY 14627, USA

W. Bowen et al. (eds.), *Single Molecule Sensing Beyond Fluorescence*,
Nanostructure Science and Technology,
https://doi.org/10.1007/978-3-030-90339-8_4

Named after the whispering gallery at St. Pauls Cathedral, London for its similar properties, an optical WGM allows photons to circulate along its edge through total internal reflection (TIR). When the photon round trip optical path length equals to the integer multiples of its wavelength, resonance occurs [10]. An important figure of merit for WGMs is the optical quality factor (Q), a measure of how long a photon can survive in the cavity. An ultra-high Q WGM resonantly builds up a substantial optical intensity in close proximity to its surface with a small amount of optical power supplied by an external source. As an illustration, a 10 milli-watt laser power continuously delivered to a 10^8-Q microcavity may establish an intracavity intensity as high as multiple tera-watts per square meter. Such high optical intensity triggers a plurality of nonlinear optical effects upon which many micron scale, low power consumption devices have been developed. To date, researchers have demonstrated several silica based microcavity structures with Q above 10^8, which include microsphere [11], microtoroid [12] and microdisk [13], etc. Among them, a microsphere, usually reflowed on a fiber tip, can reach quality factor as high as 10^9 while a microtoroid fabricated on a silicon wafer is the first on-chip device that has achieved a Q as high as 5×10^8. Further, microdisks with Q as high as 10^9 have been demonstrated with advanced lithography technologies, providing an alternative to microtoroids as an on-chip microcavity [13, 14].

Several unique features of an ultra-high Q WGM make it a perfect device for nanosensing. Firstly, a nanoparticle binding on the cavity surface causes a sudden change in cavity resonance wavelength. Therefore, by monitoring the cavity resonant wavelength change, one may sense a particle. This is called the reactive sensing method and the wavelength change can be probed by highly accurate interferometry method [4, 5, 15]. Secondly, a pair of counter propagating resonance modes exist in WGM. Their interaction yields a scattering dependent mode splitting. By detecting the change of mode splitting, one may detect scatters such as nanoparticles [5, 6, 16]. Thirdly, the strong optical intensity build-up at the WGM surface can trigger lasing, nonlinear, Raman process and short pulse generation at a threshold pump power orders of magnitude lower than the conventional methods [17–20]. More importantly, such high intensity can generate an optical force strong enough to quiver a WGM mechanically, a phenomenon known as the cavity optomechanical oscillation [21–30]. This oscillation modulates the light signal escaping from the cavity and can be interrogated through a photodetector. According to Hooke's law, a particle attached to the cavity will increase the cavity effective mass and thus cause the shift in the mechanical oscillation frequency. By monitoring this shift one can also detect the particle. These unique features of a WGM are still under-exploration for sensing.

Progress on reactive sensing has been rapid since it was first proposed in 1995 [3]. In 2008, a silica microsphere with a Q of 2.6×10^5 in water was employed to demonstrate the detection of a single Influenza A virion at a signal-to-noise ratio of 3:1 [4]. In 2011, by adopting a silica microtoroid with a reference interferometer, a ten-fold improvement of detection sensitivity was achieved [5]. The detection resolution of this approach is limited by two major factors: a thermal refractive noise that causes the cavity resonance wavelength fluctuation and the bandwidth of the resonance structures that contributes to the uncertainty on locating the resonance wavelength. A passive microtoroid typically yields a thermal refractive noise induced optical resonant frequency fluctuation of around 100 kHz and a bandwidth of several MHz in water. That limits the minimum detectable particle to be 12.5 nm in radius to date.

Typically, two approaches can be taken to improve the detection sensitivity: to enhance the signal or to reduce the detection uncertainty from noises (see Chap. 1). In the former approach, progress was made by combining the cavity with gold plasmonic nanoantennas [31–34] as discussed in detail in Chap. 2. In that case, the signal is enhanced by the conventional plasmonic structures while the cavity helps to overcome the high loss drawback of the plasmonic nanoantenna. Along the same vein, gold nanobeads, nanorods or nanoshells randomly adsorbed to a cavity were employed as a plasmonic nanoantenna. In these articles, single molecule sensitivity has been demonstrated but with orders of magnitude smaller effective detection area than conventional reactive sensing. Therefore, cavity sensing without a plasmonic antenna is still of significant importance.

On the other hand, cavity optomechanical oscillation displayed a much lower frequency noise and narrower bandwidth in liquid [35]. This makes it an attractive candidate along the sensitivity improvement through latter approach. In the past, sensing through optical or electrical driven mechanical oscillations have been demonstrated in various fields. These sensing applications typically exploit the dynamical interaction of light with a mechanical resonator together with the precision readout provided by a high quality optical microcavity. For example, a room-temperature optoelectromechanical transducer using nanomembrane was reported [36] to be capable of strong coupling of both microwave and optical fields. A GaAs-based electromechanical resonator was futher used for the study of nuclear magnetic resonance (NMR) [37]. Meanwhile, optomechanical cavities have been used for ultrabroadband and ultrasensitive magnetometry [38, 39], acceleration sensing [40–43], ultrasound sensing [44, 45], forces detection [46, 47], and potentially of quantum primary thermometry [48–50].

Cavity optomechanical sensing in liquid requires high optical quality factor and strong optical force to overcome the mechanical dissipation from the liquid viscosity, which is challenging because of large optical loss from water and solvent contaminants that could easily adhere to the cavity surface. Therefore, this approach was demonstrated mainly in gaseous environments and in cases where bulk liquid solutions flow inside the cavity so that the cavity Q will not degrade [51, 52]. To achieve optomechanical oscillation of a cavity immersed in liquid, care must be taken to prevent possible contamination. Meanwhile the cavity should be properly designed such that optical Q will not drop significantly from water absorption [35].

In this chapter we demonstrate an optomechanical sensor based on optical spring effects [53]. With this approach, sensing resolution can be enhanced by orders of magnitude compared with conventional approaches. As a result, single protein molecules can be detected at high signal-to-noise ratio (SNR).

This chapter is organised as follows, in Sect. 4.2 we first lay out the theory of cavity optomechanics and its principle for sensing while Sect. 4.3 demonstrates the coherent regenerative optomechanical oscillation of a silica microsphere immersed in heavy water (D2O). Section 4.4 further illustrates detection of single nanoparticles and protein molecules suspended in buffer solution.

4.2 Cavity Optomechanics

A spring pulling away from its equilibrium position will oscillate as its elasticity exerts a restoring force to counter react the displacement.[1] In a whispering gallery microcavity, the circulating light wave exerts an optical force to deform the cavity, causing it to oscillate mechanically in similar manner to a spring. Meanwhile, the cavity resonance wavelength changes as a result of the mechanical deformation, which in turn modulates the light intensity and the optical force. The mutual opto-mechanical interaction in the process makes the cavity a forced harmonic oscillator and this phenomenon is also known as cavity optomechanical oscillation (OMO) [54–57].

The pictorial description of this process is as follows: a single photon carries energy $E = h\nu$ and momentum $\mathbf{p} = h\nu/c\hat{\mathbf{k}}$ where h is Planck constant, ν the optical frequency, c the speed of light and $\hat{\mathbf{k}}$ the unit vector directing toward propagation. When travelling in cavity, the structural confinement from the microcavity causes the directional change of the photon momentum. Consequently the photon exerts an optical force against such changes following Newton's laws of motion. Although the amount of force by a single photon is small, the collective build-up of such forces by a large number of highly coherent photons can be strong enough to deform the cavity itself. Calculation shows that the needed photon number is easy to achieve in an ultra-high Q microcavity due to the long life time the photon can survive. In fact, when the photon wavelength is close to the cavity resonance, even a continuous supply of as little as micro watts light power to a cavity with 100 million Q can make the cavity quiver mechanically [58, 59].

Figure 4.1 illustrates such dynamic interactions between light and a cavity. Here, the cavity is a hollow Fabry-Pérot (FP) cavity with a pair of parallel placed mirrors separated by a length L. The cavity is pumped with a frequency tunable laser whose wavelength is set close to the cavity resonance wavelength. When photons travel in the cavity, they are bouncing back and forth between the pair of mirrors. If the photon round trip optical path length equals to the integer multiples of its wavelength ($\lambda_0 = 2L/m, m = 1, 2, \ldots$), optical resonance occurs. Here, m is an integer representing the longitudinal mode order of the FP cavity. The reflection at the end facets causes the photon momentum change of $\Delta p = 2|\mathbf{p}| = 2h\nu/c$. Therefore, the continuous arrival of photons at end facets exerts an optical force directing outward. The amplitude of the optical force F is determined by the number of reflected photons per unit time and can be estimated from the circulating power P_c within the cavity according to $F = 2P_c/c$. Similarly, one may obtain the optical force from photons on resonance with a whispering gallery microcavity to be $F = 2\pi n_e P_c/c$ with n_e being the effective

[1] Most of this section is reproduced/adapted with permission from Ref. [53]. Copyright 2016 ©Springer Nature.

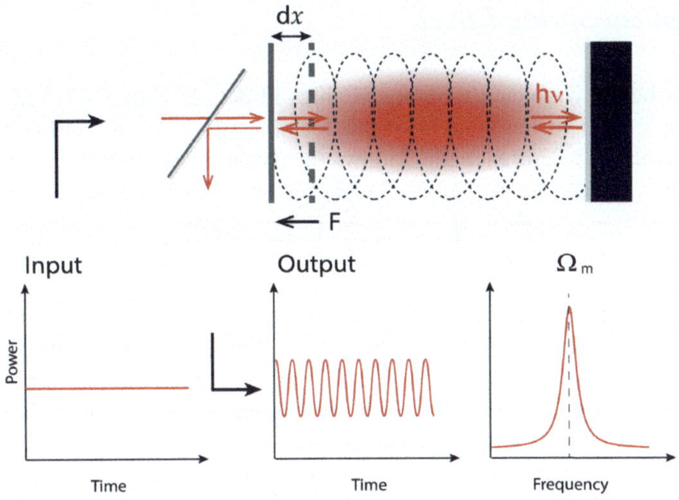

Fig. 4.1 Schematic plot showing the concept of radiation pressure. The continuous wave at a wavelength close to the cavity resonance enters a Fabry-Pérot cavity. The radiation pressure caused by the light reflecting at the mirror displaces the mirror front by an amount of dx, which modulates the output light with an oscillation frequency Ω_m' jointly determined by the natural mechanical response of the cavity and the radiation pressure

refractive index of the cavity. Further, the high intensity buildup at the cavity side wall due to the combined effects of small mode volume and high optical Q of the WGM microcavity, the resulting pressure is sufficient to deform the cavity structure. For example, an ultrahigh Q ($Q > 10^8$) WGM microcavity generates a radiation pressure above thousands of Pascal with as little as 1 mW external optical power continuously supplied to the cavity. With a sufficient power, the radiation pressure is able to push the cavity wall outward from its original position with a small radial displacement of x. Consequently, the cavity resonance shifts toward a longer wavelength because of the increased optical path length.

To describe the dynamic process, consider a pump laser operating with constant power at a fixed wavelength slightly shorter than cavity resonance (blue detuning) initially delivers light to the cavity, the radiation pressure will be established to stretch the cavity along the radial direction. Meanwhile, the increasing cavity deformation results in an increasing tensile force to counteract the radiation pressure. On the other hand, the increasing cavity circumference due to the deformation further increases the relative detuning between the cavity resonance and the pump laser wavelength and reduces radiation pressure. The combined efforts from the increasing tensile force and decreasing radiation pressure will eventually revert the stretching to compressing with tensile force starting to decrease and radiation pressure to increase. Consequently, the cavity oscillates mechanically. As portion of the intracavity light leaks out, the dynamical backaction between the optical mode and mechanical motion leads to an observable transmission modulation on the cavity output spectrum.

4.2.1 Optical Spring Effect

Here we further analyse cavity optomechanics in details with a silica microsphere as an example, and in the context of single molecule sensing. As known [60], a waveguide coupled microresonator follows the dynamic equation

$$\frac{da}{dt} = \left(j\Delta_\omega - \frac{\Gamma_t}{2}\right)a + j\sqrt{\Gamma_e}A_{in} \tag{4.1}$$

where a is the field amplitude of an optical whispering-gallery mode (WGM), normalized such that the cavity mode energy $U = |a|^2$. $A_{\rm in}$ is the amplitude of input laser field, normalized such that $P_{\rm in} = |A_{\rm in}|^2$ is the laser power at the waveguide input. $\Delta_\omega = \omega_l - \omega_0$ is the optical angular frequency detuned from the laser angular frequency ω_l to the cold cavity resonance ω_0. Γ_t is the total photon decay rate of the loaded cavity such that it relates to the loaded optical quality factor according to $Q_t = \omega_0/\Gamma_t$, and Γ_e is the extrinsic photon decay rate associated with the external waveguide coupling and inverse proportional to the coupled optical quality factor following $Q_e = \omega_0/\Gamma_e$. The decay rates Γ_t and Γ_e are related to the intrinsic photon decay rate Γ_i with $\Gamma_t = \Gamma_i + \Gamma_e$. Again, Γ_i and the intrinsic optical quality factor follows $Q_i = \omega_0/\Gamma_i$. In continuous wave (CW) mode, the steady state solution ($da/dt \equiv 0$) of Eq. (4.1) can be expressed as

$$a_0 = \frac{j\sqrt{\Gamma_e}A_{\rm in}}{\frac{\Gamma_t}{2} - j\Delta_\omega} \tag{4.2}$$

Consequently, the amplitude of the optical wave at the output of the waveguide $A_{\rm out}$ can be found as

$$A_{\rm out} = A_{\rm in} + j\sqrt{\Gamma_e}a_0 = \frac{\frac{\Gamma_i - \Gamma_e}{2} - j\Delta_\omega}{\frac{\Gamma_i + \Gamma_e}{2} - j\Delta_\omega}A_{\rm in} \tag{4.3}$$

Accordingly, one may rewrite the transmission spectrum in Chap. 2 as

$$T = \frac{|A_{\rm out}|^2}{|A_{\rm in}|^2} = 1 - \frac{\Gamma_i\Gamma_e}{\Delta_\omega^2 + (\Gamma_t/2)^2} \tag{4.4}$$

Further, we can define the power dropped to the cavity P_d as

$$\begin{aligned} P_d &= |A_{\rm in}|^2(1-T) \\ &= \frac{\Gamma_i\Gamma_e}{\Delta_\omega^2 + (\Gamma_t/2)^2}P_{\rm in} \\ &= \Gamma_i|a_0|^2 \end{aligned} \tag{4.5}$$

which indicates that P_d is expended to compensate the cavity energy dissipation $\Gamma_i|a_0|^2$.

When the circulating power is strong enough, it will produce a radiation pressure that is sufficient to drive the mechanical vibration mode of the cavity. The mechanical mode for a microsphere has a motion that modulates the device radius which in turn modulates the optical cavity resonance $\omega_0' = \omega_0 + g_{\mathrm{om}}x$ with x representing the effective mechanical displacement of the cavity radius from its cold cavity value. $g_{\mathrm{om}} = d\omega_0/dx$ is the optomechanical coupling coefficient, which scales inversely with the radius for a microsphere. A finite element simulation shows that $|g_{\mathrm{om}}|/(2\pi) \approx 6.2$ GHz/nm at a wavelength of 974 nm for a 100 μm-diameter silica microsphere. Such an optomechanical coupling modifies Eq. (4.1) by an additional detuning term $-jg_{\mathrm{om}}x$ and can be described by the following coupled equations of motion [54]:

$$\frac{da}{dt} = \left(j\Delta_\omega - jg_{\mathrm{om}}x - \frac{\Gamma_{\mathrm{t}}}{2}\right)a + j\sqrt{\Gamma_{\mathrm{e}}}A_{\mathrm{in}}, \tag{4.6}$$

$$\frac{d^2x}{dt^2} + \Gamma_{\mathrm{m}}\frac{dx}{dt} + \Omega_{\mathrm{m}}^2 x = \frac{F_{\mathrm{rad}}(t)}{m_{\mathrm{eff}}} + \frac{F_{\mathrm{L}}(t)}{m_{\mathrm{eff}}}, \tag{4.7}$$

Equation (4.6) reflects the cavity optical resonance modulation due to its mechanical motion through the optomechanical coupling coefficient. Consequently, the cavity mechanical displacement changes the optical mode by altering the laser cavity detuning according to $\Delta_\omega'(x) = \omega_1 - \omega_0(x) = \Delta_\omega - g_{\mathrm{om}}x$.

Equation (4.7) describes the mechanical oscillation of the cavity, Γ_{m}, Ω_{m}, and m_{eff} are the damping rate, intrinsic mechanical resonance frequency, and effective mass of the mechanical mode, respectively. F_{L} is the thermal Langevin force responsible for the thermal Brownian motion of the mechanical mode and F_{rad} describes the radiation pressure produced by the intra-cavity laser intensity. Note that the total photon energy in the cavity can be expressed as

$$U(x) = N\hbar\omega_0' = N\hbar(\omega_0 + g_{\mathrm{om}}x) \tag{4.8}$$

where $N = \frac{|a|^2}{\hbar\omega_0'} \approx \frac{|a|^2}{\hbar\omega_0}$ is the total number of photons in the cavity. We obtain the radiation force [61]

$$F_{\mathrm{rad}} = -\frac{dU(x)}{dx} = -N\hbar g_{\mathrm{om}} \approx -\frac{g_{\mathrm{om}}|a|^2}{\omega_0} \tag{4.9}$$

In general, $x(t)$, $a(t)$ and $F_{\mathrm{rad}}(t)$ are all functions of time. Seeking close form solutions is impossible given the highly nonlinear nature of the equation. However, when the mechanical displacement is small, under the linear approximation, we may assume that $a(t) = a_0 + \delta a(t)$ and $x(t) = x_0 + \delta x(t)$ where $|\delta a(t)| \ll |a_0|$ and $|\delta x| \ll |x_0|$. x_0 and a_0 are steady state solutions of Eqs. (4.6) and (4.7) by setting all time derivatives to zero ($da_0/dt = dx_0/dt = d^2x_0/dt^2 \equiv 0$).

$$a_0 = \frac{j\sqrt{\Gamma_{\mathrm{e}}}A_{\mathrm{in}}}{\frac{\Gamma_{\mathrm{t}}}{2} - j(\Delta_\omega - g_{\mathrm{om}}x_0)} \tag{4.10}$$

$$x_0 = -\frac{g_{\mathrm{om}}|a_0|^2}{\omega_0 m_{\mathrm{eff}}\Omega_{\mathrm{m}}^2} \tag{4.11}$$

Again, the presence of cavity optomechanics simply modifies a_0 in Eq. (4.2) by an additional static cavity resonance shift $g_{\mathrm{om}}x_0$ as the effective cavity radius stretches to $R + x_0$ in equilibrium by the radiation force. Subsequently, Eq. (4.6) can be decomposed into

$$\begin{aligned}\frac{da_0}{dt} + \frac{d\delta a(t)}{dt} &= \left[j(\Delta_\omega - g_{\mathrm{om}}(x_0 + \delta x)) - \frac{\Gamma_{\mathrm{t}}}{2}\right](a_0 + \delta a) + j\sqrt{\Gamma_{\mathrm{e}}}A_{\mathrm{in}} \\ &= \left[j(\Delta_\omega - g_{\mathrm{om}}x_0) - \frac{\Gamma_{\mathrm{t}}}{2}\right]a_0 + j\sqrt{\Gamma_{\mathrm{e}}}A_{\mathrm{in}} \\ &+ \left[j(\Delta_\omega - g_{\mathrm{om}}x_0) - \frac{\Gamma_{\mathrm{t}}}{2}\right]\delta a(t) - jg_{\mathrm{om}}a_0\delta x(t) - jg_{\mathrm{om}}\delta x(t)\delta a(t)\end{aligned} \tag{4.12}$$

Here, the first term on the left side of the equation, da_0/dt, and the first two terms on the right of the second equation sign vanish as they form the steady state equation of a_0. Further, the second order perturbation term $-jg_{\mathrm{om}}\delta x\delta a$ can be neglected, leading to a simple expression

$$\frac{d\delta a(t)}{dt} = \left(j\Delta'_\omega - \frac{\Gamma_{\mathrm{t}}}{2}\right)\delta a(t) - jg_{\mathrm{om}}a_0\delta x(t) \tag{4.13}$$

Using the the Fourier transform defined as $\delta\tilde{a}(\Omega) = \mathcal{F}\{\delta a(t)\} = \int_{-\infty}^{+\infty}\delta a(t)$ $e^{j\Omega t}dt$, Eq. (4.13) can be expressed in frequency domain

$$-j\Omega\delta\tilde{a}(\Omega) = \left(j\Delta'_\omega - \frac{\Gamma_{\mathrm{t}}}{2}\right)\delta\tilde{a}(\Omega) - jg_{\mathrm{om}}a_0\delta\tilde{x}(\Omega) \tag{4.14}$$

with $\delta\tilde{x}(\Omega) = \mathcal{F}\{\delta x(t)\}$. Consequently, we obtain the spectral response of the perturbed field as

$$\delta\tilde{a}(\Omega) = \frac{jg_{\mathrm{om}}a_0}{j(\Delta'_\omega + \Omega) - \Gamma_{\mathrm{t}}/2}\delta\tilde{x}(\Omega) \tag{4.15}$$

and its complex conjugate $\delta\tilde{a}(\Omega)$ has the form

$$\delta\tilde{a}^*(\Omega) = \frac{jg_{\mathrm{om}}a_0^*}{j(\Delta'_\omega + \Omega) + \Gamma_{\mathrm{t}}/2}\delta\tilde{x}^*(\Omega). \tag{4.16}$$

To obtain the field pattern in frequency domain, one need to find the Fourier transform of the mechanical displacement $\delta\tilde{x}(\Omega)$. To do so, we first apply the similar perturbation approach to Eq. (4.7)

$$\frac{d^2\delta x(t)}{dt^2} + \Gamma_{\mathrm{m}}\frac{d\delta x(t)}{dt} + \Omega_m^2\delta x(t) = \frac{F_{\mathrm{rad},0}}{m_{\mathrm{eff}}} - \Omega_m^2 x_0 + \frac{\delta F_{\mathrm{rad}}(t)}{m_{\mathrm{eff}}} + \frac{F_L(t)}{m_{\mathrm{eff}}} \tag{4.17}$$

Here, following Eq. (4.9), we have

$$F_{\rm rad}(t) = -\frac{g_{\rm om}|a_0 + \delta a(t)|^2}{\omega_0} \approx F_{\rm rad,0} + \delta F_{\rm rad}(t) \tag{4.18}$$

In this equation, $F_{\rm rad,0} = -\frac{g_{\rm om}|a_0|^2}{\omega_0}$ defines the new equilibrium position of the cavity boundary. The time dependent optical force perturbation is found to be $\delta F_{\rm rad}(t) = -\frac{g_{\rm om}(a_0\delta a^* + a_0^*\delta a)}{\omega_0}$, $|\delta F_{\rm rad}| \ll |F_{\rm rad,0}|$ with the second order perturbation term $-\frac{g_{\rm om}|\delta a(t)|^2}{\omega_0}$ being neglected. Note that the first two terms on the right hand side of Eq. (4.17) cancels due to Eq. (4.11), we obtain

$$\frac{d^2\delta x(t)}{dt^2} + \Gamma_{\rm m}\frac{d\delta x(t)}{dt} + \Omega_m^2\delta x(t) = \frac{\delta F_{\rm rad}(t)}{m_{\rm eff}} + \frac{F_L(t)}{m_{\rm eff}} \tag{4.19}$$

Similarly we can Fourier transform Eq. (4.19) to obtain

$$[-\Omega^2 - j\Omega\Gamma_{\rm m} + \Omega_{\rm m}^2]\delta\widetilde{x}(\Omega) = \frac{\delta \widetilde{F}_{\rm rad}(\Omega)}{m_{\rm eff}} + \frac{\widetilde{F}_L(\Omega)}{m_{\rm eff}} \tag{4.20}$$

Note in Eq. (4.7), the Langevin force $F_{\rm L}(t)$ obeys the statistical properties in the frequency domain [62]

$$\left\langle \widetilde{F}_{\rm L}(\Omega_\mu)\widetilde{F}_{\rm L}^*(\Omega_\nu)\right\rangle = m_{\rm eff}\Gamma_{\rm m}k_{\rm B}T2\pi\delta(\Omega_\mu - \Omega_\nu) \tag{4.21}$$

Here T is the temperature and $k_{\rm B}$ is the Boltzmann constant. Further, as the mechanical displacement $\delta x(t)$ is real, $\delta\widetilde{x}(\Omega) = \delta\widetilde{x}^*(-\Omega)$. One may find, according to Eq. (4.16),

$$\delta\widetilde{a}^*(-\Omega) = \frac{jg_{\rm om}a_0^*}{j(\Delta'_\omega - \Omega) + \Gamma_{\rm t}/2}\delta\widetilde{x}(\Omega) \tag{4.22}$$

Consequently,

$$\begin{aligned}
\delta\widetilde{F}_{\rm rad}(\Omega) &= \mathcal{F}\{\delta F_{\rm rad}(t)\} \\
&= -\frac{g_{\rm om}[a_0\delta\widetilde{a}^*(-\Omega) + a_0^*\delta\widetilde{a}(\Omega)]}{\omega_0} \\
&= -\frac{jg_{\rm om}^2|a_0|^2\delta\widetilde{x}(\Omega)}{\omega_0}\left[\frac{1}{j(\Delta'_\omega - \Omega) + \Gamma_{\rm t}/2} + \frac{1}{j(\Delta'_\omega + \Omega) - \Gamma_{\rm t}/2}\right] \\
&= -\frac{2g_{\rm om}^2|a_0|^2\Delta'_\omega}{\omega_0}\frac{[\Delta'^2_\omega - \Omega^2 + (\Gamma_{\rm t}/2)^2] + j\Gamma_{\rm t}\Omega}{[(\Delta'_\omega - \Omega)^2 + (\Gamma_{\rm t}/2)^2][(\Delta'_\omega + \Omega)^2 + (\Gamma_{\rm t}/2)^2]}\delta\widetilde{x}(\Omega)
\end{aligned} \tag{4.23}$$

which clearly shows that the optical force is linearly proportional to the mechanical displacement. In addition, after substituting to Eqs. (4.20), (4.23) leads to

$$\delta\widetilde{x}(\Omega) = \frac{\widetilde{F}_L(\Omega)}{m_{\text{eff}}}\left\{-\Omega^2 - j\Omega\Gamma_{\text{m}} + \Omega_m^2 + \frac{2g_{\text{om}}^2|a_0|^2\Delta'_\omega}{\omega_0 m_{\text{eff}}}\frac{[\Delta'^2_\omega - \Omega^2 + (\Gamma_{\text{t}}/2)^2] + j\Gamma_{\text{t}}\Omega}{[(\Delta'_\omega - \Omega)^2 + (\Gamma_{\text{t}}/2)^2][(\Delta'_\omega + \Omega)^2 + (\Gamma_{\text{t}}/2)^2]}\right\}^{-1}$$
$$= -\frac{\widetilde{F}_L(\Omega)}{m_{\text{eff}}}\frac{1}{\Omega'^2_{\text{m}} - \Omega^2 - j\Gamma'_{\text{m}}\Omega} \tag{4.24}$$

According to Eq. (4.5), the interaction between P_{d} and the mechanical motion modifies the dynamics of the mechanical mode, resulting in an effective mechanical damping rate Γ'_{m} and an effective mechanical frequency Ω'_{m} given by [54],

$$\Omega'^2_{\text{m}} \approx \Omega^2_{\text{m}} + \frac{2g^2_{\text{om}}P_{\text{d}}\Delta'_\omega}{m_{\text{eff}}\omega_0\Gamma_{\text{i}}}\frac{\Delta'^2_\omega - \Omega^2_{\text{m}} + (\Gamma_{\text{t}}/2)^2}{[(\Delta'_\omega + \Omega_{\text{m}})^2 + (\Gamma_{\text{t}}/2)^2][(\Delta'_\omega - \Omega_{\text{m}})^2 + (\Gamma_{\text{t}}/2)^2]} \tag{4.25}$$

$$\Gamma'_{\text{m}} \approx \Gamma_{\text{m}} - \frac{2g^2_{\text{om}}P_{\text{d}}\Delta'_\omega}{m_{\text{eff}}\omega_0\Gamma_{\text{i}}}\frac{\Gamma_{\text{t}}}{[(\Delta'_\omega + \Omega_{\text{m}})^2 + (\Gamma_{\text{t}}/2)^2][(\Delta'_\omega - \Omega_{\text{m}})^2 + (\Gamma_{\text{t}}/2)^2]}, \tag{4.26}$$

which are obtained from Eqs. (4.6) and (4.7) by treating the mechanical motion as a perturbation to the optomechanical system. Equation (4.26) shows the dependence of optomechanical amplification/cooling on the laser-cavity detuning. The mechanical damping rate Γ'_{m} is modulated by the radiation pressure according to the sign of the detuning Δ'_ω. When the pump laser is operating at a frequency ω_{l} higher than the cavity resonance $\omega'_0 = \omega_0 + g_{\text{om}}x_0$, the blue detuned pump laser ($\Delta'_\omega > 0$) leads to a decrease of the mechanical damping rate and drives the cavity into "amplification" regime. With sufficient optical power, $\Gamma'_m = 0$ and regenerative coherent oscillation or "phonon lasing" occurs. Therefore, the optical drop power threshold P^{T}_{d} for regenerative optomechanical oscillation can be found from Eq. (4.26)

$$P^{\text{T}}_{\text{d}} = \frac{m_{\text{eff}}\omega_0\Gamma_m\Gamma_{\text{i}}}{2g^2_{\text{om}}\Delta'_\omega\Gamma_{\text{t}}}\left[(\Delta'_\omega + \Omega_{\text{m}})^2 + (\Gamma_{\text{t}}/2)^2\right]\left[(\Delta'_\omega - \Omega_{\text{m}})^2 + (\Gamma_{\text{t}}/2)^2\right] \tag{4.27}$$

Consequently, Eqs. (4.25) and (4.26) can be simplified to

$$\Omega'^2_{\text{m}} \approx \Omega^2_{\text{m}} + \frac{P_{\text{d}}\Gamma_m}{P^{\text{T}}_{\text{d}}\Gamma_{\text{t}}}\left[\Delta'^2_\omega - \Omega^2_{\text{m}} + (\Gamma_{\text{t}}/2)^2\right] \tag{4.28}$$

$$\Gamma'_{\text{m}} \approx \left(1 - \frac{P_{\text{d}}}{P^{\text{T}}_{\text{d}}}\right)\Gamma_{\text{m}} \tag{4.29}$$

When $P_{\text{d}} > P^{\text{T}}_{\text{d}}$, the cavity mechanical motion (Ω'_{m}) is amplified by the radiation pressure and induces the mechanical instability. Due to the backaction of the cavity motion, the circulating light (ω_{l}) is then Doppler shifted so that has two sidebands at ($\omega_{\text{l}} - \Omega'_{\text{m}}$) and ($\omega_{\text{l}} + \Omega'_{\text{m}}$) on its spectrum [63]. Affected by the power distribution within one cavity resonance mode, the sideband with a frequency closer to the resonance has a enhanced intensity compared to the other. The unbalanced sidebands cause a net energy transfer from optical mode to mechanical motion in the blue detuning regime, since the sideband with lowered frequency photons ($\omega_l - \Omega'_m$) is enhanced by the cavity.

In general, the cavity mechanical frequency Ω_m is much smaller than effective laser cavity detuning ($\Delta'_\omega \approx \Delta_\omega$) and total cavity photon decay rate Γ_t. Therefore, Eqs. (4.26), (4.25) and (4.27) can be further simplified to

$$\Omega_m'^2 \approx \Omega_m^2 + \frac{2g_{om}^2\Gamma_e P_{in}}{m_{eff}\omega_0}\frac{\Delta_\omega}{\left[\Delta_\omega^2 + (\Gamma_t/2)^2\right]^2} \tag{4.30}$$

$$\Gamma'_m \approx \Gamma_m - \frac{2g_{om}^2\Gamma_e P_{in}}{m_{eff}\omega_0}\frac{\Delta_\omega\Gamma_t}{\left[\Delta_\omega^2 + (\Gamma_t/2)^2\right]^3} \tag{4.31}$$

$$P_d^T = \frac{m_{eff}\omega_0\Gamma_m\Gamma_i}{2g_{om}^2\Delta_\omega\Gamma_t}\left[\Delta_\omega^2 + (\Gamma_t/2)^2\right]^2, \tag{4.32}$$

With a red detuned laser on the opposite side of the cavity resonance ($\Delta'_\omega < 0$), the negative sign of the frequency detuning adds an extra effective damping rate to the intrinsic one so that further suppresses the mechanical motion of the cavity or drives the cavity to a "cooling" regime. There exists an net energy transfer from the mechanical motion to the optical mode through the optomechanics. Therefore, the cavity is considered to be cooling down by the optical force under this situation.

In summary, the dynamic backaction between the laser field and mechanical motion can be either in phase or out of phase, resulting amplifying or cooling the mechanical motion. The efficiency relies on g_{om}, the optical Q, and the laser-cavity detuning Δ'_ω. Therefore, the high optical quality and strong optomechanical coupling in a cavity would provide efficient optomechanical excitation. In addition, on the blue detuning side, the optical wave amplifies the mechanical motion and leads to a decrease of the mechanical damping rate that depends linearly on the optical power. A large enough optical power is able to boost the mechanical motion above the oscillation threshold resulting in a coherent optomechanical oscillation (OMO) with a very narrow linewidth. This underlies the principle of cavity optomechanical transduction sensing, or optical spring sensing, which will be discussed in detail later.

The dynamic interaction between light and cavity mechanics can be further interpreted by an optical spring. It is well known that the harmonic oscillation angular frequency Ω_m of a spring follows the Hooke's law $\Omega_m = \sqrt{k/m_{eff}}$ where k is the spring constant. When two spring k_1 and k_2 are connected in parallel to a load m_{eff}, they are equivalent to a spring $k = k_1 + k_2$ with eigen frequency square Ω_m^2 equals to the sum of each spring's eigen frequency ($\Omega_{1,2} = \sqrt{k_{1,2}/m_{eff}}, \Omega_m^2 = \Omega_1^2 + \Omega_2^2$. In analogy, the oscillation angular frequency of the OMO can be described by the same equation, by including the radiation pressure contribution to the spring constant, $k = k_{mech} + k_{opt}$ [64].

$$\Omega_m'^2 = \frac{k_{mech}}{m_{eff}} + \frac{k_{opt}(\Delta'_\omega, \Gamma_t)}{m_{eff}} \tag{4.33}$$

Here, $k_{\text{mech}} = m_{\text{eff}}\Omega_m^2$ is the cavity intrinsic mechanical spring constant and we call k_{opt} the optical spring constant as the optomechanical coupling behaves like another spring connect in parallel to the intrinsic one. According to Eq. (4.25), the optical spring constant can be expressed as

$$k_{\text{opt}}(\Delta'_\omega, \Gamma_t) = \frac{2g_{\text{om}}^2 P_d \Delta'_\omega}{\omega_0 \Gamma_i} \frac{{\Delta'_\omega}^2 - \Omega_m^2 + (\Gamma_t/2)^2}{[(\Delta'_\omega + \Omega_m)^2 + (\Gamma_t/2)^2][(\Delta'_\omega - \Omega_m)^2 + (\Gamma_t/2)^2]} \tag{4.34}$$

which can be further simplified, according to Eq. (4.30),

$$k_{\text{opt}}(\Delta'_\omega, \Gamma_t) \approx \frac{2g_{\text{om}}^2 \Gamma_e P_{\text{in}}}{\omega_0} \frac{\Delta_\omega}{\left[\Delta_\omega^2 + (\Gamma_t/2)^2\right]^2} \tag{4.35}$$

under the assumption that the cavity mechanical frequency is much smaller than the effective laser cavity detuning. Evidently, as the magnitude of the radiation pressure changes due to the cavity resonance detuning, the optical spring constant depends sensitively on the cavity resonance. The optical spring effect provides an effective mechanical stiffness that relies on the radiation pressure produced by the WGM, which is essential for OMO sensing described in the latter sections.

4.2.2 *Principles of Optical Spring Sensing*

When the laser wavelength λ_l is blue detuned to the cavity resonance, the optical wave can efficiently boost the mechanical motion above the threshold of regenerative coherent oscillation [54, 65], resulting in highly coherent optomechanical oscillation (OMO) with a narrow mechanical linewidth. According to Eq. (4.34), the intracavity laser field produce an effective mechanical rigidity [54], resulting the overall frequency $f_m = \Omega'_m/2\pi$ depending sensitively on the laser-cavity detuning $\Delta_\lambda = \lambda_l - \lambda'_0$ with $\lambda'_0 = 2\pi c/(\omega_0 + g_{\text{om}} x_0)$ and $\lambda_l = 2\pi c/\omega_l$. Consequently, according to Eq. (4.25), particle binding induced cavity resonance wavelength shift, $\delta\lambda$, is transduced to the frequency shift, δf_m according to $\delta f_m = -\frac{df_m}{d\Delta_\lambda}\delta\lambda$ which can be probed with an electrical spectrum analyzer (Fig. 4.2).

The minimal detectable frequency shift $(\delta f_m)_{\min}$ is determined by OMO linewidth Δf_m, which in turn determines the minimal detectable cavity resonance shift $(\delta\lambda/\lambda_0)_{\min} = \frac{\Delta f_m}{\lambda_0}/(-\frac{df_m}{d\Delta_\lambda})$. By defining an effective mechanical Q factor of coherent OMO, $Q_m^{\text{eff}} \equiv f_m/\Delta f_m$. One may derive the sensing resolution

$$\left(\frac{\delta\lambda}{\lambda_0}\right)_{\min} = \frac{1}{\eta_{\text{om}} Q_m^{\text{eff}} Q_t}, \tag{4.36}$$

where η_{om} represents the optomechanical transduction factor with a value in the order of $\eta_{\text{om}} \sim 1$. Equation (4.36) shows clearly that the sensing resolution scales not only

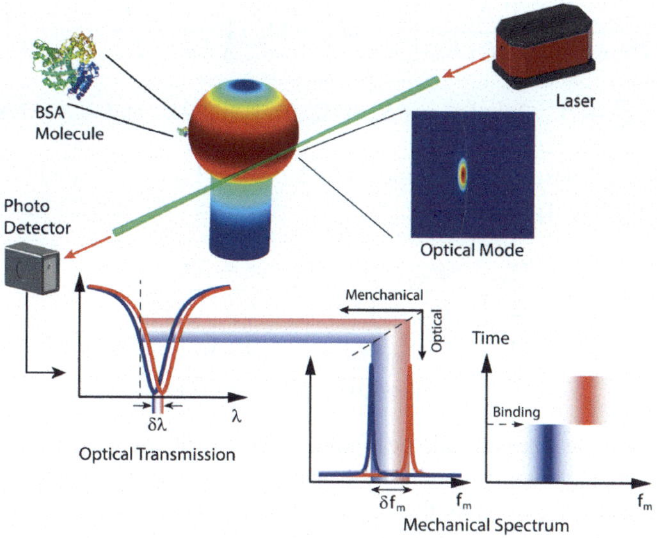

Fig. 4.2 Schematic illustrating the sensing mechanism. A protein molecule bound to an optomechanically oscillating microsphere yields an optical resonance shift $\delta\lambda$, which is transduced to a mechanical frequency shift δf_{m}. The color map on the microsphere shows the radial breathing mechanical mode simulated by the finite element method. Adapted with permission from [53] ©Springer Nature

with the optical Q of the cavity as in conventional microcavity sensors, but also with the effective mechanical Q of OMO. Consequently, the optical spring sensing is able to enhance the sensing resolution by about a factor of $Q_{\mathrm{m}}^{\mathrm{eff}}$ compared with conventional approaches without the need to scarifies the effective detection area.

4.2.3 Mechanical Frequency Shift Induced by Particle Binding

As particle adsorption at the microcavity modifies both the optical cavity resonance and the optical Q. According to the optical spring effect, the resulting OMO angular frequency shift can be found as

$$\delta\Omega'_{\mathrm{m}} = \frac{\partial\Omega'_{\mathrm{m}}}{\partial\omega_0}\delta\omega_0 + \frac{\partial\Omega'_{\mathrm{m}}}{\partial Q_{\mathrm{t}}}\delta Q_{\mathrm{t}} = -\frac{\partial\Omega'_{\mathrm{m}}}{\partial\Delta_\omega}\delta\omega_0 - \frac{\Gamma_{\mathrm{t}}}{Q_{\mathrm{t}}}\frac{\partial\Omega'_{\mathrm{m}}}{\partial\Gamma_{\mathrm{t}}}\delta Q_{\mathrm{t}}, \tag{4.37}$$

where $\delta\omega_0$ and δQ_{t} are the particle induced cavity resonance frequency and loaded cavity optical Q variation, respectively. From Eq. (4.25), we obtain

$$\frac{\partial\Omega'_{\mathrm{m}}}{\partial\Delta_\omega} = \frac{g_{\mathrm{om}}^2 P_{\mathrm{in}}\Gamma_{\mathrm{e}}}{m_{\mathrm{eff}}\omega_0\Omega'_{\mathrm{m}}L_0L_+L_-}\left[L_0 + \Omega_{\mathrm{m}}^2\left(\frac{2\Delta_\omega^2}{L_0} - 1\right) - \frac{4\Delta_\omega^2(L_0 - \Omega_{\mathrm{m}}^2)^2}{L_+L_-}\right], \tag{4.38}$$

$$\frac{\partial \Omega_m'}{\partial \Gamma_t} = \frac{g_{om}^2 P_{in} \Gamma_e \Gamma_t \Delta_\omega}{2 m_{eff} \omega_0 \Omega_m' L_0 L_+ L_-} \left[1 - (L_0 - \Omega_m^2)\left(\frac{1}{L_0} + \frac{1}{L_+} + \frac{1}{L_-}\right)\right], \tag{4.39}$$

where $L_0 \equiv \Delta_\omega^2 + (\Gamma_t/2)^2$, $L_+ \equiv (\Delta_\omega + \Omega_m)^2 + (\Gamma_t/2)^2$, and $L_- \equiv (\Delta_\omega - \Omega_m)^2 + (\Gamma_t/2)^2$.

In the sideband-unresolved regime with $\Omega_m \ll \Gamma_t$, Eqs. (4.38) and (4.39) can be simplified considerably to

$$\frac{\partial \Omega_m'}{\partial \Delta_\omega} = \frac{g_{om}^2 P_{in} \Gamma_e}{m_{eff} \omega_0 \Omega_m'} \frac{(\Gamma_t/2)^2 - 3\Delta_\omega^2}{[\Delta_\omega^2 + (\Gamma_t/2)^2]^3}, \tag{4.40}$$

$$\frac{\partial \Omega_m'}{\partial \Gamma_t} = -\frac{g_{om}^2 P_{in} \Gamma_e}{m_{eff} \omega_0 \Omega_m'} \frac{\Gamma_t \Delta_\omega}{[\Delta_\omega^2 + (\Gamma_t/2)^2]^3}. \tag{4.41}$$

When the particle size is small, the binding induced optical Q change is negligible, OMO frequency shift will be linear to the optical cavity resonance shift according to Eq. (4.37). In the cases when the particle size is large enough to cause detectable Q degradation, Eq. (4.37) introduces an additional amount of OMO shift that is nonlinear to the cavity resonance shift.

In the linear regime discussed above, the optomechanical transduction factor can be further determined by

$$\eta_{om} = \left(1 - \frac{\Omega_m^2}{\Omega_m'^2}\right) \frac{1 - 3\bar{\Delta}_\omega^2}{(1 + \bar{\Delta}_\omega^2)\bar{\Delta}_\omega}, \tag{4.42}$$

where $\bar{\Delta}_\omega = \Delta_\omega/(\frac{\Gamma_t}{2})$ is the normalized detuning with regard to the cavity resonance linewidth. When $\Omega_m' \gg \Omega_m$, η_{om} primarily depends on the laser-cavity detuning and it usually ranges from 1 to 2 for proper sensing operation.

4.2.4 Distinction from Conventional Mass Sensing

Here we provide a simple explanation to show that the optical spring sensing is distinctive from the conventional optomechanical and nanomechanical sensing. As discussed in the previous sections, the spring constant of an OMO contributes from both mechanical spring and optical spring. Consequently, the OMO frequency shift can be induced by perturbation either to the mass δm_{eff}, or to the spring constant, δk, given by

$$\frac{\delta \Omega_m'}{\Omega_m'} = \frac{1}{2}\left(\frac{\delta k}{k} - \frac{\delta m_{eff}}{m_{eff}}\right) = \frac{1}{2}\left(\frac{\delta k_{opt}}{k} + \frac{\delta k_{mech}}{k} - \frac{\delta m_{eff}}{m_{eff}}\right). \tag{4.43}$$

Conventional nanomechanical sensing relies on the last two terms of the above equation, $\frac{\delta \Omega_m'}{\Omega_m'} = \frac{1}{2}\left(\frac{\delta k_{mech}}{k} - \frac{\delta m_{eff}}{m_{eff}}\right)$, where the molecule binding either perturbs the effective mass or modifies the intrinsic mechanical spring constant of the mechanical resonator [66, 67]. Conventional OMO sensing relies on the last term, $\frac{\delta\Omega'_m}{\Omega'_m} = -\frac{1}{2}\frac{\delta m_{eff}}{m_{eff}}$, similar to nanomechanical sensing while with an optical readout [51, 68–70]. It can only detect 1 μm-diameter silica beads with a sub-picogram resolution.

In contrast, our approach takes advantage of the first term, $\frac{\delta\Omega'_m}{\Omega'_m} = \frac{1}{2}\frac{\delta k_{opt}}{k}$ where the optically induced spring depends sensitively on the laser-cavity detuning, $\delta k_{opt} = \frac{dk_{opt}}{d\omega_0}\delta\omega_0$. Consequently, the OMO frequency shift is given by (see also Eq. (4.37))

$$\frac{\delta\Omega'_m}{\Omega'_m} = \frac{1}{2k}\frac{dk_{opt}}{d\omega_0}\delta\omega_0. \tag{4.44}$$

The frequency tuning slope, $dk_{opt}/d\omega_0$, significantly amplifies the cavity resonance shift induced by molecular binding and transduces it into the OMO frequency shift. It is this very distinctive nature of optical spring that provides a sensing resolution orders of magnitude higher than conventional OMO or nanomechanical sensing.

Although previous discussions focus on the frequency domain, it can be easily converted to the wavelength domain by use of $d\omega_0/d\lambda_0 = -2\pi c/\lambda_0^2$. For example, the wavelength-dependent tuning slope can be obtained by $\partial\Omega'_m/\partial\Delta_\lambda = -(2\pi c/\lambda_0^2)(\partial\Omega'_m/\partial\Delta_\omega)$.

4.3 Experiment Demonstration of Optomechanical Oscillation in Heavy Water

The first step toward optical spring sensing is to demonstrate the OMO in an aqueous environment.[2] The challenges in such demonstration is that when a cavity is immersed in liquid, the suspension actuates strong viscous force against the cavity oscillation. This makes the optomechanical oscillation highly dissipative except in the cases where a superfluid was adopted [69]. In addition, strong optical absorption of water further reduces the optical Q of the cavity, making OMO even harder to be excited. To reduced the optical dissipation, we immerse a silica microsphere in heavy water (D2O). At a wavelength around 970 nm, the absoption coefficient of D2O is 0.0113 cm^{-1}, much lower than that of water (H2O, 0.43685 cm^{-1}).[3] Such low optical absorption helps to maintain high optical Q ($\sim 10^7$) and strong optical force in the cavity for OMO excitation.

[2] Most of this section is reproduced/adapted with permission from Ref. [35]. Copyright 2014 ©The Optical Society.

[3] The absorption coefficients are available from https://refractiveindex.info.

4.3.1 Experiment Results and Discussions

The experiment setup is shown in Fig. 4.3. An external cavity tunable laser (Newport 6300LN) operating at 970 nm wavelength range is used to drive the cavity OMO. At the laser output, a variable optical attenuator (VOA A) is inserted for power control. A 99/1 optical directional coupler (OC A) further split the light to two branches. The 99% port was connected to the input of a tapered optical fiber to actuate the OMO of silica microsphere under test. The 100 μm-diameter microsphere was fabricated by melting a silica fiber with a CO2 laser and immersed in heavy water (D2O). For precision position control, the microsphere was mounted on a nanopositioner. The transmitted optical signal from the microsphere was converted to electrical signals by a photodetector (New focus 1811) placed at the output of the same tapered fiber. The electrical signal was further split by a 50/50 electrical splitter such that the signal can be examined in both time and frequency domain with an Agilent DSO 90404A oscilloscope and a Tektronix RSA 3408B real time spectrum analyzer.

The 1% output branch of the directional coupler is connected to a reference interferometer to provide accurate optical frequency calibration in optical Q measurements. As shown in Fig. 4.3, the reference interferometer is built with a pair of optical fiber connected between two 50/50 optical couplers. The lengths of the two fibers differ by ΔL, when light enters the first coupler (OC B), its power splits evenly at the coupler outputs. The light then propagates along the fiber pair, resulting a relative time delay $\tau = \Delta L/(n_e \cdot c)$ when arriving at the inputs of the other coupler (OC C). The light will be photo mixed at OC C and converted to electrical signal by the balanced photo detector. The detector output voltage V_0 is related to the optical frequency ν and fiber delay τ through

$$\frac{V_0}{V_{0,\max}} = \cos(2\pi \nu \tau) \tag{4.45}$$

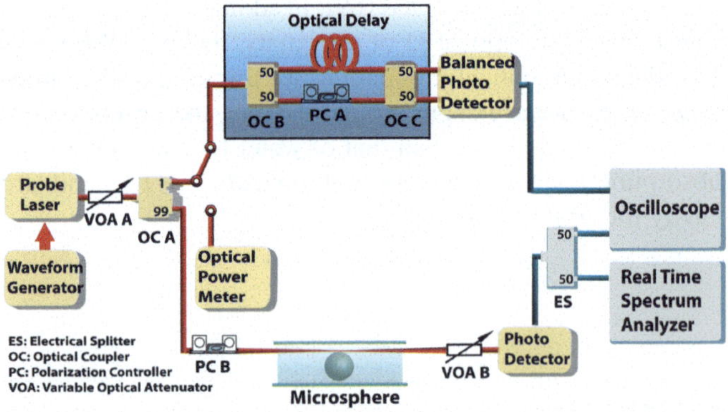

Fig. 4.3 Experiment setup for the OMO measurements. Adapted with permission from [35] ©The Optical Society

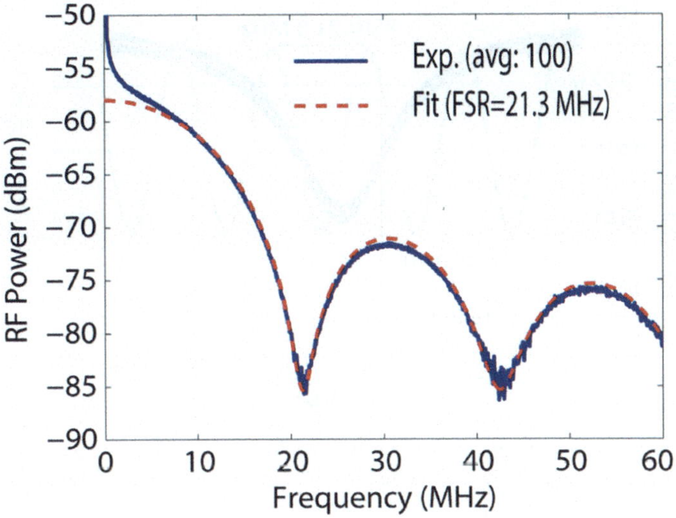

Fig. 4.4 RF power spectrum of the reference interferometer output calibrated from the dark current spectrum. Adapted with permission from [35] ©The Optical Society

with $V_{0,\max}$ being the maximum output voltage. Therefore, when τ is precisely known, V_0 maps to the optical frequency ν with high precision. To determine τ, we set the laser in CW mode and connect the photo detector output to an electrical spectrum analyzer. Figure 4.4 shows the power spectrum of the output signal after averaging over 100 measurements and calibrated by subtracting it from the 100 measurements averaged dark current spectrum. As shown, the signal spectrum displays a sinc-square shape (blue trace) and the free spectral range $\Delta f_{\rm FSR}$ of the spectrum is the inverse of the delay $\tau = 1/\Delta f_{\rm FSR}$. In this experiment, $\Delta f_{\rm FSR} = 21.3$ MHz was determined through a least square fit (red dashed line). Equivalently a time delay of $\tau = 0.0469\,\mu$s or length difference $\Delta L = 9.73$ m is obtained. After calibration, the output of the balanced detector is then connected to the oscilloscope shown in Fig. 4.3.

To measure the optical Q of the microsphere immersed in D2O, we scan the laser wavelengths around the cavity resonance while reducing its power till the thermal effect diminishes. The cavity transmission spectrum averaged over 100 measurements is displayed as the blue trace in Fig. 4.5. According to Eq. (4.4), the spectrum displays a Lorentzian shape. Meanwhile, due to Eq. (4.45), the interferometer signal (green trace) has a sinusoidal shape with a periodicity precisely equals to $\Delta_{\rm FSR}$. Therefore, using the fitted curve of the interferometer signal, we can accurately map the frequency detuning of the transmission spectrum for precision Q measurement. In this experiment, through the least square fit, we obtained an intrinsic optical Q of 1.4×10^7 when the microsphere is immersed in D2O. Under the operation coupling condition, an overall loaded optical Q of 9.8×10^6 is obtained.

To actuate OMO, we scan the wavelength of the probe laser around the resonance while gradually increasing the laser power to about 2.5 mW. With fine adjustment

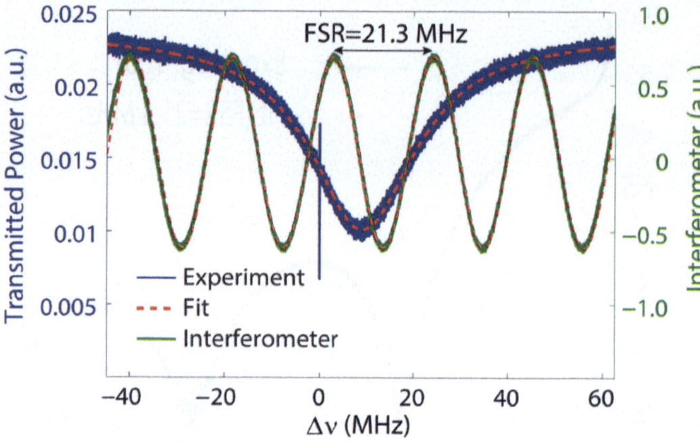

Fig. 4.5 Cavity transmitted optical power (blue trace) displays a Lorentzian shape. The laser frequency is calibrated through the transmitted signal of the reference interferometer (green trace). The red dashed lines are least square fitting results. Adapted with permission from [35] ©The Optical Society

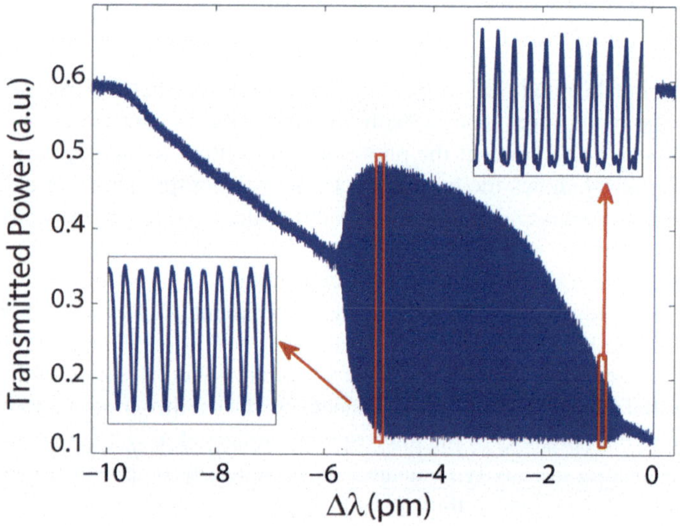

Fig. 4.6 Transmitted optical power as a function of probe laser wavelength detune. At a dropped optical power close to the threshold power, the left inset displayed a sinusoidal spectrum while at a high dropped power the spectrum displayed in the right inset was distorted by the high order harmonics. Adapted with permission from [35] ©The Optical Society

of the gap between the tapered fiber and the microsphere, OMO can be observed as shown in Fig. 4.6. As shown, the thermal broaden effect at the blue detuned regime facilitates the OMO, which helps the locking in sensing experiment discussed in later section. The insets at the bottom left of the figure shows that the OMO is single mode when the P_d is slightly above the threshold power. When the dropped power increases, however, high order harmonics occurs due to the nonlinear nature of cavity response.

To characterize OMO, we set the laser in CW mode and adjust the laser optical power using the optical attenuator (VOA A) while keeping the off-resonance transmission signal voltage at 0.6 V by adjusting the other attenuator (VOA B) accordingly. The coupler (OC A) 1% output port is switched to an optical power meter for P_d monitoring. We then scan the laser from lower wavelength toward cavity resonance till the transmission signal voltage dropped by half. At each wavelength detuning point, we measure P_d through the readings from the optical power meter. Meanwhile, transmission signal frequency spectra are collected using the spectrum analyzer and averaged every 100 traces. Figure 4.7 displays the averaged spectrum when P_d is 0.4, 1 and 1.1 mW. As shown, when P_d is at 0.4 mW, OMO is well below threshold with large spectral bandwidth of 269 kHz and the oscillation is dominated by the natural mechanical oscillation of the microsphere. When P_d increases to around the threshold power of 1 mW, the linewidth gradually reduces to 61 kHz. With further increasing P_d above threshold by as little as 0.1 to 1.1 mW, coherent regenerative optical spring oscillation rapidly narrows down the linewidth 232 Hz. As laser frequency jitter noise broadens the averaged spectrum, we examine each individual trace of the spectrum and find the narrowest linewidth to be 99 Hz, corresponding to an effective Q_m of 3,884.

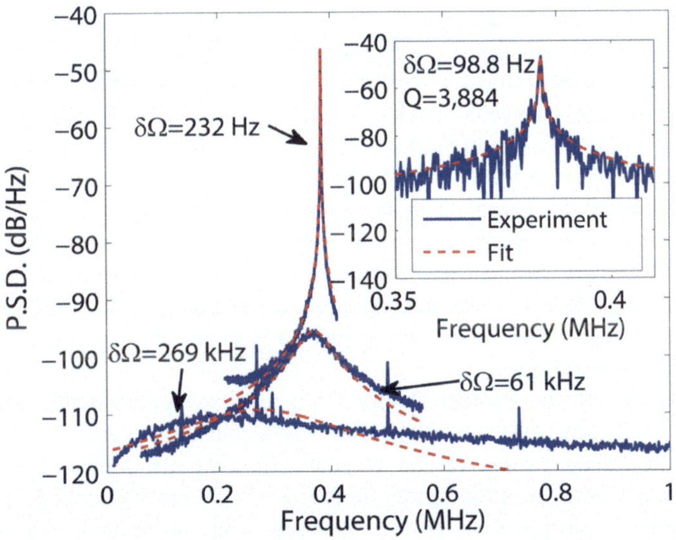

Fig. 4.7 RF spectra at dropped power of 1.1, 1 and 0.4 mW, least square fittings to the Lorentzian function indicate the linewidths of the optomechanical tones to be 232 Hz, 61 kHz and 269 kHz respectively. In the main plot, each spectrum was averaged over 100 spectral traces collected seamlessly at the same drop power level. The inset is the spectrum of single trace measurement. Adapted with permission from [35] ©The Optical Society

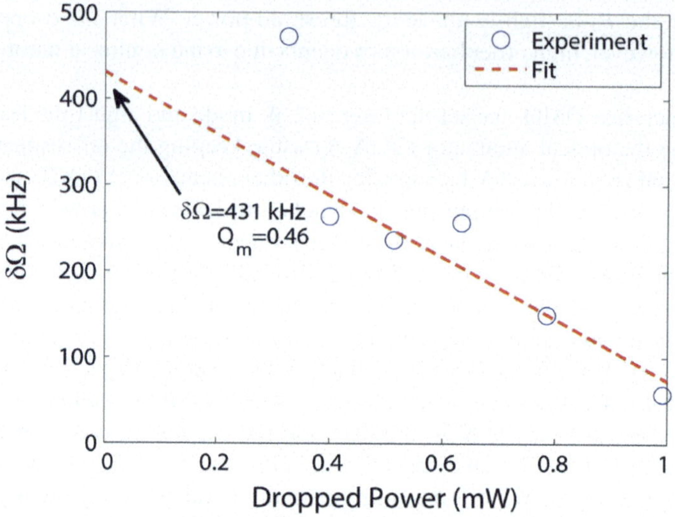

Fig. 4.8 The plot of mechanical linewidth versus dropped power, which indicates an intrinsic mechanical linewidth of 431 kHz and an effective mechanical quality factor of $Q_{\mathrm{m}} = 0.5$ through the linear extrapolation. Adapted with permission from [35] ©The Optical Society

According to Eqs. (4.28) and (4.29), at the dropped power well below threshold, Ω'_m and Γ'_m are linear to the dropped power. When optical spring diminishes at $P_{\mathrm{d}} = 0$, they equals to the intrinsic frequency and decay rate of the cavity's natural mechanical oscillation. In the inset of Fig. 4.9, we plot the OMO peak frequency vs dropped power (blue square markers). As indicated by Eq. (4.28), when below threshold and in blue detuning regime ($\Delta_\omega > 0$), the optical spring drives the peak frequency gradually from 253 kHz at $P_{\mathrm{d}} = 0.3$ mW to 370 kHz at 1.0 mW [71]. A linear extrapolation at $P_{\mathrm{d}} = 0$ (red dashed line) predicts an intrinsic mechanic frequency of $f_{\mathrm{m}} = 199$ kHz. To obtain the intrinsic decay rate Γ_{m}, we plot the linewidth (blue circles) versus P_{d} in Fig. 4.8. Again, a linear extrapolation (red dashed line) predicts an intrinsic linewidth of 431 kHz at $P_{\mathrm{d}} = 0$. Accordingly, we obtain $\Gamma_{\mathrm{m}} = 431\ (\mathrm{ms})^{-1}$, yielding a $Q_{\mathrm{m}} = 2\pi f_{\mathrm{m}} / \Gamma_{\mathrm{m}} \approx 0.5$. The low intrinsic Q is expected due to the mechanical energy dissipation to liquid.

In Fig. 4.9, we further plot the cavity mechanical energy normalized to its peak value (blue circle markers) versus P_{d}. Here, the mechanical energy is calculated by integrating the RF spectrum over the frequency interval as it denotes the oscillation displacement. As shown in the figure, at above $P_{\mathrm{d}}^{\mathrm{T}}$, the mechanical energy linearly increases with the dropped power at a rapid rate while at below the threshold, the energy grows slowly. This threshold is a typical "lasing" behaviour and confirms the observed oscillation is coherent in nature. Through a linear fit to the mechanical energy (red dashed trace), we obtain the threshold power to be 0.98 mW.

Figure 4.10 further displays as many as 24 high order harmonics within a frequency span of 10 MHz in the transmission spectrum when $P_{\mathrm{d}} = 2.6$ mW. Meanwhile the

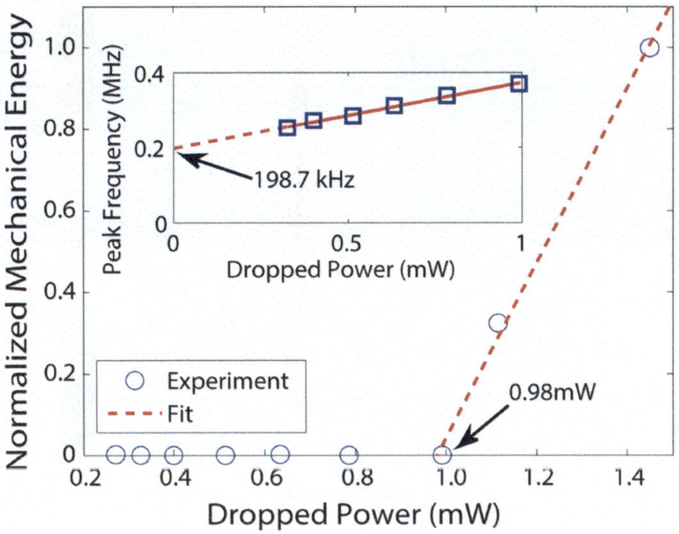

Fig. 4.9 Mechanical energy (normalized to the maximum value) as a function of the dropped power. The peak frequency as a function of the dropped power is displayed in the inset and a linear extrapolation predicts an intrinsic mechanical frequency of 198.7 kHz. Adapted with permission from [35] ©The Optical Society

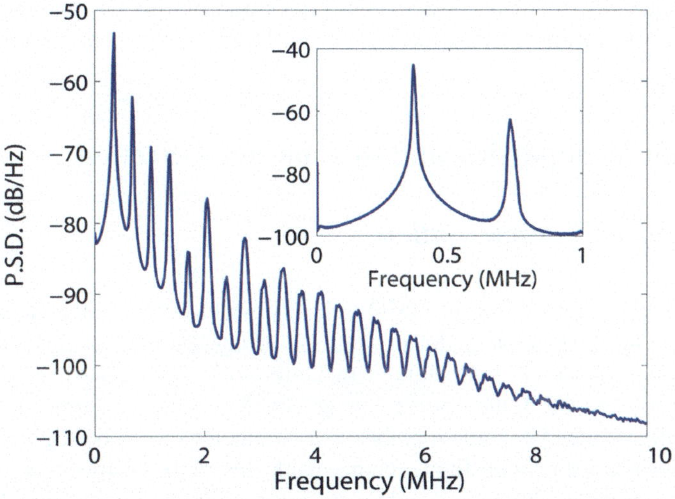

Fig. 4.10 In a separate measurement, as high as 24th order harmonics was observed in a frequency span of 10 MHz. The inset further displayed the spectrum with a frequency span set at 1 MHz. Adapted with permission from [35] ©The Optical Society

frequency doubling of the second harmonic is evident from the inset when the span is set to 1 MHz.

To characterise the oscillation stability, we collect the spectrogram in Fig. 4.11 with the peak frequency plotted as the black line. Over a time interval of 392 ms,

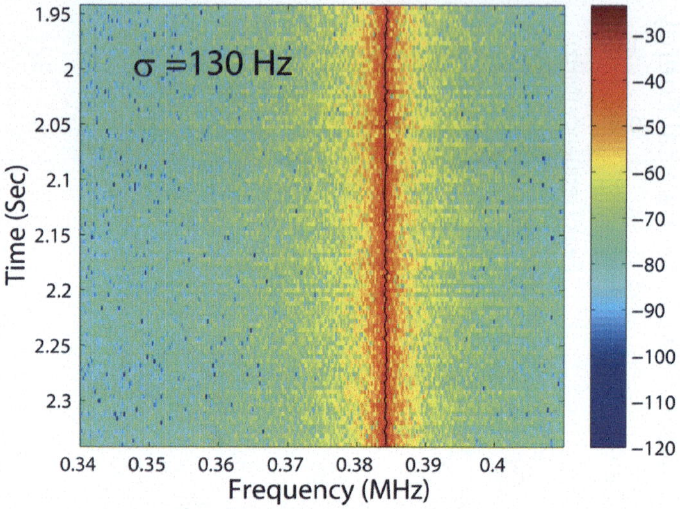

Fig. 4.11 Spectrogram of the optomechanical oscillation indicates a 130 Hz standard deviation of the oscillation peak over a time span of 392 ms. Adapted with permission from [35] ©The Optical Society

the peak frequency fluctuates at a standard deviation of 130 Hz, making it highly attractive to detect minute frequency shift caused by single protein binding events.

4.4 Single Nanoparticle and Biomolecule Detection

4.4.1 Device Characterisation

The successful demonstration of OMO in heavy water confirms the feasibility for biosensing.[4] Our next step is to detect single silica nanoparticles and Bovine Serum Albumin (BSA) proteins, both diluted in Dulbecco's Phosphate-Buffered Saline (DPBS). Similar to H2O, the optical loss of DPBS is almost 50 times higher than D2O. To mitigate the Q degradation, the cavity diameter should be slightly larger so that the majority portion of the light remains inside the cavity. On the other hand, if the size of the cavity is too large, g_{om} and reactive sensitivity will both drop. Therefore, a numerical modelling is favorable for the determination of optimal cavity size. Our A numerical modelling shows that with a diameter of around 100 μm, a microsphere will maintain its Q to around 6×10^6 when immersed in water, only degraded by half compared to that immersed in D2O. Therefore, we fabricates our silica microsphere

[4] Most of this section is reproduced/adapted with permission from Ref. [53]. Copyright 2016 ©Springer Nature.

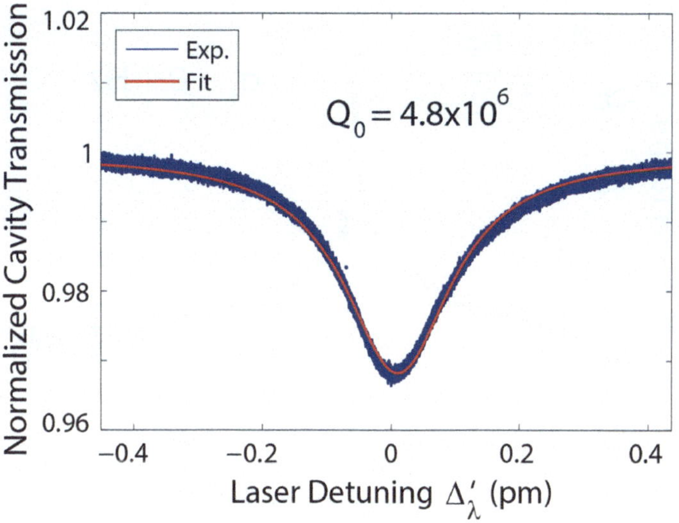

Fig. 4.12 The optical transmission spectrum of the microsphere immersed in DPBS, at a probe laser wavelength of 974 nm, with experimental data in blue and theoretical fitting in red. The input power is maintained low enough to characterize the intrinsic optical property of device, which exhibits an intrinsic optical Q of 4.8×10^6. Adapted with permission from [53] ©Springer Nature

Fig. 4.13 The optical transmission spectrum at an input laser power of 8.5 mW. The coherent OMO was excited with a threshold power of 3.0 mW dropped into the cavity. Adapted with permission from [53] ©Springer Nature

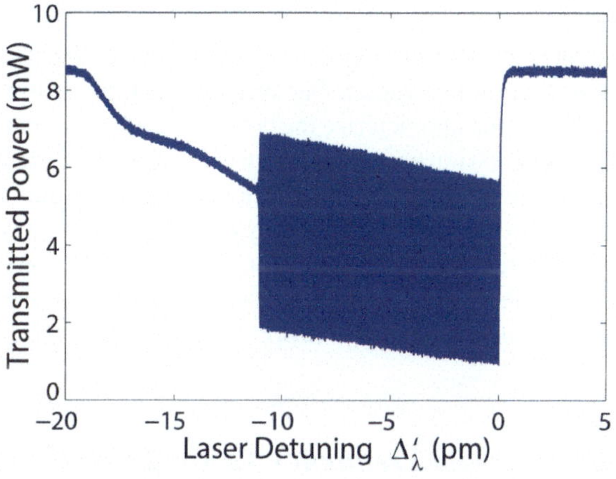

to around that diameter. Further, we measure the optical Q using the identical setup described in previous section. In our experiment, an intrinsic optical Q of 4.8×10^6 (Fig. 4.12) was obtained. The close agreement between measurement and numerical simulation confirms that our sample preparation and measurement procedure virtually eliminates all excessive contaminants that could degrade the Q.

With such a high optical Q, OMO should be observable with a reasonably higher threshold power. The transmission spectrum displayed in Fig. 4.13 shows that strong

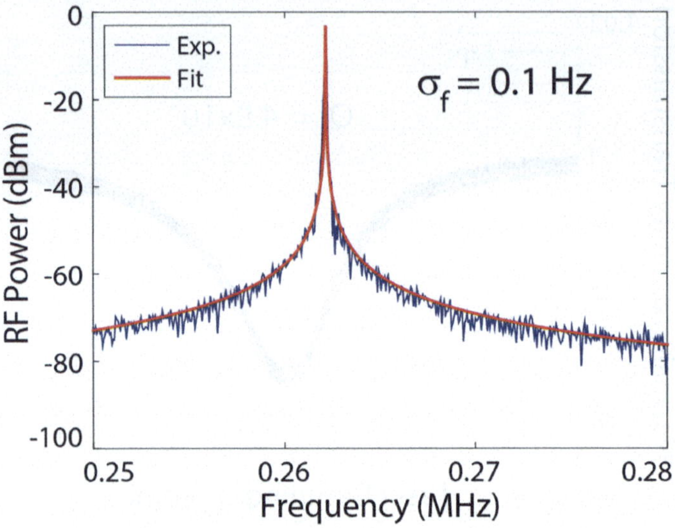

Fig. 4.14 The detailed spectrum of the fundamental oscillation tone, with experimental data in blue and theoretical fitting in red. The OMO exhibits a full-width at half maximum of 0.1 Hz, corresponding to an effective mechanical Q of 2.6×10^6. Adapted with permission from [53] ©Springer Nature

OMO is observable above a P_d^T of 3.0 mW. It is also worth mentioning that the OMO starts near the middle point of the resonance dip, which is corresponding to the high transduction rate regime because of the slope on the Lorentzian function.

To further characterize the coherent OMO, we set the laser to CW mode and record the transmission power spectrum use a procedure identical to that in D2O experiments as shown in Fig. 4.14. Through a least square fitting to a Lorentzian shape, we found the OMO peaked at a frequency of 262 kHz with a linewidth of 0.1 Hz. In this case, the effective mechanical Q becomes 2.6×10^6 was obtained, almost three orders of magnitude higher than that recorded in D2O. Similar to D2O experiments, Fig. 4.15 shows that OMO at high dropped power also leads to a harmonic comb [71, 72].

4.4.2 OMO Frequency Versus Laser-Cavity Detuning

To verify the sensing principle, we operates the laser in constant power while increasing its wavelength step by step from off-resonance toward the cavity resonance. At each wavelength, the transmitted signals are simultaneously recorded by the oscilloscope and the realtime electrical spectrum analyzer. Figure 4.16 shows the OMO spectrum at all wavelength points where the strong optical spring effect results in an OMO frequency dependent on the laser-cavity detuning. To confirm Eq. (4.30), we plot the OMO peak frequency versus laser-cavity detuning $\Delta_\lambda = \lambda - \lambda_0$ with

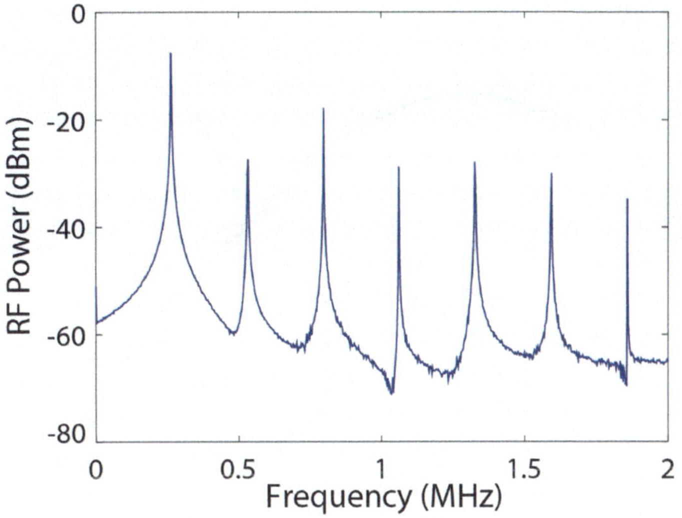

Fig. 4.15 An example of the power spectral density of the cavity transmission. The fundamental oscillation frequency is located at 262 kHz, with 6 high-order harmonics clearly visible on the spectrum. Adapted with permission from [53] ©Springer Nature

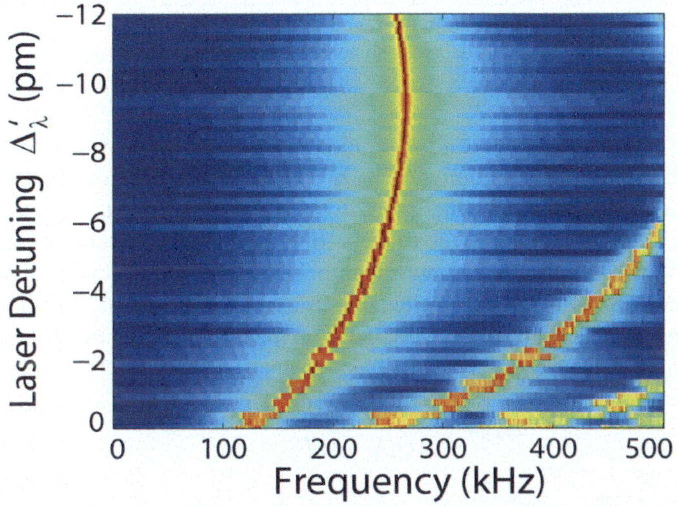

Fig. 4.16 Spectrogram of cavity transmitted signal as a function of laser wavelength detuning Δ'_λ (see Fig. 4.17 for the meaning of Δ'_λ), showing the detuning dependent mechanical frequency. The proportional frequency variations at the second and third harmonics are clearly visible. Every spectrum was averaged over 5 traces acquired continuously. Adapted with permission from [53] ©Springer Nature

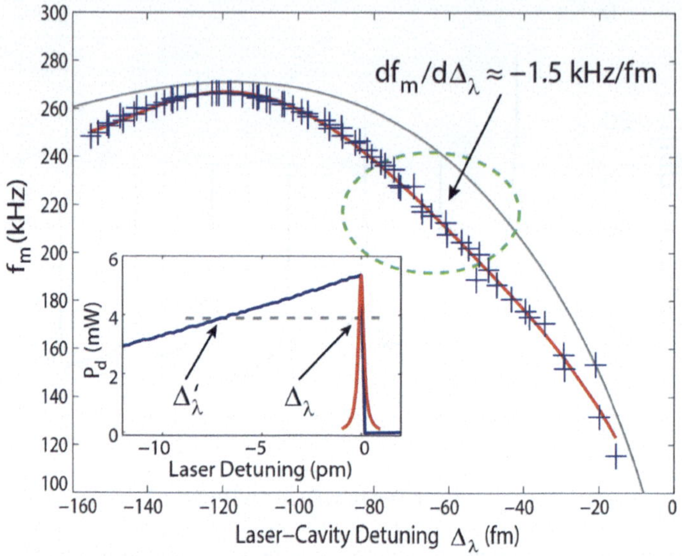

Fig. 4.17 The OMO frequency as a function of laser-cavity wavelength detuning. The blue crosses show the experimental data and the grey curve shows the theory. The red curve is a polynomial fitting to the experimental data. The dashed circle indicates the operating regime for the particle and molecule sensing, with a frequency tuning slope of $df_m/d\Delta_\lambda = -1.5$ kHz/fm at a laser-cavity detuning of $\Delta_\lambda = -70$ fm. Inset: Recorded dropped optical power as a function of laser wavelength detuning. This curve was used to obtain the real laser-cavity wavelength detuning $\Delta_\lambda = \lambda_l - \lambda_0$ where λ_l is the laser wavelength. Adapted with permission from [53] ©Springer Nature

λ being the laser wavelength and λ_0 cold cavity resonance wavelength. Note that in Fig. 4.16, laser-cavity detuning $\Delta'_\lambda = \lambda - \lambda'_0$ is computed from the wavelength of the operation point to that where the dropped power returns to zero as shown in inset of Fig. 4.17 (blue trace). Due to the strong thermal broadening in ultra-high Q cavity, the hot cavity resonance wavelength λ'_0 deviates significantly from λ_0 and P_d versus laser detuning is not Lorentzian as expected according to Eq. (4.5) (red trace). Therefore, Δ_λ needs to be extracted from the measured Δ'_λ using the following equation

$$\Delta_\lambda = -\frac{\lambda_0}{2Q_\mathrm{t}}\sqrt{\frac{P_\mathrm{d}(0) - P_\mathrm{d}(\Delta'_\lambda)}{P_\mathrm{d}(\Delta'_\lambda)}}, \tag{4.46}$$

where $P_\mathrm{d}(0)$ is the optical power dropped to the cavity when the probe laser is on-resonance.

As shown in Fig. 4.17, when the laser wavelength increases from far below cold cavity resonance, OMO frequency (blue plus markers) increases to a peak value of 267 kHz at $\Delta_\lambda \approx -119$ fm and then decreases quickly. A large and close-to-uniform tuning slope of $df_m/d\Delta_\lambda \approx -1.5$ kHz/fm is found around $\Delta_\lambda \approx -70$ fm in the green dashed circle. When operating in that regime, every 1-fm cavity resonance wavelength shift induced by a particle binding event can be transduced to an OMO frequency change of about 1.5 kHz, almost four orders of magnitude larger than the linewidth of optomechanical oscillation. Figure 4.17 clearly shows that our transducing sensing resolution is 10^4 times higher than conventional cavity reactive sensing with the same optical Q [73] and even 50 times higher than that with a cavity of 10^8 Q [5]. It is also worth mentioning that the higher order harmonics oscillation frequency is multiple times larger than the OMO fundamental frequency (Fig. 4.16). They can also be applied to detect particle bindings. In practice, the larger frequency shifts on the higher-order harmonics allows us to use a coarser resolution bandwidth and shorter acquisition time for spectral measurements and reduces considerably the excessive detection noises that usually accumulates over long time interval.

In this figure, we also plot the theoretical prediction according to Eq. (4.30) as grey curve. Further, Eq. (4.30) predicts the peak OMO frequency occurs at $\Delta_\lambda^{max} = -\lambda_0/(2\sqrt{3}Q_t)$. Assuming the cavity is at critical coupling condition, $Q_t = Q_0/2 = 2.4 \times 10^6$, we estimate $\Delta_\lambda^{max} = -117$ fm, which is in close agreement with the experiment observation. The slight discrepancy showing in the figure is due to that the theoretic model is derived from perturbation theory while the strong OMO amplitude in our experiment does not fully satisfy the underline approximation. Nevertheless, Fig. 4.17 clearly shows that OMO frequency is dominated by the optical spring instead of the natural mechanical oscillation of the cavity.

4.4.3 Silica Nanoparticles Detection

This subsection illustrates the results of sensing experiments on single silica nanoparticles and protein molecules. For highest resolution, we set the laser-cavity detuning within the dashed circle of Fig. 4.17. In our first set of experiments, we dilute silica nanobeads in DPBS to various concentrations and deliver them to the microsphere in the order from low to high concentrations. Similar to reactive sensing, a particle binding event introduces a sudden cavity resonance wavelength shift, which transduces to the OMO frequency shift that can be detected from the transmission spectra. In our experiment, the OMO spectra are recorded seamlessly by the spectrum analyzer to form spectrograms. We then locate the OMO peak frequencies by fitting each spectrum in the spectrogram to a Lorentzian function. A step finder algorithm is further applied to the peak frequency versus time curve to detect sudden changes of OMO frequency from particle binding. Figure 4.18a–d display typical binding signals of 11.6, 25, 50, and 85 nm radius silica nanobeads. Here all spectra are recorded at the fundamental frequency of OMO except for the case of 11.6 nm beads where the third harmonic is recorded for higher signal-to-noise ratio as mentioned before. As shown in Fig. 4.18a, a 11.6 nm bead binding event yields a step of 1.3 ± 0.1 kHz, or equivalent to 0.43 ± 0.03 kHz step at its fundamental tone with an SNR of 13. As the transduce slope is negative ($\sim$1.5kHz/fm), a positive frequency step implies an increase of the cavity resonance wavelength, which suggests a 11.6 nm silica bead binds to the surface of the cavity. Figures 4.18b–d show frequency steps of

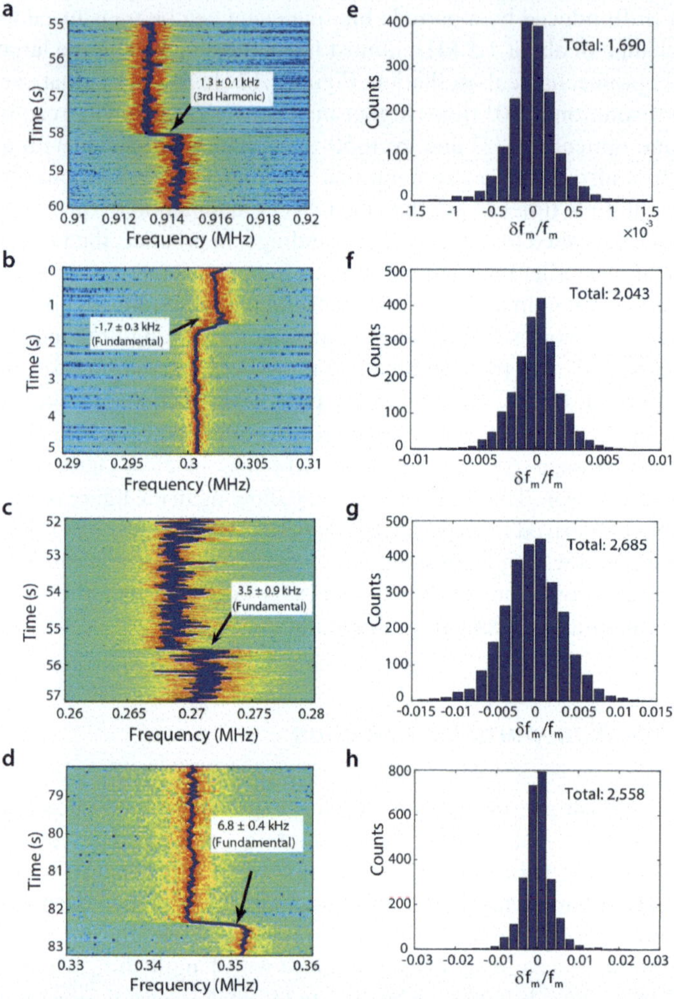

Fig. 4.18 **a–d** Typical mechanical spectrograms for the binding events of silica beads with average radii of 11.6, 25, 50, and 85 nm, where **a** shows that of third harmonic and **b–d** show those of the fundamental oscillation frequency. **e–h** The histograms of the normalized frequency steps $\delta f_m / f_m$. Adapted with permission from [53] ©Springer Nature

-1.7 ± 0.3 kHz, 3.5 ± 0.9 kHz, and 6.8 ± 0.4 kHz of the 25, 50 and 85 nm beads respectively, where all positive frequency steps suggest binding events while negative steps unbinding. According to the numerical simulation, direct contribution of particle binding to m_{eff} is negligible since the masses of the nanobeads ($\sim 0.01 - 5$ fg depending on particle radius) are more than 9 orders of magnitude smaller than the effective mass of the cavity ($\sim 1\,\mu$g). Therefor, the OMO frequency shifts by particle bindings are overwhelmingly from the optical spring transduction.

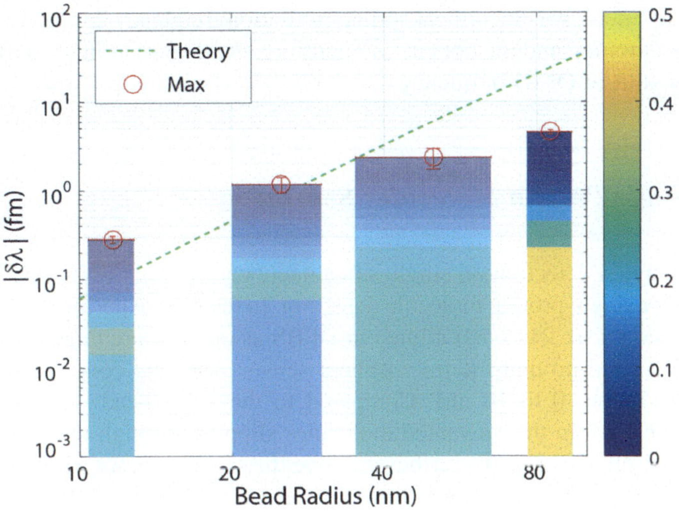

Fig. 4.19 The corresponding cavity resonance shifts induced by the particle binding as a function of bead radius. The color bars show the probability density functions of the recorded cavity resonance wavelength shifts induced by particle binding, where the bar width indicates the standard deviation of the bead size (provided by the manufacturer) and the color map indicates the magnitude of probability density. The red circles indicate the recorded maximum wavelength shifts of the cavity resonance. The dashed curve shows the theoretical prediction. Adapted with permission from [53] ©Springer Nature

To investigate the statistical properties of binding, we collected 500, 121, 521, 389, 1, 335, 415, and 758, 728 spectra in 11.6, 25, 50, and 85 nm beads expereiments. Among them, 1, 690, 2, 043, 2, 685, and 2, 558 frequency steps whose SNR exceeding unity are identified. Figure 4.18e–h show the histograms of the frequency steps normalized to OMO peak frequencies, $\delta f_m/f_m$. Among all frequency steps, we also found maximum OMO frequency shifts of $\delta f_m/f_m = (1.4 \pm 0.4) \times 10^{-3}$, $(-7.8 \pm 1.5) \times 10^{-3}$, $(1.3 \pm 0.3) \times 10^{-2}$, and $(-2.3 \pm 0.6) \times 10^{-2}$ for 11.6 nm, 25 nm, 50 nm, and 85 nm radius beads. Further, using the transduction rate $d f_m/d\Delta_\lambda = -1.5$ kHz/fm, we estimated the corresponding cavity resonance wavelength shifts $\delta\lambda$ and plot the probability density function of their absolute values as color bars in Fig. 4.19. The maximum wavelength shifts of $|\delta\lambda|/\lambda_0 = 2.6 \times 10^{-10}$, 1.2×10^{-9}, 2.4×10^{-9}, and 4.6×10^{-9} for the four sizes of beads are also plotted as red circles in the figure.

Further, we numerically computed maximum wavelength steps using perturbation method and plot them as the green dashed line [74]. A comparison to the experiment results (red circles) shows that our experiments agree with the numerical results on small beads (11.6 and 25 nm) while at larger bead radii (50 and 85 nm), the experiment values are consistently smaller. This is because large bead binding on the cavity degrades the cavity optical Q noticeably. For example, a 85 nm bead binding to the equator of a 100 μm microsphere would cause the optical Q droop to 2.7×10^4.

According to Eq. (4.30), the optical spring oscillation frequency depends on both the laser-cavity detuning and the optical Q. Therefore, the degradation of cavity Q leads to a smaller shift of OMO frequency.

4.4.4 Single Protein Molecules Sensing

The demonstrated detection on single silica nanobeads at high SNR establishes the confidence for single protein molecule detection. In the next experiment, we selected BSA (Biovision, Cat. 2119-10) diluted in DPBS as our sensing target. The suspension was delivered around the microsphere sensor, with the concentration gradually increased from 0 to 10 nM. Compared to the silica beads experiments, the excessive noises from the unwanted molecules slipping through the cavity surface were significantly reduced. To further improve the sensitivity, we again monitor the third order harmonic of frequency steps for detection. Figure 4.20a shows a maximum frequency step of -0.67 ± 0.04 kHz with an SNR of 16.8, corresponding to a step of -0.22 ± 0.01 kHz at the fundamental oscillation tone. In this experiment, we recorded a total of 145, 407 spectra and found 1, 785 frequency steps. Figure 4.20b displays the histogram of the normalized frequency steps with the max-

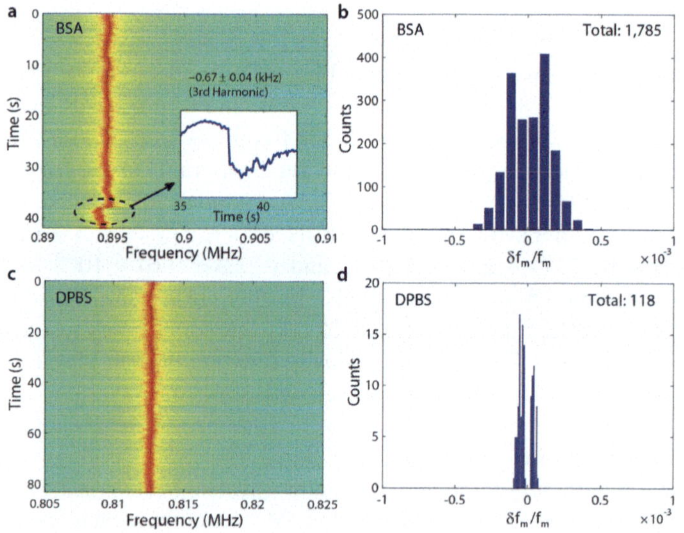

Fig. 4.20 **a** A typical mechanical spectrogram recorded at the third harmonic of the oscillation tone, capturing the event of a BSA protein molecule detaching from the silica microsphere surface at 38 s, with a clear frequency step (inset) of -0.67 ± 0.04 kHz. **b** The histogram of the normalized frequency steps. **c** A mechanical spectrogram in the absence of protein molecules. **d** The histogram of the normalized frequency steps. Adapted with permission from [53] ©Springer Nature

imum step of $\delta f_m/f_m = (-7.6 \pm 0.4) \times 10^{-4}$, corresponding to a cavity resonance shift of $|\delta\lambda| = 0.15$ fm.

As a control experiment, we also collected spectra from pure DPBS. The spectrogram shown in Fig. 4.20c indicates the OMO is stable. Further, the histogram of all 118 steps found by our program displayed in Fig. 4.20d shows the baseline noise is well below the signal detected in BSA experiments. This observation clearly proves the capability of sensing single BSA molecules with a molecule weight of 66 kDa. By assuming the resonance shift is proportional to the mass (or equivalently, to the volume) of the protein [75], we derive that our current set-up is capable of detecting proteins as small as 3.9 kDa with an SNR above unity.

The double digit SNR for BSA molecule detection and the predicted 3.9 kDa sensitivity demonstrated the astounding sensitivity of optical spring sensors. Note that improvements can be further made to significantly increase the sensitivity. For example, in our current setup, the external cavity laser used in our experiments produces large frequency jitter noise that is transduced to OMO frequency noise and limits the detection sensitivity. With the future adoption of advanced frequency locking circuitry can further improve the sensing resolution by $\sim$100 times to around $\delta\lambda/\lambda_0 \sim 10^{-14}$. On the other hand, the optical Q can be increased to above 10^8 if a visible laser is employed [5], which would further improve the sensing resolution by more than one order of magnitude. Further, the proposed approach does not needs additional sensing element such as plasmonic nanoantenna [31–33] attached to the device. Therefore, it is capable of utilizing the entire effective sensing area offered by a whispering-gallery microcavity which is more than five orders of magnitude larger than that with a plasmonic strucutre. Moreover, an incorporation of a plasmonic nanoantenna can enhance the cavity resonance wavelength shift by orders of magnitude and consequently transducing to the OMO frequency shift by the same order of magnitude. These future improvements would enable detecting small molecules and atoms with a mass down to sub-Dalton level in cryogenic environment, with a great potential for dramatically advancing the capability of sensing to an unprecedented level. In particular, as the molecule binding occurs during the coherent mechanical motion of the sensor, controlling the motion pattern of the coherent OMO (amplitude, phase, time waveform and so on) may function as a unique paradigm to study/control the mechanical properties of molecule binding and unbinding. This, in combination with certain functionalization of the sensor surface [76] and with implementation of potentially versatile optomechanical motions [54], may offer a unique multifunctional biomolecule toolbox that is not only able to observe cellular machineries at work, but also to selectively manipulate single-molecule interactions.

Additionally, although we focus here on the particle and molecule sensing, the demonstrated optical spring sensing principle can be applied for other physical sensing applications [77], such as inertial sensing [78, 79], electromagnetic field sensing [80], gas sensing [81] and so on, which are based upon sensitive detection of optical cavity resonance shifts induced by external physical perturbations. Therefore, we expect the demonstrated optical spring to be of great promise for broad applications beyond the particle and molecule sensing itself.

References

1. M. S. Hanay, S. Kelber, A. K. Naik, D. Chi, S. Hentz, E. C. Bullard, E. Colinet, L. Duraffourg, and M. L. Roukes. Single-protein nanomechanical mass spectrometry in real time. *Nature Nanotech.*, 7:602–608, 2012.
2. Marek Piliarik and Vahid Sandoghdar. Direct optical sensing of single unlabelled proteins and super-resolution imaging of their binding sites. *Nature Communications*, 5:4495, 2013.
3. A. Serpengüzel, G. Griffel, and S. Arnold. Excitation of resonances of microspheres on an optical fiber. *Opt. Lett.*, 20(7):654–656, Apr. 1995.
4. F. Vollmer, S. Arnold, and D. Keng. Single virus detection from the reactive shift of a whispering-gallery mode. *Proceedings of the National Academy of Sciences*, 105(52):20701–20704, 2008.
5. Tao Lu, Hansuek Lee, Tong Chen, Steven Herchak, Ji-Hun Kim, Scott E. Fraser, Richard C. Flagan, and Kerry Vahala. High sensitivity nanoparticle detection using optical microcavities. *Proceedings of the National Academy of Sciences*, 108(15):5976–5979, 2011.
6. JG Zhu, SK Ozdemir, YF Xiao, L Li, LN He, DR Chen, and L Yang. On-chip single nanoparticle detection and sizing by mode splitting in an ultrahigh-Q microresonator. *Nature Photonics*, 4(1):46–49, January 2010.
7. Robert W. Boyd and John E. Heebner. Sensitive disk resonator photonic biosensor. *Appl. Opt.*, 40(31):5742–5747, Nov. 2001.
8. Joachim Knittel, Jon D. Swaim, David L. McAuslan, George A. Brawley, and Warwick P. Bowen. Back-scatter based whispering gallery mode sensing. *Scientific Reports*, 3:2974, Aug. 2013.
9. Saverio Avino, Anika Krause, Rosa Zullo, Antonio Giorgini, Pietro Malara, Paolo De Natale, Hans Peter Loock, and Gianluca Gagliardi. Direct sensing in liquids using whispering-gallery-mode droplet resonators. *Advanced Optical Materials*, 2(12):1155–1159, 2014.
10. KJ Vahala. Optical microcavities. *Nature*, 424(6950):839–846, 8 2003.
11. M. L. Gorodetsky, A. A. Savchenkov, and V. S. Ilchenko. Ultimate Q of optical microsphere resonators. *Optics Letters*, 21(7):453–455, Apr. 1996.
12. DK Armani, TJ Kippenberg, SM Spillane, and KJ Vahala. Ultra-high-Q toroid microcavity on a chip. *Nature*, 421(6926):925–928, Feb. 27 2003.
13. Hansuek Lee, Tong Chen, Jiang Li, Ki Youl Yang, Seokmin Jeon, Oskar Painter, and Kerry J. Vahala. Chemically etched ultrahigh-Q wedge-resonator on a silicon chip. *Nat. Photonics*, 6(6):369–373, June 2012.
14. Lue Wu, Heming Wang, Qifan Yang, Qingxin Ji, Boqiang Shen, Chengying Bao, Maodong Gao, and Kerry Vahala. Greater than one billion Q factor for on-chip microresonators. *Optics Letters*, 45(18):5129–5131, 2020.
15. Xudong Fan, Ian M. White, Siyka I. Shopova, Hongying Zhu, Jonathan D. Suter, and Yuze Sun. Sensitive optical biosensors for unlabeled targets: A review. *Analytica Chimica Acta*, 620(1-2):8 – 26, 2008.
16. Tao Lu, Tsu-Te Judith Su, Kerry J. Vahala, and Scott Fraser. Split frequency sensing methods and systems. *US Patent*, pages US 8,593,638 B2, November 2013. preliminary filing in Oct. 2008.
17. Po Dong, Long Chen, Qianfan Xu, and Michal Lipson. On-chip generation of high-intensity short optical pulses using dynamic microcavities. *Opt. Lett.*, 34(15):2315–2317, Aug. 2009.
18. Jiang Li, Hansuek Lee, and Kerry J. Vahala. Low-noise Brillouin laser on a chip at 1064 nm. *Opt. Lett.*, 39(2):287–290, Jan. 2014.
19. Eric P. Ostby and Kerry J. Vahala. Yb-doped glass microcavity laser operation in water. *Optics Letters*, 34(8):1153–1155, 2009.
20. Ivan S. Grudinin, Andrey B. Matsko, and Lute Maleki. Brillouin lasing with a CaF_2 whispering gallery mode resonator. *Phys. Rev. Lett.*, 102:043902, Jan. 2009.
21. H. Rokhsari, T. Kippenberg, T. Carmon, and K.J. Vahala. Radiation-pressure-driven micro-mechanical oscillator. *Opt. Express*, 13(14):5293–5301, Jul. 2005.

22. T. J. Kippenberg, H. Rokhsari, T. Carmon, A. Scherer, and K. J. Vahala. Analysis of radiation-pressure induced mechanical oscillation of an optical microcavity. *Physical Review Letters*, 95:033901, Jul. 2005.
23. Tal Carmon, Hossein Rokhsari, Lan Yang, Tobias J. Kippenberg, and Kerry J. Vahala. Temporal behavior of radiation-pressure-induced vibrations of an optical microcavity phonon mode. *Physical Review Letters*, 94:223902, June 2005.
24. Matt Eichenfield, Ryan Camacho, Jasper Chan, Kerry J. Vahala, and Oskar Painter. A picogram- and nanometre-scale photonic-crystal optomechanical cavity. *Nature*, 459:550–555, May 2009.
25. Siddharth Tallur, Suresh Sridaran, and Sunil A. Bhave. A monolithic radiation-pressure driven, low phase noise silicon nitride opto-mechanical oscillator. *Opt. Express*, 19(24):24522–24529, Nov. 2011.
26. Michael A. Taylor, Alex Szorkovszky, Joachim Knittel, Kwan H. Lee, Terry G. McRae, and Warwick P. Bowen. Cavity optoelectromechanical regenerative amplification. *Opt. Express*, 20(12):12742–12751, June 2012.
27. Wei C. Jiang, Xiyuan Lu, Jidong Zhang, and Qiang Lin. High-frequency silicon optomechanical oscillator with an ultra low threshold. *Opt. Express*, 20(14):15991–15996, July 2012.
28. Xiankai Sun, King Y. Fong, Chi Xiong, Wolfram H. P. Pernice, and Hong X. Tang. Ghz optomechanical resonators with high mechanical Q factor in air. *Opt. Express*, 19(22):22316–22321, Oct. 2011.
29. Yang Deng, Fenfei Liu, Zayd C. Leseman, and Mani Hossein-Zadeh. Thermo-optomechanical oscillator for sensing applications. *Opt. Express*, 21(4):4653–4664, Feb. 2013.
30. Kyu Hyun Kim, Gaurav Bahl, Wonsuk Lee, Jing Liu, Matthew Tomes, Xudong Fan, and Tal Carmon. Cavity optomechanics on a microfluidic resonator with water and viscous liquids. *Light: Science & Applications*, 2, 2013.
31. S. I. Shopova, R. Rajmangal, S. Holler, and S. Arnold. Plasmonic enhancement of a whispering-gallery-mode biosensor for single nanoparticle detection. *Appl. Phys. Lett.*, 98(24):243104, 2011.
32. Venkata R. Dantham, Stephen Holler, Curtis Barbre, David Keng, Vasily Kolchenko, and Stephen Arnold. Label-free detection of single protein using a nanoplasmonic-photonic hybrid microcavity. *Nano Letters*, 13(7):3347–3351, 2013.
33. Martin D. Baaske, Matthew R. Foreman, and Frank Vollmer. Single-molecule nucleic acid interactions monitored on a label-free microcavity biosensor platform. *Nature Nanotechnology*, 9:933–939, 2014.
34. Bumki Min, Eric Ostby, Volker Sorger, Erick Ulin-Avila, Lan Yang, Xiang Zhang, and Kerry Vahala. High-Q surface plasmon polariton whispering gallery microcavity. *Nature*, 457(7228):455–458, January 2009.
35. Wenyan Yu, Wei C. Jiang, Qiang Lin, and Tao Lu. Coherent optomechanical oscillation of a silica microsphere in an aqueous environment. *Opt. Express*, 22(18):21421–21426, Sept. 2014.
36. T Bagci, A Simonsen, S Schmid, L G Villanueva, E Zeuthen, J Appel, J M Taylor, A S⊘rensen, K Usami, A Schliesser, and E S Polzik. Optical detection of radio waves through a nanomechanical transducer. *Nature*, 507(7490):81–5, 3 2014.
37. Yuma Okazaki, Imran Mahboob, Koji Onomitsu, Satoshi Sasaki, Shuji Nakamura, Nobu-Hisa Kaneko, and Hiroshi Yamaguchi. Dynamical coupling between a nuclear spin ensemble and electromechanical phonons. *Nature Communications*, 9:2993, 2018.
38. Bei-Bei Li, George Brawley, Hamish Greenall, Stefan Forstner, Eoin Sheridan, Halina Rubinsztein-Dunlop, and Warwick P. Bowen. Ultrabroadband and sensitive cavity optomechanical magnetometry. *Photon. Res.*, 8(7):1064–1071, Jul. 2020.
39. S. Forstner, S. Prams, J. Knittel, E.D. van Ooijen, J.D. Swaim, G.I. Harris, A. Szorkovszky, W.P. Bowen, and J. Rubinsztein-Dunlop. Cavity Optomechanical Magnetometer. *Phys. Rev. Lett.***108**, 120801, 2012.
40. Alexander G. Krause, Martin Winger, Tim D. Blasius, Qiang Lin, and Oskar Painter. A high-resolution microchip optomechanical accelerometer. *Nature Photonics*, 6:768–772, 2012.
41. F. Zhou, Y. Bao, R. Madugani, D.A. Long, J.J. Gorman, and T.W. LeBrun. Broadband thermomechanically limited sensing with an optomechanical accelerometer. *Optica*, 8:350–356, 2021.

42. F. Guzman Cervantes, L. Kumanchik, J. Pratt, and J.M. Taylor. High sensitivity optomechanical reference accelerometer over 10 kHz. *Applied Physics Letters* **104**, 221111, 2014.
43. O. Gerberding, F.G. Cervantes, J. Melcher, J.R. Pratt, and J.M. Taylor. Optomechanical reference accelerometer. *Metrologia* **52**, 654, 2015.
44. Sahar Basiri-Esfahani, Ardalan Armin, Stefan Forstner, and Warwick P. Bowen. Precision ultrasound sensing on a chip. *Nature Communications*, 10:132, 2019.
45. W.J. Westerveld, M. Mahmud-Ul-Hasan, R. Shnaiderman, V. Ntziachristos, X. Rottenberg, S. Severi, and V. Rochus. Sensitive, small, broadband and scalable optomechanical ultrasound sensor in silicon photonics. *Nature Photonics*, **15**, 341–345 2021
46. E. Gavartin, P. Verlot, and T.J. Kippenberg. A hybrid on-chip optomechanical transducer for ultrasensitive force measurements. *Nature Nanotechnology* **7**, 509–514, 2012.
47. G.I. Harris, D.L. McAuslan, T.M. Stace, A.C. Doherty, and W.P. Bowen. Minimum requirements for feedback enhanced force sensing. *Physical Review Letters* **111**, 103603, 2013.
48. T. P. Purdy, K. E. Grutter, K. Srinivasan, and J. M. Taylor. Quantum correlations from a room-temperature optomechanical cavity. *Science*, 356(6344):1265–1268, 2017.
49. T. Purdy, P.L. Yu, N. Kampel, R. Peterson, K. Cicak, R. Simmonds, and C. Regal. Optomechanical Raman-ratio thermometry. *Physical Review A* **92** 031802, 2015.
50. R. Singh, and T.P. Purdy. Detecting Acoustic Blackbody Radiation with an Optomechanical Antenna. *Physical Review Letters* **125**, 120603, 2020.
51. Fenfei Liu, Seyedhamidreza Alaie, Zayd C. Leseman, and Mani Hossein-Zadeh. Sub-pg mass sensing and measurement with an optomechanical oscillator. *Opt. Express*, 21(17):19555–19567, Aug. 2013.
52. G. Bahl, Kyu H. Kim, W. Lee, J. Liu, X. Fan, and T. Carmon. Brillouin cavity optomechanics with microfluiuidic devices. *Nature Communications*, 4:1994, 2013.
53. Wenyan Yu, Wei C. Jiang, Qiang Lin, and Tao Lu. Cavity optomechanical spring sensing of single molecules. *Nature Communications*, 7:12311, 2016.
54. Markus Aspelmeyer, Tobias J. Kippenberg, and Florian Marquardt. Cavity optomechanics. *Rev. Mod. Phys.*, 86:1391–1452, Dec. 2014.
55. Ivan Favero and Khaled Karrai. Optomechanics of deformable optical cavities. *Nature Photon.*, 3:201–205, Mar. 2009.
56. Dries Van Thourhout and Joris Roels. Optomechanical device actuation through the optical gradient force. *Nature Photonics*, 4(4):211–217, Apr. 2010.
57. F. Marquardt and S. M. Girvin. Optomechanics. *Physics*, 2:40, 2009.
58. V.B. Braginsky, S.E. Strigin, and S.P. Vyatchanin. Parametric oscillatory instability in Fabry-Perot interferometer. *Physics Letters A*, 287:331–338, September 2001.
59. Vladimir B Braginsky, Sergey E Strigin, and Sergey P Vyatchanin. Analysis of parametric oscillatory instability in power recycled LIGO interferometer. *Physics Letters A*, 305:111 – 124, 2002.
60. H. A. Haus, *Waves and fields in optoelectronics,* Prentice Hall, Inc., 1984.
61. Qiang Lin, Jessie Rosenberg, Xiaoshun Jiang, Kerry J. Vahala, and Oskar Painter. Mechanical oscillation and cooling actuated by the optical gradient force. *Phys. Rev. Lett.*, 103:103601, Aug. 2009.
62. W. Coffey, Y.P. Kalmykov, and J.T. Waldron. *The Langevin Equation: With Applications to Stochastic Problems in Physics, Chemistry, and Electrical Engineering*. Series in contemporary chemical physics. World Scientific, 2004.
63. Tobias J. Kippenberg and Kerry J. Vahala. Cavity opto-mechanics. *Optics Express*, 15(25):17172–17205, Dec 2007.
64. Mani Hossein-Zadeh and Kerry J. Vahala. Observation of optical spring effect in a microtoroidal optomechanical resonator. *Optics Letters*, 32(12):1611–1613, June 2007.
65. T. J. Kippenberg and K. J. Vahala. Cavity optomechanics: Back-action at the mesoscale. *Science*, 321(5893):1172–1176, 2008.
66. P. S. Waggoner and H. G. Craighead. Micro- and nanomechanical sensors for envinronmental, chemical, and biological detection. *Lab on a Chip*, 7:1238, 2007.

67. J. L. Arlett, E. B. Myers, and M. L. Roukes. Comparative advantages of mechanical biosensors. *Nature Nanotech.*, 6:203–215, 2011.
68. Fenfei Liu and M. Hossein-Zadeh. Mass sensing with optomechanical oscillation. *Sensors Journal, IEEE*, 13(1):146–147, Jan. 2013.
69. S. Forstner, S. Prams, J. Knittel, E. D. van Ooijen, J. D. Swaim, G. I. Harris, A. Szorkovszky, W. P. Bowen, and H. Rubinsztein-Dunlop. Cavity optomechanical magnetometer. *Phys. Rev. Lett.*, 108:120801, Mar. 2012.
70. K. H. Kim and X. Fan. Surface sensitive microfluidic optomechanical ring resonator sensors. *Appl. Phys. Lett.*, 105:191101, 2014.
71. Mani Hossein-Zadeh, Hossein Rokhsari, Ali Hajimiri, and Kerry J. Vahala. Characterization of a radiation-pressure-driven micromechanical oscillator. *Physical Review A*, 74:023813, Aug. 2006.
72. Chi Xiong, Xiankai Sun, King Y. Fong, and Hong X. Tang. Integrated high frequency aluminum nitride optomechanical resonators. *Appl. Phy*, 100:171111, 2012.
73. Ian M. White and Xudong Fan. On the performance quantification of resonant refractive index sensors. *Opt. Express*, 16(2):1020–1028, Jan. 2008.
74. S. Arnold, M. Khoshsima, I. Teraoka, S. Holler, and F. Vollmer. Shift of whispering-gallery modes in microspheres by protein absorption. 28(4):272–274, February 2003.
75. Iwao Teraoka, Stephen Arnold, and Frank Vollmer. Perturbation approach to resonance shifts of whispering-gallery modes in a dielectric microsphere as a probe of a surrounding medium. *J. Opt. Soc. Am. B*, 20(9):1937–1946, Sep. 2003.
76. F. Vollmer and S. Arnold. Whispering-gallery-mode biosening: label-free detection down to single molecules. *Nature Methods*, 5:591–596, 2008.
77. Matthew R. Foreman, Jon D. Swaim, and Frank Vollmer. Whispering gallery mode sensors. *Adv. Opt. Photon.*, 7(2):168–240, June 2015.
78. Caterina Ciminelli, Francesco Dell'Olio, Carlo E. Campanella, and Mario N. Armenise. Photonic technologies for angular velocity sensing. *Adv. Opt. Photon.*, 2(3):370–404, Sept. 2010.
79. Alexander G. Krause, Martin Winger, Tim D. Blasius, Qiang Lin, and Oskar Painter. A high-resolution microchip optomechanical accelerometer. *Nature Photon.*, 6:768–772, 2012.
80. Tindaro Ioppolo, Ulas Ayaz, and M. Volkan Ötügen. Tuning of whispering gallery modes of spherical resonators using an external electric field. *Opt. Express*, 17(19):16465–16479, Sept. 2009.
81. Nebiyu A. Yebo, Sreeprasanth Pulinthanathu Sree, Elisabeth Levrau, Christophe Detavernier, Zeger Hens, Johan A. Martens, and Roel Baets. Selective and reversible ammonia gas detection with nanoporous film functionalized silicon photonic micro-ring resonator. *Opt. Express*, 20(11):11855–11862, May 2012.

Chapter 5
Quantum Optical Theories of Molecular Optomechanics

Mohsen Kamandar Dezfouli and Stephen Hughes

Abstract We present several quantum optics models to describe regimes of molecular optomechanics for single molecules coupled to plasmonic cavity systems and show how this relates to surface enhanced Raman spectroscopy (SERS). We first present a general-medium open system picture of single molecule SERS (or molecular optomechanics) in a system-bath approximation, and show how the detected Raman spectrum originates from an interplay of nonlinear light generation and propagation. We apply this theory to study several different resonator systems whose cavity modes are described through a quasinormal theory of open cavities, and identify important limits of commonly used electric-field rules for Raman signal enhancement. Second, in the good cavity limit, we present a quantum optics picture of off-resonant SERS to explore molecular optomechanics in the sideband-resolved regime, where the cavity-emitted spectrum results in anharmonic resonances. This latter regime exploits hybrid metal-dielectric resonators, which yield deep sub-wavelength plasmonic confinement and dielectric-like quality factors. Finally, we present a generalized master equation approach to describe resonant SERS in the strong coupling regime of cavity-QED, while also in the ultrastrong vibrational coupling regime, and show several applications of the theory, including signatures of phonon-dressed polaritons in the emission spectrum.

M. Kamandar Dezfouli
Advanced Electronics and Photonics Research Centre, National Research Council Canada, 1200 Montreal Rd., Ottawa, ON K1A 0R6, Canada
e-mail: Mohsen.KamandarDezfouli@nrc-cnrc.gc.ca

S. Hughes (✉)
Department of Physics, Engineering Physics and Astronomy, Queen's University, Kingston, ON K7L 3N6, Canada
e-mail: shughes@queensu.ca

W. Bowen et al. (eds.), *Single Molecule Sensing Beyond Fluorescence*, Nanostructure Science and Technology,
https://doi.org/10.1007/978-3-030-90339-8_5

5.1 Introduction

Photons interacting with molecules can induce spontaneous Raman scattering [1], causing light scattering at phonon-shifted frequencies with respect to the excitation laser frequency, manifesting in nonlinear Stokes (red shifted) and anti-Stokes (blue shifted) signals (see also Chap. 2.12). While most Raman experiments involve very small scattering cross-sections, surface-enhanced Raman spectroscopy (SERS) with metal nanoparticles (MNPs) can enhance the Raman emission from molecules by many orders of magnitude [2–11], leading to applications in sensing technologies and fundamental quantum optics. Furthermore, localized surface resonances from MNPs facilitate extreme spatial confinement of the electromagnetic fields, well below the diffraction limit, which has enabled SERS to emerge as a powerful tool for identifying different molecules and proteins down to the single molecule level [5, 6, 11–13], and even observation of single atomic ions [14] as discussed in Chap. 2. There are also significant advances being made to explore the fundamental precision limits of biosensing due to the quantization of light, e.g., using evanescent biosensors for single-molecule detection [15] (see Chap. 1). Single molecules and engineered optical fields can give rise to extremely rich regimes for light-matter interactions and quantum optics in open systems, but one in which losses and molecular back-action can form a significant part of the system.

As a first approximation, SERS can be viewed as a simple enhanced nonlinear scattering process, and one commonly adopted scaling rule assumes the total Raman signal scales with the field enhancement at the laser frequency, ω_L, as well as field enhancement at the Raman frequency, ω_R, yielding a total enhancement factor of [9]

$$\mathrm{EF}\left(\omega_R, \omega_L\right) \propto \frac{\left|E\left(\omega_R\right)\right|^2\left|E\left(\omega_L\right)\right|^2}{\left|E_0\left(\omega_R\right)\right|^2\left|E_0\left(\omega_L\right)\right|^2}, \tag{5.1}$$

which is termed the E^4 electric-field law for SERS enhancement. Many theoretical approaches have been used to describe SERS in plasmonic environments [16–23], including density functional theory [18, 21]. Recent work has described how Raman scattering of single molecules in MNP cavities can also be seen as a model for molecular optomechanics [24–26], induced by an intrinsic optomechanical coupling between the localized surface plasmon resonance and the molecular vibrations. Figure 5.1 shows a molecular energy diagram for a single molecule coupled to an optical cavity mode (e.g., from a localized plasmon), and depicts how it can be viewed in the same way as a stereotypical optomechanical coupling problem with a 1D cavity mode in the presence of a vibrating mirror [24].

There has also been much interest in using plasmonic structures to study emerging regimes of quantum plasmonics [27–32], including pulsed molecular optomechanics [33]. Pulsed optomechanics has also been used to explore quantum optomechanics beyond the quantum coherent oscillation regime [34]. However, strong optomechanical coupling at the few photon regime, which allows nonlinear quantum optical effects such as the single photon blockade [35, 36], remains largely unexplored in

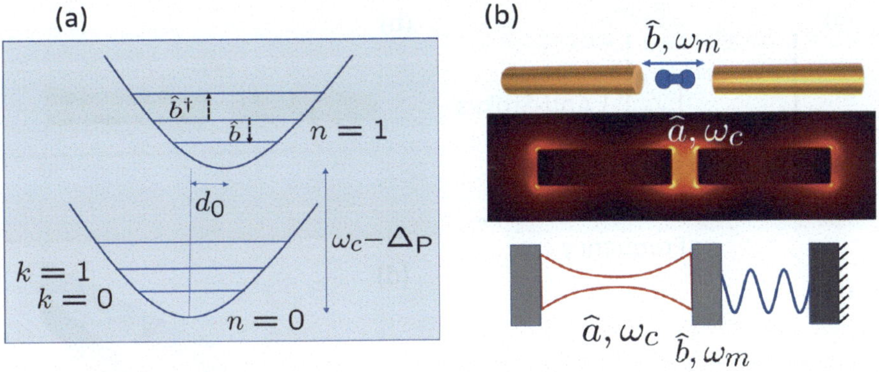

Fig. 5.1 **a** Molecular energy level diagram showing two photon manifolds ($n = 0, 1$) with phonon states (k), where the $n = 1$ photon manifold has a normalized displacement (d_0) and a polaronic frequency shift ($-\Delta_P$). **b** Metal dimer with an oscillating molecule, represented through $\hat{b}$, ω_m, and a cavity mode profile of the localized plasmon mode (formally a quasinormal mode, represented with $\hat{a}$, ω_c), and an analogue optomechanics system using a simple 1D cavity with a vibrating mirror

the context of SERS, and new ideas with hybrid resonator systems are now emerging [37, 38].

Various photonic systems can be used to enhance SERS and molecular optomechanics, including plasmonic resonators [12, 13, 30, 42, 43], waveguide geometries [44–46], and hybrid material systems [47–50]. A few example systems are shown schematically in Fig. 5.2. In order to be able to describe a wide range of different photonic systems, we need accurate and flexible theoretical formalisms.

In this chapter, we present several theoretical techniques for studying molecular optomechanics and single molecule SERS for plasmonic cavity systems, with a common theme of using formalisms from the viewpoint of open system quantum optics. We first present a quantum optical theory of SERS in a system-bath approximation, and show how the Raman spectrum stems from an interplay of nonlinear light generation and light propagation. We also identify the limits of the E^4 electric-field rule for Raman signal enhancement for low pump powers, and derive a different E^8 rule at higher pump powers. We derive a quantum master equation where the general photonic system (e.g., lossy and inhomogeneous) is treated as a photonic bath. The photonic bath function is implicitly contained in the electric-field Green function. We then use this theory to investigate a selection of resonator systems whose cavity modes are conveniently described through quasinormal modes (QNMs) [51, 52]. Second, we study molecular optomechanics in a strong vibrational coupling regime, where the cavity modes are quantized at the system level, without any form of linearization for the optomechanical coupling. Although the significant dissipation inherent to MNPs can prevent accessing such a regime, hybrid plasmonic devices, consisting of dielectric and metal parts, offer extra design flexibility in terms of the resonance line shapes and cavity mode properties [41, 47, 48, 53, 54], where the system-bath theo-

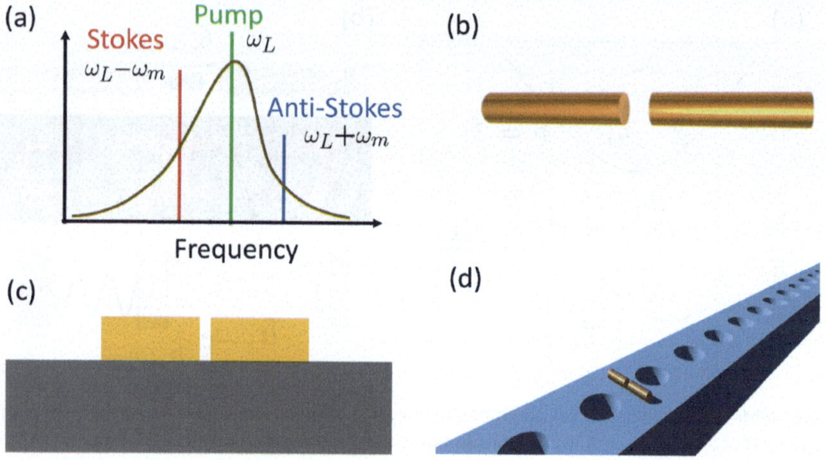

Fig. 5.2 **a** Schematic of the main resonances involved in SERS where a single molecule of interest is coupled to a plasmonic resonator or any general photonic environment. The broad (and in general non-Lorentzian) plasmonic response is shown where the Stokes and anti-Stokes signal are enhanced according to a frequency dependent coupling strength. **b–d** Schematics of example plasmonic structures, including a metal-dimer nanoantenna [39], a metal slot waveguide [40], and a hybrid metal-dielectric cavity [41]. For further discussion of the latter, see Chap. 2

ries break down. We describe how such a system can yield *anharmonic* emission lines in the sideband-resolved regime of the cavity-emitted spectrum. Finally, we explore resonant SERS in the strong coupling regime of cavity-QED [55]. In this latter study, we use a more advanced generalized master equation that includes realistic baths for the cavity mode, vibrational mode, and the resonant two level system (TLS). We then explore how phonon interactions modify the cavity-QED polariton states. We also show the impact of TLS pure dephasing and demonstrate the clear failure of using standard master equations without properly accounting for the system-bath coupling dynamics.

5.2 Quantum Optics Model of Single Molecule SERS Using a System-Bath Master Equation Approach

In this section, we describe an open-system quantum optics model to derive analytical expressions for the detected Stokes and anti-Stokes intensities of single molecules coupled to a general photonic medium [26]. The theory applies to a wide range of photonic geometries, including cavities and waveguides. We adopt a master equation approach [56–58], where the photon Green function appears directly, which includes the full frequency dependence of the medium, including light propagation and quenching effects.

Importantly, the theory uses a quantum field theory that is valid for any inhomogeneous and lossy medium [59, 60]. We include the vibrational mode (or phonon mode) at the system level and trace out the field operators in a bath approximation; however, we still treat the bath interaction in a self-consistent way. Using the master equation, and the quantum field theory, we then derive a simple and transparent analytical formula for the emitted SERS spectrum, and we use this to connect to approximate SERS theories in the literature, and point out their limitations.

We subsequently apply the theory to model single molecule SERS from several resonator structures, including a metal dimer resonator made of two gold cylindrical nanorods (which has good quantum efficiency [39]) as shown in Fig. 5.1b (and also Fig. 5.2b). We then study a more complex hybrid photonic-plasmonic system, where the dimer is on top of a photonic crystal nanobeam cavity (see Fig. 5.2d). The optical cavity modes of these resonator systems are obtained using a powerful QNM theory, which allows the Green function to be obtained analytically in terms of a QNM expansion.

5.2.1 *System Hamiltonian, Photonic Green Function and Interaction Hamiltonian for the Raman Induced Dipole*

Here we introduce the system Hamiltonian, including the molecular vibrations of a given frequency, the photon bath of the photonic medium, and discuss the Raman interaction between phonons and the electromagnetic fields. We then present the quantized electric field operator for an arbitrary plasmonic/photonic environment, the photonic Green function, as well as the induced dipole through the Raman process.

The molecular vibrational mode is treated as a quantized harmonic oscillator, with the corresponding Hamiltonian (cf. Fig. 5.1b)

$$H_{\rm m} = \hbar\omega_m \hat{b}^\dagger \hat{b}, \tag{5.2}$$

where $\hat{b}$ and $\hat{b}^\dagger$ satisfy Bosonic commutation rules, $\left[\hat{b}, \hat{b}^\dagger\right] = 1$, and ω_m is the vibrational mode frequency. For simplicity, we assume that there is only one vibrational mode (or phonon mode) of interest, though the theory can easily include multiple vibrational modes. The Hamiltonian for the fields, which is valid for an arbitrary inhomogeneous and lossy medium, is expressed in terms of a continuum of field operators, $\hat{\mathbf{f}}(\mathbf{r}, \omega)$, with [59, 60]

$$H_{\rm ph} = \hbar \int d\mathbf{r} \int_0^\infty d\omega\, \omega \hat{\mathbf{f}}^\dagger(\mathbf{r}, \omega) \hat{\mathbf{f}}(\mathbf{r}, \omega), \tag{5.3}$$

where

$$\left[\hat{\mathbf{f}}(\mathbf{r}, \omega), \hat{\mathbf{f}}^\dagger(\mathbf{r}', \omega')\right] = \mathbf{1}\delta(\mathbf{r} - \mathbf{r}')\delta(\omega - \omega'). \tag{5.4}$$

For any general medium described by the complex permittivity function $\epsilon(\mathbf{r}, \omega)$, the electric-field operator is [59, 60]

$$\hat{\mathbf{E}}(\mathbf{r}, \omega) = \int d\mathbf{r}' \mathbf{G}(\mathbf{r}, \mathbf{r}'; \omega) \cdot \sqrt{\frac{\hbar}{\pi \epsilon_0} \text{Im}\{\epsilon(\mathbf{r}', \omega)\}} \hat{\mathbf{f}}(\mathbf{r}', \omega), \tag{5.5}$$

where the two space-point Green function satisfies

$$\nabla \times \nabla \times \mathbf{G}(\mathbf{r}, \mathbf{r}'; \omega) - \frac{\omega^2}{c^2} \epsilon(\mathbf{r}, \omega) \mathbf{G}(\mathbf{r}, \mathbf{r}'; \omega) = \frac{\omega^2}{c^2} \mathbf{1} \delta(\mathbf{r} - \mathbf{r}'), \tag{5.6}$$

subjected to appropriate boundary conditions.

Assuming a dipole interaction between the fields and the molecule, the induced Raman polarization is

$$\hat{\mathbf{p}}_R = \hat{\boldsymbol{\alpha}}_R \cdot \hat{\mathbf{E}}(\mathbf{r}_m), \tag{5.7}$$

where $\hat{\boldsymbol{\alpha}}_R$ is the Raman polarizability tensor of the molecule, which can be expressed in terms of the Raman tensor of the molecule, $\mathbf{R}$, through [9]

$$\hat{\boldsymbol{\alpha}}_R = \mathbf{R} \sqrt{\frac{\hbar}{2\omega_m}} \left(\hat{b} + \hat{b}^\dagger\right), \tag{5.8}$$

and $\hat{\mathbf{E}}(\mathbf{r}_m)$ is the total electric field operator at the molecule location $\mathbf{r}_m$,

$$\hat{\mathbf{E}}(\mathbf{r}) = \int_0^\infty d\omega \, \hat{\mathbf{E}}(\mathbf{r}, \omega) + \text{H.c.} \tag{5.9}$$

Subsequently, the interaction Hamiltonian is

$$H_{\text{int}} = -\hat{\mathbf{p}}_R \cdot \hat{\mathbf{E}}(\mathbf{r}_m) = -\sqrt{\frac{\hbar}{2\omega_m}} \left(\hat{b} + \hat{b}^\dagger\right) \hat{\mathbf{E}}(\mathbf{r}_m) \cdot \mathbf{R} \cdot \hat{\mathbf{E}}(\mathbf{r}_m). \tag{5.10}$$

Note the general structure of the total Hamiltonian is similar in form to the simple coupled-mode approaches [25, 61], which we discuss in more detail later.

5.2.2 Pump Field Enhancement Simplified Interaction Hamiltonian

We now derive the pump field enhancement term and use that to obtain a simplified version of the interaction Hamiltonian.

For the optical pump term, we assume a continuous wave (CW) drive, treated as a c number, with amplitude $\mathbf{F}_0$ and frequency ω_L. This classical pump model is equivalent to the common linearization procedures [25, 62]. Accounting for scattering from the pump field, the effective field at the molecule position is

$$\mathbf{F}_\mathrm{p}(\mathbf{r}_m, t) = \eta \left[\mathbf{F}_0(\mathbf{r}_m) e^{-i\omega_L t} + \mathbf{F}_0^*(\mathbf{r}_m) e^{i\omega_L t}\right], \tag{5.11}$$

where η is the field enhancement factor, obtained from Green function theory:

$$\eta = 1 + \frac{\int_V [\epsilon(\mathbf{r}, \omega_L) - \epsilon_B] \mathbf{n} \cdot \mathbf{G}(\mathbf{r}_m, \mathbf{r}; \omega_L) \cdot \mathbf{F}_0(\mathbf{r}) d\mathbf{r}}{\mathbf{n} \cdot \mathbf{F}_0(\mathbf{r}_m)}, \tag{5.12}$$

where V is the volume of the scattering geometry, ϵ_B is the background medium dielectric constant, and $\mathbf{n}$ is a unit vector that represents the direction of the induced Raman dipole (which will depend on the dominant Raman tensor element, R_{nn}). We highlight that η depends on the molecule location ($\mathbf{r}_m$) as well as the laser frequency (ω_L).

The interaction Hamiltonian can then be written, within a rotating-wave approximation, as

$$\begin{aligned} H_\mathrm{int} = & -R_{nn}\eta\sqrt{\frac{\hbar}{2\omega_m}}\left(\hat{b} + \hat{b}^\dagger\right)\left\{\int d\omega \int d^3\mathbf{r}\left[\mathbf{n} \cdot \mathbf{F}_0^*(\mathbf{r}_m)\right] e^{i\omega_L t}\right. \\ & \left. \times\, \mathbf{n} \cdot \mathbf{G}(\mathbf{r}_m, \mathbf{r}; \omega) \cdot \sqrt{\frac{\hbar}{\pi\epsilon_0}\mathrm{Im}\{\epsilon(\mathbf{r}, \omega)\}}\, \hat{\mathbf{f}}(\mathbf{r}, \omega) + \mathrm{H.c.}\right\}. \end{aligned} \tag{5.13}$$

5.2.3 General Quantum Master Equation Using a Photonic Bath Approximation

Introducing a time-local master equation in the interaction picture (tilde represents the interaction picture), and using a Born-Markov approximation, we start from [56, 57]

$$\frac{\partial\tilde{\rho}(t)}{\partial t} = -\frac{1}{\hbar^2}\int_0^t d\tau\, \mathrm{Tr}_\mathrm{R}\left\{\left[\tilde{H}_\mathrm{int}(t), \left[\tilde{H}_\mathrm{int}(t-\tau), \tilde{\rho}(t)\rho_\mathrm{R}\right]\right]\right\}, \tag{5.14}$$

where $\tilde{\rho}$ is the reduced density matrix in the basis of molecular vibrations, ρ_R is the plasmonic bath density operator, and $\tilde{H}_\mathrm{int}$ is the interaction Hamiltonian in the interaction picture. We will also assume the following bath approximations:

$$\mathrm{Tr}_{\mathrm{R}}\left\{\hat{\mathbf{f}}^{\dagger}(\mathbf{r},\omega)\hat{\mathbf{f}}(\mathbf{r}',\omega')\rho_{\mathrm{R}}\right\}=0,$$
$$\mathrm{Tr}_{\mathrm{R}}\left\{\hat{\mathbf{f}}(\mathbf{r},\omega)\hat{\mathbf{f}}^{\dagger}(\mathbf{r}',\omega')\rho_{\mathrm{R}}\right\}=\mathbf{1}\delta(\mathbf{r}-\mathbf{r}')\delta(\omega-\omega'). \tag{5.15}$$

If we now transform the Hamiltonian (5.13) to the interaction picture, and substitute into the master Eq. (5.14), we obtain the reduced density matrix of the system,[1]

$$\begin{aligned}\frac{\partial\tilde{\rho}}{\partial t}&=J_{\mathrm{ph}}(\omega_L+\omega_m)\left(2\hat{b}\tilde{\rho}\hat{b}^{\dagger}-\hat{b}^{\dagger}\hat{b}\tilde{\rho}-\tilde{\rho}\hat{b}^{\dagger}\hat{b}\right)\\&+J_{\mathrm{ph}}(\omega_L-\omega_m)\left(2\hat{b}^{\dagger}\tilde{\rho}\hat{b}-\hat{b}\hat{b}^{\dagger}\tilde{\rho}-\tilde{\rho}\hat{b}\hat{b}^{\dagger}\right),\end{aligned} \tag{5.16}$$

where $J_{\mathrm{ph}}(\omega)$ is defined through

$$J_{\mathrm{ph}}(\omega)\equiv\frac{R_{nn}^2|\eta|^2|\mathbf{n}\cdot\mathbf{F}_0|^2}{2\epsilon_0\omega_m}\mathrm{Im}\{G_{nn}(\mathbf{r}_m,\mathbf{r}_m;\omega)\}, \tag{5.17}$$

which includes the effects of the Raman tensor element, the medium local density of states (LDOS), and the enhancement factor of the pump field. Note that J_{ph} has the same units as γ (a rate).

In Eq. (5.16), the first term represents the Stokes signal (emitted at $\omega_L-\omega_m$), and is proportional to $J_{\mathrm{ph}}(\omega_L+\omega_m)$; while the second term represent the anti-Stokes signal (emitted at $\omega_L+\omega_m$), which is proportional to $J_{\mathrm{ph}}(\omega_L-\omega_m)$. Consequently, the Stokes (anti-Stokes) emission depends on the plasmonic enhancement through the projected LDOS at the anti-Stokes (Stokes) frequency. This may seem contrary to the common wisdom that SERS involves a field enhancement at the corresponding frequencies for both Stokes and anti-Stokes signals, but the actual detected signal also depends on light propagation effects as well as pump enhancement effects (at the laser frequency), as will be made clear later. Note also that while the unusual mixed LDOS features may seem surprising, similar effects often occur in the domain of quantum nonlinear optics, e.g., a similar non-trivial LDOS coupling has been noted with the Mollow triplet spectrum emitted from TLSs coupled to metal resonators, where the center peak width of the Mollow triplet depends only on the LDOS at the two side peaks and not on the LDOS at the laser frequency [63].

To obtain the full master equation, we also account for the coupling of the Raman vibrations to the thermal environmental, by using additional Lindblad terms associated with thermal dissipation and incoherent pumping [25]. We then derive the following master equation for the total molecule-photonic system:

[1] To obtain Eq. (5.16) from Eq. (5.14), the upper limit in the time integral has been extended to infinity, as the time scale for system changes are much longer than the bath correlation time.

$$\begin{aligned}\frac{\partial \rho}{\partial t} &= -i\omega_m \left[\hat{b}^\dagger\hat{b}, \rho\right] + J_{\rm ph}\left(\omega_L + \omega_m\right)\left(2\hat{b}\rho\hat{b}^\dagger - \hat{b}^\dagger\hat{b}\rho - \rho\hat{b}^\dagger\hat{b}\right) \\ &+ J_{\rm ph}\left(\omega_L - \omega_m\right)\left(2\hat{b}^\dagger\rho\hat{b} - \hat{b}\hat{b}^\dagger\rho - \rho\hat{b}\hat{b}^\dagger\right) \\ &+ \gamma_m\left(\bar{n}^{\rm th} + 1\right)\left(2\hat{b}\rho\hat{b}^\dagger - \hat{b}^\dagger\hat{b}\rho - \rho\hat{b}^\dagger\hat{b}\right) + \gamma_m\bar{n}^{\rm th}\left(2\hat{b}^\dagger\rho\hat{b} - \hat{b}\hat{b}^\dagger\rho - \rho\hat{b}\hat{b}^\dagger\right),\end{aligned} \tag{5.18}$$

where the thermal population of the vibrational mode is

$$\bar{n}^{\rm th} = \frac{1}{\exp\left(\hbar\omega_m/k_B T\right) - 1}, \tag{5.19}$$

where T the bath temperature, and k_B the Boltzmann constant. Note that the density operator has now been transformed back to the usual Schrödinger picture.

5.2.4 Analytical Expression for the SERS Spectrum

Using the general medium master equation, Eq. (5.18), we now seek to derive an analytical expression for the detected spectrum for the case of a CW excited molecular-cavity system.

The detected spectrum, at position $\mathbf{r}_{\rm D}$, is obtained from

$$S\left(\mathbf{r}_{\rm D}, \omega\right) \equiv \left\langle \hat{\mathbf{E}}^\dagger\left(\mathbf{r}_{\rm D}, \omega\right) \cdot \hat{\mathbf{E}}\left(\mathbf{r}_{\rm D}, \omega\right)\right\rangle, \tag{5.20}$$

where we assume a point detector. The electric field operator includes the effects of propagation through the Green function, and is obtained from [64]

$$\hat{\mathbf{E}}\left(\mathbf{r}_{\rm D}, \omega\right) = \frac{1}{\epsilon_0}\mathbf{G}\left(\mathbf{r}_{\rm D}, \mathbf{r}_m; \omega\right) \cdot \hat{\mathbf{p}}_R\left(\mathbf{r}_m; \omega\right). \tag{5.21}$$

Using the electric field operator and the molecular polarization equations introduced earlier, we derive the detected spectrum as

$$\begin{aligned} S\left(\mathbf{r}_{\rm D}, \omega\right) &= \frac{\hbar R_{nn}^2 \left|\eta\right|^2 \left|\mathbf{n}\cdot\mathbf{F}_0\right|^2}{2\omega_m \epsilon_0^2} \left|\mathbf{G}\left(\mathbf{r}_{\rm D}, \mathbf{r}_m; \omega\right)\cdot\mathbf{n}\right|^2 S_0\left(\omega\right), \\ &\equiv A(\eta, \mathbf{F}_0)\left|\mathbf{G}\left(\mathbf{r}_{\rm D}, \mathbf{r}_m; \omega\right)\cdot\mathbf{n}\right|^2 S_0\left(\omega\right), \end{aligned} \tag{5.22}$$

where $A(\eta, \mathbf{F}_0)$ accounts for the enhancement factor, the two-space-point Green function accounts for light propagation (from the molecule to the detector), and the third term, $S_0\left(\omega\right) = S_0^{\rm st}\left(\omega\right) + S_0^{\rm as}\left(\omega\right)$, is the *emitted spectrum*, that includes a sum from the Stokes process, $S_0^{\rm st}\left(\omega\right)$, and the anti-Stokes process, $S_0^{\rm as}\left(\omega\right)$.

The emission contributions are obtained from [63, 65][2]

$$S_0^{\rm st}(\omega) \equiv \left\langle \hat{b}(\omega)\hat{b}^\dagger(\omega)\right\rangle = {\rm Re}\left\{\int_0^\infty d\tau e^{i(\omega-\omega_L)\tau}\left\langle \hat{b}(t)\hat{b}^\dagger(t+\tau)\right\rangle\right\}, \tag{5.23}$$

and

$$S_0^{\rm as}(\omega) \equiv \left\langle \hat{b}^\dagger(\omega)\hat{b}(\omega)\right\rangle = {\rm Re}\left\{\int_0^\infty d\tau e^{i(\omega-\omega_L)\tau}\left\langle \hat{b}^\dagger(t)\hat{b}(t+\tau)\right\rangle\right\}. \tag{5.24}$$

Using the quantum regression theorem [56], and the general master equation, Eq. (5.18), these can be solved analytically to give:

$$S_0^{\rm st}(\omega) = {\rm Re}\left\{\frac{i\left[\gamma_m\left(\bar{n}^{\rm th}+1\right)+J_{\rm ph}\left(\omega_L+\omega_m\right)\right]}{\left[\omega-\left(\omega_L-\omega_m\right)+i\left(\gamma_m+\Delta J_{\rm ph}\right)\right]\left(\gamma_m+\Delta J_{\rm ph}\right)}\right\}, \tag{5.25}$$

$$S_0^{\rm as}(\omega) = {\rm Re}\left\{\frac{i\left[\gamma_m\bar{n}^{\rm th}+J_{\rm ph}\left(\omega_L-\omega_m\right)\right]}{\left[\omega-\left(\omega_L+\omega_m\right)+i\left(\gamma_m+\Delta J_{\rm ph}\right)\right]\left(\gamma_m+\Delta J_{\rm ph}\right)}\right\}, \tag{5.26}$$

where $\Delta J_{\rm ph} = J_{\rm ph}(\omega_L+\omega_m) - J_{\rm ph}(\omega_L-\omega_m)$.

We next discuss some general trends from the analytical formulas, Eqs. (5.22) (detected spectrum), and (5.25)–(5.26) (emitted Stokes and anti-Stokes spectra), including the role of an increasing pump power. For example, let us consider the Stokes emission given in Eq. (5.25): In the low pump limit, the $J_{\rm ph}$ contribution will be relatively small, and the maximum Stokes generation is simply

$$S_0^{\rm st}(\omega_L-\omega_m) \approxeq \frac{\bar{n}^{\rm th}+1}{\gamma_m}, \tag{5.27}$$

which is independent of $J_{\rm ph}$. However, in the strong pump limit, assuming $\Delta J_{\rm ph}$ is small compared to γ_m, the Stokes emission is then

$$S_0^{\rm st}(\omega_L-\omega_m) \approxeq \frac{J_{\rm ph}(\omega_L+\omega_m)}{\gamma_m^2}, \tag{5.28}$$

which depends on the pump power as well as the LDOS (see Eq. (5.17)). Similar arguments can be made for the anti-Stokes signal shown with Eq. (5.26).

Interestingly, we see that the Green function enters our formalism twice: once at the emission stage and once for light propagation to the detection point. The former effect samples the LDOS as the dressed resonances, while the latter is a linear propagation effect, which also includes quenching (at the excitation laser frequency). These combined effects approximately recover the expected E^4 enhancement rule

[2] For this section, we have neglected $\langle\hat{b}(\omega)\hat{b}(\omega)\rangle$ and $\langle\hat{b}^\dagger(\omega)\hat{b}^\dagger(\omega)\rangle$ terms as they are associated with higher-order Raman intensities.

for SERS at low pump powers, but also lead to a E^8 enhancement rule at high pump power.

5.2.5 Coupled Mode Quantum Optomechanical Model with Simple Lindblad Decay Processes

It is useful to connect our general medium approach to simpler coupled mode formalisms that have been used in the literature, e.g., Refs. [25, 61]. The two-mode system Hamiltonian, including a cavity pump term with Rabi frequency Ω, is

$$H_s = \hbar\omega_c \hat{a}^\dagger \hat{a} + \hbar\omega_m \hat{b}^\dagger \hat{b} + \hbar g\, \hat{a}^\dagger \hat{a} \left(\hat{b} + \hat{b}^\dagger\right) + i\hbar\Omega \left(\hat{a}\, e^{i\omega_L t} - \hat{a}^\dagger e^{i\omega_L t}\right), \quad (5.29)$$

where $\hat{a}$ and $\hat{a}^\dagger$ are the MNP cavity mode operators and g is the cavity-molecule optomechanical coupling factor, assumed now to be for a single cavity mode.

The corresponding master equation, including the cavity decay rate γ_c (half width at half maximum, see next subsection on QNMs for more details), is

$$\begin{aligned}\frac{d\rho}{dt} &= -\frac{i}{\hbar}[H_s, \rho] + \gamma_c \left(2\hat{a}\rho\hat{a}^\dagger - \hat{a}^\dagger\hat{a}\rho - \rho\hat{a}^\dagger\hat{a}\right)\\ &+ \gamma_m \left(\bar{n}^{\text{th}} + 1\right)\left(2\hat{b}\rho\hat{b}^\dagger - \hat{b}^\dagger\hat{b}\rho - \rho\hat{b}^\dagger\hat{b}\right) + \gamma_m \bar{n}^{\text{th}} \left(2\hat{b}^\dagger\rho\hat{b} - \hat{b}\hat{b}^\dagger\rho - \rho\hat{b}\hat{b}^\dagger\right),\end{aligned} \quad (5.30)$$

which can be solved numerically in a basis of N cavity (plasmon) photons.

The cavity emitted spectrum is computed from

$$S(\omega) = \omega^4\, \text{Re}\left[\int_0^\infty d\tau\, e^{i(\omega-\omega_L)\tau} \left\langle \hat{a}^\dagger(t)\, \hat{a}(t+\tau)\right\rangle\right], \quad (5.31)$$

where the quantum regression theorem can be used to compute the two-time correlation function $\langle \hat{a}^\dagger(t)\hat{a}(t+\tau)\rangle$. Alternatively, using a linearization procedure [25, 66], one can readily solve the equations analytically, which is the approach we follow in the numerical results Sect. 5.2.7 when comparing with our general master equation solution.

We note the two mode spectrum here does not include the full propagation effects from the lossy medium, and the additional ω^4 term is used to account for a simple dipole propagation argument for far field detection. Clearly, our arbitrary medium approach is much more general, as it can also be used to model the detected spectrum as a function of distance, including the fuell spatial regimes in the near field, and it is not restricted to simple cavity modes.

5.2.6 *Quasinormal Modes, Green Function Expansions, and Purcell Factors for Two Example Resonators*

Here we briefly discuss the use of QNMs and the QNM expansion technique for the Green function, which that can be used to describe a general open cavity system. Extensions of the arguments given below to other plasmonic structures such as waveguides and slabs is straightforward, when the appropriate alternative Green function expansion is used. For example, a detailed study of SERS enhancement using plasmonic slot waveguides using this theory is presented in Ref. [40], which employs a waveguide Green function approach. We will also show how these QNM properties can be used to rigorously connect to the medium LDOS and the Purcell factor of an embedded quantum emitter or dipole.

The electric-field QNMs, $\tilde{\mathbf{f}}_\mu(\mathbf{r})$, are solutions to the Helmholtz equation,

$$\nabla \times \nabla \times \tilde{\mathbf{f}}_\mu(\mathbf{r}) - \left(\frac{\tilde{\omega}_\mu}{c}\right)^2 \epsilon(\mathbf{r}, \tilde{\omega}_\mu)\, \tilde{\mathbf{f}}_\mu(\mathbf{r}) = 0, \tag{5.32}$$

where $\tilde{\omega}_\mu = \omega_\mu - i\gamma_\mu$ is the complex eigenfrequency, and $\epsilon(\mathbf{r}, \omega)$ is the dielectric function, which is in general complex. The open boundary conditions ensure the Silver-Müller radiation condition [67], and the quality factor of each resonance is $Q_\mu = \omega_\mu/2\gamma_\mu$. Note that the complex part of the eigenfrequency accounts for both radiative losses as well as nonradiative losses (such as Ohmic losses). The normalization of the QNMs can be done in different ways [68–70], and additional care is needed as the QNMs spatially diverge because of the open boundary conditions, which is a consequence of solving a non-Hermitian eigenvalue problem.

To obtain a general definition of the Purcell factor for a point dipole emitter in an arbitrary medium (i.e., the enhanced spontaneous emission rate), we require the Green function, defined in Eq. (5.6). The normalized QNMs can be used to construct the Green function for locations near (or within) the scattering geometry through [71–73]

$$\mathbf{G}(\mathbf{r}, \mathbf{r}_0, \omega) = \sum_\mu A_\mu(\omega)\, \tilde{\mathbf{f}}_\mu(\mathbf{r})\, \tilde{\mathbf{f}}_\mu(\mathbf{r}_0), \tag{5.33}$$

with $A_\mu(\omega) = \omega^2/2\tilde{\omega}_\mu\left(\tilde{\omega}_\mu - \omega\right)$ or $A_\mu(\omega) = \omega/2(\tilde{\omega}_\mu - \omega)$; note that these are related through a sum relationship [73].

Considering a dipole emitter at location $\mathbf{r}_0$, with dipole moment $\mathbf{d}$ $(= d\,\mathbf{n}_d)$, then the classical spontaneous emission rate is [51]

$$\Gamma(\mathbf{r}_0, \omega) = \frac{2}{\hbar\epsilon_0}\mathbf{d} \cdot \mathrm{Im}\{\mathbf{G}(\mathbf{r}_0, \mathbf{r}_0, \omega)\} \cdot \mathbf{d}, \tag{5.34}$$

and the generalized Purcell factor is [51, 74]

$$F_{\rm P}(\mathbf{r}_0, \omega) = 1 + \frac{\Gamma_{\rm total}(\mathbf{r}_0, \omega)}{\Gamma_0(\mathbf{r}_0, \omega)}, \tag{5.35}$$

where $\Gamma_0(\mathbf{r}_0, \omega) = \frac{2}{\hbar\epsilon_0}\mathbf{d} \cdot \text{Im}\{\mathbf{G}_0(\mathbf{r}_0, \mathbf{r}_0, \omega)\} \cdot \mathbf{d}$, and $\mathbf{G}_0$ is the Green function for a homogeneous medium. Note that the factor of 1 is included for convenience, and whose origin is from the background Green function contribution for dipoles outside the resonator [72].

For the cavity structures investigated below, the QNMs can be obtained from an efficient dipole scattering approach in complex frequency [75], which is essentially an inverse Green function approach. We implement such an approach in COMSOL. The total Green function can also be obtained numerically from the full dipole response (namely, without any modal approximation), which we also carry out in COMSOL to check the accuracy of the QNM expansions and mode approximations.

As a first resonator example, we consider a nanorod gold dimer where each nanorod is a cylinder with radius $r_{\rm r} = 10$ nm and height $h_{\rm r} = 80$ nm; see Fig. 5.3a for a top down schematic. The dimer is assumed to be in free space and the Drude model is used to model its dispersive and lossy behavior through

$$\epsilon(\omega) = 1 - \frac{\omega_p^2}{\omega(\omega + i\gamma_p)}, \tag{5.36}$$

where the plasmon frequency and the collisional decay rate are $\omega_p = 8.29$ eV[3] and $\gamma_p = 0.09$ eV, respectively. In the same figure, we show the computed spatial profile of the QNM where a significant field mode is formed between the 20 nm gap. We also plot the corresponding Purcell factor for a y-polarized dipole emitter at the center of the dimer gap. We clearly see a single QNM behavior over a wide range of frequencies centered at $\omega_c = 1.78$ eV, with a corresponding quality factor of $Q = 13$. We stress that the blue solid line employed Eq. (5.35) with no fitting parameters at all, and the red dashed line is a full numerical dipole result (with no mode approximations). The analytical QNM theory is thus quantitatively accurate [26, 41].

Next, we consider the same dimer on top of a photonic crystal nanobeam cavity, as shown in Fig. 5.3b. The nanobeam cavity, studied in [41], is modeled as silicon-nitride with a refractive index of $n = 2.04$, where the height is $h = 200$ nm, and the width is $w = 367$ nm. This design uses a mirror section as well as a taper section [41], to yield a high quality factor of $Q = 3 \times 10^5$, with resonance frequency $\omega_c = 1.62$ eV. As a consequence of the coupling between the dimer and photonic crystal cavity, a strong hybridization of the individual modes occur. The resonance frequencies (real part) of the two hybridized QNMs are $\omega_1 = 1.64$ eV and $\omega_2 = 1.61$ eV, with the corresponding quality factors of $Q_1 = 15$ and $Q_2 = 55$, respectively.

In Fig. 5.3b, we show the spatial profile of the first QNM with a lower Q, where mixing of the individual modes (MNP and photonic crystal parts) is seen. A similar feature is also observed for the other QNM with the higher Q. When both of the

[3] We use units of $\hbar = 1$ if quoting ω (or rates) in eV, to maintain consistency with the graphs shown in this section.

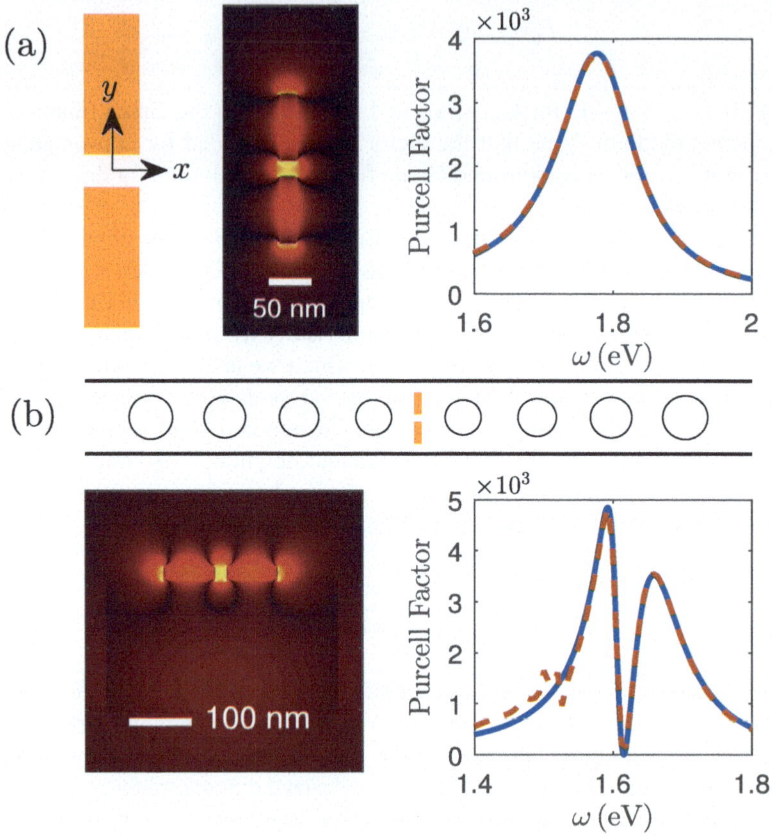

Fig. 5.3 **a** Two cylindrical gold nanorods which form a single mode plasmonic resonance. The three sub-figures show the dimer schematic, projected on xy plane, the spatial map of the dimer QNM, and the Purcell factor for a y-polarized dipole in the middle of the dimer gap. **b** The same dimer is now placed on top of a photonic crystal nanobeam cavity and 5 nm away from the surface of the beam to form a hybrid system. The three sub-figures show the top view of the hybrid device, projected to xy plane, the spatial map of one of the system QNMs, and the Purcell factor, again with a y-polarized dipole in the middle of the dimer gap. The blue-solid curve is calculated using the QNM expansion of the Green function where the red-dashed curve is the full numerical dipole calculations. **Reproduced from with permission from Ref.** [26]. **Copyright (2017) American Chemical Society**

hybrid QNMs are used, an accurate representation of the system LDOS is obtained over a wide range of frequencies as shown in the same figure. For the hybrid LDOS, we obtain a significant interference regime between the two main QNMs, which manifests in a strong and non-trivial modification of the spontaneous emission rate [41].

It is worth mentioning that the QNMs for both of these resonator systems (dimers and hybrid-dimers) have recently been quantized [76] (namely, quantized while fully accounting for the QNM losses), with a remarkably good agreement with the semiclassical results; this work also highlights regimes where the usual quantized mode theories (such as the dissipative Jaynes-Cummings models) can completely fail, e.g., for describing the Fano resonance feature of the hybrid coupling regime [76]. Features unique to a fully quantized QNM model have also been recently explored, well beyond the bad cavity limit [77].

We can now use the analytical QNM theory to connect to the the E^4 scaling rule for the SERS spectrum. From Eq. (5.22), we first consider the enhancement at the laser pump frequency through η. We consider a single QNM expansion of the Green function, Eq. (5.33), using $\tilde{\mathbf{f}}(\mathbf{r})$. Assuming an initial pump field that is polarized along $\mathbf{n}$, η becomes

$$\eta = A(\omega_L)\left[\mathbf{n}\cdot\tilde{\mathbf{f}}(\mathbf{r}_m)\right]\left\{\int\left[\epsilon(\mathbf{r},\omega_L)-\epsilon_B\right]\tilde{\mathbf{f}}(\mathbf{r})\,d\mathbf{r}\right\}, \tag{5.37}$$

where $A(\omega_L) \approx Q$ near resonance, and the spatial integration is performed over the cavity resonator. Using the modal field value at the molecule location, $\tilde{\mathbf{f}}(\mathbf{r}_m)$, in Eq. (5.22), we obtain the expected squared dependence of the SERS spectrum at the laser frequency.

Finally, we also consider the propagation effects from the term $|\mathbf{G}(\mathbf{r}_\mathrm{D}, \mathbf{r}_\mathrm{m}; \omega)\cdot\mathbf{n}|^2$, and again use a single QNM expansion to obtain

$$\mathbf{G}(\mathbf{r}_\mathrm{D}, \mathbf{r}_\mathrm{m}; \omega)\cdot\mathbf{n} = A(\omega)\,\tilde{\mathbf{f}}(\mathbf{r}_\mathrm{D})\left[\tilde{\mathbf{f}}(\mathbf{r}_m)\cdot\mathbf{n}\right], \tag{5.38}$$

where $\tilde{\mathbf{f}}(\mathbf{r}_\mathrm{D})$ QNM is the detection point, and $\tilde{\mathbf{f}}(\mathbf{r}_m)\cdot\mathbf{n}$ is the projected QNM value at the molecule location. Combining these enhancement and propagation effects together, then indeed we recover the approximate $|\tilde{\mathbf{f}}(\mathbf{r}_m)|^4$ scaling rule for the detected SERS spectrum.

Similar arguments can be made for the LDOS at the molecule location, $\mathrm{Im}\{G_{nn}(\mathbf{r}_\mathrm{m}, \mathbf{r}_\mathrm{m}; \omega)\}$, where we find that J_ph, defined in Eq. (5.17), is also proportional to $|\tilde{\mathbf{f}}(\mathbf{r}_m)|^4$. Thus, for sufficiently high pump powers, we now find that the Raman signal scales approximately with $|\tilde{\mathbf{f}}(\mathbf{r}_m)|^8$ instead.

5.2.7 *Numerical Results for the Cavity-Emitted SERS Spectrum from Single Molecules*

We next present example numerical results of our SERS theory using the QNMs and Green functions for the resonators discussed above. For the molecule, we use a vibrational mode similar to the R6G molecule at $\omega_m = 160\,\mathrm{meV}$ [78], with a dissipation rate of $\gamma_m = 1.6$ meV.

The spectrum obtained from Eq. (5.22) (along with Eqs. (5.25) and (5.26) as input) is shown in Fig. 5.4 for three different excitation conditions; in all cases, we show the emitted spectrum (using $S_0(\omega)$, left column) and the detected spectrum (using

$S(\mathbf{R}, \omega)$, when the detector is at $x = 500\,\text{nm}$ away form the dimer, right column); the results are qualitatively the same for larger propagation distances.[4] The three cases shown in Fig. 5.4 are: (a)/(d) $\omega_L = \omega_c$, where the pump laser is on-resonance with the plasmonic mode; (b)/(e) $\omega_L = \omega_c + \omega_m$, where the laser pump is detuned to the blue side of the plasmonic resonance by exactly the frequency of R6G vibration; and (c)/(f), $\omega_L = \omega_c - \omega_m$. These calculations assume room temperature conditions, where $\bar{n}^{\text{th}} = 0.002$, and the pump intensity is fixed for all cases.

As discussed in the theory section above, the spectral properties of the emitted Raman signals can be quite different to the detected spectrum. The differences partly depend on the operating frequency for the pump field and the changes of the LDOS between the laser drive frequency and the Raman sidebands. For example, by comparing Fig. 5.4a and d, when the system is pumped on resonance with the plasmonic mode, the propagation effect to the far-field has more enhancement for the anti-Stokes signal compared to the Stokes signal. This difference becomes even more pronounced when one excites the system in a red detuning configuration, where the anti-Stokes emission exploits the maximum enhancement from the MNP environment.

Next, in Fig. 5.5, we study the effects of laser detuning from the plasmonic resonance in more detail. We first show the two different theory results in Fig. 5.5a, b, where the Stokes and anti-Stokes signals are calculated as follows: (a) using the simple quantum optomechanical description shown in Sect. 5.2.5; (b) using our analytical expression for the detected spectrum given in Eq. (5.22), where the gold dimer Green function is computed using the QNM expansion of Eq. (5.33). The agreement between the two methods is clearly qualitatively good, since the response of the system happens to be very similar to a Lorentzian line shape; but there are still quantitative differences, in part from the differences to our estimation of the pump enhancement effect, which involves computing η from Eq. (5.12). To help clarify this further, we note that there is an excellent agreement between the prediction of our model and the coupled-mode quantum optomechanical model of Ref. [25], if η was estimated using the plasmonic LDOS rather than using Eq. (5.12). This response is obviously a limitation of the coupled mode formalism.

We also consider three different pump intensity regimes in Fig. 5.5c. For the R6G molecule in the gold dimer, an increasing pump field reshapes the anti-Stokes signal so that a single peak feature at a different spectral location is observed. For the same pump values, small changes to the general shape of the Stokes signal also occur. Similar effects were noted in Ref. [25], due to nonlinear interactions. We highlight again that the J_{ph} terms in the master equations are proportional to square of the pump field, in comparison to the thermal dissipation terms. We also note that the pump values used in Fig. 5.5 are rather high, with $2\epsilon_0 c\,|\mathbf{F}_0|^2 = 1.3 \times 10^5\,\text{W}/\mu\text{m}^2$, which is mainly due to the lower plasmonic enhancement achieved in the gold dimer system.

---

[4] Note that at sufficiently large distances away from the resonator, QNMs can be used in conjunction with the background propagator to obtain the renormalized QNM fields, which are non-divergent [72]; renormalized QNMs can also be obtained using near-field to far-field transforms [79].

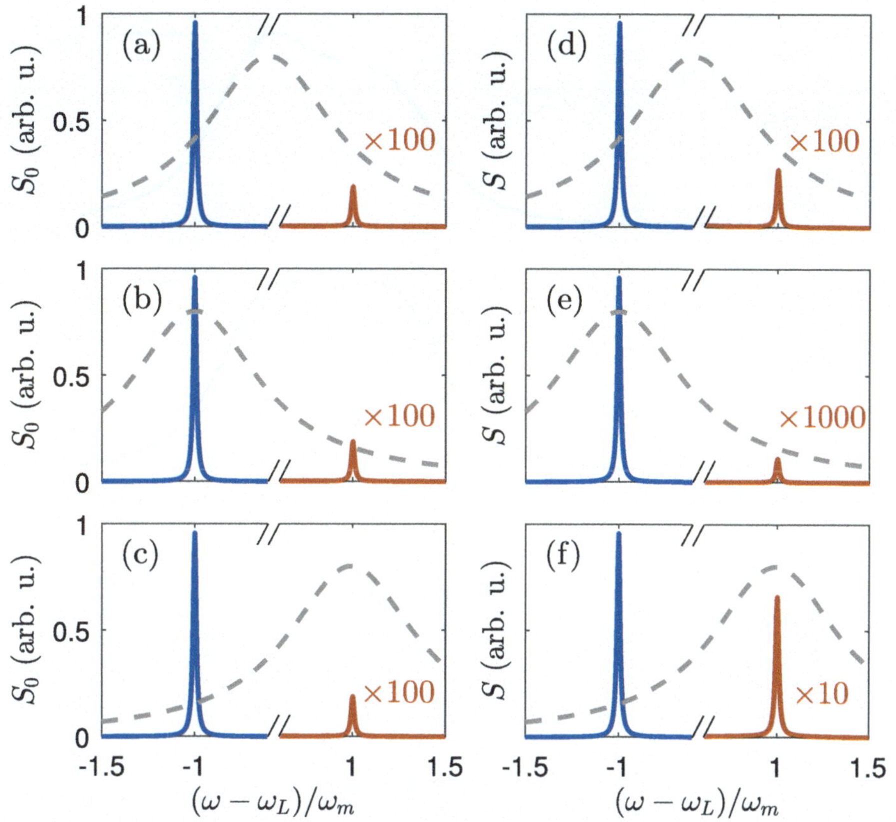

Fig. 5.4 Raman induced spectra calculated for a R6G molecule coupled to the gold dimer, using Eq. (5.22) for three different laser pump frequencies: **a/d** where $\omega_L = \omega_c$, **b/e** where $\omega_L = \omega_c + \omega_m$ and **c/f** where $\omega_L = \omega_c - \omega_m$. The plots on the left only show the $S_0(\omega)$ (emitted spectrum), where in contrast, on the right, the full far-field (detected) spectrum is plotted. The same pump power, $2\epsilon_0 c\,|\mathbf{F}_0|^2 = 130\,\mathrm{mW}/\mu\mathrm{m}^2$, and the temperature, $T = 300$ K, is assumed in all calculations. The gray-dashed line on the background of all plots indicates the LDOS profile of the plasmonic resonance, and the magnifying factors are only applied to the anti-Stokes intensities. **Reproduced from with permission from Ref.** [26]. **Copyright (2017) American Chemical Society**

Next, we consider a more complex resonator where the simple coupled mode approaches would clearly fail, and use the hybrid photonic-plasmonic structure discussed in Sect. 5.2.6. Using Eq. (5.22), and the QNM representation of the system Green function for the hybrid system through expansion of Eq. (5.33), we plot the Stokes (blue-solid) and anti-Stokes (red-dashed) detected intensities for different pump detuning in Fig. 5.6. These results are shows in two different ways: (a) the detuning with respect to the first hybrid resonance at $\omega_1 = 1.64$ eV, which is associated with the lower Q mode, and is more dimer-like (or plasmon-like); (b) detuning with respect to the second hybrid resonance at $\omega_2 = 1.61$ eV, which is associated with the higher Q mode, and is less dimer-like. For clarity, we also show the projected

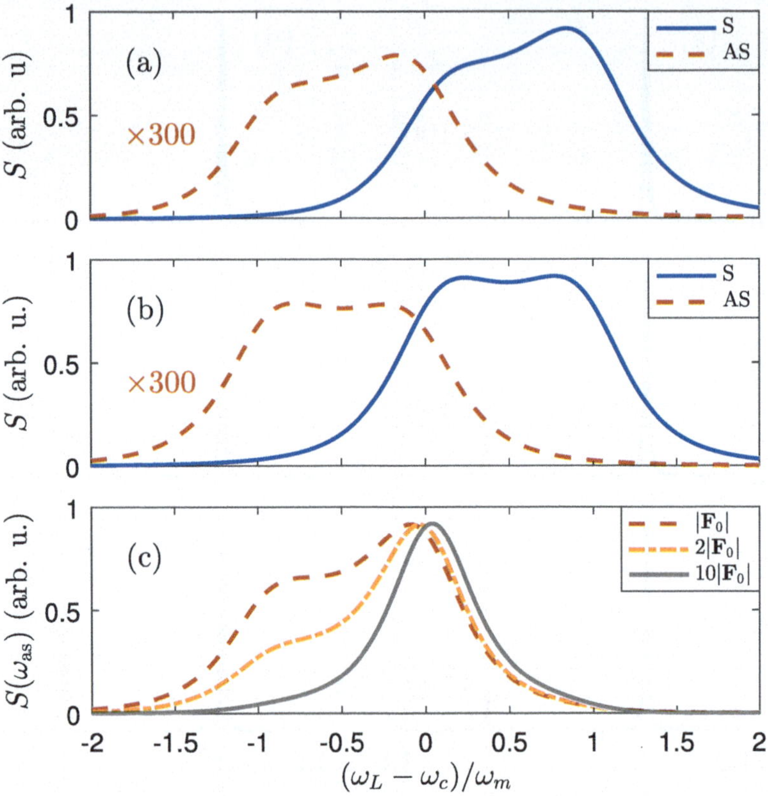

Fig. 5.5 Plot of the computed Stokes and anti-Stokes intensity peaks calculated as a function of laser detuning when: **a** The plasmonic dimer line shape is assumed to have a Lorentzian line shape and the quantum optomechanical model of Sect. 5.2.5 is used where the pump power is set to $\Omega = 1$ eV. **b** The full gold dimer response is used through inclusion of the system Green function in Eq. (5.22) and the pump power is set to $2\epsilon_0 c\,|\mathbf{F}_0|^2 = 130\,\mathrm{mW}/\mu\mathrm{m}^2$. **c** The effects of increasing the pump intensity on the anti-Stokes detectable intensities when going from red-dashed to gray-solid using the general medium master equation. **Reproduced from with permission from Ref.** [26]. **Copyright (2017) American Chemical Society**

LDOS with a gray dotted-dashed line, which is suppressed when near zero detuning because of the QNM interference effects; thus, the Raman intensities follow the same trend. We also see that there seems to be three high intensity resonances available for both Stokes and anti-Stokes spectra. The maximum detection is now obtained near one ω_m to the right (left) for the Stokes (anti-Stokes) in Fig. 5.6b, where the detuning is with respect to the ω_2. Thus, for this hybrid system, the SERS signals is mainly coupled to the QNM frequency ω_2 that is primarily due to the nanobeam cavity resonance rather than the gold dimer.

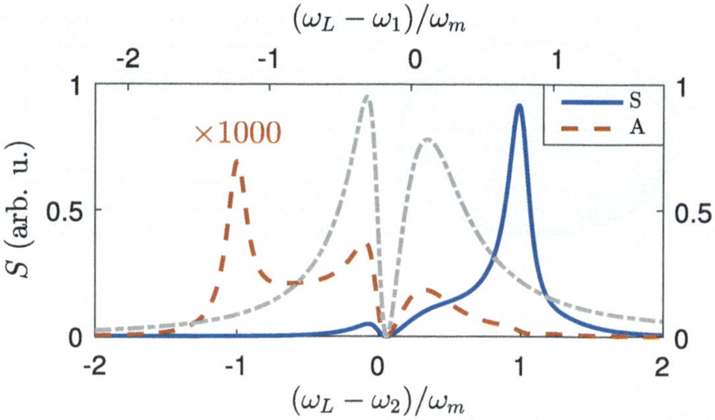

Fig. 5.6 Stokes (blue-solid curve) and anti-Stokes (red-dashed curve) detected intensities for a simulated R6G molecule when coupled to the hybrid cavity device at exactly the middle of the dimer gap and oriented along the y axis. The photonic LDOS is shown in gray in the background for comparison. The top and bottom axes measure the laser detuning with respect to ω_1 and ω_2, respectively. **Reproduced from with permission from Ref.** [26]. **Copyright (2017) American Chemical Society**

Finally, we note that the general bath model here also allows accurate and rapid calculation of both the Stokes and anti-Stokes intensities as a function of the spatial location for both for the molecule and the detector [26].

5.3 Molecular Optomechanics in the Sideband-Resolved Strong Coupling Regime Using Hybrid Metal-Dielectric Cavity Modes

Having presented a general theory of molecular SERS that can be explained with perturbative techniques in open system quantum optics, we next explore the non-perturbative regime, where the plasmonic cavity mode must now be treated self-consistency and at the system level (good cavity regime). In such a regime, we can more fully explore quantum back-action effects between a MNP cavity mode and the molecular vibrational mode. To do this, we will extend the theory into more advanced master equation techniques, and introduce a similar hybrid metal-dielectric system (schematically shown in Fig. 5.7), one that can probe nonlinear quantum Raman peaks, which are otherwise obscured by the usual MNP dissipation rates [37].

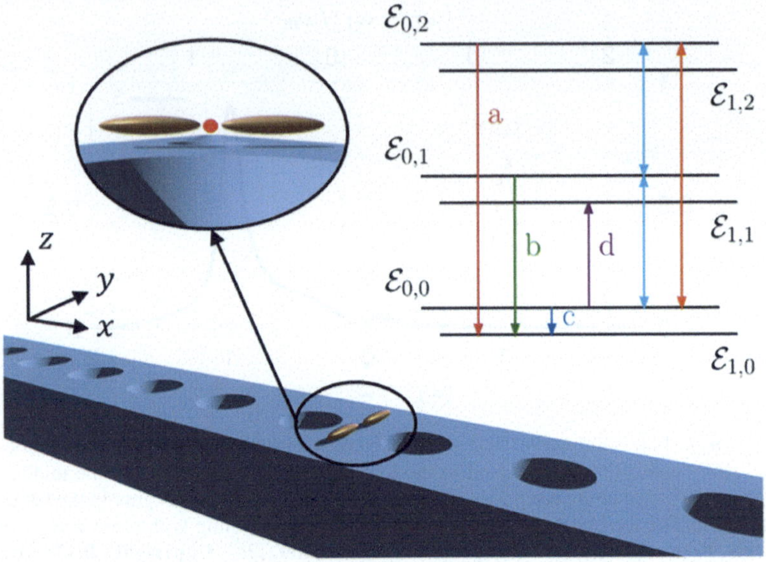

Fig. 5.7 Schematic of another hybrid metal-dielectric device with a MNP dimer (now with elliptical rods) coupled to a photonic crystal nanobeam cavity; the inset shows a closeup of the metal dimer where molecules (shown by a red filled circle) can be located. A partial energy diagram of the coupled molecule-cavity system is also shown, where red/green/blue/purple show four transition lines for the cavity emitted SERS spectrum. For example, the cyan and orange arrows show standard harmonic transitions associated with first order and second order Raman (Stokes/anti-Stokes), respectively. **Reproduced from with permission from Ref.** [37]. **Copyright (2019) American Chemical Society**

5.3.1 *Optomechanical System Hamiltonian and the Dressed States*

The optomechanical interaction, without any form of linearization, including the cavity pump term, and in the interaction picture [66], is

$$H_s = \hbar\Delta\, \hat{a}^\dagger\hat{a} + \hbar\omega_m\, \hat{b}^\dagger\hat{b} - \hbar g\, \hat{a}^\dagger\hat{a}\left(\hat{b}^\dagger + \hat{b}\right) + \hbar\Omega\left(\hat{a}^\dagger + \hat{a}\right), \tag{5.39}$$

where $\Delta = \omega_c - \omega_L$, ω_m is again the frequency of molecular vibrational mode, and Ω is the Rabi frequency of the optical cavity mode. Note that cavity operator terms $\hat{a}\hat{a}$ and $\hat{a}^\dagger\hat{a}^\dagger$ (dynamical Casimir effects) can be ignored here, as $\omega_c \gg \omega_m$ [80, 81]. The optomechanical coupling factor is defined from [24]

$$g = (\hbar R_m/2\omega_m)^{-1/2}\,\frac{\omega_c}{\epsilon_0 V_c}, \tag{5.40}$$

with R_m the Raman activity[5] that depends on the vibrational mode of interest, and V_c as the effective mode volume of the cavity mode.

The optomechanical coupling term in Eq. (5.39) is appropriate for describing off-resonant Raman interactions. For resonant interactions, as discussed later, the plasmonic MNP also interacts with electronic (two-level) vibrational degrees of freedom, yielding the Raman term [82] $\omega_m \sqrt{S_{\rm HR}}\, \hat{\sigma}^+\hat{\sigma}^-(\hat{b}+\hat{b}^\dagger)$, where $\hat{\sigma}^+, \hat{\sigma}^-$ are the Pauli operators and $S_{\rm HR}$ is the Huang-Rhys parameter [83], which accounts for the phonon displacement between the ground and excited electronic states. For certain excitation conditions, resonant Raman effects may be treated phenomenologically [9], resulting in an effective increase of the off-resonant interaction above. However, this is likely only a good approximation for weak pumping fields (and/or sufficiently broad resonances), where the Fermionic states behave as harmonic oscillator states.[6] For our studies below, we concentrate on the off-resonant Raman interactions but also use Raman cross sections that can increased using resonant Raman interactions. Resonant Raman effects are discussed in Sect. 5.4.

With optical pumping, the SERS Hamiltonian (Eq. (5.39)) can be solved analytically, and the eigenenergies take the form [84]:

$$\mathcal{E}_{n,k} = n\hbar\Delta + k\hbar\omega_m - n^2\frac{\hbar g^2}{\omega_m}, \tag{5.41}$$

with the corresponding eigenstates,

$$\left|\Psi_{n,k}\right\rangle = \hat{D}^\dagger\left(\frac{gn}{\omega_m}\right)|n,k\rangle\,, \tag{5.42}$$

where $\hat{D}$ is the displacement operator. We see that the phonon states are displaced as: $\hat{b} \to \hat{b} - d_0\hat{a}^\dagger\hat{a}$, where $d_0 = g/\omega_m$ is the normalized displacement. Figure 5.1a shows these energy states as well as the normalized displacement. Note that the polaron shift of the $n = 1$ cavity manifold (which is a red shift) is $\Delta_{\rm P} = g^2/\omega_m = d_0\omega_m$.

The optomechanical coupling thus yields photon manifolds that contain phonon sub-levels, and the sub-level splitting depends on the photon number state. To resolve the lowest-order anharmonic levels ($n = 1$ photon manifold), we require $g^2/\omega_m > \kappa$. However, one also requires $\kappa < \omega_m$, to be in the sideband resolved regime [85]. These optomechanical states now involve interactions beyond the rotating-wave approximation for the mechanical mode, where one also finds that $g/\omega_m > 0.1$, which is the regime of ultrastrong coupling (USC) [82, 86–89] for the photon-phonon interactions. Figure 5.7 show six of the optomechanical energy levels that can be resolved in the emitted spectrum, in the presence of cavity pumping. For $\Delta = 0$, the first

[5] Note that the Raman activity is related to the elements of the Raman tensor, but they have different units.

[6] A more detailed discussion between these two models, also from a polaron viewpoint, is given in the next section.

three energy levels for $n = 0$, are $\mathcal{E}_{0,0} = 0$, $\mathcal{E}_{0,1} = \hbar\omega_m$ and $\mathcal{E}_{0,2} = 2\hbar\omega_m$ for the ground state, first order and second order Raman states – these relate to the standard Stokes and anti-Stokes resonances. However, the first three energy levels for the $n = 1$ photon manifold contain the anharmonic side-bands, $\mathcal{E}_{1,0} = -\hbar g^2/\omega_m$, $\mathcal{E}_{1,1} = \hbar\omega_m - \hbar g^2/\omega_m$ and $\mathcal{E}_{1,2} = 2\hbar\omega_m - \hbar g^2/\omega_m$.

5.3.2 *Quantum Master Equations for Exploring Molecular Optomechanics in the Sideband-Resolved Regime*

We can first employ a standard master equation approach [56, 57], including cavity and mechanical bath interactions [25, 26]:

$$\frac{d\rho(t)}{dt} = -\frac{i}{\hbar}[H_{\rm S}, \rho(t)] + \frac{\kappa}{2}\mathcal{D}[\hat{a}]\rho(t) + \frac{\gamma_m\left(\bar{n}^{\rm th}+1\right)}{2}\mathcal{D}[\hat{b}]\rho(t) + \frac{\gamma_m\bar{n}^{\rm th}}{2}\mathcal{D}[\hat{b}^\dagger]\rho(t), \tag{5.43}$$

where κ is the cavity decay rate, γ_m is the vibrational decay rate, $\bar{n}^{\rm th}$ is again the thermal population of the vibrational mode, and the Lindblad superoperator $\mathcal{D}$ is defined via:

$$\mathcal{D}[\hat{O}]\rho(t) = 2\hat{O}\rho(t)\hat{O}^\dagger - \hat{O}^\dagger\hat{O}\rho(t) - \rho(t)\hat{O}^\dagger\hat{O}. \tag{5.44}$$

Note that the decay rates here are defined with a factor of 2 greater than the ones used in Eq. (5.30) (to be consistent with a full-width half-maximum of a spectral line).

A general problem with the standard master equation is that it neglects internal coupling effects between the system operators [90], since the correct bath interaction should occur at the dressed resonances of the system. This internal coupling interaction has been used to highlight various effects in quantum optics, e.g., Mollow triplets with MNP resonators interacting with two level atoms [63], circuit QED [91], and general regimes of USC physics [81]. Neglecting the influence of the weak pumping field, then $\hat{b}$ and $\hat{a}$ can be solved solved analytically (without dissipation). Neglecting terms that oscillate at $\exp(\pm i\omega_m t)$ and $\exp(\pm i2\omega_m t)$ (in the interaction picture), and treating the phonon bath spectral function as Ohmic, the modified master equation is [88]

$$\begin{aligned}\frac{d\rho(t)}{dt} &= -\frac{i}{\hbar}[H_{\rm s}, \rho(t)] + \frac{\kappa}{2}\mathcal{D}[a]\rho(t) + \frac{\gamma_m\left(\bar{n}^{\rm th}+1\right)}{2}\mathcal{D}[\hat{b} - d_0\hat{a}^\dagger\hat{a}]\rho(t)\\ &+ \frac{\gamma_m\bar{n}^{\rm th}}{2}\mathcal{D}[\hat{b}^\dagger - d_0\hat{a}^\dagger\hat{a}]\rho(t) + \frac{2\gamma_m k_{\rm B}{\rm T}}{\hbar\omega_m}d_0^2\,\mathcal{D}[\hat{a}^\dagger\hat{a}]\rho(t).\end{aligned} \tag{5.45}$$

While there is no modification on the final cavity decay terms,[7] the mechanical operator terms are displaced in the dissipators. There is also a bath-induced pure dephasing process that is found to be negligible in the regimes studied below, though it becomes relevant at sufficiently high temperatures and for larger d_0.

Below we will show solutions first with the standard master equation of Eq. (5.43), and then also carry out a comparison with the modified master equation, Eq. (5.45). For numerical calculations, we again use the QuTiP package [92, 93] under Python.

5.3.3 *Plasmonic Quasinormal Modes with High Quality Factors and Large Optomechanical Coupling Rates*

We next discuss a hybrid cavity system to explore the sideband-resolved regime. Similar to the previous section, we consider a MNP dimer that is top-coupled to a photonic crystal nanobeam cavity again (see Fig. 5.7). However, here we improve the design to yield a stronger light-matter coupling, which requires a larger Q and smaller V_c); this is achieved by adjusting the gap size, the aspect ratio, and the shape of the dimer as well as its spacing from the nanobeam. We also deliberately keep the photonic crystal-like mode *off resonance* from the main plasmon mode, so it can be approximately treated as a single QNM and maintain a suitably large Q factor. Similar metal-dielectric hybrid structures are discussed in Ref. [94].

We consider a gold dimer that is now made of two ellipsoids, each 60 nm long and 15 nm wide. A small gap, 0.5–5 nm in between them is used to create a pronounced field for coupling to the molecule. These gap sizes are within the range where electron tunneling effects are negligible [95]. Also, we note that using a fully three-dimensional nonlocal QNM theory at the level of a hydrodynamical model [96], we have verified that nonlocal effects simply blueshift the low Q mode by a few percent. Therefore, the MNP is modeled again using a local Drude theory.

The classical QNM calculations are performed as before, where the mode parameters are used to estimate the effective mode volumes at the emitter location as well as the quality factors. As described earlier, the QNM complex eigenfrequencies are defined from [51] $\tilde{\omega}_c = \omega_c - i\gamma_c$, where $\kappa = 2\gamma_c$, $Q = \omega_c/\kappa$ and the effective mode volume is obtained from the normalized QNM spatial profile at dimer gap center [51], $V_c = \mathrm{Re}\{1/\epsilon_\mathrm{B}\tilde{\mathbf{f}}^2(\mathbf{r}_0)\}$, where $\tilde{\mathbf{f}}^2(\mathbf{r}_0)$ is the normalized QNM.

To gain insight into the role of the dimer gap, we model both metal dimer structures, and dimers on top of photonic crystal cavity beams (see Fig. 5.7). Further details of the QNM calculations are discussed in Ref. [37]. The nanobeam cavity has a refractive index of $n = 2.04$ (as before), with height $h = 200$ nm, and width $w = 367$ nm. A dielectric cavity region that is 126 nm long is created in the middle of the nanobeam, with a taper section and then a mirror section on either side. The

[7] However, there is an effect if one also treats the cavity dissipation in the form of an Ohmic bath function, as shown in the next section. For this current derivation, the cavity bath was assumed to be spectrally flat over the frequencies of interest [88].

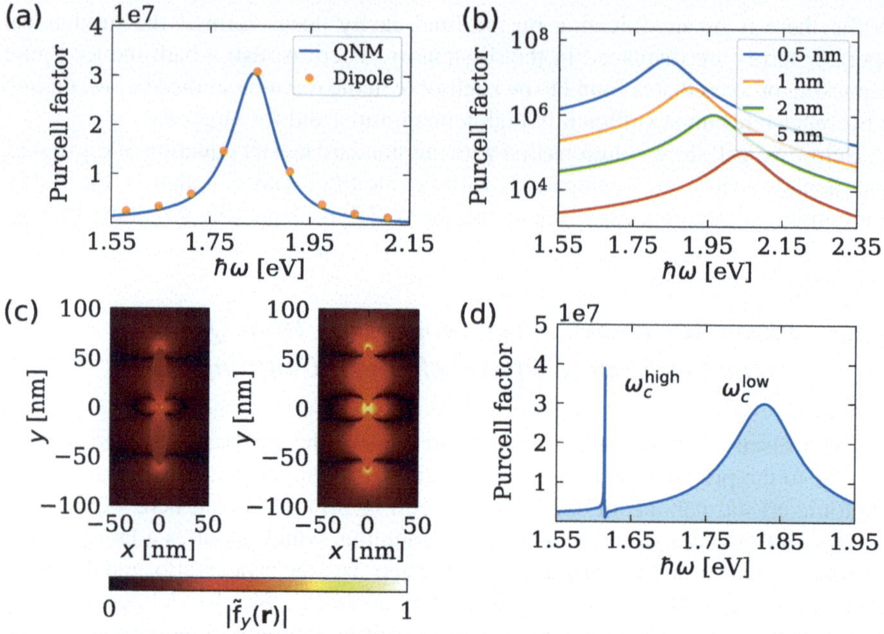

Fig. 5.8 **a** Purcell factor calculations for the smaller gaps size of 0.5 nm using the analytical QNM theory (solid curve) and fully vectorial solutions to Maxwell's equations from a dipole excitation (symbols). **b** QNM calculated Purcell factors for dimer designs with different gap sizes. **c** Calculated mode profile of two gold dimers with different gaps, 0.5 nm (left) and 5 nm (right). **d** Purcell factor of a dipole emitter placed inside the plasmonic gap of the hybrid device, where a sharp high Q mode (dielectric-like) is present next to the broader low Q mode (plasmon-like). **Reproduced (with modification) from with permission from Ref.** [37]. **Copyright (2019) American Chemical Society**

taper section is made of 7 holes linearly increasing from 68 nm to 86 nm in radius, and from 264 nm to 299 nm in spacing. The mirror section consists of 7 more holes having the same radius of $r = 86$ nm and the same spacing of $a = 306$ nm. For the main mode of interest for this cavity design, we obtained the effective mode volume at the beam center inside dielectric region to be $V_c = 0.078\,\lambda^3$, with the corresponding quality factor of $Q = 3 \times 10^5$.

Figure 5.8 shows how the MNP dimer response changes as a function of gap size (a-c), and also displays the hybrid modes for the smallest gap size on a photonic crystal cavity beam cavity (d). Figure 5.8c shows the near-field mode profile for gap sizes of 0.5 nm and 5 nm, respectively. In Fig. 5.8a, we confirm the accuracy of the QNM theory for accurately capturing the system Purcell factor; even for this very small gap sizes of 0.5 nm, a single QNM evidently gives an excellent fit compared to full numerical dipole simulations. Figure 5.8b shows how the QNM computed Purcell factors [51] change for several dimer designs with different gap sizes, ranging from

0.5 nm to 5 nm. The smaller dimer gaps cause a rapid decrease of the effective mode volume and a spectral redshift.

These hybrid MNP-dielectric modes can trap molecules that couple to extremely localized fields, while maintaining a sufficiently large Q factor in the presence of dielectric mode coupling. Figure 5.8d shows the total Purcell factors for a dipole emitter that is (again) oriented along the dimer axis when on top of the photonic crystal cavity. Over the wide spectral range of 400 meV, only two modes contribute dominantly, which are the two hybrid cavity modes [41]. The resonant frequencies of the hybrid modes are $\hbar\omega_c^{\rm HQ} = 1.61$ eV and $\hbar\omega_c^{\rm LQ} = 1.83$ eV, with corresponding quality factors of $Q^{\rm HQ} = 3500$ ($\hbar\kappa = 23.3$ meV) and $Q^{\rm LQ} = 17$ ($\hbar\kappa = 108$ meV), respectively. The effective mode volumes for the two hybrid modes are $V_c^{\rm HQ} = 5.36 \times 10^{-6}\,\lambda_c^3$ and $V_c^{\rm LQ} = 4.54 \times 10^{-8}\,\lambda_c^3$. Note that, as discussed before, the high Q mode has an extremely small mode volume inherited from the plasmonic dimer structure.

Turning now to the molecular vibrational mode, we consider a low frequency oscillation with resonance energy $\hbar\omega_m = 10$ meV, with a Raman activity of $R_m = 10^3$ Å^{4}amu^{-1}; these values are within the range of values reported in the literature, e.g., for single-walled carbon nanotubes [24, 97]. We also consider a mechanical quality factor of $Q_m = \omega_m/\gamma_m = 100$ for the molecular vibrational mode, though note the temperature dependence will increase the vibrational linewidth, through $\bar{n}^{\rm th}$, as shown in Eq. (5.51). For this vibrational mode, the two hybrid optical modes yield $\hbar g^{\rm HQ} = 0.1$ meV and $\hbar g^{\rm LQ} = 20$ meV. As discussed earlier, even though the low Q mode offers a much larger coupling factor, it fails to exploit the strong coupling interaction for sideband resolution, since one requires $\kappa < \omega_m$. We note that these estimates assume standard off-resonant SERS, and with resonant enhancement effects, the effective coupling factor can likely be enhanced by more than $\times 1000$ when resonant Raman regimes are used, and so we also consider several values of the coupling factor within the range: $\hbar g^{\rm HQ} = 0.1 - 4$ meV [37].

5.3.4 Cavity Emitted Spectrum and Population Dynamics in the Sideband-Resolved Regime

Employing the good- cavity master equations, Eqs. (5.43) and (5.45), and the quantum regression theorem [56], we calculate the cavity-emitted (incoherent) spectrum from:

$$S(\omega) \equiv \text{Re}\left\{\int_0^\infty dt\, e^{i(\omega-\omega_L)t}\ \left[\langle \hat{a}^\dagger(t)\,\hat{a}(0)\rangle_{\rm ss} - \langle a^\dagger\rangle_{\rm ss}\langle \hat{a}\rangle_{\rm ss}\right]\right\}, \tag{5.46}$$

where the expectation values are computed in *steady state*, and the latter contribution subtracts off the coherent contribution. The population dynamics of the cavity mode population is obtained from $n_c(t) = \langle \hat{a}^\dagger\hat{a}\rangle(t)$.

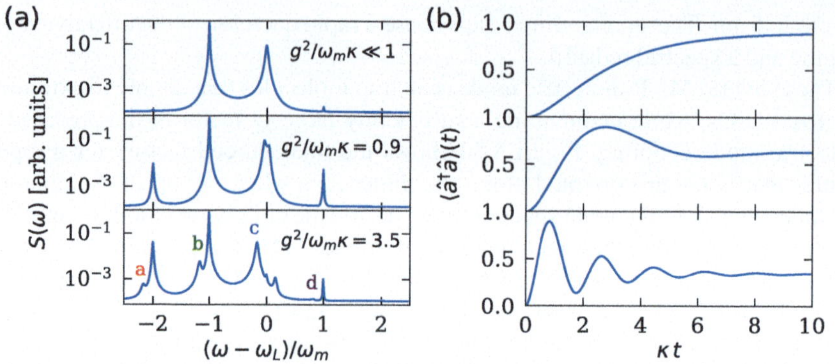

Fig. 5.9 **a** Cavity emitted spectrum and **b** cavity photon population versus time, for $\hbar g = 0.1, 2, 4$ meV, top to bottom, respectively. The temperature is $T = 4$ K, and for the larger values of g, we have $d_0 = g/\omega_m = 0.4$. which is also in the (vibrational) USC regime. The first (lowest) g value is for the hybrid high Q design shown earlier, with the estimated non-resonant Raman configuration. Here, the cavity decay rate is $\hbar\kappa = 0.46$ meV, the frequency of vibration is $\hbar\omega_m = 10$ meV, and the Rabi energy is $\hbar\Omega = 0.1$ meV. Note that $\omega_m \gg \kappa$, which is required for sideband resolution. **Reproduced (with modification) from with permission from Ref.** [37]. **Copyright (2019) American Chemical Society**

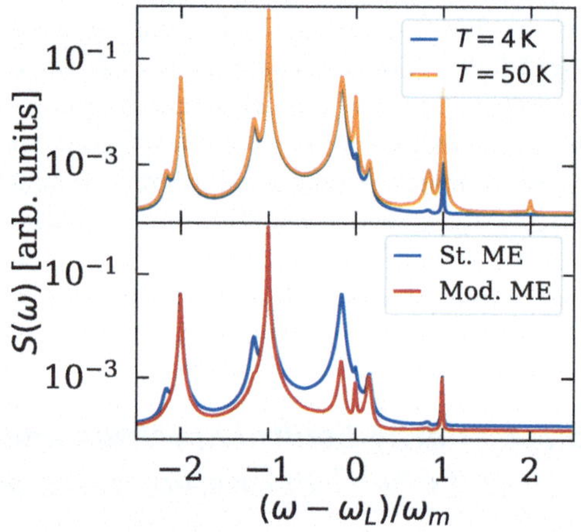

Fig. 5.10 Cavity emitted spectrum for different cavity-laser detunings $\Delta = -\omega_m, \omega_m, g^2/\omega_m$ from top to bottom, respectively. **Reproduced (with modification) from with permission from Ref.** [37]

In Figs. 5.9a, b, we show the cavity-emitted spectrum and corresponding cavity mode population dynamics, for the three optomechanical coupling rates: $\hbar g^{\mathrm{HQ}} = 0.1, 2, 4$ meV. The simulations in Figs. 5.9a, b use the standard master equation (Eq. (5.51)), with a bath temperature of $T = 4$ K. By increasing g, we see significant spectral shifts of the cavity mode resonance and the emergence of the Raman side-peaks (Fig. 5.9a). In Fig. 5.9b, the corresponding cavity mode populations also become non-trivial as one enter a strong coupling regime. Referring to the schematic energy diagram of Fig. 5.7, the usual Raman emissions (first-order and second-order) correspond to single phonon step and two phonon step jumps on the molecule (har-

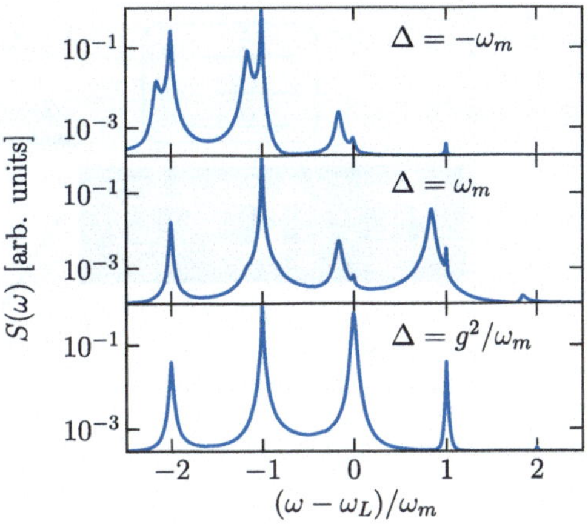

Fig. 5.11 Influence of temperature (top, 4 K versus 50 K) and the modified Lindbald dissipation terms (bottom, Eq. (5.43) versus Eq. (5.45)) on the strong coupling spectrum for $\hbar g = 4$ meV case at $T = 4$ K. **Reproduced (with modification) from with permission from Ref.** [37]. **Copyright (2019) American Chemical Society**

monic) ladder. The spectral sidebands, however, are mediated from the anharmonic energy levels when there is a sufficiently strong optomechanical coupling.

We also investigate the role of increasing temperature on the emission spectrum, as shown in Fig. 5.11 (top, for $T = 4,\ 50$ K). The spectrum shows that increasing the temperature primarily affects the anti-Stokes emissions. Even at room temperature, the thermal phonon populations for the vibration frequency of the ω_m=10 meV is about $n^{\text{th}} \approx 2$, causing a further increase of the anti-Stokes emissions as well as extra broadening. In Fig. 5.11 (lower), we also show the effect of the modified dissipation terms given in Eq. (5.45), where the modified Lindblad terms of the modified (improved) master equation introduce additional dissipation terms that obscures some of the side-peaks, consistent with the results in Ref. [88].

In Fig. 5.10, we also show the effect of the laser-cavity detuning, Δ, on the emitted spectrum, using three different detunings: $\Delta = -\omega_m,\ \omega_m,\ g^2/\omega_m$. The first two detuning values effectively lead to a different enhancement depending on how far the excitation laser is from the cavity resonance, either on the Stokes or anti-Stokes peaks. However, the third case tunes into the exact polaron-shifted resonance condition, where the anharmonic transitions are no longer visible.

5.4 Resonant Raman Scattering in the Strong Coupling Regime of Cavity-QED

In this final Raman configuration, we discuss some recent ideas with resonant SERS, again for single molecules, when also strongly coupled to a MNP cavity mode in the cavity-QED regime [55], which is shown schematically in Fig. 5.12. For resonant cavity-QED schemes, where MNP modes are resonantly coupled to the molecular electronic levels, some signatures of vacuum Rabi splitting has been experimentally observed [30]. To understand such systems theoretically, one must couple the physics

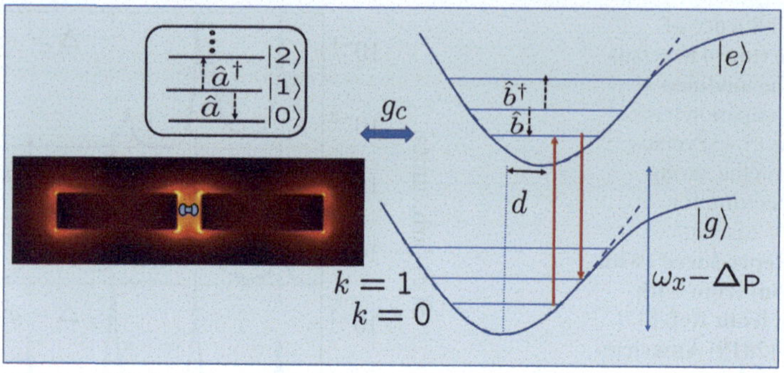

Fig. 5.12 Simple schematic of a plasmonic field hot-spot coupled to a vibrating molecule and the cavity mode ladder states in the regime of resonant SERS. At the right, we show the two electronic manifolds, each containing a subset of phonon levels ($k = 1, 2, \ldots$) separated by ω_m, and coupled to the cavity mode through g_c; the excited manifold is shifted down in frequency from the bare exciton resonance (ω_x) through the polaron shift, $\Delta_P = d^2\omega_m$, and shifted in space by the normalized displacement d. This graph is similar to Fig. 5.1, but now we also have coupling to electronic levels for the resonant TLS, described by Pauli operators, $\hat{\sigma}^+$ and $\hat{\sigma}^-$ (see text)

of resonant SERS and traditional strong coupling physics, where the cavity mode cannot be treated as a bath, and the effects of system-bath coupling can be much more subtle.

Several theoretical approaches have been developed to explore resonant Raman interactions, with various approximations. Reference [99] explored a bad cavity limit to study the interplay between Mollow triplet physics and vibrational coupling, and found that the Mollow resonances can be split under certain conditions; their theory was based on the standard master equation and they also included pure dephasing processes phenomenologically for the TLS at elevated temperatures. Recently, Ref. [38] studied USC effects in molecular cavity-QED using the quantum Rabi model as well as interactions between phonons and excitations as the system level. They solved model Hamiltonians and dynamical coupling in a one photon subspace, though without accounting for dissipation. They also showed how the time-averaged occupations depends on the strength of the optomechanical coupling rates. Recently, Ref. [100] studied resonant SERS in the good cavity limit, presenting useful analytical solutions and numerical solutions from a standard master equation approach. Below we will present a generalized master equation to describe resonant SERS [9], which rigorously includes spectral baths for the cavity and vibrational degrees of freedom, as well as a pure dephasing bath for the resonant two-level system. We also demonstrate the clear failure of using a standard master equation in a regime that involves strong cavity coupling and ultrastrong phonon-exciton coupling.

5.4.1 *Polaronic Picture and Connection Between Off-Resonant SERS and Resonant SERS*

To make a closer connection between off-resonant SERS and resonant SERS, it is useful to consider the system Hamiltonians in a polaronic picture.

For the off-resonant SERS Hamiltonian, the polaron transform yields a $\hat{a}^\dagger\hat{a}$-dependent displacement of the mechanical resonator: $\hat{S} = d_0 a^\dagger \hat{a}(\hat{b}^\dagger - \hat{b})$. The polaron transformed system Hamiltonian. for off-resonant SERS, $H_s' = e^{\hat{S}} H_s e^{-\hat{S}}$. is then

$$H_s' = \hbar(\omega_c - \Delta_P)\hat{a}^\dagger\hat{a} + \hbar\omega_m \hat{b}^\dagger\hat{b} - \hbar\Delta_P\hat{a}^\dagger\hat{a}^\dagger\hat{a}\hat{a}, \tag{5.47}$$

where $\hat{a} \to \hat{a}\hat{X}$, and the displacement operator is $\hat{X} = \exp[d_0(\hat{b} - \hat{b}^\dagger)]$. In the polaron frame, the effective cavity resonance is shifted down by $\Delta_P = g_{om}^2/\omega_m = d_0^2\omega_m$ (see Fig. 5.1), as also shown from the eigenvalues of the optomechanical coupling problem. Note, the Kerr-like term (4 operators) causes a nonlinear (photon-photon) dependence on photon number. However, if term can be neglected, then one can write

$$H_s' \approx \hbar(\omega_c - \Delta_P)\hat{a}^\dagger\hat{a} + \hbar\omega_m \hat{b}^\dagger\hat{b}, \tag{5.48}$$

which clearly has a very simple form and interpretation.

For *resonant* Raman interactions (see Fig. 5.12), the system Hamiltonian is

$$H_s = \hbar\omega_c\hat{a}^\dagger\hat{a} + \hbar\omega_m\hat{b}\hat{b}^\dagger + \hbar\omega_x\hat{\sigma}^+\hat{\sigma}^- + \hbar d\omega_m\hat{\sigma}^+\hat{\sigma}^-\left(\hat{b}^\dagger + \hat{b}\right) + \hbar g_c(\hat{\sigma}^+\hat{a} + \hat{a}^\dagger\hat{\sigma}^-), \tag{5.49}$$

where the latter term is the Jaynes-Cummings term. Here we assume $g_c \ll \omega_c$, so can neglect USC effects related to the TLS-cavity coupling [38]; the TLS has a resonance exciton energy ω_x, and g_c is the exciton-cavity coupling rate (we use the 'c' subscript to avoid confusion with the g used in the off-resonant SERS scheme). The polaron transformed system Hamiltonian, for resonant SERS, now using $\hat{S}{=}d\hat{\sigma}^+\hat{\sigma}^-(\hat{b}^\dagger - \hat{b})$ and $\hat{X} = \exp(d(\hat{b} - \hat{b}^\dagger))$, is

$$H_s' = \hbar\omega_c\hat{a}^\dagger\hat{a} + \hbar\omega_m\hat{b}\hat{b}^\dagger + \hbar(\omega_x - \Delta_P)\hat{\sigma}^+\hat{\sigma}^- + \hbar g_c(\hat{\sigma}^+\hat{a}\hat{X} + \hat{a}^\dagger\hat{X}^\dagger\hat{\sigma}^-). \tag{5.50}$$

The equivalence between the off-resonant and resonant SERS can then be made, by invoking a harmonic oscillator approximation for the TLS in a bad cavity limit. Thus, in such a regime, the resonant SERS polaron transformed Hamiltonian then becomes identical in form to the off-resonance case, apart from the Kerr-like term (which vanishes for Fermions). Therefore, a one simply replaces $\hat{a}$, $\hat{a}^\dagger$ with $\hat{\sigma}^-$, $\hat{\sigma}^+$, and identifies $d_0 = d$. Clearly, when Fermionic behavior becomes important in the two state system, e.g., Mollow physics and strong coupling between the cavity mode and TLS, then one must use the Pauli operators to describe the resonant Raman.

5.4.2 Standard and Generalized Master Equations for Resonant SERS

Including cavity and mechanical bath interactions, the standard master equation for resonant SERS is

$$\frac{d\rho(t)}{dt} = -\frac{i}{\hbar}\left[H_s + H_{\text{pump}}, \rho(t)\right] + \frac{\kappa}{2}\mathcal{D}[\hat{a}]\rho(t) + \frac{\gamma_\phi}{2}\mathcal{D}[\hat{\sigma}^+\hat{\sigma}^-]\rho(t)$$

$$+ \frac{\gamma_m\left(\bar{n}^{\text{th}}+1\right)}{2}\mathcal{D}[\hat{b}]\rho(t) + \frac{\gamma_m \bar{n}^{\text{th}}}{2}\mathcal{D}[\hat{b}^\dagger]\rho(t), \quad (5.51)$$

where $\gamma_\phi = \Gamma_\phi(0)$ (see below) is the pure dephasing rate of the TLS, H_s is the system Hamiltonian in the interaction picture in a rotating frame at the laser frequency,

$$H_s = \hbar(\omega_c - \omega_L)\hat{a}^\dagger\hat{a} + \hbar\omega_m\hat{b}\hat{b}^\dagger + \hbar(\omega_x - \omega_L)\hat{\sigma}^+\hat{\sigma}^- + \hbar d\omega_m\hat{\sigma}^+\hat{\sigma}^-\left(\hat{b}^\dagger + \hat{b}\right)$$

$$+ \hbar g_c(\hat{\sigma}^+\hat{a} + \hat{a}^\dagger\hat{\sigma}^-), \quad (5.52)$$

and H_{pump} is the CW pump term for the cavity mode:

$$H_{\text{pump}} = \Omega(\hat{a} + \hat{a}^\dagger). \quad (5.53)$$

As highlighted earlier, the standard master equation will generally fail in such regimes, since it neglects internal coupling between the system operators when deriving the dissipators [63, 81, 90, 91], and these interactions can be significant for MNP-molecular systems. We also have the additional problem here of including the exciton-cavity coupling, and one has to introduce a more general pure dephasing bath function, which is known experimentally [101] and theoretically [102] to introduce pronounced spectral asymmetries. Thus for a suitably large g_c and d, one needs a more general master equation, where the dissipators are derived with respect to the dressed states of the hybrid system. However, in contrast to our earlier SERS theory with a generalized bath function, we are now including the MNP cavity mode at the system level, as well as a resonant exciton, and we need to include an appropriate bath function for pure dephasing as well.

To address this problem rigorously, we exploit a generalized master equation approach [81], which first computes the eigenstates and eigenenergies of the system Hamiltonian, and account for the fact that the bath interactions occur as the dressed resonances of the system, The generalized master equation for the resonant Raman scheme can be written as [9]

$$\frac{\mathrm{d}}{\mathrm{dt}}\rho = -\frac{i}{\hbar}[H_\mathrm{s} + H_\mathrm{pump}, \rho] + \mathcal{L}_\mathrm{G}\rho + \mathcal{L}^\phi_\mathrm{G}\rho, \tag{5.54}$$

where the cavity and phonon baths terms are included through the term

$$\begin{aligned}\mathcal{L}_\mathrm{G}\rho = \frac{1}{2}\sum_{\alpha=c,m}\sum_{\omega,\omega'>0}\\ &\Gamma_\alpha(\omega)(1+\bar{n}^\mathrm{th}_\alpha(\omega))[\hat{x}^+(\omega)\rho\hat{x}^-(\omega') - \hat{x}^-(\omega')\hat{x}^+(\omega)\rho]\\ &+\Gamma_\alpha(\omega')(1+\bar{n}^\mathrm{th}_\alpha(\omega'))[\hat{x}^+(\omega)\rho\hat{x}^-(\omega') - \rho\hat{x}^-(\omega')\hat{x}^+(\omega)]\\ &+\Gamma_\alpha(\omega)\bar{n}^\mathrm{th}_\alpha(\omega)[\hat{x}^-(\omega')\rho\hat{x}^+(\omega) - \rho\hat{x}^+(\omega)x^-(\omega')]\\ &+\Gamma_\alpha(\omega')\bar{n}^\mathrm{th}_\alpha(\omega')[\hat{x}^-(\omega')\rho\hat{x}^+(\omega) - \hat{x}^+(\omega)\hat{x}^-(\omega')\rho]\\ &+\Gamma'_\alpha(T)[2\hat{x}^0_\alpha\rho\hat{x}^0_\alpha - \hat{x}^0_\alpha\hat{x}^0_\alpha\rho - \rho\hat{x}^0_\alpha\hat{x}^0_\alpha],\end{aligned} \tag{5.55}$$

and we neglect counter-rotating wave terms. It is also important to note that we do not make any secular approximations.

The dressed-state operators, solved in a basis of energy eigenstates with respect to H_s, are defined through

$$\hat{x}^+_c(\omega) = \sum_{j,k>j}(\langle j|\hat{a}+\hat{a}^\dagger|k\rangle)\,|j\rangle\,\langle k|\,, \tag{5.56}$$

$$\hat{x}^+_m(\omega) = \sum_{j,k>j}\left(\left\langle j|\hat{b}+\hat{b}^\dagger|k\right\rangle\right)|j\rangle\,\langle k|\,, \tag{5.57}$$

$$\hat{x}^+_c = \sum_j(\langle j|\hat{a}+\hat{a}^\dagger|j\rangle)\,|j\rangle\,\langle j|\,,\quad \hat{x}^+_m = \sum_j\left(\left\langle j|\hat{b}+\hat{b}^\dagger|j\right\rangle\right)|j\rangle\,\langle j|\,, \tag{5.58}$$

where $\omega = \omega_k - \omega_j > 0$ and $\hat{x}^- = (\hat{x}^+)^\dagger$.

For both the cavity and phonon baths, we have assumed Ohmic bath functions, $J_\alpha(\omega) = \Gamma_\alpha\omega/2\pi\omega_\alpha$, and the decay rates are then defined from [81]

$$\Gamma_\alpha(\omega) = \frac{\gamma_\alpha\omega}{\omega_\alpha}, \tag{5.59}$$

$$\Gamma'_\alpha(T) = \frac{\gamma_\alpha T}{\omega_\alpha}, \tag{5.60}$$

where the latter is a bath-induced pure dephasing term. At low temperatures, and using MNP cavity parameters, these are typically negligible, though at room temperature the vibrational pure dephasing becomes important. Similar terms were noted before for the off-resonant SERS scheme, but now we also include a more general model for the cavity bath.

In addition to the above phonon and cavity bath interaction terms, the pure dephasing term associated with the TLS, e.g., through electron-phonon interactions, becomes

$$
\begin{aligned}
\mathcal{L}_{\mathrm{G}}^{\phi}\rho = \frac{1}{2}\sum_{\omega,\omega'>0} & \Gamma_{\phi}^{\downarrow}(\omega)[\hat{x}_{\mathrm{a}}^{+}(\omega)\rho\hat{x}_{\mathrm{a}}^{-}(\omega') - \hat{x}_{\mathrm{a}}^{-}(\omega')\hat{x}_{\mathrm{a}}^{+}(\omega)\rho] \\
& + \Gamma_{\phi}^{\downarrow}(\omega')[\hat{x}_{\mathrm{a}}^{+}(\omega)\rho\hat{x}_{\mathrm{a}}^{-}(\omega') - \hat{x}_{\mathrm{a}}^{-}(\omega')\hat{x}_{\mathrm{a}}^{+}(\omega)\rho] \\
& + \Gamma_{\phi}^{\uparrow}(\omega)[\hat{x}_{\mathrm{a}}^{-}(\omega)\rho\hat{x}_{\mathrm{a}}^{+}(\omega') - \hat{x}_{\mathrm{a}}^{+}(\omega')\hat{x}_{\mathrm{a}}^{-}(\omega)\rho] \\
& + \Gamma_{\phi}^{\uparrow}(\omega')[\hat{x}_{\mathrm{a}}^{-}(\omega)\rho\hat{x}_{\mathrm{a}}^{+}(\omega') - \hat{x}_{\mathrm{a}}^{+}(\omega')\hat{x}_{\mathrm{a}}^{-}(\omega)\rho] \\
& + \Gamma_{\phi}(0)[2\hat{x}_{\mathrm{a}}^{+}\rho\hat{x}_{\mathrm{a}}^{-} - \hat{x}_{\mathrm{a}}^{-}\hat{x}_{\mathrm{a}}^{+}\rho - \rho\hat{x}_{\mathrm{a}}^{+}\hat{x}_{\mathrm{a}}^{-}], \qquad (5.61)
\end{aligned}
$$

where for clarity, we define the TLS pure dephasing rates implicitly including $\bar{n}^{\mathrm{th}}(\omega)$ (upwards transition) and $(1+\bar{n}^{\mathrm{th}}(\omega))$ (downward transition) to better explain the physics of the asymmetric baths coupling:

$$
\begin{aligned}
\Gamma^{\uparrow}(\omega) &= 2\pi J_{\phi}(\omega)[1+\bar{n}^{\mathrm{th}}(\omega)], \quad \omega \geq 0, \\
\Gamma^{\downarrow}(\omega) &= 2\pi J_{\phi}(-\omega)\bar{n}^{\mathrm{th}}(-\omega), \quad \omega < 0. \qquad (5.62)
\end{aligned}
$$

The TLS dressed state operators for molecular pure dephasing are defined through:

$$
\hat{x}_{\mathrm{a}}^{+}(\omega) = \sum_{j,k>j} (\langle j|\hat{\sigma}^{+}\hat{\sigma}^{-}|k\rangle)\,|j\rangle\,\langle k|, \qquad (5.63)
$$

$$
\hat{x}_{\mathrm{a}}^{+} = \sum_{j} (\langle j|\hat{\sigma}^{+}\hat{\sigma}^{-}|j\rangle)\,|j\rangle\,\langle j|, \qquad (5.64)
$$

Note that in the standard master equation, only the final term in Eq. (5.61) would be used for TLS pure dephasing. We use the 'a' subscript here to represent a pure dephasing bath the resonant atom or TLS.

For the pure dephasing spectral function, we use an Ohmic form with a cut-off frequency [82],

$$
J_{\phi}(\omega) = \eta_{\phi}\omega e^{-\left(\frac{\omega}{\omega_{\mathrm{cut}}}\right)^{2}}, \qquad (5.65)
$$

where $\omega_{\mathrm{cut}} = 160$ meV is the cut-off frequency, and η_{ϕ} is the coupling strength which we define from having $\gamma_{\phi} = \Gamma_{\phi}(0) = 10$ meV at room temperature (and we assume this value scales linearly with temperature).

In the time domain, it is also useful to define the bath correlation function:

$$
\phi(\tau) = \int_{0}^{\infty} d\omega J_{\phi}(\omega)[(\bar{n}^{\mathrm{th}}(\omega)+1)e^{-i\omega\tau} + \bar{n}^{\mathrm{th}}(\omega)e^{i\omega\tau}], \qquad (5.66)
$$

which in the high temperature limit ($\hbar\omega_{\rm cut} \gg k_B T$), behaves as $\phi(\tau) \propto \exp(-\omega_{\rm cut}^2 \tau^2/4)$. Thus, larger cut off frequencies cause ultrafast interactions with the bath, which can have a significant influence on the emitted spectra.

Equations (5.55)–(5.61) describe a completely generalized dissipator which correctly considers dissipation from the hybrid light-matter system to the coupling baths [81]. Finally, in the interaction picture at the laser frequency ω_L, the pump term also has to be included in the dressed-state basis, so that

$$H_{\rm pump} = \Omega(\hat{x}_c^+ + \hat{x}_c^-), \tag{5.67}$$

which is included after diagonalizing the density matrix from the solution from the resonant SERS $H_{\rm s}$ (Eq. (5.49)). Numerically, we obtain an appropriately large number of energy states from a basic of n photons, m photons and the TLS, and truncate to the lowest N levels in the dressed-state basis, and check that this truncation is numerically conserved for each problem studied below. This results in significant memory savings when the interaction strengths are large, as the correct energy level are ensured even from a much smaller truncated basis.

5.4.3 Numerical Results for Resonant SERS in the Strong Coupling Regime

To exemplify the power of the theory, we next study the effects of SERS in the strong coupling regime of cavity-QED. For the MNP cavity mode, we use $\hbar\kappa = 100$ meV, and $\hbar\omega_{\rm c} = 1.7$ eV. We will also explore the phonon dressing of strong exciton-cavity coupling with a relatively low vibrational frequency of $\hbar\omega_m = 20$ meV. The influence of larger vibrational frequencies are shown elsewhere [55], and can lead to much richer spectra with simultaneous cavity coupling.

To help better understand the role and need for pure dephasing of the TLS, we show in Fig. 5.13 the pure dephasing rates as a function of bath frequency for two different temperatures. We have also depicted some example dressed-state energies for the strong coupling problem, where $g_c = 5\omega_m = \kappa$. Typically, since $g_c > \kappa/2$, we expect a symmetric Rabi splitting at $\pm 5\omega_m$ if the drive strength is not too strong. However, because the dephasing bath is much larger at higher frequencies, phonon absorption processes will dominate phonon emission at low temperatures, and may cause a spectral asymmetry, especially at $T = 4$ K. Thus it is important to include such features, which is well known also for semiconductor quantum dots [103–105].

We are now ready to study *simultaneous strong cavity coupling and ultrastrong vibrational coupling*, with parameters $g_c = 5\omega_m = \kappa$ and $d = 0.2$. Larger values of d in strong cavity coupling regimes are shown in Ref. [55]. Figure 5.14a shows the energy levels without dissipation as a function of g_c, for a selection of the first 6 phonon levels in the $n = 1$ photon subspace.[8] At $g = 5\omega_{\rm m}$, the $m = 0$ lower polariton

[8] We reduce this number for visual clarity, but use 15 phonon levels in the spectral calculations.

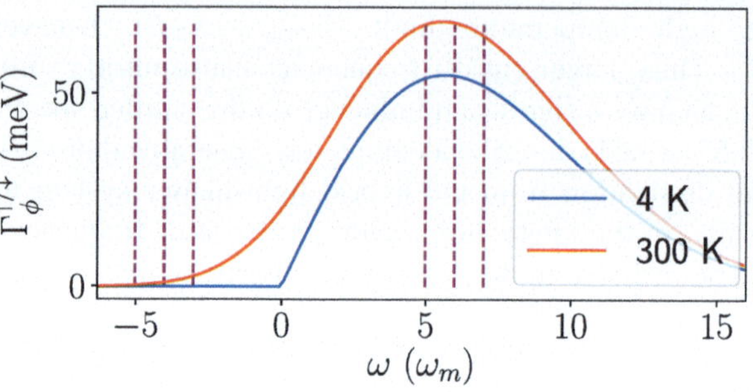

Fig. 5.13 Pure dephasing rates mediated from TLS interactions with the non-Markovian spectral function, $J_\phi(\omega)$ (Eq. (5.65)), where $\Gamma_\phi^\uparrow$ occurs for $\omega > 0$ and $\Gamma_\phi^\uparrow$ occurs for $\omega < 0$, shown for two different temperatures. The standard master equation would use the zero phonon value, namely $\Gamma_\phi(0)$, which is negligible at low temperatures. The magenta vertical dashed lines show some example on-resonance dressed state resonances associated with strong cavity coupling, with $g_c = 5\omega_\mathrm{m}$, separated by the vibrational (phonon) mode frequency

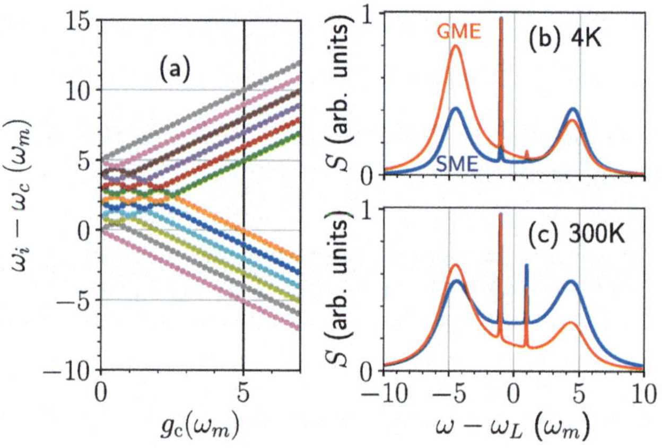

Fig. 5.14 **a** Eigenfrequencies of the system Hamiltonian, Eq. (5.49), for the $n = 1$ photon manifold as a function of g_c with $d_0 = 0.2$. **b** Cavity emitted spectra with a coherent drive at $g_c = 5\omega_\mathrm{m}$ and 4 K, with $\hbar\kappa = 100$ meV and $g_c = \kappa$, showing the standard master equation (Eq. (5.51), blue curve) and generalized master equation solution (Eq. (5.54), red curve). **c** Cavity emitted spectra at 300 K. Figure based on the results from Ref. [55]

state shifts down to $-5\omega_\mathrm{m}$ as expected, with the phonon states split by exactly one phonon frequency; the same shifts (but upwards) happens to the upper polariton states. In the $n = 0$ photon subspace, we simply obtain constant energy levels split by ω_m, with the zero phonon residing at $\omega = 0$. Since photon transitions can take place from any of these excited phonon states, in general the splitting of cavity emitted photons will always be less than g_c ($5\omega_m$), even when $\kappa \ll g_c$.

To study these energy states in the presence of resonance Raman scattering, we compute the cavity emitted spectrum with the CW driving field $\Omega = 0.5\omega_m$, using the same techniques as before (see Eq. (5.46)). Figure 5.14b shows the results at a temperature of 4 K, showing the first Stokes and anti-Stokes sidebands at $\pm\omega_m$, and polariton peaks (from cavity-exciton coupling) around $\pm 4.5\omega_{\rm m} < g_c$; these resonance energies are not increased with smaller κ and are due to the collective summation of the phonon dressed states contributing to the total emission linewidth.

Interestingly, we also see a pronounced asymmetry in the generalized master equation solution calculations, which is mainly coming from the spectral properties of the pure dephasing bath. This shows that even though the zero phonon line has negligible dephasing, the properties of the bath at the dressed resonances can have a substantial influence on the oscillator strengths. The bath-induced resonances cause downward transitions between the upper polariton and lower polariton states, while the transitions from the lower to higher states are negligible at low temperature. Such an effect cannot be predicted by the SME regardless of what Γ_ϕ value is used. Specifically, $J_\phi(\omega) > J_\phi(-\omega)$, which breaks the detailed balance of the diagonal dephasing process in the SME [62, 91].

Figure 5.14c shows room temperature results (300 K), which shows that these effects survive at elevated temperatures, and that the bath-induced asymmetry is still visible but less pronounced in the generalized master equation, since now there is a larger probability also for bath-induced upwards transitions between the polariton states. The anti-Stokes Raman transition is also much more visible as one might expect.

5.5 Conclusions and Outlook

We have presented several different theoretical approaches to model SERS and molecular optomechanics for single molecules coupled to plasmonic cavity systems, from the framework of open system quantum optics. These regimes are made possible by the tremendous progress over the years with MNP fabrication and spectroscopic techniques related to trapping and positioning single molecules in plasmonic cavities and other photonic structures. While MNP cavities naturally come with very lossy modes (low Q), they also yield regimes of extreme spatial confinement, and with suitable cavity designs, it is possible to have both high Q and very low mode volumes. Emerging hybrid quantum systems also motivate the need for new theories to understand emerging light-matter and optomechanical coupling effects in these extreme confinement regimes, which allow one to explore fundamentally new regimes in quantum plasmonics, where traditional theories in quantum optics can fail.

In the first part, we presented an intuitive Green function formalism that can be used to model SERS using plasmonic systems of arbitrary geometry. A general medium master equation was derived where the plasmonic environment is treated as a photonic bath and the system dynamics are projected onto the basis of the molecular vibrations. While the plasmonic degrees of freedom are traced out, the exact medium Green function is self-consistently included to account for the plasmonic

LDOS characteristics at the dressed-state resonances and for describing important light propagation effects. This approach describes the nonlinear light generation, enhancement, and quenching effects of light detection during SERS. Using this general medium approach, analytical expressions were derived for the molecular SERS spectrum near plasmonic environments which can be used to quantify the Stokes and anti-Stokes emission intensities. We also discussed how the dynamics of the SERS process is quite different in the emission process compared to light propagation and detection. The induced Raman polarization emits photons at Stokes frequencies that are proportional to the plasmonic enhancement (LDOS) at the co-existing anti-Stokes frequency, and vice-versa for the anti-Stokes frequencies. However, the plasmonic LDOS also comes into play differently at the light propagation stage, such that both the Stokes and anti-Stokes photons are enhanced at their respective frequencies. Our model can easily explore SERS as a function of laser detuning, pump power, spatial location of the molecule/detector, and the medium LDOS. We also demonstrated how the Green function response and LDOS can be efficiently described in terms of QNMs, which allow one to model SERS from various molecular positions and detector positions in a very efficient way. Examples were shown for a gold dimer cavity and a more complex hybrid photonic-plasmonic cavity system.

In the second part, we explored the regime of molecular optomechanics in a strong vibrational coupling regime (requiring multiple phonons and cavity dynamics to be included self-consistently), where a strong modification of the usual SERS spectrum is obtained because of the influence of higher lying quantum states, which have an *anharmonic* level spacing g^2/ω_m. These quantum states can be spectrally resolved if $g^2/\omega_m > \kappa$ and $\kappa < \omega_m$ is satisfied, which is typically not possible with broadband plasmonic resonators. However, our cavity design uses a hybrid metal-dielectric system where a MNP dimer is on top of a photonic crystal nanobeam cavity, which enables very small effective mode volumes as well as a sufficiently high quality factor ($Q^{\rm HQ} = 3500$). The high Q (small κ) and large optomechanical g are two essential criteria to probe the strong coupling anharmonic ladder states of the optomechanical system. While our designs use extremely small gap antennas, the prospect of using larger Raman active and resonant Raman processes in molecules indicates that emerging experiments in quantum plasmonic systems are not too far off reaching such a regime. There is also prospects for exploring similar regimes with new designs of very small mode dielectric systems [106, 107].

In the third and final part, we introduced a theory of resonant SERS using a generalized master equation approach, which treats system-bath coupling rigorously for the cavity, phonon (molecular vibration), and TLS interactions. We showed explicitly how the system eigenenergies are affected by an increasing exciton cavity coupling rate, and explored how phonon dressing affects the cavity emitted spectra in the strong coupling regime of cavity-QED. We also demonstrated the important role of including a pure dephasing bath, which results in pronounced spectral asymmetries, which is caused by the relatively large spectral splitting in the strong coupling regime. We described a rich regime of phonon-dressed polaritons, which coupled strong cavity QED effects with ultrastrong exciton-phonon interactions.

With the continued experimental and theoretical progress in single molecular sensing, trapping, quantum plasmonics, and MNP cavity designs, we anticipate continued improvements in both dielectric and plasmonic systems, as well as hybrid plasmon-dielectric cavities, opening up a wide range of effects in and applications in molecular cavity QED and plasmonic cavity QED in general. Since MNP environments are inherently very lossy, we also believe that future theoretical developments will benefit significantly from a quantized QNM approach, where unique features can appear such as quantum mechanical loss-induced coupling between classically orthogonal modes [76, 77].

Acknowledgements We acknowledge Queen's University, the Canadian Foundation for Innovation (CFI), and the Natural Sciences and Engineering Research Council of Canada (NSERC) for financial support, as well as CMC Microsystems for the provision of COMSOL Multiphysics. We gratefully acknowledge Reuven Gordon for collaborations and discussions on the hybrid cavity design and molecular optomechanics work in the off-resonant strong vibrational coupling regime. SH thanks Salvatore Savasta and Franco Nori for collaborations and discussions on the resonant Raman work.

References

1. C. V. Raman and K. S. Krishnan. A new type of secondary radiation. *Nature*, 121:501–502, March 1928.
2. Surbhi Lal, Stephan Link, and Naomi J. Halas. Nano-optics from sensing to waveguiding. *Nature Photonics*, 1:641, 2007.
3. Katherine A. Willets and Richard P. Van Duyne. Localized surface plasmon resonance spectroscopy and sensing. *Annual Review of Physical Chemistry*, 58(1):267–297, 2007.
4. Janina Kneipp, Harald Kneipp, and Katrin Kneipp. SERS-a single-molecule and nanoscale tool for bioanalytics. *Chem. Soc. Rev.*, 37:1052–1060, 2008.
5. Shuming Nie and Steven R. Emory. Probing single molecules and single nanoparticles by surface-enhanced Raman scattering. *Science*, 275(5303):1102–1106, 1997.
6. Katrin Kneipp, Yang Wang, Harald Kneipp, Lev T. Perelman, Irving Itzkan, Ramachandra R. Dasari, and Michael S. Feld. Single molecule detection using surface-enhanced Raman scattering (SERS). *Phys. Rev. Lett.*, 78:1667–1670, Mar 1997.
7. Alexandre G. Brolo, Erin Arctander, Reuven Gordon, Brian Leathem, and Karen L. Kavanagh. Nanohole-enhanced Raman scattering. *Nano Letters*, 4(10):2015–2018, October 2004.
8. Stefan Alexander Maier. *Plasmonics: Fundamentals and Applications*. Springer, 2007.
9. Eric C. Le Ru and Pablo G. Etchegoin. *Principles of Surface-Enhanced Raman Spectroscopy*. Elsevier, 2009.
10. Aftab Ahmed and Reuven Gordon. Single molecule directivity enhanced Raman scattering using nanoantennas. *Nano Letters*, 12(5):2625–2630, April 2012.
11. Eric C. Le Ru and Pablo G. Etchegoin. Single-molecule surface-enhanced Raman spectroscopy. *Annual Review of Physical Chemistry*, 63(1):65–87, 2012.
12. R. Zhang, Y. Zhang, Z. C. Dong, S. Jiang, C. Zhang, L. G. Chen, L. Zhang, Y. Liao, J. Aizpurua, Y. Luo, J. L. Yang, and J. G. Hou. Chemical mapping of a single molecule by plasmon-enhanced Raman scattering. *Nature*, 498:82–86, 2012.
13. Steven Yampolsky, Dmitry A. Fishman, Shirshendu Dey, Eero Hulkko, Mayukh Banik, Eric O. Potma, and Vartkess A. Apkarian. Seeing a single molecule vibrate through time-resolved coherent anti-Stokes Raman scattering. *Nat Photon*, 8:650–656, 2014.

14. Martin D. Baaske and Frank Vollmer. Optical observation of single atomic ions interacting with plasmonic nanorods in aqueous solution. *Nature Photonics*, 10:733, 2016.
15. N. P. Mauranyapin, L. S. Madsen, M. A. Taylor, M. Waleed, and W. P. Bowen. Evanescent single-molecule biosensing with quantum-limited precision. *Nature Photonics*, 11(8):477–481, August 2017.
16. Wenhua Gu, Hyungsoo Choi, and Kyekyoon (Kevin) Kim. A quantum mechanical theory for single molecule-single nanoparticle surface enhanced raman scattering. *The Journal of Physical Chemistry A*, 111(33):8121–8125, August 2007.
17. Anne Myers Kelley. A molecular spectroscopic view of surface plasmon enhanced resonance Raman scattering. *The Journal of Chemical Physics*, 128(22):224702, 2008.
18. David J. Masiello and George C. Schatz. Many-body theory of surface-enhanced Raman scattering. *Phys. Rev. A*, 78:042505, Oct 2008.
19. David J. Masiello and George C. Schatz. On the linear response and scattering of an interacting molecule-metal system. *The Journal of Chemical Physics*, 132(6):064102, 2010.
20. Aida Delfan, Marco Liscidini, and John E. Sipe. Surface enhanced Raman scattering in the presence of multilayer dielectric structures. *J. Opt. Soc. Am. B*, 29(8):1863–1874, Aug 2012.
21. John L. Payton, Seth M. Morton, Justin E. Moore, and Lasse Jensen. A hybrid atomistic electrodynamics-quantum mechanical approach for simulating surface-enhanced Raman scattering. *Accounts of Chemical Research*, 47(1):88–99, 2014.
22. John R. Lombardi and Ronald L. Birke. Theory of surface-enhanced Raman scattering in semiconductors. *The Journal of Physical Chemistry C*, 118(20):11120–11130, 2014.
23. Niclas S. Mueller, Sebastian Heeg, and Stephanie Reich. Surface-enhanced Raman scattering as a higher-order raman process. *Phys. Rev. A*, 94:023813, Aug 2016.
24. Philippe Roelli, Christophe Galland, Nicolas Piro, and Tobias J. Kippenberg. Molecular cavity optomechanics as a theory of plasmon-enhanced Raman scattering. *Nat Nano*, 11:164–169, Feb 2016.
25. Mikolaj K. Schmidt, Ruben Esteban, Alejandro González-Tudela, Geza Giedke, and Javier Aizpurua. Quantum mechanical description of Raman scattering from molecules in plasmonic cavities. *ACS Nano*, 10(6):6291–6298, 2016.
26. Mohsen Kamandar Dezfouli and Stephen Hughes. Quantum optics model of surface-enhanced Raman spectroscopy for arbitrarily shaped plasmonic resonators. *ACS Photonics*, 4(5):1245–1256, 2017.
27. M. S. Tame, K. R. McEnery, Ş. K. Özdemir, J. Lee, S. A. Maier, and M. S. Kim. Quantum plasmonics. *Nature Physics*, 9(6):329–340, June 2013.
28. Armin Regler, Konrad Schraml, Anna A. Lyamkina, Matthias Spiegl, Kai Müller, Jelena Vuckovic, Jonathan J. Finley, and Michael Kaniber. Emission redistribution from a quantum dot-bowtie nanoantenna. *Journal of Nanophotonics*, 10(3):033509, 2016.
29. Tao Cai, Subhojit Dutta, Shahriar Aghaeimeibodi, Zhili Yang, Sanghee Nah, John T. Fourkas, and Edo Waks. Coupling emission from single localized defects in two-dimensional semiconductor to surface plasmon polaritons. *Nano Letters*, 17(11):6564–6568, 2017.
30. Rohit Chikkaraddy, Bart de Nijs, Felix Benz, Steven J. Barrow, Oren A. Scherman, Edina Rosta, Angela Demetriadou, Peter Fox, Ortwin Hess, and Jeremy J. Baumberg. Single-molecule strong coupling at room temperature in plasmonic nanocavities. *Nature*, 535:127–130, 2016.
31. C. Belacel, B. Habert, F. Bigourdan, F. Marquier, J.-P. Hugonin, S. Michaelis de Vasconcellos, X. Lafosse, L. Coolen, C. Schwob, C. Javaux, B. Dubertret, J.-J. Greffet, P. Senellart, and A. Maitre. Controlling spontaneous emission with plasmonic optical patch antennas. *Nano Letters*, 13(4):1516–1521, 2013.
32. Gleb M. Akselrod, Christos Argyropoulos, Thang B. Hoang, Cristian Ciracì, Chao Fang, Jiani Huang, David R. Smith, and Maiken H. Mikkelsen. Probing the mechanisms of large Purcell enhancement in plasmonic nanoantennas. *Nat Photon*, 8:835–840, 2014.
33. Anna Lombardi, Mikołaj K. Schmidt, Lee Weller, William M. Deacon, Felix Benz, Bart de Nijs, Javier Aizpurua, and Jeremy J. Baumberg. Pulsed molecular optomechanics in plasmonic nanocavities: From nonlinear vibrational instabilities to bond-breaking. *Phys. Rev. X*, 8:011016, Feb 2018.

34. Kiran E. Khosla, George A. Brawley, Michael R. Vanner, and Warwick P. Bowen. Quantum optomechanics beyond the quantum coherent oscillation regime. *Optica*, 4(11):1382, November 2017.
35. P. Rabl. Photon blockade effect in optomechanical systems. *Phys. Rev. Lett.*, 107:063601, Aug 2011.
36. A. Nunnenkamp, K. Børkje, and S. M. Girvin. Single-photon optomechanics. *Phys. Rev. Lett.*, 107:063602, Aug 2011.
37. Mohsen Kamandar Dezfouli, Reuven Gordon, and Stephen Hughes. Molecular optomechanics in the anharmonic cavity-QED regime using hybrid metal–dielectric cavity modes. *ACS Photonics*, 6(6):1400–1408, May 2019.
38. Jorge Calvo, David Zueco, and Luis Martin-Moreno. Ultrastrong coupling effects in molecular cavity QED. *Nanophotonics*, 9(2):277–281, February 2020.
39. Rong-Chun Ge and S. Hughes. Design of an efficient single photon source from a metallic nanorod dimer: a quasi-normal mode finite-difference time-domain approach. *Opt. Lett.*, 39(14):4235–4238, Jul 2014.
40. Herman M. K. Wong, Mohsen Kamandar Dezfouli, Lu Sun, Stephen Hughes, and Amr S. Helmy. Nanoscale plasmonic slot waveguides for enhanced Raman spectroscopy. *Phys. Rev. B*, 98:085124, Aug 2018.
41. Mohsen Kamandar Dezfouli, Reuven Gordon, and Stephen Hughes. Modal theory of modified spontaneous emission of a quantum emitter in a hybrid plasmonic photonic-crystal cavity system. *Phys. Rev. A*, 95:013846, Jan 2017.
42. P. Alonso-González, P. Albella, M. Schnell, J. Chen, F. Huth, A. Garcia-Etxarri, F. Casanova, F. Golmar, L. Arzubiaga, L.E. Hueso, J. Aizpurua, and R. Hillenbrand. Resolving the electromagnetic mechanism of surface-enhanced light scattering at single hot spots. *Nature Communications*, 3:684, 2012.
43. Felix Benz, Mikolaj K. Schmidt, Alexander Dreismann, Rohit Chikkaraddy, Yao Zhang, Angela Demetriadou, Cloudy Carnegie, Hamid Ohadi, Bart de Nijs, Ruben Esteban, Javier Aizpurua, and Jeremy J. Baumberg. Single-molecule optomechanics in "picocavities". *Science*, 354(6313):726–729, 2016.
44. Yingzhou Huang, Yurui Fang, Zhenglong Zhang, Ling Zhu, and Mengtao Sun. Nanowire-supported plasmonic waveguide for remote excitation of surface-enhanced Raman scattering. *Light: Science & Applications*, 3:e199, 2014.
45. Stephan J. P. Kress, Felipe V. Antolinez, Patrizia Richner, Sriharsha V. Jayanti, David K. Kim, Ferry Prins, Andreas Riedinger, Maximilian P. C. Fischer, Stefan Meyer, Kevin M. McPeak, Dimos Poulikakos, and David J. Norris. Wedge waveguides and resonators for quantum plasmonics. *Nano Letters*, 15(9):6267–6275, 2015.
46. Changhyoup Lee, Frederik Dieleman, Jinhyoung Lee, Carsten Rockstuhl, Stefan A. Maier, and Mark Tame. Quantum plasmonic sensing: Beyond the shot-noise and diffraction limit. *ACS Photonics*, 3(6):992–999, 2016.
47. Frédéric Peyskens, Ashim Dhakal, Pol Van Dorpe, Nicolas Le Thomas, and Roel Baets. Surface enhanced Raman spectroscopy using a single mode nanophotonic-plasmonic platform. *ACS Photonics*, 3(1):102–108, 2016.
48. Michael Barth, Stefan Schietinger, Sabine Fischer, Jan Becker, Nils Nüsse, Thomas Aichele, Bernd Löchel, Carsten Sönnichsen, and Oliver Benson. Nanoassembled plasmonic-photonic hybrid cavity for tailored light-matter coupling. *Nano Letters*, 10(3):891–895, 2010.
49. Martin D. Baaske and Frank Vollmer. Optical observation of single atomic ions interacting with plasmonic nanorods in aqueous solution. *Nat Photon*, 10:733–739, 2016.
50. Ishita Mukherjee, Ghazal Hajisalem, and Reuven Gordon. One-step integration of metal nanoparticle in photonic crystal nanobeam cavity. *Opt. Express*, 19(23):22462–22469, Nov 2011.
51. Philip Trøst Kristensen and Stephen Hughes. Modes and mode volumes of leaky optical cavities and plasmonic nanoresonators. *ACS Photonics*, 1(1):2–10, 2014.
52. Philippe Lalanne, Wei Yan, Kevin Vynck, Christophe Sauvan, and Jean-Paul Hugonin. Light interaction with photonic and plasmonic resonances. *Laser & Photonics Reviews*, 12(5):1700113, 2018.

53. Ali El Eter, Thierry Grosjean, Pierre Viktorovitch, Xavier Letartre, Taha Benyattou, and Fadi I. Baida. Huge light-enhancement by coupling a bowtie nano-antenna's plasmonic resonance to a photonic crystal mode. *Opt. Express*, 22(12):14464–14472, Jun 2014.
54. Hugo M. Doeleman, Ewold Verhagen, and A. Femius Koenderink. Antenna-cavity hybrids: Matching polar opposites for Purcell enhancements at any linewidth. *ACS Photonics*, 3(10):1943–1951, 2016.
55. Stephen Hughes, Alessio Settineri, Salvatore Savasta, and Franco Nori. Resonant Raman scattering of single molecules under simultaneous strong cavity coupling and ultrastrong optomechanical coupling in plasmonic resonators: phonon-dressed polaritons, Phys. Rev. B 104, 045431, 2021.
56. Howard J. Carmichael. *Statistical Methods in Quantum Optics I*. Springer, 2002.
57. Heinz-Peter Breuer and Francesco Petruccione. *The Theory of Open Quantum Systems*. Oxford, 2007.
58. Anna Kowalewska-Kudlaszyk and Ryszard Tanas. Generalized master equation for a two-level atom in a strong field and tailored reservoirs. *Journal of Modern Optics*, 48(2):347–370, 2001.
59. Ho Trung Dung, Ludwig Knöll, and Dirk-Gunnar Welsch. Three-dimensional quantization of the electromagnetic field in dispersive and absorbing inhomogeneous dielectrics. *Phys. Rev. A*, 57:3931–3942, May 1998.
60. L. G. Suttorp and A. J. van Wonderen. Fano diagonalization of a polariton model for an inhomogeneous absorptive dielectric. *EPL (Europhysics Letters)*, 67(5):766, 2004.
61. Philippe Roelli, Nicolas Piro, and Tobias J Kippenberg. Molecular cavity optomechanics as a theory of plasmon-enhanced Raman scattering. *Nat Nano*, 11:164–169, 2016.
62. A. A. Clerk, M. H. Devoret, S. M. Girvin, Florian Marquardt, and R. J. Schoelkopf. Introduction to quantum noise, measurement, and amplification. *Rev. Mod. Phys.*, 82:1155–1208, Apr 2010.
63. Rong-Chun Ge, C. Van Vlack, P. Yao, Jeff. F. Young, and S. Hughes. Accessing quantum nanoplasmonics in a hybrid quantum dot–metal nanosystem: Mollow triplet of a quantum dot near a metal nanoparticle. *Physical Review B*, 87(20):205425, may 2013.
64. Peijun Yao, C. Van Vlack, A. Reza, M. Patterson, M. M. Dignam, and S. Hughes. Ultrahigh Purcell factors and Lamb shifts in slow-light metamaterial waveguides. *Phys. Rev. B*, 80:195106, Nov 2009.
65. Marlan O. Scully and M. Suhail Zubairy. *Quantum Optics*. Cambridge University Press, 1997.
66. Markus Aspelmeyer, Tobias J. Kippenberg, and Florian Marquardt. Cavity optomechanics. *Rev. Mod. Phys.*, 86:1391–1452, Dec 2014.
67. Philip Trøst Kristensen, Rong-Chun Ge, and Stephen Hughes. Normalization of quasinormal modes in leaky optical cavities and plasmonic resonators. *Phys. Rev. A*, 92:053810, Nov 2015.
68. P. T. Kristensen, C. Van Vlack, and S. Hughes. Generalized effective mode volume for leaky optical cavities. *Opt. Lett.*, 37(10):1649–1651, May 2012.
69. C. Sauvan, J. P. Hugonin, I. S. Maksymov, and P. Lalanne. Theory of the spontaneous optical emission of nanosize photonic and plasmon resonators. *Phys. Rev. Lett.*, 110:237401, Jun 2013.
70. E. A. Muljarov, W. Langbein, and R. Zimmermann. Brillouin-wigner perturbation theory in open electromagnetic systems. *EPL*, 92(5):50010, 2010.
71. P. T. Leung, S. Y. Liu, and K. Young. Completeness and orthogonality of quasinormal modes in leaky optical cavities. *Physical Review A*, 49(4):3057–3067, April 1994.
72. Rong-Chun Ge, Philip Trøst Kristensen, Jeff F Young, and Stephen Hughes. Quasinormal mode approach to modelling light-emission and propagation in nanoplasmonics. *New Journal of Physics*, 16(11):113048, 2014.
73. Philip Trøst Kristensen, Kathrin Herrmann, Francesco Intravaia, and Kurt Busch. Modeling electromagnetic resonators using quasinormal modes. *Advances in Optics and Photonics*, 12(3):612, 2020.
74. Pascal Anger, Palash Bharadwaj, and Lukas Novotny. Enhancement and quenching of single-molecule fluorescence. *Phys. Rev. Lett.*, 96:113002, Mar 2006.

75. Q. Bai, M. Perrin, C. Sauvan, J-P Hugonin, and P. Lalanne. Efficient and intuitive method for the analysis of light scattering by a resonant nanostructure. *Opt. Express*, 21(22):27371–27382, Nov 2013.
76. Sebastian Franke, Stephen Hughes, Mohsen Kamandar Dezfouli, Philip Trøst Kristensen, Kurt Busch, Andreas Knorr, and Marten Richter. Quantization of quasinormal modes for open cavities and plasmonic cavity quantum electrodynamics. *Phys. Rev. Lett.*, 122:213901, May 2019.
77. Sebastian Franke, Marten Richter, Juanjuan Ren, Andreas Knorr, and Stephen Hughes. Quantized quasinormal-mode description of nonlinear cavity-QED effects from coupled resonators with a Fano-like resonance. *Phys. Rev. Research*, 2:033456, Sep 2020.
78. Hiroyuki Watanabe, Norihiko Hayazawa, Yasushi Inouye, and Satoshi Kawata. DFT vibrational calculations of Rhodamine 6G adsorbed on silver: Analysis of tip-enhanced Raman spectroscopy. *The Journal of Physical Chemistry B*, 109(11):5012–5020, 2005.
79. Juanjuan Ren, Sebastian Franke, Andreas Knorr, Marten Richter, and Stephen Hughes. Near-field to far-field transformations of optical quasinormal modes and efficient calculation of quantized quasinormal modes for open cavities and plasmonic resonators. *Physical Review B*, 101(20), May 2020.
80. Vincenzo Macrì, Alessandro Ridolfo, Omar Di Stefano, Anton Frisk Kockum, Franco Nori, and Salvatore Savasta. Nonperturbative dynamical Casimir effect in optomechanical systems: Vacuum Casimir-Rabi splittings. *Phys. Rev. X*, 8:011031, Feb 2018.
81. Alessio Settineri, Vincenzo Macrí, Alessandro Ridolfo, Omar Di Stefano, Anton Frisk Kockum, Franco Nori, and Salvatore Savasta. Dissipation and thermal noise in hybrid quantum systems in the ultrastrong-coupling regime. *Physical Review A*, 98(5), November 2018.
82. Javier del Pino, Johannes Feist, and F. J. Garcia-Vidal. Signatures of vibrational strong coupling in Raman scattering. *The Journal of Physical Chemistry C*, 119(52):29132–29137, dec 2015.
83. K. Huang and A. Rhys. Theory of light absorption and non-radiative transitions in F -centres. *Proceedings of the Royal Society of London. Series A. Mathematical and Physical Sciences*, 204(1078):406–423, December 1950.
84. Girish S. Agarwal. *Quantum Optics*. Cambridge, 2012.
85. Lukas Neumeier, Tracy E. Northup, and Darrick E. Chang. Reaching the optomechanical strong-coupling regime with a single atom in a cavity. *Physical Review A*, 97(6):063857, jun 2018.
86. Anton Frisk Kockum, Adam Miranowicz, Simone De Liberato, Salvatore Savasta, and Franco Nori. Ultrastrong coupling between light and matter. *Nature Reviews Physics*, 1(1):19–40, January 2019.
87. T. Niemczyk, F. Deppe, H. Huebl, E. P. Menzel, F. Hocke, M. J. Schwarz, J. J. Garcia-Ripoll, D. Zueco, T. Hümmer, E. Solano, A. Marx, and R. Gross. Circuit quantum electrodynamics in the ultrastrong-coupling regime. *Nature Physics*, 6(10):772–776, Jul 2010.
88. Dan Hu, Shang-Yu Huang, Jie-Qiao Liao, Lin Tian, and Hsi-Sheng Goan. Quantum coherence in ultrastrong optomechanics. *Phys. Rev. A*, 91:013812, Jan 2015.
89. P. Forn-Díaz, L. Lamata, E. Rico, J. Kono, and E. Solano. Ultrastrong coupling regimes of light-matter interaction. *Rev. Mod. Phys.*, 91:025005, Jun 2019.
90. H. J. Carmichael and D. F. Walls. Master equation for strongly interacting systems. *Journal of Physics A: Mathematical, Nuclear and General*, 6(10):1552–1564, Oct 1973.
91. Félix Beaudoin, Jay M. Gambetta, and A. Blais. Dissipation and ultrastrong coupling in circuit QED. *Physical Review A*, 84(4):043832, oct 2011.
92. J. R. Johansson, P. D. Nation, and Franco Nori. Qutip: An open-source Python framework for the dynamics of open quantum systems. *Computer Physics Communications*, 183(8):1760 – 1772, 2012.
93. J. R. Johansson, P. D. Nation, and Franco Nori. Qutip 2: A Python framework for the dynamics of open quantum systems. *Computer Physics Communications*, 184(4):1234 – 1240, 2013.
94. Isabelle M. Palstra, Hugo M. Doeleman, and A. Femius Koenderink. Hybrid cavity-antenna systems for quantum optics outside the cryostat? *Nanophotonics*, 8(9):1513–1531, May 2019.

95. Wenqi Zhu, Ruben Esteban, Andrei G. Borisov, Jeremy J. Baumberg, Peter Nordlander, Henri J. Lezec, and Kenneth B. Aizpurua, Javier an Crozier. Quantum mechanical effects in plasmonic structures with subnanometre gaps. *Nature Communications*, 7:11495, June 2016.
96. Mohsen Kamandar Dezfouli, Christos Tserkezis, N. Asger Mortensen, and Stephen Hughes. Nonlocal quasinormal modes for arbitrarily shaped three-dimensional plasmonic resonators. *Optica*, 4(12):1503–1509, Dec 2017.
97. Sergei Lebedkin, Carolin Blum, Ninette Stürzl, Frank Hennrich, and Manfred M. Kappes. A low-wavenumber-extended confocal Raman microscope with very high laser excitation line discrimination. *Review of Scientific Instruments*, 82(1):013705, 2011.
98. Maurizio Montagna and Roberto Dusi. Raman scattering from small spherical particles. *Phys. Rev. B*, 52:10080–10089, Oct 1995.
99. Tomáš Neuman, Ruben Esteban, Geza Giedke, Mikołaj K. Schmidt, and Javier Aizpurua. Quantum description of surface-enhanced resonant Raman scattering within a hybrid-optomechanical model. *Phys. Rev. A*, 100:043422, Oct 2019.
100. Tomáš Neuman and Javier Aizpurua and Ruben Esteban. Quantum theory of surface-enhanced resonant Raman scattering (SERRS) of molecules in strongly coupled plasmon–exciton systems. *Nanophotonics*, 121:295–308, Feb 2020.
101. Jino George, Shaojun Wang, Thibault Chervy, Antoine Canaguier-Durand, Gael Schaeffer, Jean-Marie Lehn, James A. Hutchison, Cyriaque Genet, and Thomas W. Ebbesen. Ultrastrong coupling of molecular materials: spectroscopy and dynamics. *Faraday Discussions*, 178:281–294, 2015.
102. Tomáš Neuman and Javier Aizpurua. Origin of the asymmetric light emission from molecular exciton–polaritons. *Optica*, 5(10):1247, October 2018.
103. S. Weiler, A. Ulhaq, S. M. Ulrich, D. Richter, M. Jetter, P. Michler, C. Roy, and S. Hughes. Phonon-assisted incoherent excitation of a quantum dot and its emission properties. *Phys. Rev. B*, 86:241304, Dec 2012.
104. Kaushik Roy-Choudhury and Stephen Hughes. Quantum theory of the emission spectrum from quantum dots coupled to structured photonic reservoirs and acoustic phonons. *Phys. Rev. B*, 92:205406, Nov 2015.
105. Ahsan Nazir and Dara P S McCutcheon. Modelling exciton–phonon interactions in optically driven quantum dots. *Journal of Physics: Condensed Matter*, 28(10):103002, February 2016.
106. Shuren Hu and Sharon M. Weiss. Design of photonic crystal cavities for extreme light concentration. *ACS Photonics*, 3(9):1647–1653, 2016.
107. Hyeongrak Choi, Mikkel Heuck, and Dirk Englund. Self-similar nanocavity design with ultrasmall mode volume for single-photon nonlinearities. *Phys. Rev. Lett.*, 118:223605, May 2017.

Part III
Biomolecular Manipulation

The manipulation and detection of biomolecules by electromagnetic fields is the topic of the third part of this book, particularly using dielectrophoresis (DEP) and optical tweezers. While physically the same, the main difference between dielectrophoresis and optical tweezers is the frequency of operation—dielectrophoresis operates in the radiofrequency regime, whereas optical tweezers are in the UV-NIR regime. To reach down the scale of single biomolecules, nanoscale features are used. In DEP, these are electrodes. DEP has the advantage of not requiring an external laser to manipulate biomolecules; however, the detection of biomolecules still often requires fluorescence. In the future, one may envision that simple changes in local dielectric environment could be directly detected through capacitance changes. The theory, geometries, and practicalities of DEP manipulation are outlined in Chap. 6. For optical tweezers, the nanostructures can be photonic crystals, resonators, plasmonic particles, or apertures in metal films. A brief introduction to trapping single molecules is outlined in Chap. 7. Apertures in metal films allow for trapping and detecting the presence of biomolecules by simply measuring changes in the aperture transmission. This 'dielectric loading' effect provides a useful way to monitor proteins and their interactions without labels or tethers, as detailed in Chap. 8. Finally, one may consider extending these approaches to detect and manipulate chiral molecules, since chirality plays an important role in biomolecular function—emerging approaches towards this end-goal are outlined in Chap. 9.

Chapter 6
Dielectrophoresis of Single Molecules

Binoy Paulose Nadappuram, Christopher T. Ertsgaard, Joshua B. Edel, and Sang-Hyun Oh

Abstract The ability to manipulate individual molecules with high spatio-temporal resolution is critical for understanding the structural, functional and compositional features of various biological molecules. Since Phol and Hank demonstrated single entity manipulation using dielectrophoresis (DEP), many researchers have attempted to extend the capability of this technique to develop easy and inexpensive methods for trapping and sorting of single molecules. Many of these devices can be easily fabricated using the existing micro/nanofabrication methods standard to the semiconductor industry. To date, DEP based platforms were reported for the precise manipulation of a range of biomolecules including nucleic acids and proteins. This chapter aims to provide an overview of nanoscale DEP and introduce the different device configurations and experimental strategies reported for single molecule manipulation.

6.1 Introduction

As detection schemes move towards single particle and single molecules, analyte dimensions approach the sub-micron and sub-nanometer regimes. Often various nanotechnology techniques can be employed to more tightly focus measurement signals to volumes comparable in size for efficient measurement transduction. However, often with the reduction now of sensing volumes, the next challenge posed is

B. Paulose Nadappuram · J. B. Edel
Imperial College London, London, UK
e-mail: b.paulose@imperial.ac.uk

J. B. Edel
e-mail: joshua.edel@imperial.ac.uk

C. T. Ertsgaard · S.-H. Oh (✉)
University of Minnesota, Minneapolis, USA
e-mail: sang@umn.edu

C. T. Ertsgaard
e-mail: ertsg003@umn.edu

W. Bowen et al. (eds.), *Single Molecule Sensing Beyond Fluorescence*,
Nanostructure Science and Technology,
https://doi.org/10.1007/978-3-030-90339-8_6

delivering target analytes to these volumes. Common techniques consist of centrifugation and sedimentation, convective stirring/vortexing, and/or electromagnetic (EM) tags (e.g. for electrophoresis or magnetic separation). However, most of these techniques require a laboratory setting and/or relevant and expertise to facilitate. Therefore, more commonly simple diffusion is used in which analytes follow a random walk (i.e. Brownian motion) to the sensing surface. In this regard, detection times can be quite long (e.g. hours or days) and ultimately results in random placement of the analyte on the sensing surface [1, 2].

Consequently, active trapping and positioning of suspended analytes in solution poses great advantage. Namely, it can accelerate detection times and precisely place target analytes to interfaces with the most sensitive regions of nanostructures and sensing elements. This is especially advantageous when the delivery technique is simple, portable, and cheap to operate. Dielectrophoresis (DEP) offers these advantageous by utilizing radio-frequency (RF) signals to manipulate unlabeled dielectric particles. By using non-uniform electric fields (E-fields), unbalanced forces on the induced dipole causes particle movement toward or away from the increasing E-field gradient. Capable of manipulating unlabeled particles, this contrasts EM tagging and electrophoresis which manipulates only a subset of particles (i.e. bearing a net charge) and/or requires chemical tagging steps. Operating within the RF regime, this enables DEP to utilize digital power sources which may be more readily available to a broader community compared to laboratory specific techniques (e.g. centrifugation or vortexing). Further, DEP electrodes can typically be constructed using standard microfabrication processes that are well developed in the complementary metal oxide semiconductor (CMOS) technology. This mature infrastructure promotes a more direct path to commercialization as it does not require intricate assembly or optical alignment compared to micro-electromechanical systems (MEMS) based convective stirring or optical trapping.

In the following chapter, a conceptual description of the DEP force and its derivation will be provided. An overview of its application for single particle manipulation and integration with various sensing techniques will be explored, followed by a discussion on the limits to the technique and the future vision for this powerful technology.

6.2 Theory

In order to appreciate the origin of the DEP force, a conceptual description is helpful to lay the foundation and provide intuition for future design and optimization. While electrophoresis refers to the movement and/or separation of charged species in the presence of an electric field (E-field), **E**, dielectrophoresis is the "uncharged" equivalent–in which dielectric particles can be manipulated in the presence of an E-field gradient, $\nabla\mathbf{E}$. This is felt as a macroscopic DEP force, that will be derived below.

This derivation will consist of first defining a complex dielectric permittivity to describe general dielectric materials with some conductive components and their frequency response to an oscillating E-field followed by the description of an "effective" polarizability and subsequent Clausius-Mossotti factor (CMF). Then, a derivation of the DEP force will be made using the Kelvin Polarization force conceptually outlined below. Finally, a description of a trapping volume will be outlined, defined as the volume in which the DEP force is able to manipulate a given particle.

6.2.1 Conceptual Description

Two point charges will exhibit a mutual force (either repulsion or attraction) due to Coulomb interaction. More practically, if a conductive surface is now held at a given voltage potential (i.e. an accumulation of surface charges on an electrode), a corresponding electric field (E-field) will extend radially and terminate at some ground reference frame. A test charge now placed within this E-field will again experience a force (i.e. Lorentz Force) in the direction towards opposite charge, see Fig. 6.1.

However, if now instead a complex particle which generally will contain both positive charge (e.g. atomic nuclei) and negative (e.g. electron electrode or free electrons), these individual charges within the composite particle can respond with opposite responses and induce more complex responses from the particle as a whole. For instance, if the net charge of the particle zero (i.e. an even balance/distribution of positive and negative charge throughout) then these individual charges will polarize across the volume of the particle exhibiting equal and opposite forces resulting in a net zero force induces on the particle. This scenario is often the case for most polarizable materials (i.e. dielectrics). While there are multiple configuration of charge distribution about a given particle (i.e. n-pole), the most common and useful for this discussion is the dipole, in which an asymmetry of opposite charge exists across only one spatial axis. Therefore, a dipole moment, $\mathbf{p}$ [units: Cm], can be defined simply as the product between the magnitude of charge, $|q|$, and their distance of separation, $\mathbf{r}$, see Eq. 6.1.

$$\begin{aligned} \mathbf{p} &= |q|\,\mathbf{r} \\ &= \alpha\,\mathbf{E} \end{aligned} \tag{6.1}$$

The particle's polarizability, α [units: Cm^2/V], then is simply a proportionality relating how "able" the said particle can form a dipole moment in the presence of an E-field, $\mathbf{E}$. Consequently, a larger dipole moment for a given E-field will indicate a more polarizable particle. Adding to the complexity, this polarizability term, α, depends on many factors including atomic composition, size, shape and orientation, electrical properties of the interfacial ambient environment and frequency at which the E-field oscillates [3–5]. This will be developed further below.

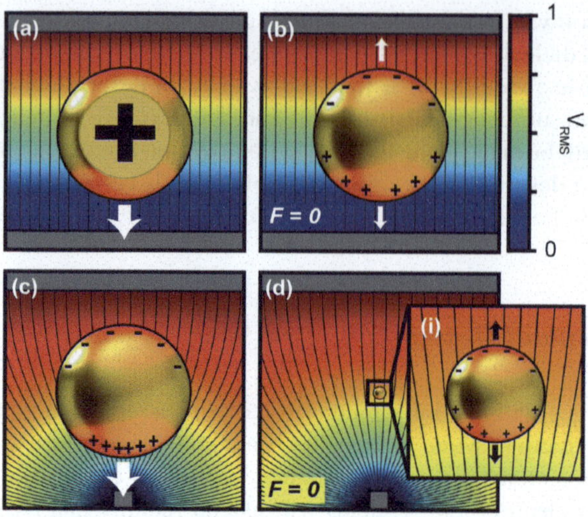

Fig. 6.1 **Polarization force with E-field gradients**: A 1 V_{RMS} is applied across two electrodes (gray) and a surface plot is used to illustrate the drop in potential across the electrodes. The electric field (E-field) lines are drawn in black. **a** A charged particle within a uniform E-field feels a net force towards opposite charge due to Coulomb interaction. **b** A dielectric or polarizable material with zero net charge will feel no net force within a uniform E-field due to equal and opposite attraction at the terminals of its dipole. **c** It can, however, experience a net force within a spatially varying electric field (i.e. $\nabla \mathbf{E} \neq 0$), due to an uneven concentration of charge at the terminals of its dipole. This is known as the Polarization force (Eq. 6.2). **d** This net force is diminished as the particle size is reduced-e.g. 10× smaller shown in inset **(i)**. This in part is due to the electric field lines locally appearing uniform (i.e. $\nabla \mathbf{E} \approx 0$) resulting in a near zero net force on the particle, similar to Panel **(b)**

As mentioned earlier, if the dielectric particle is placed within a uniform E-field, the dipole will polarize and result in equal but opposite forces on the bound charges. This yields a zero macroscopic force on the particle. However, if the E-field instead is spatially non-uniform across the particle, an unequal distribution of charge can occur as the particle polarizes. Consequently, this results in an unequal balance of force on either end of the dipole and thus a net macroscopic force felt by the particle. This is known as the polarization force and was developed extensively by Lord Kelvin in the 19th century. Therefore, intuitively a particle that is more polarizable will experience a larger polarization force compared one that is less polarizable. Likewise, the greater the non-uniformity of the E-field across the volume of the particle, the larger the polarization force. This spatial variation in the E-field is described quantitatively as the gradient of the E-field-i.e. the rate-of-change in the E-field over a given distance. Thus, a greater change in E-field across the distance/volume of the particle, the larger the asymmetry in charge distribution within the dipole and thus a greater polarization force results. Therefore, by taking these components, intuitively we can now describe this Kelvin polarization force density, $\mathbf{F_K}$, quantitatively [6]:

$$\mathbf{F_K} = \mathbf{P} \cdot \nabla \mathbf{E} \tag{6.2}$$

where, **P** [C/m^2], is the particle's polarization defined simply as the effective dipole moment, **p**, per volume.

The reader may already begin to appreciate some of the dilemmas that arise as one seeks to use the DEP force to trap and manipulate smaller/single molecule particles in polar water solution. First, as the the particle volume is reduced, the extent of the E-field gradient across the particle is reduced. This is analogous to the mathematical concept when approximating the derivative of an arbitrary function as a linear/straight line with a small spatial step, dx. Similarly, the E-fields will begin to appear more "uniform" across the particle's volume as it is reduced in size. Additionally, the ambient solution in which the particle is suspended will typically also be a dielectric and can polarize. Water molecules carry a fixed dipole moment and thus respond to the Kelvin polarization force as well. The frequency of operation then becomes a critical parameter as particles larger than water molecules will typically polarize more slowly or less effectively at higher frequencies. Therefore, the particle size and frequency of operation will become an important consideration when considering DEP actuation.

6.2.2 *Polarizability and Clausius-Mossotti Factor (CMF)*

As mentioned in Sect. 6.2.1, the polarizability, α, of a given particle depends on multiple factors including geometry, frequency of E-field oscillation (with angular frequency, ω) and the electrical properties of the particle and ambient surrounding environment. The constitutive law governing these properties of interest will be the dielectric permittivity, ε, and conductivity, σ, of the particle and surrounding solution. Since nearly all real materials will contain both a polarizable dielectric property and conductive component, it is helpful to characterize the materials' electrical properties using a single complex dielectric permittivity, $\widetilde{\varepsilon}(\omega)$, see Eq. 6.3.

$$\widetilde{\varepsilon} = \varepsilon - \imath \frac{\sigma}{\omega} \tag{6.3}$$

Since DEP typically operates within the low-frequency (e.g. RF regime), ε and σ are simply the material's DC dielectric permittivity and conductivity constants and $\imath$ is the imaginary unit. This complex term, $\widetilde{\varepsilon}$, can then capture both its dielectric and conductive components and frequency response in a single term [3–5].

An effective polarizability, $\widetilde{\alpha}$ [units: C/Vm], then will be defined per unit volume of the particle containing both the complex permittivity of the particle, $\widetilde{\varepsilon}_p$, and the surrounding liquid medium, $\widetilde{\varepsilon}_L$–both frequency, ω, dependent.

In addition to the electric properties of the particle and solution, the geometry of the particle also plays a significant role in its effective polarizability. For instance, an ellipsoidal or elongated particle will feel a larger polarization force due to a greater spatial extent across the E-field gradient as compared to a spherical counterpart. Often analytes are approximated as spheres (classic spherical "horse" physics approach) to simplify derivation. This in fact is a fair approach as many particles are spherical in nature and/or are spherical shells (e.g. membrane found cells, exosomes, viral capsids, etc.). Additionally, the spherical derivation will serve as a conservative "upper-bound" on the DEP force as geometrically it is the least polarizable shape (i.e. the 3D shape with smallest surface area and largest symmetry) and thus will be the shape with the smallest DEP force [3, 4]. Therefore, the polarizability for a sphere and spherical shell will be discussed.

The effective polarizability for a sphere has been derived in extensive detail in other works [3, 4] and is provided below-written in a form to include the Clausius-Mossotti factor (CMF), $\widetilde{f}_{cm}(\omega)$ [4], see Eq. 6.4.

$$\widetilde{\alpha} = 3\,\varepsilon_L\,\widetilde{f}_{cm}(\omega) \tag{6.4}$$

The, ε_L, is the real dielectric constant of the surrounding medium/liquid. All of the complex components have been captured by the CMF for simplicity. A sphere with a complex dielectric permittivity, $\widetilde{\varepsilon}_p$ (defined using Eq. 6.3), surrounded in a medium with a complex dielectric permittivity, $\widetilde{\varepsilon}_L$, the classic CMF equation becomes the following:

$$\widetilde{f}_{cm}(\omega) = \frac{\widetilde{\varepsilon}_p - \widetilde{\varepsilon}_L}{\widetilde{\varepsilon}_p + 2\,\widetilde{\varepsilon}_L} \tag{6.5}$$

When considering a spherical shell particle (e.g. cell, virus, functionalized nanoparticle, etc.), a similar effective polarizability will result. The significant difference is a change in the complex dielectric permittivity of the particle, $\widetilde{\varepsilon}_p$, which becomes more complicated due to a core with a different material and polarizability response. Therefore, Eq. 6.5 will instead replace the solid sphere complex permittivity, $\widetilde{\varepsilon}_p$ with it composite shell counterpart, i.e. $\widetilde{\varepsilon}_p \rightarrow \widetilde{\varepsilon}_{sp}$, defined as the following [3, 4]:

$$\widetilde{\varepsilon}_{sp} = \widetilde{\varepsilon}_s \left(\frac{\Gamma^3 + 2\widetilde{M}^{sc}}{\Gamma^3 - \widetilde{M}^{sc}} \right) \tag{6.6}$$

$$\widetilde{M}^{sc} = \frac{\widetilde{\varepsilon}_c - \widetilde{\varepsilon}_s}{\widetilde{\varepsilon}_c + 2\,\widetilde{\varepsilon}_s} \tag{6.7}$$

$$\Gamma = \frac{a}{a - k} \tag{6.8}$$

where, $\widetilde{\varepsilon}_c$, and, $\widetilde{\varepsilon}_s$, is the complex dielectric permittivity of the particle's core and shell, respectively. Additionally, a, and, k, is the radius of the particle and the thickness of its shell/membrane layer, respectively.

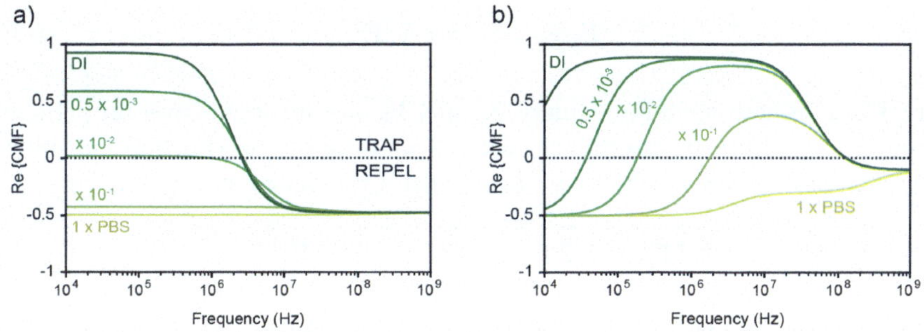

Fig. 6.2 Clausius–Mossotti factor (CMF): Example plot of the real CMF for different particles in 1×PBS, 0.1×PBS, 0.01×PBS, 0.0005×PBS buffer ($\varepsilon_L = 80\ \varepsilon_0$, $\sigma = 1.5$ S/m) and DI water ($\varepsilon_L = 80\ \varepsilon_0$, $\sigma = 4 \times 10^{-4}$ S/m). Positive values indicate frequencies for pDEP (or trapping) and negative values indicate nDEP (or repelling) from the E-field gradient. **a** CMF for solid polystyrene ($\varepsilon_p = 2.56\ \varepsilon_0$, $\sigma - 160 \times 10^{-4}$ S/m) spheres (Eq. 6.5). **b** CMF for a vesicle with a 1 μm radius (core: $\varepsilon_c = 60\ \varepsilon_0$, $\sigma = 0.5$ S/m) with a 5 nm shell/membrane ($\varepsilon_s = 10\ \varepsilon_0$, $\sigma = 1 \times 10^{-8}$ S/m) using the shell CMF model (Eq. 6.6)

Using Eq. 6.3, it is clear that the CMF will be frequency dependent with the conductivity values, σ, dominating the magnitude of the complex dielectric permittivity at low frequencies and the dielectric constants, ε, dominating the complex magnitude at high frequencies. Therefore, the real component of the CMF (i.e. Eq. 6.5) can change as a function of ω, and will range from -1 to 1. Ultimately, the sign on this value will then determine the direction of force exerted on the particle and thus two regimes of operation will exist. The condition in which the real part of the CMF is a positive value, will herein be referred to as positive DEP (pDEP), and conversely negative DEP (nDEP) if the opposite is true. Examples of the CMF plotted for solid spheres composed of different materials and a different sized shelled vesicles is provided in Fig. 6.2.

6.2.3 Dielectrophoresis (DEP) Force Derivation

Starting with the Kelvin Polarization force density (Eq. 6.2) the polarization vector, **P**, can be written in terms of the effective polarizability, $\widetilde{\alpha}$ per unit volume (Eq. 6.4) using the dipole relation (Eq. 6.1):

$$\mathbf{F_K} = \widetilde{\alpha}\ \mathbf{E} \cdot \nabla\mathbf{E} \tag{6.9}$$

Equation 6.9 can then be written more compactly using the following vector property,[1] see Eq. 6.10

[1] $\mathbf{A} \cdot \nabla(\mathbf{A}) = \frac{1}{2}\nabla(\mathbf{A} \cdot \mathbf{A})$, where **A** is a vector.

$$\mathbf{F}_{DEP} = \frac{1}{2}\,\widetilde{\alpha}\,v\,\nabla|\mathbf{E}|^2 \tag{6.10}$$

Here, $\mathbf{F}_{DEP}$, is now a real force [units: N] after $\mathbf{F}_{\mathbf{K}}$ was integrated over the particle's volume, v. Therefore, by inserting the derived polarizability, $\widetilde{\alpha}$, (Eq. 6.4) and volume of a sphere, Eq. 6.10 is cast into the classic DEP force equation:

$$\boxed{\mathbf{F}_{\text{DEP}} = 2\pi\,\varepsilon_L\,a^3\,Re\left\{\widetilde{f_{cm}}\right\}\,\nabla|\mathbf{E}|^2} \tag{6.11}$$

In Eq. 6.11, the real part of the CMF was explicitly used to express the real force exerted on the particle with a radius, a. The E-field is assumed either DC or an RMS AC value.

A couple of important observations can be made from Eq. 6.11. First, the magnitude of the DEP force depends on the gradient of the E-field squared. Increasing the DEP force can be realized by either increasing the E-field strength itself (by increasing the electric potential) resulting in a squared response in the DEP force, or utilizing a geometry that promotes larger spatially varying E-field lines. This will motivate the use of nanogap spaced electrodes in the following sections to boost the gradient term without introducing more power into the system. This comes at the cost of reduced far-field manipulation (i.e. the trapping volume is decreased) due to more spatially confined E-fields. Next, as mentioned in Sect. 6.2.2, the sign of $\widetilde{f_{cm}}$ will determines the direction the sphere moves, either towards or way from the E-field gradient. Therefore, careful choice in the operating frequency is needed depending on the desired application and manipulation of the particle. Lastly, the DEP force has a cubic response on the size/radius, a, of the particle. Therefore, smaller particles will experience a significantly reduced DEP force compared to their larger counterparts with all things being equal. For example, by reducing the particle size by half, the DEP force is reduced 8-fold. This motivates a discussion on the effective trapping volumes, i.e. the volume surrounding the E-field gradient in which a particle can be effectively actuated via DEP forces, see Sect. 6.2.4.

6.2.4 DEP Trapping Volume

The DEP trapping volume is generally defined as the volume in which the DEP force (Eq. 6.11) can "sufficiently" manipulate a given particle. Since the magnitude of $|\mathbf{E}|^2$ will diminish the farther the particle's position is from the electrodes, the resulting DEP force will decay at a similar rate of $\sim|\mathbf{r}|^{-2}$, where, $\mathbf{r}$ is some position vector from the electrodes. Further, for a given voltage, the position, $\mathbf{r}$, at which the particle can be manipulated will greatly depend on its radius, a. Therefore, the trapping volume will grow with a squared response to increasing voltage and will be unique for a given particle size.

Once a solution is at equilibrium particles are evenly disbursed within a solution and their resulting movement is stochastic (i.e. Brownian motion), governed my thermal energy. Therefore, classically the trapping volume is defined as the extent in which the DEP force can overcome a 1D thermal motion (either a radial step towards or way from the gap), assuming the electrodes are infinity long in one direction, defined as [4, 7–9]:

$$|\mathbf{F}_{\mathrm{DEP}}| > \frac{k_b\,T}{2a} \tag{6.12}$$

where T is the temperature of the solution and, k_b, is Boltzmann's constant. Using Eqs. 6.11 and 6.12, the volume in which $\nabla|\mathbf{E}|$ satisfies Eq. 6.13 will define the trapping volume in which particles entering the volume will then rapidly be accelerated either toward or away from the E-field gradient depending on the sign of the CMF, i.e. $Re\left\{\widetilde{f}_{cm}(\omega)\right\}$.

$$\boxed{\nabla|\mathbf{E}|^2 > \frac{k_b\,T}{4\pi\ \varepsilon_L\ a^4\ Re\left\{\widetilde{f}_{cm}\right\}}} \tag{6.13}$$

Noteworthy observation, smaller particles exhibit stronger thermal responses, and thus the trapping volume in Eq. 6.13 was reduced by another factor of the particle size, a, that further exacerbates the challenge of DEP trapping small particles. The reducing effect on the trapping volume as a function of the particle size is illustrated using coplanar electrodes, see Fig. 6.3.

Lastly, it is often useful to consider the amount of time necessary for a certain particle concentration is reached for discernible detection when utilizing DEP in biosensing applications. With a DEP trapping volume defined (Eq. 6.13), a sensing

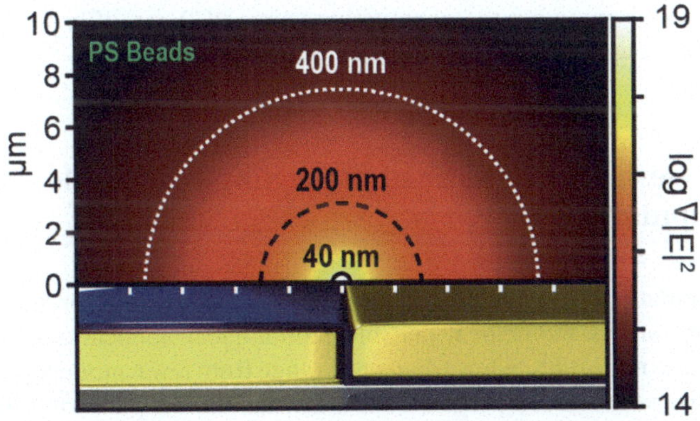

Fig. 6.3 DEP trapping radius: A 2D electroquasistatic simulation of the spatial distribution of $\nabla|E|^2$ for a 10 nm conformal Al_2O_3 layer (purple) that defines the electrode gap between coplanar electrodes. Simulation was ran for a 1 V_{RMS} (1 MHz signal) electric potential and is assumed the coplanar electrodes are infinite into and out of the page. The radial distance in which the beads will be trapped due to the DEP force overcoming the thermal force, F_{th}, of Brownian motion (Eq. 6.13) are defined for polystyrene (PS) spheres with diameters of 40, 200, and 400 nm in DI water

surface is now effectively projected onto the surface area of this volume where particles that pass through the surface of this volume with pDEP will be accelerated to the sensor. Therefore, Eq. 6.15 can be used to predict the relative speed DEP will accelerate the time of detection.

Considering a sensing area in which analytes will stick upon collision (e.g. chemical binding or DEP trapping) that is semi-infinite, the diffusion driven surface coverage per area as a function of time has been derived and is provided below (ignoring surface saturation effects) [10]:

$$\Pi(t) = 2\, C_B \sqrt{\frac{Dt}{\pi}} \tag{6.14}$$

where C_B is the equilibrium bulk concentration, D is the Stokes-Einstein diffusion constant and t is the time for the sensing area to reach a coverage of: $\Pi(t)$ [units: m^{-2}]. Now, if analyte detection can be made at some limit-of-detection surface coverage, Π_{LOD}, the time, t_d, for diffusion to fill the sensing surface to this value is:

$$t_d = \frac{\pi}{4D}\left(\frac{\Pi_{\text{LOD}}}{C_B}\right)^2 \tag{6.15}$$

The DEP force can then reduces the time for detection, t_d, by effectively decreasing the limit-of-detection, Π_{LOD}, due to the increased sensing surface area. For example, if the limit of detection was one particle within a diffraction-limited laser spot (waist $= 1\ \mu\text{m}$), the $\Pi_{\text{LOD}} \approx 1/\pi(0.5\ \mu\text{m})^2$, where the area of the laser spot defined the sensing area. Now, if DEP is used with a cylindrical trapping volume (radius $= 1\ \mu\text{m}$), the projected sensing area is expanded resulting in a 4-fold reduction in the limit of detection: $\Pi_{\text{LOD}} \approx 1/\pi(1\ \mu\text{m})^2$ and thus 16× faster detection. In this way, Eq. 6.15 can be used to approximate the quantitative gains DEP can offer for accelerated sensing applications.

6.3 DEP Design and Fabrication for Single Molecule Manipulation

Precise manipulation of single molecules such as DNA, RNA and protein are highly important for various biomedical and biotechnological applications [11–14]. To manipulate these molecules, whose diameters range from few nanometers to micrometers, researchers have developed various micro-/nanofluidic platforms that employs methods including electrophoresis, electroosmosis, electrofusion, electrowetting, and dielectrophoresis (DEP) [15, 16]. Among these methods, DEP has been widely reported for sorting, trapping, concentrating, and general manipulation of single molecules [17–19]. Typically, these devices are coupled to microfluidics, either in a basic configuration or more complex geometries up to full lab-on-a-chip systems. Often DEP is coupled with other AC electrokinetic phenomena to achieve additional capabilities (e.g. electroosmosis). Besides single molecule manipulation, these devices are also widely employed for the manipulation of single particle such as cell, virus etc. A comprehensive review of single particle manipulation using DEP is not attempted here but can be found elsewhere [3, 20].

A typical DEP device structure used for single molecule manipulation consists of at least two metal electrodes with a gap size from 100 nm down to 10 nm between them. More often an array of such electrodes was employed to scale up the device efficiency. A major bottleneck in DEP manipulation of single molecules is the fabrication of metal and/or insulating structures capable of generating the high electric field gradients required for trapping and concentrating sub-micron/nano sized particles—such as biomolecules. At these scales, particles exhibit high thermal forces due to Brownian movement that must be overcome. Since F_{DEP} is proportional to $V^2 L^{-3}$, (where V is the applied voltage and L is the distance between electrodes), a high dielectric trapping force can be achieved either by increasing the voltage between the electrodes or by reducing the distance between electrodes. Increasing applied voltage could lead to unwanted heat generation, bubble formation, and electrochemical reactions (e.g. hydrolysis) and hence is not desirable. On the other hand, shrinking the separation between two electrodes can significantly increase the force for a given bias voltage. Hitherto, most common DEP devices reported for single molecule manipulation are nanogaps between at least two adjacent planar electrode structures [21–23], although there are several reports on using nanopipettes for DEP trapping [24–26].

Different methods including electron beam lithography, electrochemical plating, Atomic layer deposition (ALD), focused ion beam (FIB) lithography, shadow mask evaporation, scanning probe microscopy lithography, on-wire lithography, and formation of mechanical break junctions were reported to fabricate DEP devices for single molecule manipulation [23, 30]. In many cases, different methods are often combined to obtain the desired device design. Figure 1.4 shows the different DEP device designs used for single molecule manipulation. The first design is a common four electrode assembly, called quadrupole electrode configuration [3, 31–34] (or sometimes called polynomial electrode system because the electrode potential in these system at any point is defined by a polynomial that obeys Laplace's equation) which are used for DEP trapping while maintaining the capability for electrorotation for additional characterization of molecules (Fig. 6.4a). Molecules are trapped due to positive DEP between the electrodes where the electric field gradient is high (between AC/ground pairs) and negative DEP causes them to accumulate towards the center of the structure. Generally, this design consists of a pair of two electrodes (AC/ground), but most common one employs four electrodes or the quadrupole for more stable trapping. Versions of this configuration were reported for trapping and manipulation of oligonucleotides [35], proteins [33] and single entities such as cells and viruses [3]. The design in Fig. 6.4b is an example of an AC/ground pair with a sharp end electrode assembly reported for trapping small molecules such as proteins [9]. These devices are fabricated mainly using electron beam lithography that offers precise control of the electrode design. The small interelectrode distance realized in these devices enables the generation of large electric field gradients required for the manipulation of small biomolecules. The interdigitated device design depicted in Fig. 6.4c is

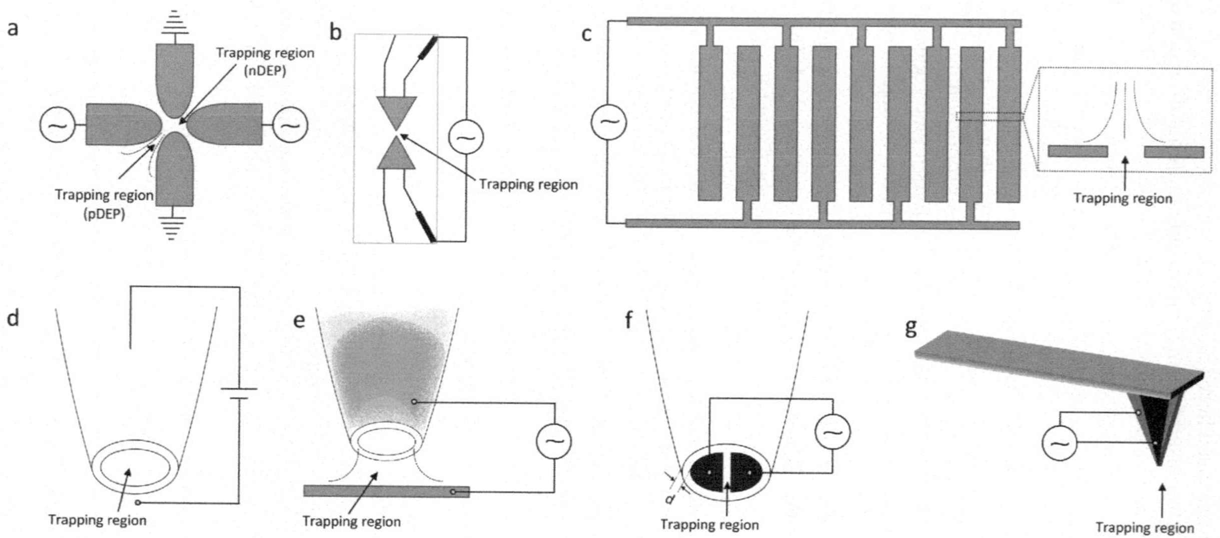

Fig. 6.4 Schematic representations of various DEP devise designs Used for DEP single molecule manipulation: **a** Quadrupole DEP device, where positive DEP (pDEP) induces trapping of molecules between the AC/ground electrode pairs and negative DEP (nDEP) concentrates molecules at the centre of the structure [27]. **b** The DEP device comprised of a planar gold electrode fabricated using electron beam lithography [9]. A pair of triangular electrodes with a radius of curvature less than 60 nm and a small interelectrode distance realised in this structure helps to achieve a maximum field gradient while keeping the heating effect and electrohydrodynamic fluid flow to a minimum. **c** An interdigitated electrode assembly. Inset: vertical view of trapping region between electrodes [27]. **d** Schematic of nanopipette based DEP trapping device used for molecule trapping via DC-DEP [28]. **e** Nanopipette based DEP-nanopore device for trapping and sensing of analyte molecules from ultra-low concentration solutions [25]. **f** Design of DEP nanotweezer used for single molecule manipulation and single cell analysis [24]. **g** Schematic of AFM-tip based DEP device for DEP manipulation of molecules [29]

commonly used for separation and concentration of biomolecules in flow-through devices [21]. By coupling with other techniques such as acoustic waves [36] with interdigitated electrode devices, molecules can be preconcentrated in these systems prior to focusing them to specific flow channels or locations by using DEP. Other variants of interdigitated devices such as castellated or sinusoidally corrugated electrodes assemblies were also developed for precise manipulation of single entities [37, 38].

Nanopipette based DEP devices were reported for single molecule manipulation. The nanopipettes used for this application (Fig. 6.4d) generally has an internal diameter of tip of 100–150 nm, sufficient to create electric field strengths required for trapping single protein molecules. Using a physiological buffer to retain protein integrity, Clarke et al. measured protein conductivity and demonstrated the trapping of two proteins, protein G, and immunoglobulin G, and a maximum of 3000-fold protein concentration via positive DEP with reversible protein accumulation [28, 39]. In a later study Edel et al. reported another nanopipette based DEP device incorporating a nanopore for trapping of DNA molecules from ultra-low concentrations at the nanopore for their subsequent detection via resistive pulse measurement [25]. A schematic of this device is shown in Fig. 6.4e. To adapt the nanopipettes (25 nm diameter) for use in DEP trapping, a thin 5 nm layer of Au was deposited onto the area surrounding the pipette tip. With the gold electrode in proximity to the pore, high gradient forces can be generated at the pipette tip (nanopore) by placing it close to a metal surface and applying an AC field between the gold coated at the pipette and the metal surface to trap and concentrate analyte molecules at the nanopore. A recent study from the same group reported another nanopipette based DEP device comprised of two nanoelectrodes spaced few nanometers apart to trap and manipulate single molecules from both solutions and living cells (Fig. 6.4f) [24]. The application of an AC voltage across these electrodes creates an electric field gradient, as high as $10^{28}\,V^2m^{-3}$ near the electrode gap. With such high field gradients, single DNA molecules well below 200 base pairs could be trapped.

The AFM tip-based DEP device proposed by Wickramasinghe et al. enabled its incorporation into a scanning probe platform for spatially resolved manipulation of biomolecules within living cells, Fig. 6.4g [29]. These devices were fabricated using commercially available conical, highly doped (resistivity 4^{-6} Ωcm) silicon AFM probes by growing a 20 nm thick layer of thermal SiO_2 on them to electrically insulate the entire silicon probe including the AFM cantilever and handling chip. This was followed by the deposition of a Ti/Pt (10/20 nm) layer onto the probe tip by electron beam evaporation. The DEP electrodes were then formed by polishing the end of the Pt coated tip until the doped silicon tip is just exposed. Application of an AC field across the electrodes provided the electric field gradient required for biomolecule trapping. Detailed reviews focusing on fabrication of nanoscale DEP devices can be found in these references [22, 23].

6.4 Dielectrophoretic Manipulation of Single Molecules

Single molecule sensing is a rapidly expanding technology for the detection and quantification of a wide range of biomolecules including nucleic acids and proteins. Many single-molecule detection technologies, such as nanopore sensing [40] and field-effect biosensing [41] architectures offer significant opportunities for the advancement of rapid Point-of-Care (POC) diagnostics. Nevertheless, the dominant mechanism of capture and detection of the analyte in these sensors is diffusion-limited, making it difficult to concentrate and perform high-throughput detection of ultra-dilute samples [25]. Ever since Pohl and Hank first demonstrated the possibility of employing DEP for selective separation of single entities [42], many researchers attempted to extend the capability of this platform to break the diffusion barrier to increase the selectivity and sensitivity of single molecule techniques. Hitherto, DEP was reported for the manipulation of a range of biomolecules including DNA, RNA and proteins [3, 17, 43]. However, as will be highlighted in the following sections, DEP manipulation of single molecules is at varying stages of development with each method requiring continued research to reach its full potential.

6.4.1 Dielectrophoretic Manipulation of DNA and RNA

The ability to manipulate DNA molecules by exploiting its induced dipole moment offers exciting possibilities for the development of novel DNA separation techniques and their controlled movement for DNA diagnostics to a wider use in genomics and proteomics. Experiments over the past three decades, have demonstrated ways to orient, stretch, transport, and trap DNA using nonuniform fields generated in microfabricated devices [3, 44–48].

In the first reported study on DEP behavior of DNA, Washizu showed that lambda phage DNA (λ DNA) suspended in deionized water attracted towards the high field regions of DEP electrodes upon application of a 1 MHz voltage bias [49]. A follow up study from the same group demonstrated the DEP induced stretching and alignment of single λ DNA molecule in solution [48]. Subsequent studies using different electrode designs investigated the dielectric and DEP behavior of DNA.

An exciting application of the earlier research on DEP effect on DNA can be seen in the study by Yamamoto et al. where a novel method for the space-resolved dissection (molecular surgery) of deoxyribonucleic acid (DNA) using electrostatic molecular manipulation is demonstrated [46]. Unlike in the conventional restriction digestion of DNA where DNA-cutting enzymes (Restriction enzymes, e.g. DNase) and DNA are mixed in the reaction buffer and the restriction process is depending on stochastic chances, the presented method offered a reproducible cutting of DNA at any desired position along the DNA molecule by its physical manipulation using DEP. To realize this space-resolved fragmentation, DEP was employed to stretch and anchor the molecule on a solid surface. Besides these applications, a wide range of reports on

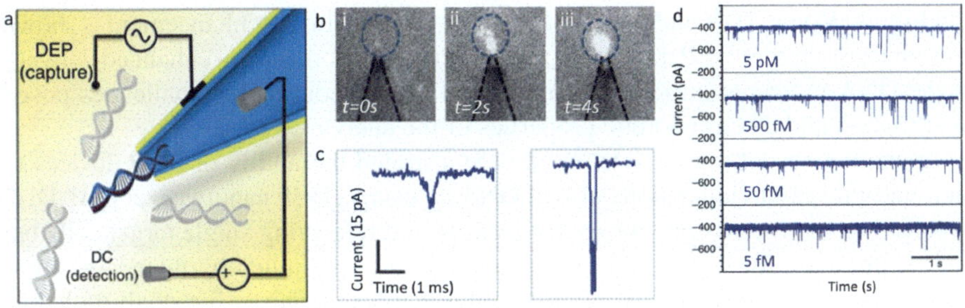

Fig. 6.5 DEP-Nanopore device for single molecule detection from ultra-low concentration solutions: **a** Schematic of the nanopipette DEP-nanopore device showing DNA being threaded through the tip of the device. **b** Fluorescent micrograph of DEP trapping showing the progression of DNA trapping at the tip of the nanopipette at different time intervals. **c** Typical current modulation associated with single DNA molecule translocation recoded across the nanopore. **d** Current traces obtained using DEP as a pre-concentration step for four different concentrations (5 pM, 500 fM, 50 fM and 5 fM). A bias of $|\Delta V| = 500$ mV was applied across the nanopore for all the experiments. A single molecule translocation event is seen as spikes in the current trace. Figures reprinted in part with permission from Freedman et al. [25] Copyright 2016, Springer Nature

DNA fragments sorting and concentration using various electrode configurations and DEP characterization of different DNA molecules in low and high conductivity media have also been reported. Comprehensive reviews on these developments can be seen in these references [3, 44].

As mentioned in the previous section, an important aspect of single molecule manipulation is to overcome the diffusion barrier in delivering analyte molecules into the sensing regions of single molecule detection platforms—such as nanopore sensors. This is particularly important for detecting ultra-low concentration analytes such as cell-free circulating DNAs for early cancer diagnosis. To address this challenge, Edel et al. demonstrated a simple, yet powerful, method based on coupling single-molecule dielectrophoretic trapping to nanopore sensing for trapping and delivering DNA from ultra-diluted sample solution to the sensing region for fast and easy single molecule detection [25]. A schematic of the experimental set-up used for this study is show in Fig. 6.5a. Incorporation of a DEP trap to the nanopore was achieved by adding a 5 nm layer of gold onto the area surrounding the nanopore (tip of the nanopipette). Typical nanopore experiments use a constant DC bias to prompt the translocation of the analyte molecule through the nanopore while monitoring the current across the nanopore. The DC current modulation resulting from the translocation of an analyte molecule across the nanopore is used to identify and characterize the translocating molecule (Fig. 6.5c). In this study, an AC voltage (10–20 V and 0.5–4 MHz) was applied to the metallized layer surrounding the nanopipette for dielectrophoretic trapping and delivery of DNA (Fig. 6.5b) to the nanopore followed by a DC voltage which translocates the molecules across the nanopore for sensing. The AC and DC voltages are applied to the system using the two Au electrodes and two Ag/AgCl electrodes, respectively. By using this set-up, high sensitivity detection in concentrations as low as 5 fM was achieved using optimized DEP trapping conditions (Vpp = 20 V and 1 MHz), Fig. 6.5d. The sensitivity enhancement is shown to stem from a larger DEP trapping volume in combination with an enhanced delivery of the DNA molecule into the sensing region by overcoming the challenges posed by diffusion and electrophoretic properties of the analytes.

In another recent report, Edel et al. demonstrated the ability of DEP devices to pick-and-place individual molecules of DNA by using a DEP nanotweezer [24]. DEP tweezers were reported previously for picking and relocating single target cells but their application for single molecule manipulation was limited by their large footprint and the difficulty in generating high DEP forces required for small molecule manipulation. The DEP nanotweezer, on the other hand, was fabricated easily and inexpensively from dual barrel capillaries by pulling them into sharp end pipettes and pyrolytically depositing carbon inside the nanopipette to form two co-planar nanoscale electrodes separated by ca. 20 nm septum to generate high DEP forces required for single molecule manipulation (Fig. 6.6a). This nanotweezer was then employed in conjunction with an XYZ positioning platform to perform 'pick-and-place'-type measurements in which single molecules were trapped, moved at a velocity as high as 30 μms^{-1} and then released at a specific location. This was demonstrated for DNA molecules of which a single molecule was traced using an image tracking algorithm to follow the trajectory of the molecule from capture (Fig. 6.6b), movement in the x–y plane (Fig. 6.6 panels c and d) and subsequent release (Fig. 6.6e). Apart from using the DEP nanotweezers for performing single molecule manipulation in solutions, they were also employed for trapping and extracting DNA, RNA and individual organelles from living cells without affecting their viability.

One of the most important aspects in designing a DEP device for single molecule manipulation is the voltage required for its operation. Most of the DEP devices employed from DNA manipulation requires a high voltage for generating sufficiently large electric field gradients [51]. Because of the sophisticated electronic instrumentation required for generating these high voltages, it is of utmost importance to devise strategies to reduce input voltage requirements to facilitate their deployment in ana-

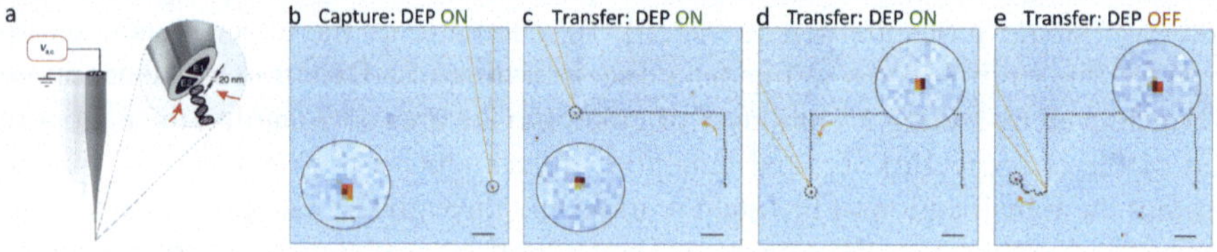

Fig. 6.6 DEP nanotweezer for single molecule manipulation: **a** Schematic of a typical DEP nanotweezer. Application of an AC voltage on the nanotweezer generates a highly localized electric field gradient that is suitable for targeted molecular trapping in solution or inside a cell. Panels **b** to **e** Pick-and-place of single molecule DNA using the DEP nanotweezer. The DNA molecule is captured at the nanotweezer tip by turning on the AC field (**b**), the captured single molecule is transferred from one position to another by moving the nanotweezer using a micromanipulator with the AC field kept on (panels(**c**) and (**d**)) and then released by turning off the DEP (**e**). Scale bars (right), 10 μm; inset, 2 μm [24]

lytical platforms. Additionally, the use of high voltages in microfluidic channels filled with conductive solutions generate significant Joule heating affecting the device sensitivity. Thus, high input-voltages represent an obstacle toward creating fully integrated, stand-alone platforms that can be used outside of a laboratory. To overcome this challenge, Barik et al. developed a Graphene-edge dielectrophoretic tweezers for low-voltage trapping of biomolecules [50]. A combination of ALD, Photolithography and electron beam deposition was used to fabricate this DEP device, in addition to the atomically sharp edges of monolayer graphene to generate singular electrical field gradients for trapping DNA molecules. A schematic of the device design is shown in Fig. 6.7a where a metal electrode (Ti/Pd) was first patterned using photolithography onto a thick layer of SiO_2 (300 nm) thermally grown on a Si wafer. This was followed by a combination of reactive ion etc.hing and wet etc.hing. An 8 nm thick dielectric layer was deposited on top of the Pd electrodes with HfO_2 using atomic layer deposition (ALD). Then, single-layer graphene grown by chemical vapor deposition (CVD) was transferred onto the wafer and etched into rectangular patterns. Electrical contacts (Cr/Au) were patterned on top of the graphene layer via photolithography and electron beam deposition. The bottom electrode and these top contact electrodes were arranged in an interdigitated fashion to minimize series resistance and to produce more exposed edges to efficiently trap the DNA molecules. To demonstrate the utility of the graphene electrodes in capturing biomolecules, DEP trapping of 10 kbp and 500 bp DNA molecules tagged with YOYO-1 dye at a final concentration of 10 pM in a 10 μM KCl solution was performed. Trapping and releasing of 10 kbp DNA was observed at 1 MHz and 10 MHz, respectively (Fig. 6.7b panels i to vii). 500 bp DNA molecules were found to trap along the graphene edge at 2.5 V and 1 MHz frequency. This approach is scalable and highly reproducible at the wafer scale. Additionally, the interdigitated electrode arrays allowed large coverage selective trapping and positioning of these DNA molecules precisely along the atomically sharp edges of graphene under the same low-voltage operation due to the extreme DEP forces generated from the atomic graphene electrode flake.

High-aspect ratio, coplanar nanogaps have also been demonstrated for large surface coverage with low-volt DEP; this configuration was especially optimal for integration with anamorphic optics for rapid, real-time vibrational spectroscopic imaging of molecular cargo within liposome capsules [52]. Moreover, the extreme coplanar nanogaps were fabricated using a novel atomic layer lithography technique that comprises the peeling of adhesive tape for creating the extreme aspect ratio (e.g. <10 nm electrode gap maintained over several millimeters in length) coplanar electrodes [7, 53]. The entire coplanar nanogap can then be excited simultaneously for Raman imaging using a laser line (via anamorphic optics) in which DEP precisely positions the analytes within the excitation and imaging path for readout using an imaging spectrometer (Fig. 6.8). The nanogap separation offered 1 V amplitude trapping and active passing between segmented parallel structures of 70 nm liposomes containing 4-Mercaptopyridine (4-MPY), a small (0.1 kDa) Raman active carbon ring molecule. Bare gold nanoparticles (AuNP) spiked into solution would simultaneously be trapped (over a wide range of operating frequencies) to offer surface enhanced Raman spectroscopy of the 4-MPY molecular cargo. Referred to as

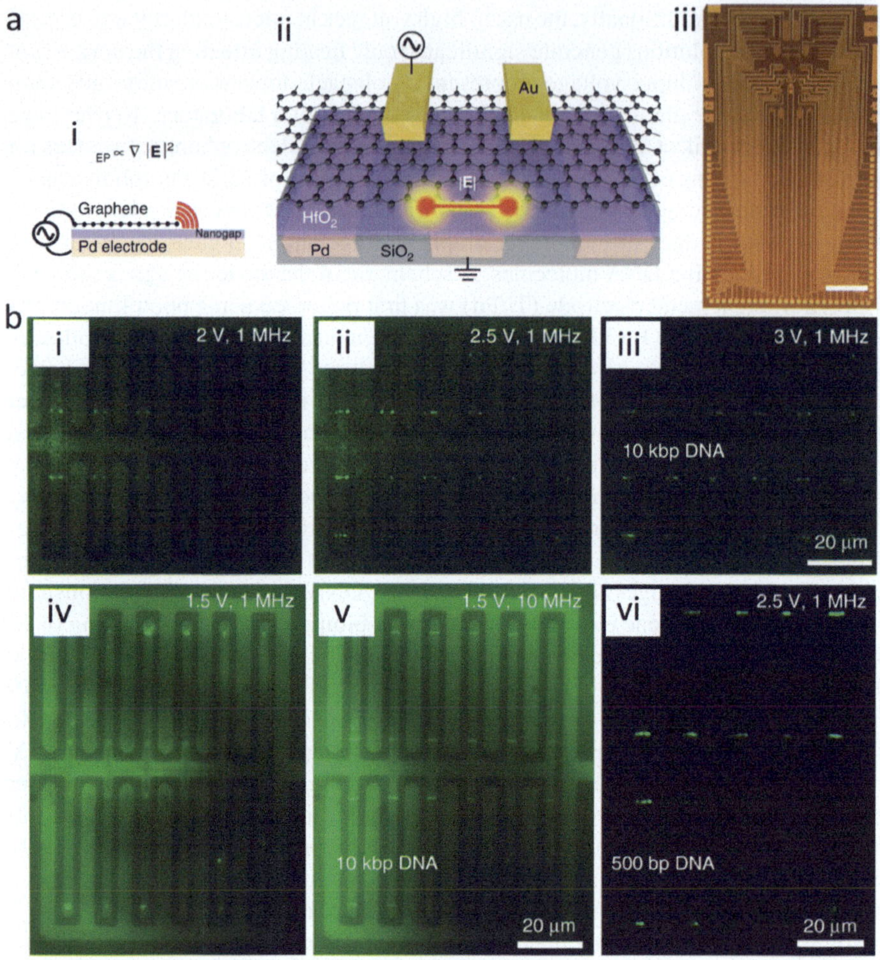

Fig. 6.7 Atomic scale graphene electrode for low-volt, large surface coverage DNA positioning. a Fabrication of a graphene-edge dielectrophoretic tweezers. **(i)** Schematic of the device. The edge of graphene could provide the smallest possible radius of curvature necessary for generating high electric field gradients. **(ii)** An illustration showing the region of strongest electric field gradient is generated at the intersection of the edge of the graphene by applying an AC bias between the graphene contact electrode (gold) and palladium gate electrode. **(iii)** A photograph of the chip. Scale bar: 2 mm. **b** DEP manipulation of DNA molecules. Panels **(i)** to **(iii)** Trapping and releasing of 10 kbp DNA was observed at 1 MHz. Increasing the voltage amplitude from 2 to 2.5 to 3 V, traps more DNA molecules at the trapping sites. Panels **(iv)** and **(v)** At a higher solution conductivity (1 nM DNA in 1 mM KCl, 0.93 mS/cm), DNA localization vs the tight trapping phenomenon is observed at a higher frequency range. **(ix)** 500 bp DNA molecules (threshold $\nabla |E|^2$ an order of magnitude higher than that of the 10 kbp DNA molecules) were trapped along the graphene edge at 2.5 V and 1 MHz frequency. Figures reprinted in part with permission from Barik et al. [50] Copyright 2017, Springer Nature

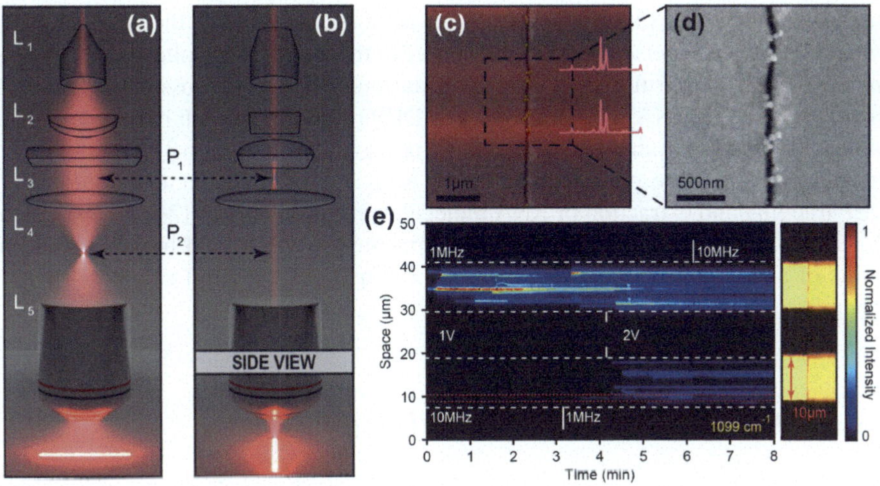

Fig. 6.8 Coplanar nanogap DEP for trapping, Raman And Imaging Line (TRAIL) biosensing of molecular cargo: **a–b** Anamorphic optics are used to generate a laser line excitation for Raman imaging. A Powell lens (L_1) and two cylindrical lenses (L_2, L_3) elongate a laser spot along one dimension (**a**) and creates the laser line image at P_1. A spherical lens (L_4) then collimates the short dimension (**b**) at P_2, located at the back-focal plane of a microscope objective. **c–d** A scanning electron microscope image of gold nanoparticles (AuNPs) trapped along the coplanar nanogap via DEP (**d**) with the locations of the observed surface-enhanced Raman spectroscopy (SERS) overlaid (**c**) of 4-MPY attached to the trapped AuNPs. **e** Two independent DEP devices are independently operated for real-time positive and negative control TRAIL chemical imaging of liposomes carrying 4-MPY cargo. Bare AuNPs spiked into the solution provides the necessary SERS in which the 1099 cm^{-1} peak is plotted. The top device operates using a 1 V amplitude, 1 MHz signal and generates pDEP for trapping both the liposomes and AuNPs. The bottom device initially operates at 1 V, 10 MHz which generates nDEP for the liposomes and thus no SERS is observed. Once the bottom device is switched to 1 MHz, SERS spectra is then observed as liposomes carrying 4-MPY are trapped. Inset contains a microscope image of the two independent DEP devices spatial locations. Figures reprinted in part with permission from Ertsgaard et al. Copyright 2018, American Chemical Society [52]

Trapping Raman And Imaging Line (TRAIL), this method offered label-free (i.e. no chemical tagging necessary) for both rapid collection and detection of the target analytes that was 100 × faster than a similar point-scan imaging system using the same dwell time and resolution [52]. By combining DEP with spectroscopic imaging, the TRAIL platform can provide high-information datasets, containing spatial, temporal, and spectroscopic with relative intensity information regarding molecular cargo in biocapsules (Fig. 6.8). This could be especially advantageous for the characterizing of exosomes–extracellular vesicles that are often low in concentration and known to vary in size and molecular content, for early diagnosis of pathogenic behavior [52].

Unlike the large number of DEP literature on manipulation of DNA, very few applications of DEP for manipulation of RNA was reported thus far. Giruad et al. examined the DEP behavior of 16 and 23 s subunits of E. Coli rRNA using interdigitated microelectrodes [54]. In their study the authors characterized the frequency and voltage dependence on the trapping efficiency of the RNAs. The voltage dependence measured at 1 MHz ruled out any significant contribution of possible permanent dipole moment of RNA molecules on their DEP response. In each bias, these RNA molecules showed maximum DEP trapping (positive DEP) between 3 kHz and 1 MHz and a negative DEP response above 9 MHz. More recently, Edel et al. demonstrated DEP trapping and extraction of RNA molecules from living cells [24]. Even though the technology is in its infancy, DEP manipulation of RNA hold high potential in terms of analytical applications especially for spatially resolved single cell RNA mapping.

6.4.2 Dielectrophoretic Manipulation of Protein

Selective trapping and manipulation of proteins is essential for diagnostic applications as well as for understanding their structural and functional features. Hitherto, many examples of DEP devices for trapping and manipulation of protein have been reported in the literature [27]. A comprehensive review of dielectrophoretic trapping can be seen in these references [27, 55]. Among the examples of DEP manipulation of proteins reported thus far, the seminal work by Washizu et al. deserves special mentioning [37]. Up until then, DEP was mainly employed for manipulation of micrometer-sized particles such as biological cells, since DEP manipulation of proteins requires relatively high electric field gradients to overcome the high Brownian motion. Nevertheless, by using an interdigitated sinusoidally corrugated DEP device, Washizu et al. demonstrated the DEP trapping of Avidin, concanavalin A, chymotripsinogen A and Ribonuclease A and attempted to perform 'dielectrophoretic chromatography' to separate biopolymers based on their polarizability. Nearly a decade later, Asokan et al., demonstrated another example of DEP manipulation of actin-myosin systems using a quadrupole electrode system [56]. Figure 6.9a shows the actin filaments trapped at high field region in the quadrupole electrode with the filaments oriented parallel to the field lines.

Another study by Holzel et al. demonstrating the trapping of freely diffusing R-phycoerythrin protein deserves special attention, partially because of the improvement in device design that enabled the generation of relatively high electric field gradients [9]. The DEP device was comprised of planar gold electrodes prepared on a low n-doped silicon substrate. To achieve a maximum field gradient while keeping the heating effects and electrohydrodynamic fluid flow to a minimum, a pair of triangular electrodes with a radius of curvature of less than 60 nm and an interelectrode distance of 500 nm was fabricated using electron beam lithography. Figure 6.9b shows an electron micrograph of a typical DEP device used in this study. R-phycoerythrin was chosen as the analyte molecule because of its intense autofluorescence and the visualization of the trapping was achieved by recording the fluorescence intensity at the electrode surface. Applications of a sinusoidal signal of 10 V (root mean square) at 1 MHz resulted in fluorescing spots appearing independently at both electrode tips

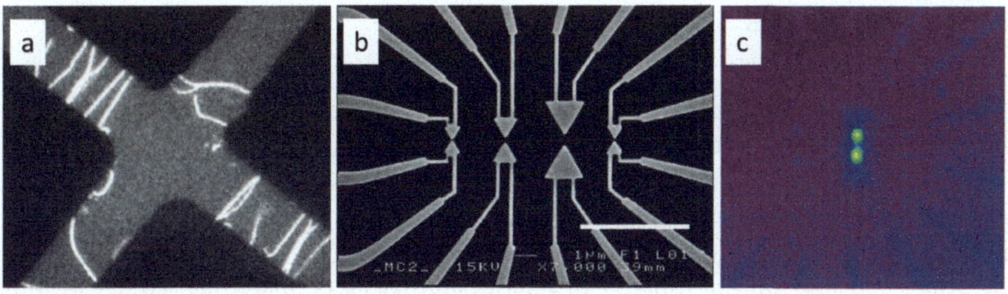

Fig. 6.9 Dielectrophoretic manipulation of protein a Fluorescent micrograph showing actin filaments collected at the high field regions of a quadrupole electrode at an applied frequency of 1 MHz and 7 V. The polarized filaments orient themselves parallel to the field lines. Figures reprinted in part with permission from Asokan et al. [56] Copyright 2003, American Chemical Society. **b** Electron micrograph of DEP device used for trapping freely suspended R-phycoerythrin molecules. Scale bar 5 μm. The largest electrode pair with 500 nm gap width was used for DEP trapping of the protein molecule. **c** Fluorescence micrograph recorded 10 s after applying a voltage bias and subsequent DEP trapping. The fluorescent spot at the electrode tip is due to the accumulation of R-phycoerythrin molecules. Figures reprinted in part with permission from Holzel et al. [9] Copyright 2005, American Physical Society

within about 10 s Fig. 6.9c). When the voltage was turned off, the fluorescent intensity decreased as the molecules freely diffused away from the electrode tip. Subsequent studies have demonstrated the DEP trapping and manipulation of a wide range of protein molecules including streptavidin, immunoglobulins and amyloid beta fibrils. An updated list of list of proteins captured or trapped using DEP devices can be seen in this reference [55].

6.5 Discussion

While many advantages exist with using DEP for particle manipulation, it is important to understand its limitations and how it can be optimized in the future. The first consideration is the operating frequency. It was provided in Sect. 6.2 that the CMF determines the frequency response of the particle and that the sign on the real part of Eq. 6.5 indicates whether the particle feels pDEP where the particle feels a force towards the E-field gradient (i.e. trapping) or nDEP where the particle is repelled from the E-field gradient. Composite particles (e.g. a shell) can have complex frequency responses that limit the range of frequency for either pDEP or nDEP (Fig. 6.2). Two additional constraints can exist which can further limit the operating frequency and/or whether DEP actuation can be feasible.

6.5.1 *Electrolysis*

The first is electrolysis and/or hydrolysis. Electrolysis refers to the oxidation or reduction (REDOX) of chemical species within liquid bodies that is activated under a voltage potential. This is called hydrolysis if the REDOX occurs with specifically water molecules–generating hydrogen and oxygen gas. Hydrolysis can occur with DC voltages as low as 1.23 V [57]. Often passivization layers are used to mitigate this effect and prevent the transfer of charge and REDOX reactions. However, this can be problematic for DEP by consuming most of the strong E-field (especially for nanogap DEP) within the passivation layer itself and thus reducing DEP actuation.

Another approach is to increase the operating frequency of the AC signal to reduce the time for REDOX reactions. With sufficiently high operating frequency, the electrolytic reactions can be reversed before adequate separation of the products can accumulate (e.g. bubbles and gas formation) [58, 59]. Therefore, a balance must be kept between the voltage, current, frequency, and passivization layer used. As a reference, frequencies around 100 kHz, using 1 W of power (under 6–7 V) and no passivation layer was at the threshold of negligible bubble formation as observed under high-speed camera [58]. This then motivates using the highest operating frequency available that still satisfies the CMF for the desired DEP actuation.

6.5.2 *RC Roll-Off*

The second limit on the operating frequency is RC roll-off [60, 61]. This is an important consideration for nanogap DEP which will typically have larger device capacitance and/or transparent and thin DEP electrodes which will typically have larger series resistance. In principle, the DEP device behaves like a low-pass filter with an upper bound cutoff frequency that must be considered. The DEP electrodes will possess a capacitance, C, that must be given adequate time to fully "fill" or charge for actuation. If the signal oscillates with too high of a frequency, only a fraction of the voltage signal is dropped over the device resulting in poor DEP actuation.

Therefore, the operating frequency should be kept below the circuit's cutoff frequency (Eq. 6.16) for all applications and is an upper bound for efficient power usage for liquid actuation.

$$\omega < \frac{1}{RC} \tag{6.16}$$

In Eq. 6.16, ω, is the angular operating frequency and R, is the entire circuit's series resistance (and/or sheet resistance for 2D materials). Thin, transparent, or 2D material conductors and semi-conductors used as DEP electrodes will often large resistance which can reduce the cutoff frequency. Further, RF power sources are typically issued a standard output impedance of 50 Ω. Therefore, this too will pose a limit on the design of the DEP device and/or max frequency that can be used. Since nanogap

DEP will have relatively larger capacitance due to a small electrode nanogap, g (see Equation for a parallel plate capacitor) the electrodes should then be designed with minimal area of overlap, A, (e.g. coplanar electrodes) to mitigate RC roll-off.

$$C = \frac{\varepsilon A}{g} \tag{6.17}$$

6.5.3 *Joule Heating*

Lastly, as EM power is dissipated either as Ohmic current in ionic solutions or displacement current in dielectric, the dissipation of this power often is converted into heat (i.e. Joule heating) of the sample. The change in temperature, ΔT, of the solution scales with the power dissipated (i.e. $\propto V^2$) and can be approximated as follows [7, 62]:

$$\Delta T \approx \frac{\sigma V^2}{2k} \tag{6.18}$$

where σ is the electrical conductivity of the solution and k is the thermal conductivity of the solution. Therefore, to mitigate the effects of Joule heating increasing the E-field gradient (rather than the operating voltage) is desirable and is a main advantage of using nanogap DEP.

6.6 Future Outlook

DEP has proven to be a versatile method for trapping, concentration, sorting and manipulation of single molecules including nucleic acids and proteins. Single molecule manipulation using DEP is straightforward and fully tuneable by altering the electric field strength and the medium conductivity. Hitherto, DEP devices for single molecule manipulation are mainly composed of planar electrode geometries that can be readily integrated into microfluidic platforms. Nevertheless, there are several reports on using nanopipettes for DEP manipulation of single molecules. Ever since the first demonstration of DEP trapping of DNA, the huge amount of research on single molecule manipulation using DEP has resulted in the development of a variety of device configurations and experimental strategies. The mechanical robustness and portability makes these platforms ideal candidates for using them with various point-of-care applications in clinical diagnostics. Although much progress has been made in the last three decades, the DEP single molecule manipulation systems can be further improved in many ways and adopted for new applications such as continuous live cell sampling, early detection of disease biomarkers (e.g. exosome characterization) and the structural investigation of biomolecules.

References

1. Paul E. Sheehan and Lloyd J. Whitman. Detection limits for nanoscale biosensors. *Nano Letters*, 5(4):803–807, apr 2005.
2. Todd M Squires, Robert J Messinger, and Scott R Manalis. Making it stick: convection, reaction and diffusion in surface-based biosensors. *Nature Biotechnology*, 26(4):417–426, apr 2008.
3. Ronald R. Pethig. *Dielectrophoresis: Theory, Methodology and Biological Applications*. WILEY, 2017.
4. Hywel Morgan. *AC electrokinetics : colloids and nanoparticles*. Research Studies Press, Philadelphia, PA, 2003.
5. Parikha Mehrotra, Baibhab Chatterjee, and Shreyas Sen. EM-wave biosensors: A review of RF, microwave, mm-wave and optical sensing. *Sensors*, 19(5):1013, feb 2019.
6. Hermann Haus. *Electromagnetic fields and energy*. Prentice Hall, Englewood Cliffs, New Jersey, 1989.
7. Avijit Barik, Xiaoshu Chen, and Sang-Hyun Oh. Ultralow-power electronic trapping of nanoparticles with sub-10 nm gold nanogap electrodes. *Nano Letters*, 16(10):6317–6324, sep 2016.
8. Herbert Pohl. *Dielectrophoresis : the behavior of neutral matter in nonuniform electric fields*. Cambridge University Press, Cambridge New York, 1978.
9. Ralph Holzel, Nils Calander, Zackary Chiragwandi, Magnus Willander, and Frank F. Bier. Trapping single molecules by dielectrophoresis. *Physical Review Letters*, 95(12), sep 2005.
10. D. Brynn Hibbert, J. Justin Gooding, and Paul Erokhin. Kinetics of irreversible adsorption with diffusion: application to biomolecule immobilization. *Langmuir*, 18(5):1770–1776, mar 2002.
11. Susan K Sarkar. Single molecule biophysics. In *Single Molecule Biophysics and Poisson Process Approach to Statistical Mechanics*. IOP Publishing, 2016.
12. Deyu Zhao, Siyun Liu, and Ying Gao. Single-molecule manipulation and detection. *Acta Biochimica et Biophysica Sinica*, 50(3):231–237, jan 2018.
13. Christopher P. Calderon, Nolan C. Harris, Ching-Hwa Kiang, and Dennis D. Cox. Analyzing single-molecule manipulation experiments. *Journal of Molecular Recognition*, 22(5):356–362, may 2009.
14. Ulrich Bockelmann. Single-molecule manipulation of nucleic acids. *Current Opinion in Structural Biology*, 14(3):368–373, jun 2004.
15. Cheng Qian, Haibo Huang, Liguo Chen, Xiangpeng Li, Zunbiao Ge, Tao Chen, Zhan Yang, and Lining Sun. Dielectrophoresis for bioparticle manipulation. *International Journal of Molecular Sciences*, 15(10):18281–18309, oct 2014.
16. P.K. Wong, T.-H. Wang, J.H. Deval, and C.-M. Ho. Electrokinetics in micro devices for biotechnology applications. *IEEE/ASME Transactions on Mechatronics*, 9(2):366–376, jun 2004.
17. Talukder Z. Jubery, Soumya K. Srivastava, and Prashanta Dutta. Dielectrophoretic separation of bioparticles in microdevices: A review. *Electrophoresis*, 35(5):691–713, feb 2014.
18. Daihyun Kim, Mukul Sonker, and Alexandra Ros. Dielectrophoresis: From molecular to micrometer-scale analytes. *Analytical Chemistry*, 91(1):277–295, nov 2018.
19. Nurhaslina Abd Rahman, Fatimah Ibrahim, and Bashar Yafouz. Dielectrophoresis for biomedical sciences applications: A review. *Sensors*, 17(3):449, feb 2017.
20. Linbo Liu, Ke Chen, Nan Xiang, and Zhonghua Ni. Dielectrophoretic manipulation of nanomaterials: A review. *Electrophoresis*, 40(6):873–889, oct 2018.
21. Khashayar Khoshmanesh, Saeid Nahavandi, Sara Baratchi, Arnan Mitchell, and Kourosh Kalantar-zadeh. Dielectrophoretic platforms for bio-microfluidic systems. *Biosensors and Bioelectronics*, 26(5):1800–1814, jan 2011.
22. Anton Kuzyk. Dielectrophoresis at the nanoscale. *Electrophoresis*, 7(10217), jul 2011.
23. M Li, W H Li, J Zhang, G Alici, and W Wen. A review of microfabrication techniques and dielectrophoretic microdevices for particle manipulation and separation. *Journal of Physics D: Applied Physics*, 47(6):063001, jan 2014.

24. Binoy Paulose Nadappuram, Paolo Cadinu, Avijit Barik, Alexander J. Ainscough, Michael J. Devine, Minkyung Kang, Jorge Gonzalez-Garcia, Josef T. Kittler, Keith R. Willison, Ramon Vilar, Paolo Actis, Beata Wojciak-Stothard, Sang-Hyun Oh, Aleksandar P. Ivanov, and Joshua B. Edel. Nanoscale tweezers for single-cell biopsies. *Nature Nanotechnology*, 14(1):80–88, dec 2018.
25. Kevin J. Freedman, Lauren M. Otto, Aleksandar P. Ivanov, Avijit Barik, Sang-Hyun Oh, and Joshua B. Edel. Nanopore sensing at ultra-low concentrations using single-molecule dielectrophoretic trapping. *Nature Communications*, 7(10217), jan 2016.
26. Kevin J. Freedman, Colin R. Crick, Pablo Albella, Avijit Barik, Aleksandar P. Ivanov, Stefan A. Maier, Sang-Hyun Oh, and Joshua B. Edel. On-demand surface- and tip-enhanced raman spectroscopy using dielectrophoretic trapping and nanopore sensing. *ACS Photonics*, 3(6):1036–1044, may 2016.
27. Asuka Nakano and Alexandra Ros. Protein dielectrophoresis: Advances, challenges, and applications. *Electrophoresis*, 34(7):1085–1096, apr 2013.
28. Richard W. Clarke, Samuel S. White, Dejian Zhou, Liming Ying, and David Klenerman. Trapping of proteins under physiological conditions in a nanopipette. *Angewandte Chemie*, 117(24):3813–3816, jun 2005.
29. D. Nawarathna, T. Turan, and H. Kumar Wickramasinghe. Selective probing of mRNA expression levels within a living cell. *Applied Physics Letters*, 95(8):083117, aug 2009.
30. Tao Li, Wenping Hu, and Daoben Zhu. Nanogap electrodes. *Advanced Materials*, 22(2):286–300, jan 2010.
31. Ling-Sheng Jang, Pao-Hua Huang, and Kung-Chieh Lan. Single-cell trapping utilizing negative dielectrophoretic quadrupole and microwell electrodes. *Biosensors and Bioelectronics*, 24(12):3637–3644, aug 2009.
32. Satoshi Tsukahara, Katsuyuki Yamanaka, and Hitoshi Watarai. Flow fractionation of microparticles under a dielectrophoretic field in a quadrupole electrode capillary. *Analytical Chemistry*, 73(23):5661–5668, dec 2001.
33. Eva-Maria Laux, Frank F. Bier, and Ralph Hölzel. Electrode-based AC electrokinetics of proteins: A mini-review. *Bioelectrochemistry*, 120:76–82, apr 2018.
34. Lifeng Zheng, James P. Brody, and Peter J. Burke. Electronic manipulation of DNA, proteins, and nanoparticles for potential circuit assembly. *Biosensors and Bioelectronics*, 20(3):606–619, oct 2004.
35. Satoshi Tsukahara, Katsuyuki Yamanaka, and Hitoshi Watarai. Dielectrophoretic behavior of single DNA in planar and capillary quadrupole microelectrodes. *Chemistry Letters*, 30(3):250–251, mar 2001.
36. R. Ghayour, Y. Hojjat, M.R. Karafi, and H. Sadeghiyan. Development of a hybrid DEP-SAW device for trapping/sensing target cells. *Applied Acoustics*, 141:355–361, dec 2018.
37. M. Washizu, S. Suzuki, O. Kurosawa, T. Nishizaka, and T. Shinohara. Molecular dielectrophoresis of biopolymers. *IEEE Transactions on Industry Applications*, 30(4):835–843, 1994.
38. R Pethig, Ying Huang, Xiao bo Wang, and J P H Burt. Positive and negative dielectrophoretic collection of colloidal particles using interdigitated castellated microelectrodes. *Journal of Physics D: Applied Physics*, 25(5):881–888, may 1992.
39. Richard W. Clarke, Samuel S. White, Dejian Zhou, Liming Ying, and David Klenerman. Trapping of proteins under physiological conditions in a nanopipette. *Angewandte Chemie International Edition*, 44(24):3747–3750, jun 2005.
40. Liang Xue, Hirohito Yamazaki, Ren Ren, Meni Wanunu, Aleksandar P. Ivanov, and Joshua B. Edel. Solid-state nanopore sensors. *Nature Reviews Materials*, 5(12):931–951, sep 2020.
41. Yu-Cheng Syu, Wei-En Hsu, and Chih-Ting Lin. Review—field-effect transistor biosensing: Devices and clinical applications. *ECS Journal of Solid State Science and Technology*, 7(7):Q3196–Q3207, 2018.
42. H. A. Pohl and I. Hawk. Separation of living and dead cells by dielectrophoresis. *Science*, 152(3722):647–649, apr 1966.
43. Seungjun Lee, Seong Min Roh, Eunji Lee, Yejin Park, Byung Chul Lee, Youngeun Kwon, Hye Jin Kim, and Jinsik Kim. Applications of converged various forces for detection of

biomolecules and novelty of dielectrophoretic force in the applications. *Sensors*, 20(11):3242, jun 2020.
44. R. Holzel. Dielectric and dielectrophoretic properties of DNA. *IET Nanobiotechnology*, 3(2):28, 2009.
45. S Nedelcu and J H P Watson. Size separation of DNA molecules by pulsed electric field dielectrophoresis. *Journal of Physics D: Applied Physics*, 37(15):2197–2204, jul 2004.
46. T. Yamamoto, O. Kurosawa, H. Kabata, N. Shimamoto, and M. Washizu. Molecular surgery of DNA based on electrostatic micromanipulation. *IEEE Transactions on Industry Applications*, 36(4):1010–1017, 2000.
47. M. Washizu, O. Kurosawa, I. Arai, S. Suzuki, and N. Shimamoto. Applications of electrostatic stretch-and-positioning of DNA. In *Conference Record of the 1993 IEEE Industry Applications Conference Twenty-Eighth IAS Annual Meeting*. IEEE, 1993.
48. M. Washizu and O. Kurosawa. Electrostatic manipulation of DNA in microfabricated structures. *IEEE Transactions on Industry Applications*, 26(6):1165–1172, 1990.
49. Masao Washizu. Electrostatic manipulation of biological objects. *Journal of Electrostatics*, 25(1):109–123, jun 1990.
50. Avijit Barik, Yao Zhang, Roberto Grassi, Binoy Paulose Nadappuram, Joshua B. Edel, Tony Low, Steven J. Koester, and Sang-Hyun Oh. Graphene-edge dielectrophoretic tweezers for trapping of biomolecules. *Nature Communications*, 8(1867), nov 2017.
51. Cinthia J. Ramirez-Murillo, J. Martin de los Santos-Ramirez, and Victor H. Perez-Gonzalez. Toward low-voltage dielectrophoresis-based microfluidic systems: A review. *Electrophoresis*, nov 2020.
52. Christopher T. Ertsgaard, Nathan J. Wittenberg, Daniel J. Klemme, Avijit Barik, Wei-Chuan Shih, and Sang-Hyun Oh. Integrated nanogap platform for sub-volt dielectrophoretic trapping and real-time raman imaging of biological nanoparticles. *Nano Letters*, 18(9):5946–5953, aug 2018.
53. Xiaoshu Chen, Hyeong-Ryeol Park, Matthew Pelton, Xianji Piao, Nathan C. Lindquist, Hyungsoon Im, Yun Jung Kim, Jae Sung Ahn, Kwang Jun Ahn, Namkyoo Park, Dai-Sik Kim, and Sang-Hyun Oh. Atomic layer lithography of wafer-scale nanogap arrays for extreme confinement of electromagnetic waves. *Nature Communications*, 4(1), sep 2013.
54. Gerard Giraud, Ronald Pethig, Holger Schulze, Grace Henihan, Jonathan G. Terry, Anoop Menachery, Ilenia Ciani, Damion Corrigan, Colin J. Campbell, Andrew R. Mount, Peter Ghazal, Anthony J. Walton, Jason Crain, and Till T. Bachmann. Dielectrophoretic manipulation of ribosomal RNA. *Biomicrofluidics*, 5(2):024116, jun 2011.
55. Mark A. Hayes. Dielectrophoresis of proteins: experimental data and evolving theory. *Analytical and Bioanalytical Chemistry*, 412(16):3801–3811, apr 2020.
56. Sreeja B. Asokan, L. Jawerth, R. Lloyd Carroll, R. E. Cheney, S. Washburn, and R. Superfine. Two-dimensional manipulation and orientation of actin-myosin systems with dielectrophoresis. *Nano Letters*, 3(4):431–437, apr 2003.
57. Mingyong Wang, Zhi Wang, Xuzhong Gong, and Zhancheng Guo. The intensification technologies to water electrolysis for hydrogen production – a review. *Renewable and Sustainable Energy Reviews*, 29:573–588, jan 2014.
58. A. V. Postnikov, I. V. Uvarov, M. V. Lokhanin, and V. B. Svetovoy. Highly energetic phenomena in water electrolysis. *Scientific Reports*, 6(1):39381, dec 2016.
59. Naohiro Shimizu, Souzaburo Hotta, Takayuki Sekiya, and Osamu Oda. A novel method of hydrogen generation by water electrolysis using an ultra-short-pulse power supply. *Journal of Applied Electrochemistry*, 36(4):419–423, dec 2005.
60. Shih-Kang Fan, Po-Wen Huang, Tsu-Te Wang, and Yu-Hao Peng. Cross-scale electric manipulations of cells and droplets by frequency-modulated dielectrophoresis and electrowetting. *Lab on a Chip*, 8(8):1325, 2008.
61. Michael B. Sano, Alireza Salmanzadeh, and Rafael V. Davalos. Multilayer contactless dielectrophoresis: Theoretical considerations. *Electrophoresis*, 33(13):1938–1946, jul 2012.
62. A Castellanos, A Ramos, A González, N G Green, and H Morgan. Electrohydrodynamics and dielectrophoresis in microsystems: scaling laws. *Journal of Physics D: Applied Physics*, 36(20):2584–2597, oct 2003.

Chapter 7
Optical Trapping of Single Molecules

Joshua Kolbow, Nathan Lindquist, and Sang-Hyun Oh

Abstract Since their invention in the 1970s, optical trapping techniques have become well-known and effective methods for trapping particles on the micrometer and nanometer scale. However, conventional optical trapping methods are unable to trap particles as small as single molecules without using high optical powers that would undoubtedly damage the sample. A variety of methods for single-molecule trapping have been proposed and developed; this chapter discusses these methods and their applications. The instrumentation and fabrication methods for performing single-molecule trapping are also discussed.

7.1 Introduction

Because light carries momentum, when light collides with an object it transfers some of its momentum to that object, exerting a force. This force is known as radiation pressure. In the 1970s, Arthur Ashkin discovered that it was possible to utilize radiation pressure to manipulate the position of micron-scale particles [1]. A laser beam, when tightly focused, can produce radiation pressure high enough to generate what is known as an optical potential well at the center of the focus. The result is that nearby particles experience a force pulling them toward the center of the focus and, if strong enough, holding them stationary there.

During his foundational experiments, Ashkin used focused laser light to counteract the Brownian motion of polystyrene microspheres ranging from 10 μm to 25 μm in size suspended in water [2]. He called this device a single-beam gradient force optical

J. Kolbow · S.-H. Oh (✉)
University of Minnesota, Kenneth H. Keller Hall, 200 Union St SE, Minneapolis, MN 55455, USA
e-mail: sang@umn.edu

J. Kolbow
e-mail: kolbo007@umn.edu

N. Lindquist
Bethel University, 3900 Bethel Dr, St. Paul, MN 55112, USA
e-mail: n-lindquist@bethel.edu

W. Bowen et al. (eds.), *Single Molecule Sensing Beyond Fluorescence*,
Nanostructure Science and Technology,
https://doi.org/10.1007/978-3-030-90339-8_7

trap. These experiments demonstrated that optical methods alone can be used to trap particles on these small scales. Ever since, plentiful research has been devoted to utilizing and improving optical trapping techniques. In particular, it has found use in microbiological research because of its ability to trap particles within biological fluids for analysis. Optical trapping techniques have even been used to trap and cool atoms to near absolute zero [3].

7.1.1 Theory

The basic theory behind optical trapping lies in the interaction between the electromagnetic field generated by light and the matter of the trapped particle. When placed in an electromagnetic field, an object will experience a force, **F**, that can be written in terms of the Maxwell Stress Tensor, **T**, as follows [4]:

$$< \mathbf{F} >= \int_{\sigma} < \hat{\mathbf{T}} > \cdot \mathbf{n} ds \tag{7.1}$$

where ds is an infinitesimal surface element and **n** is the unit vector normal to ds. This equation can be expanded to be generally written as the sum of three forces. The first is the gradient force which, in the context of optical trapping, pulls objects toward points of local maxima in the electric field magnitude. The second force is the primary scattering force, pulling objects in the direction of light propagation. The third force is a secondary scattering force; for the purposes of optical trapping, this force is generally significantly smaller than the other two and, thus, can be ignored [4]. This expanded equation can be written as follows [5]:

$$< \mathbf{F} >= \frac{1}{4}\mathcal{R}[\alpha]\nabla|\mathbf{E}|^2 + \sigma\frac{1}{2}\mathcal{R}[\frac{1}{c}\mathbf{E}\times\mathbf{H}^*] + \sigma\frac{1}{2}\mathcal{R}[i\frac{\epsilon_0}{k_0}(\mathbf{E}\cdot\nabla)\mathbf{E}^*] \tag{7.2}$$

where **F** is the force on the object, α is the polarizability of the object, c is the speed of light, **E** is the electric field, **H** is the magnetic field, k_0 is the wavenumber, ϵ_0 is the permittivity of free space, and $\sigma \equiv \frac{k_0 Im[\alpha]}{\epsilon_0}$.

The behavior of these forces is significantly impacted by the size of the object relative to the wavelength of the light carrying the field. In conventional optical trapping, generally the object size is on the same order of magnitude as the wavelength of the trapping light, at the boundary of Mie theory (for objects significantly larger than the wavelength) and Rayleigh theory (for objects significantly smaller than the wavelength). Single-molecule trapping, however, deals with particles lying well within the Rayleigh regime. For such small particles, the magnitudes of the gradient force and trapping force can be rewritten as follows [6]:

$$F_{scat} = \frac{I_0}{c}\frac{128\pi^5 r^6}{3\lambda^4}(\frac{N^2-1}{N^2+2})^2 n \tag{7.3}$$

$$F_{grad} = -\frac{n^3 r^3}{2}\left(\frac{N^2-1}{N^2-2}\right)^2 \nabla |\mathbf{E}|^2 \tag{7.4}$$

where I_0 is the intensity of the trapping laser, r is the radius of the object, λ is the wavelength of the incident light, N is the ratio of the refractive index of the object to that of the surrounding medium, n is the refractive index of the object, $\mathbf{E}$ is the electric field.

As can be discerned from Eq. 7.4, the only way to increase the optical trapping force without altering the particle to be trapped or the environment around the particle is to increase the gradient of the electric field magnitude. This can be done in two ways. The first is to increase the overall magnitude of the electric field. The simplest way to achieve this is to increase the power of the trapping laser. The second way is to localize the electric field in a smaller volume. This is difficult to achieve because the spot size of a laser beam at its focus is subject to the diffraction limit. Specifically, the radius of the spot at the focus is given by this equation [7]:

$$r = 1.22\frac{\lambda}{2NA} \tag{7.5}$$

where λ is the wavelength of the light and NA is the numerical aperture of the focusing lens. There are a variety of factors that limit these values that are discussed in Sect. 7.3.2.

Equations 7.3 and 7.4 also make it clear that, in most cases of trapping objects in the Rayleigh regime, the gradient force will dominate over the scattering force. This is because the gradient force is proportional to r^3 while the scattering force is proportional to r^6. Thus, decreasing the size of the object decreases the scattering force much more than the gradient force. Because the gradient force is dominant, optical traps can be understood as optical potential wells located within the potential landscape created by a light field. The optical potential created by a light field can be described at any point $\mathbf{r}_0$ in terms of the gradient force, $\mathbf{F}$, by the equation [8]:

$$U(\mathbf{r}_0) = -\int_{\infty}^{\mathbf{r}_0} \mathbf{F}(\mathbf{r}) \cdot d\mathbf{r} \tag{7.6}$$

The strength or stiffness of an optical trap can be understood as the depth of the optical potential well. In order for an optical trap to be effective, the depth of the potential well must exceed the kinetic energy of the object, which is approximately equal to its thermal energy kT where k is Boltzmann's constant and T is the object's temperature (Fig. 7.1).

The depth of the potential well is directly related to the gradient force on an object and, therefore, is similarly proportional to the cube of the radius of the object. Thus, the depth of the potential well decreases significantly with the size of the particle. This is why optical trapping gets significantly more difficult when trying to trap particles on the nanometer scale. In order to account for this, the electric field

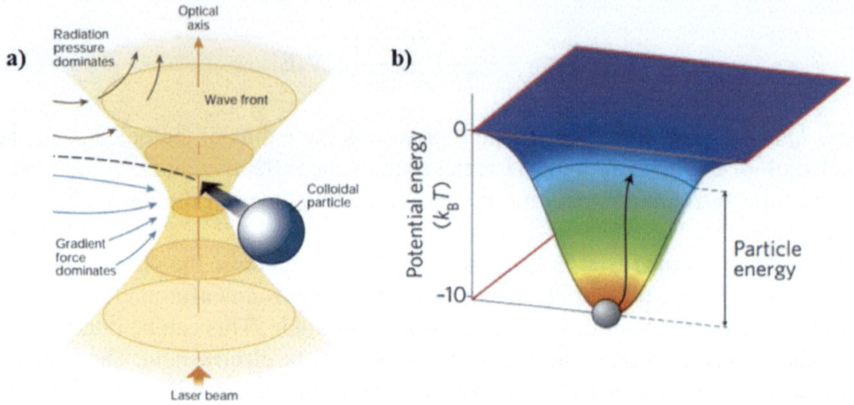

Fig. 7.1 **a** In optical trapping, colloidal particles near the focus of the laser beam are pulled to the center of the focus by the gradient force. Reprinted with permission from [9] ©Springer Nature. **b** Particles within an optical trap are unable to exit unless their kinetic energy exceeds the depth of the potential well created by the trap. Adapted with permission from [10] ©Springer Nature

intensity gradient needs to be significantly increased to trap particles at these small sizes.

7.1.2 Single-Molecule Trapping

Due to the value of being a contactless and precise method by which to manipulate particles in fluids, optical tweezers have been of particular interest to the research of biological samples. Being able to selectively trap and move particles of interest within a biological sample so that they can be analyzed individually is a valuable asset. However, in order to trap particles as small as single molecules, conventional tweezers would need to drastically increase the laser power being used, with laser powers on the scale of 1 W being required to trap particles around 10 nm in size [2]. This is generally unacceptable for trapping in biological samples because laser powers of this level will damage biological materials. Furthermore, such high laser powers generate a significant amount of heat and a high temperature gradient; this can induce thermophoresis that can cause particles to escape from the trap. Instead, other methods must be used to increase the trapping force for biological material.

Due to these limitations, early methods for the trapping of single molecules were indirect methods. It was possible to manipulate single molecules with conventional optical tweezers by first binding the molecule to a much larger particle (such as a silica bead) and then trapping said larger particle. A variety of experiments have been conducted using these methods. In particular, the measurement of biological forces within protein binding and folding processes [11, 12]. DNA stretching experiments have also been performed via indirect trapping techniques [13, 14]. These techniques, while effective, are not always useful and often complex because of the necessity of binding mechanisms to tether the molecule to the larger particle.

In order to realize the nano-optical traps necessary for the trapping of single molecules, a variety of methods have been proposed. In order to increase the trapping force, there are two primary phenomena commonly utilized by these methods. The first of these is the use of nanostructures. When light is incident on these nanostructures, the electric field is influenced by the shape of the structure. These nanostructures can induce resonances in the electric field leading to much higher magnitudes than the the focused light on its own. Furthermore, nanostructures can significantly decrease the trapping volume by localizing the electric field maxima within and at the edges of these nanostructure. These highly localized maxima are commonly known as "hotspots." As discussed earlier, both of these effects will lead to an increase in the trapping force.

The other phenomena that many nano-optical trapping techniques utilize is plasmonics. When light is incident on a metallic surface, the electric field causes oscillations of the free electrons at the surface. These oscillations are known as surface plasmons. These plasmons propagate along the surface from where the light is incident, generating electric fields at the surface as they go. Plasmons are useful for trapping techniques because they can be localized far beyond the diffraction limit of light [15]. Furthermore, the electric field magnitude generated by these plasmons drops off exponentially from the surface. As such, the field gradients generated by plasmons can be orders of magnitude higher than those generated by the incident light source.

7.2 Single-Molecule Trapping Techniques

A variety of techniques have been proposed and implemented to enable single-molecule optical trapping. In this section we will provide an overview of a few of these techniques and the results they have achieved.

7.2.1 Plasmonic Nanostructure-Based Traps

Some of the earliest nano-optical traps were achieved by combining the two phenomena described in Sect. 7.1.2. By designing metallic nanostructures specifically to induce plasmonic resonance, it is possible to produce high-intensity sub-wavelength trapping volumes and thereby achieve significantly higher trapping forces than conventional optical traps. There are a variety of nanostructures that have been pursued and are capable of accomplishing this (Fig. 7.2).

One of the first instances of such a trapping device was realized in 2008 by Grigorenko et al. [16] and consisted of an array of gold nanopillars in closely spaced pairs on a glass surface. These nanopillars, when illuminated, each had their own

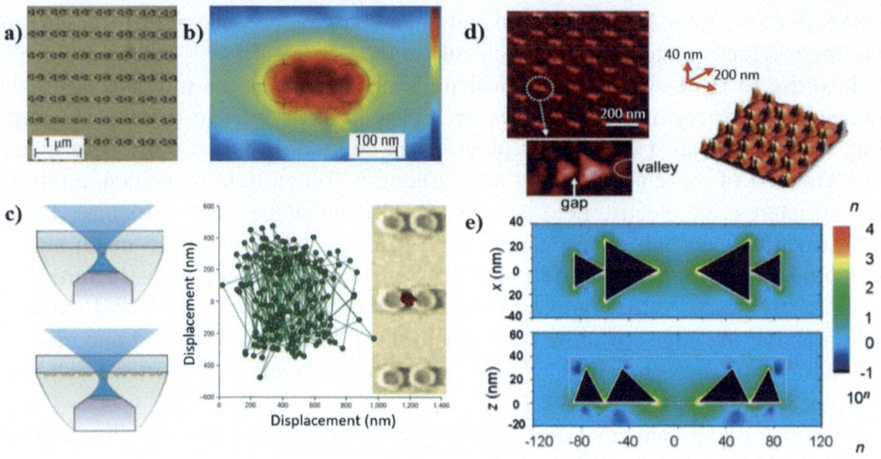

Fig. 7.2 **a** Micrograph of the array of nanopillar pairs used by Grigorenko et al. Adapted with permission from [16] ©Springer Nature. **b** Simulation of the excited electric field power of a nanopillar pair. Adapted with permission from [16] ©Springer Nature. **c** By focusing light onto a nanopillar pair as shown in the schematics on the left, a 200nm bead was trapped in the gap between the pillars and its motion was recorded in the graph on the right. Adapted with permission from [16] ©Springer Nature. **d** Scanning electron micrograph (left) and three-dimensional atomic force micrograph (right) of the double nanopyramid structures used by Tsuboi et al. Adapted with permission from [17] ©American Chemical Society. **e** Simulation of the ratio of the induced electric field intensity to the incident electric field intensity when the double nanopyramid structures are illuminated by laser light. Adapted with permission from [17] ©American Chemical Society

plasmon resonance. When placed with a small gap between them, the electromagnetic interaction between two nanopillars alters the individual resonance of each, forming two new resonances for the pair of nanopillars. This results in an intensity peak at the gap between the pillars that is significantly higher than the incident beam or the field around a lone nanopillar and is also well below the diffraction limit in volume. Ultimately, they reported that the average trapping force produced by this system was approximately 10 times higher in magnitude than a similar conventional optical trap would produce. Thus, this trap was able to efficiently trap particles with sizes well below the wavelength of the incident light. This experiment was only scratching the surface of the potential of these devices.

In the following years a plethora of research groups fabricated their own plasmonic nanostructure-based traps. Many of these showed significant improvements on earlier designs. Other designs utilized structures such as nanoblocks [18] and nanopyramids. One such design by Tsuboi et al. [17] utilized gold nanopyramids that were tightly grouped in pairs similar to the nanopillar design. This design, however, offered much greater field enhancement and, thus, the ability to trap much smaller particles and at lower incident light intensities. It was demonstrated that this design was capable of enhancing the incident electric field intensity by as much as four orders of magnitude and stably trap quantum dots as small as 10 nm in diameter. In a later experiment, they demonstrated that this structure could be utilized for the trapping and micropatterning of DNA [19]. Plasmonic nanostructure devices also can be integrated with heat sinks to avoid the potential problem of thermophoresis from heating due to high intensity peaks [20].

Nanoapertures are another type of nanostructure that have been extensively used for nanoscale trapping. Apertures such as double nanoholes and nanobowties in particular have been utilized in a variety of single-molecule trapping experiments. These structures are particularly effective compared to other nanostructures due to their ability to utilize a phenomena known as self-induced back-action to improve trap stability. This phenomena and the devices that utilize them are discussed in detail in Sect. 7.2.2. Saleh and Dionne suggested that a coaxial nanoaperture could be a powerful trapping structure, specifically by tapering the thickness of the aperture it could be capable of trapping particles under 2 nm in size while using reasonably low trapping powers. Zhao et al. demonstrated that coaxial nanostructures and circularly polarized trapping light could be used to selectively trap sub 20 nm chiral molecules based on the handedness of their chirality [21].

One of the primary drawbacks of these plasmonic nanostructure-based traps is that their ability to manipulate the position of a trapped particle is severely limited because the trap is located at a stationary nanostructure. A few setups have shown a limited ability to manipulate the position of a trapped particle around the surface of a nanostructure by utilizing the effects of polarization on plasmon propagation to control the location of the trap [20]. A common method by which conventional optical tweezers manipulate particles is by using holography to shape the trapping beam and adjust the position of the trap. Holographic beam-shaping can similarly be used to control the position of plasmonic traps. Huft et al. [22] utilized holography to control the phase of a ring-shaped beam of light incident on a bull's eye nanostructure. In doing so, they demonstrated control over the location of positive interference of the plasmons (the trapping location) in the region at the center of the structure. While allowing for more arbitrary movement, manipulation was still limited to the confines of the structure.

One method by which the manipulation capabilities of these structures has been significantly increased is by attaching these structures to probe tips. A common method of doing this is to place a plasmonic nanoaperture at the tip of a near-field scanning optical microscopy (NSOM) probe [23]. Berthelot et al. [24] demonstrated that such a setup was capable of stably trapping particles as small as 20 nm in diameter and arbitrarily moving them through the sample.

7.2.2 Self-Induced Back-Action Traps

While nanoaperture structures allow for significant increases in optical trapping forces through plasmonic amplification alone, there is also another effect present when using such structures. This effect is reliant on the fact that the presence of a particle within an aperture has a significant impact on the transmission of light

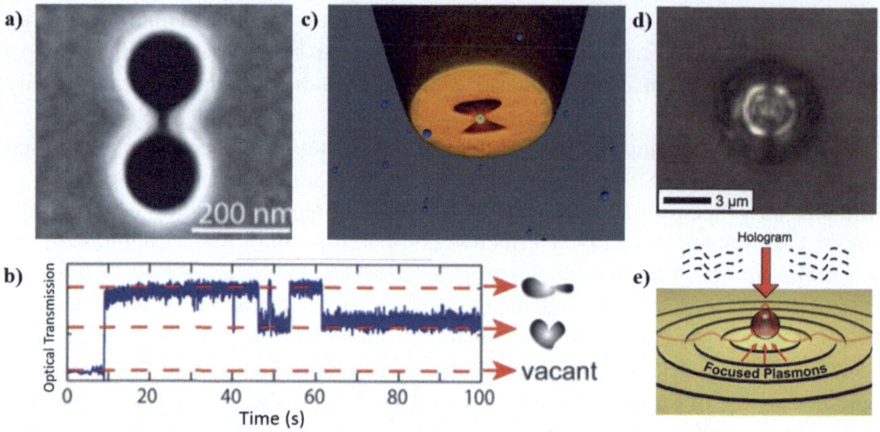

Fig. 7.3 **a** Scanning electron microscope image of the double nanohole aperture used by Pang and Gordon. Adapted with permission from [25] ©American Chemical Society. **b** Graph of the optical transmission through a double nanohole aperture during the trapping of a BSA molecule. The trapping event can be seen clearly as well as the unfolding of the protein within the trap. Adapted with permission from [25] ©American Chemical Society. **c** A schematic of a bowtie nanoaperture at the tip of an NSOM probe. Adapted with permission from [24] ©Springer Nature. **d** Microscope image of a laser beam shaped into a ring via hologram and focused onto a bull's eye nanostructure for the experiment done by Huft et al. Adapted with permission from [22] ©American Chemical Society. **e** Illustration of the constructive interference of plasmons at the center of the bull's eye structure in the Huft et al. experiment. Adapted with permission from [22] ©American Chemical Society

through said aperture. If the particle being trapped within the aperture has a higher index of refraction then water, the transmission through the aperture will increase. This increase in transmission is equivalent to an increase in the amount of photon momentum travelling through the aperture. Similarly, the particle moving out of the aperture will decrease the amount of photon momentum travelling through the aperture. Newton's third law dictates that a restorative force will act in opposition to either of these momentum changes. Essentially, this restorative force will act opposite to the movement of any particle within the aperture. This phenomenon is known as self-induced back-action (SIBA) and significantly increases the efficacy and efficiency of plasmonic nanoaperture traps (Fig. 7.3).

Juan et al. [26] investigated the benefits of SIBA in gold single nanohole apertures. They found that SIBA actually had a greater impact on the trapping efficiency than the Rayleigh gradient force that accounts for the entirety of trapping in most other techniques. In fact, it was determined that the introduction of SIBA allowed for stable trapping at optical powers an order of magnitude lower than other methods allowed. This is particularly impactful for single-molecule trapping because it enables trapping significantly smaller particles without increasing the optical intensity, significantly reducing the risk of photothermal damage to the molecule or thermophoretic effects destabilising the trap.

Pang and Gordon [25] demonstrated the power of SIBA traps using a double-nanohole aperture in a gold film. Using this structure, they were able to trap a single BSA molecule with a radius of 3.4 nm at the boundary between the two nanoholes. Furthermore, they were able to stably trap these molecules with laser powers as low as 3.5 mW, which is quite low for most optical trapping setups. A subsequent experiment demonstrated that it was possible to perform Raman spectroscopy on particles trapped in this structure [27]. Another aspect of SIBA traps that was demonstrated by this experiment was that the significant impact a particle has on transmission through the aperture can serve as an extremely sensitive sensor with a high signal to noise ratio. Because they are nanoaperture-based, SIBA traps can similarly be attached to the end of probe tips to allow manipulation of the position of trapped particles [24].

7.2.3 Waveguide-Based Traps

Another method proposed to overcome the diffraction limit to increase the trapping force is by using optical waveguides that are sub-wavelength in size. By using such waveguides it is possible to condense the electromagnetic energy of light into volumes smaller than is allowed by the diffraction limit. However, most of this energy is contained within the waveguide itself and, thus, is not helpful for trapping purposes. However, structures known as slot waveguides are able to confine the electromagnetic energy of light within a liquid core by surrounding that core with materials of high refractive index. Thus, slot waveguide based optical traps are able to trap particles within the core of the waveguide.

Yang et al. [28] demonstrated the efficacy of such a method. Their slot waveguide was capable of achieving trapping volumes on the scale of 10 s of nm, thus being able to trap sub-100 nm particles across the length of the waveguide. Furthermore, the trapped particles were simultaneously transported along the length of the waveguide by the propagation of light within the waveguide. Such a system is potentially a powerful tool as it can serve as an optically-driven microfluidic channel which particles can be pulled into and transported throughout the system.

Waveguides have also been used in coordination with photonic crystal resonators to produce stable nanoscale optical traps. Mandal et al. [29] demonstrated this by coupling a waveguide to a silicon photonic crystal resonator producing a stationary interference pattern within the resonator. This interference pattern resulted in an electric field that was both high intensity and tightly confined in a sub-100 nm volume. This setup was able to stably trap sub-100 nm particles. Chen et al. demonstrated that a similar device could be used to trap individual Wilson disease proteins [30].

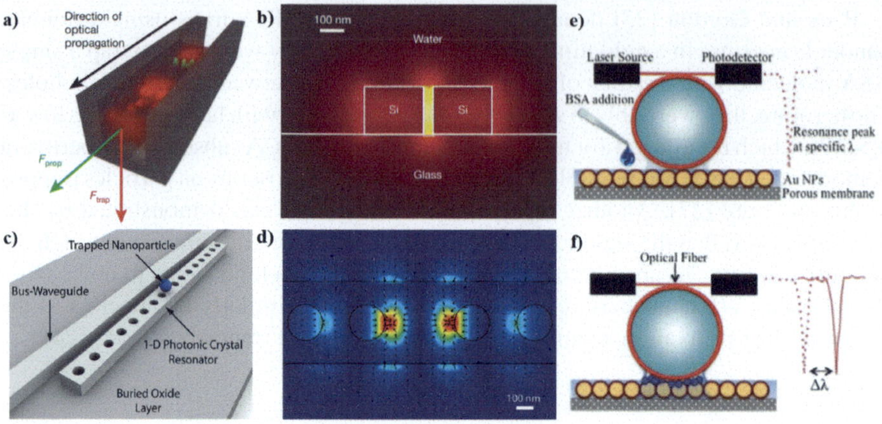

Fig. 7.4 **a** Schematic illustrating the trapping and transport of nanoparticles within a slot waveguide. Adapted with permission from [28] ©Springer Nature. **b** Illustration of the electric mode responsible for trapping within a slot waveguide. Adapted with permission from [28] ©Springer Nature. **c** Schematic of the waveguide-integrated photonic crystal resonator trap used by Mandal et al. Adapted with permission from [29] ©American Chemical Society. **d** Simulation of the electrical field and optical forces present within the architecture used by Mandal et al. Adapted with permission from [29] ©American Chemical Society. **e** Diagram of the WGM resonator optical trapping setup. Adapted with permission from [31] ©John Wiley and Sons. **f** BSA molecules are trapped at plasmonic hotspots and cause a shift in the resonance peak detected at the photodetector. Adapted with permission from [31] ©John Wiley and Sons

7.2.4 Whispering Gallery Mode Based Traps

Whispering gallery mode (WGM) resonators are devices wherein a laser source is coupled to a microsphere (often glass or silica) where the light is contained at the surface of the sphere via total internal reflection. Due to the high Q-factors that these devices are able to achieve, these devices are able to detect minute wavelength shifts in optical spectra. For this reason, they have been used as extremely effective biosensors. By coupling these devices to plasmonic structures, it is possible to create nanoscale optical traps (Fig. 7.4).

Such a device was first realized by coupling a silica microsphere WGM resonator to a random array of gold nanoparticles [31]. The electric field at the surface of the WGM resonator excited plasmons within the 55 nm diameter nanoparticles. The result was a random array of sub-wavelength plasmonic hotspots that each act as an optical trap. Santiago et al. created such a device and demonstrated the optical trapping of BSA molecules at these plasmonic hotspots. The integrated optical trapping allows this device to achieve significantly higher sensitivities than similar WGM biosensors. A recently proposed WGM device based on a silica double-toroid structure with a nanoscale gap between the toroids [32]. Simulations suggest that this structure should be able to stably trap sub-10 nm particles.

7.3 Fabrication and Instrumentation

Beyond the underlying theory, single-molecule optical trapping experiments rely on a variety of fabrication techniques, optical components, and other instruments. In this section we provide an overview of the fabrication techniques and instrumentation that enable single-molecule optical trapping.

7.3.1 Fabrication of Plasmonic and Resonant Nanostructures

Because so many single-molecule optical trapping techniques rely on nanostructures and the potential landscapes that they are capable of producing, fabrication methods are a vital step in the development of these techniques. Many of the previously discussed methods relied on the creation of structures with features on the nanometer scale. Furthermore, the propagation of surface plasmons across a metallic surface relies on the smoothness of said surface, with atomic-scale defects being capable of disturbing plasmon propagation and the field it produces. The development and improvement of fabrication methods capable of producing small feature sizes and atomically-flat surfaces has similarly enhanced the capabilities of these single-molecule trapping techniques.

A common approach to fabricating these structures is by utilizing top-down lithography techniques of which there are two primary varieties: electron beam (E-beam) lithogrpahy and focused ion beam (FIB) lithography. FIB lithography, uses a beam of charged ions as a method of etching through materials. As such, many of the plasmonic nano-apertures (negative-type structures) discussed previously were constructed via FIB lithography [24]. In contrast, positive-type structures such as nanopillars [16, 20] and nanoblocks [18] tend to be fabricated using E-beam lithography. E-beam lithography is an extremely versatile technique because it does not rely on photomasks and instead can expose resist in arbitrary patterns. It is also capable of producing the nanometer-scale feature sizes that many nano-optical traps require. E-beam lithography is often used with positive resists such as PMMA followed by a metal lift-off process.

For plasmonic nanostructures, there are a variety of methods that can be used to create atomically-flat surfaces. A common method is to use e-beam lithography to pattern an atomically-flat substrate such as silicon to use as a template for the structure [20, 22]. Metal can then be deposited onto this patterned substrate and subsequently removed by adhering the metal to another, unpatterned, substrate. The patterned substrate can then be reused to fabricate more structures. This process is known as template-stripping. Another valuable method is atomic layer deposition (ALD). ALD is a process characterized by deposition of metallic material by one layer of atoms at a time. This allows for the fabrication of layers and, thus, features with sub-nm resolutions and atomically-flat surfaces.

Though these two techniques are the most common, there are a variety of other techniques that have been utilized for optical trapping purposes. Of particular interest are methods that are able to fabricate many nanostructures in parallel. One such method is colloidal lithography. It was recently shown that by randomly attaching nanospheres to a surface by evaporating a solution of these colloidal nanospheres atop the surface and subsequently depositing metal onto the surface it is possible to create single and double nanohole apertures [33]. The nanopyramid structures used by Tsuboi et al. [17] were fabricated using a similar nanosphere lithography technique known as angle-resolved nanosphere lithography (AR NSL).

7.3.2 Instrumentation

At the most basic level, an optical trapping setup only requires a trapping laser and lens to focus the laser to a diffraction-limited spot.

There are a variety of factors that affect the choice of trapping laser for an optical trapping setup. In general, optical trapping setups have incorporated lasers with wavelengths in the red or infrared regime, though some have reported achieving higher trapping efficiencies using green light [34]. Infrared lasers, despite their longer wavelengths leading to larger trapping volumes, have been commonly used in the optical trapping of biological material because biological material tends to have a much lower absorption coefficient in the infrared regime. Infrared lasers, therefore, are much less likely to damage such material even at the high intensities required for optical trapping. Most commonly, continuous-wave lasers are used to ensure constant and stable trapping; however, in some cases it is preferable to use a pulsed laser because it is possible to achieve significantly higher electric field intensities, and thus larger trapping forces, momentarily without significantly increasing the photothermal effects on the sample. Shoji et al. demonstrated that when trapping biological material using plasmonic nanostructures using a pulsed laser can prevent the material from becoming fixed on the surface of the nanostructure [19].

The choice of focusing lens can also have a significant impact on the efficacy of the trap. In particular, choosing a lens with a high numerical aperture can significantly decrease the spot size at the focus (as can be seen in Eq. 7.5). Generally, a microscope objective is used as the focusing lens. These lenses are readily available and allow optical tweezing setups to be integrated into existing microscope setups. This is very useful for simultaneous observation of the sample during trapping. In particular, oil immersion objectives are highly valuable for optical tweezing as they allow numerical apertures between 1 and 1.4, whereas 1 is the theoretical maximum for conventional objectives.

While these are the two primary components of every optical trapping setups, there are a variety of other components that are commonly and preferentially used. Most often, the trapping laser is sent through a series of optical elements to align and collimate the laser. Ideally, the trapping laser is expanded and collimated such that it fills the back aperture of the focusing lens; this guarantees that the laser beam is focused to a tight diffraction-limited spot [35]. Other optical elements can be used to perform beam steering. By translating the beam steering elements, it is possible to deflect the focus of the trapping beam within the sample, moving the trap and any trapped particles along with it. The beam may also be reflected off of or transmitted through a spatial light modulator to enable holographic beam-shaping to control the position and shape of the trap at the sample.

Most often, being able to observe the trap during operation is an important component of an optical trapping setup. For this purpose, a secondary light source is used for illumination. Some setups utilize the white light source often already incorporated into microscope setups to illuminate the sample, while other setups use a secondary illumination laser. A secondary illumination laser is commonly used when the particles to be trapped are fluorescent or fluorescently tagged, as this can make observing even particles on the nm scale trivial. Regardless of what secondary illumination source is chosen, it is usually necessary to use a dichroic beamsplitter or bandpass filter to separate the trapping laser light from the illumination light before observation because the high intensity of the focused trapping laser can make observing the trapped particle difficult or potentially even cause damage to the imaging device.

There are a variety of instruments that can be used to image or sense trapping events. Some setups, such as the one shown in Fig. 7.5, are able to detect trapping events by simply measuring the transmission of the trapping laser through the sample. This is method is particularly effective in nanoaperture traps due to the large effect

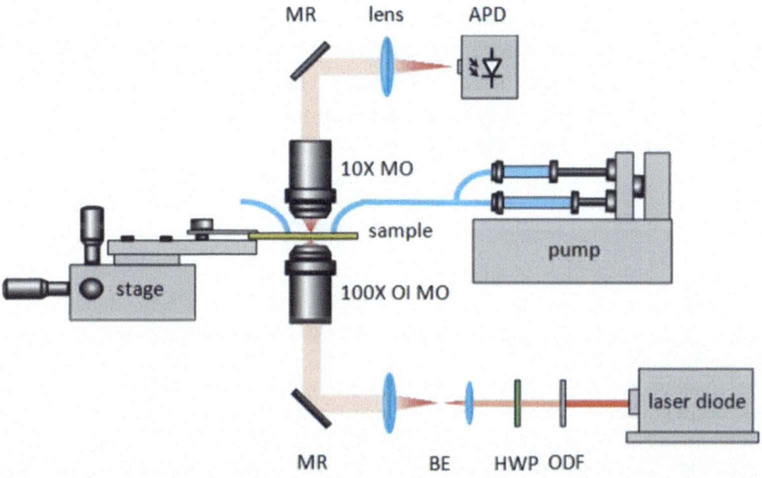

Fig. 7.5 A fairly orthodox example of an optical tweezing setup. The laser passes through an optical density filter (ODF) to limit the power, a half-wave plate to rotate the polarization of the beam(HWP), and a beam expander (BE) before being direct by a mirror (MR) into the focusing oil immersion microscope objective (OI MO) which focuses the beam onto the sample. The beam then passes through another microscope objective (MO) to finally be focused on an avalanche photodetector (APD) to measure the transmitted light. Adapted with permission from [36] ©The Optical Society

nanoparticles present in the aperture have on transmission through the aperture, as discussed in detail in Sect. 7.2.2. Similar results can be achieved with other resonant plasmonic nanostructures since nearby particles will affect the resonant modes within the structure, though the effect will be considerably less and thus must be measured using sensitive devices such as an avalanche photodiode or electron multiplying CCD. In contrast, when illuminating the sample with a secondary source, observation can simply be done using a CCD camera. It is also often desirable to be able to quantify the position of the trapped particle in order to quantify the stiffness of the trap. Processing the images from a CCD camera using position tracking software (e.g. rapidSTORM) makes acquiring this information fairly straightforward.

7.4 Outlook

Optical tweezers have proven themselves to be a valuable tool in research in microbiology and throughout the field of nanotechnology. The last two decades have seen significant strives made in improving optical trapping techniques and pushing them into the trapping of nanoscale particles. As can be seen throughout this chapter, there are a variety of paths and techniques that have been and continue to be explored for the trapping of single-molecules. Optical trapping can significantly increase the sensitivity and effectiveness of biosensors and spectroscopy techniques. As microfabrication technology improves, the capabilities of these techniques continue to improve with it, opening up new opportunities in the pursuit and utilization of single-molecule optical trapping techniques. Furthermore, the continued growth of interest in the development of lab-on-a-chip devices drives an increased need for effective and non-invasive methods by which to transport material throughout these devices. Optical tweezers can surely be an effective tool through which to do so.

References

1. Arthur Ashkin. Acceleration and trapping of particles by radiation pressure. *Physical Review Letters*, 24(4):156, 1970.
2. Arthur Ashkin, James M Dziedzic, John E Bjorkholm, and Steven Chu. Observation of a single-beam gradient force optical trap for dielectric particles. *Optics Letters*, 11(5):288–290, 1986.
3. Steven Chu, Leo Hollberg, John E Bjorkholm, Alex Cable, and Arthur Ashkin. Three-dimensional viscous confinement and cooling of atoms by resonance radiation pressure. *Physical Review Letters*, 55(1):48, 1985.
4. Lukas Novotny and Bert Hecht. *Principles of nano-optics*. Cambridge university press, 2012.
5. Silvia Albaladejo, Manuel I Marqués, Marine Laroche, and Juan José Sáenz. Scattering forces from the curl of the spin angular momentum of a light field. *Physical Review Letters*, 102(11):113602, 2009.
6. Justin E Molloy and Miles J Padgett. Lights, action: optical tweezers. *Contemporary Physics*, 43(4):241–258, 2002.

7. E. Hecht. *Optics*. Pearson, 2014.
8. Lukas Novotny, Randy X Bian, and X Sunney Xie. Theory of nanometric optical tweezers. *Physical Review Letters*, 79(4):645, 1997.
9. David G Grier. A revolution in optical manipulation. *Nature*, 424(6950):810–816, 2003.
10. Mathieu L Juan, Maurizio Righini, Romain Quidant. Plasmon nano-optical tweezers. *Nature Physics*, 5(6):349–356, 2011.
11. Zhiqun Xi, Ying Gao, George Sirinakis, Honglian Guo, and Yongli Zhang. Single-molecule observation of helix staggering, sliding, and coiled coil misfolding. *Proceedings of the National Academy of Sciences*, 109(15):5711–5716, 2012.
12. Ciro Cecconi, Elizabeth A Shank, Frederick W Dahlquist, Susan Marqusee, and Carlos Bustamante. Protein-dna chimeras for single molecule mechanical folding studies with the optical tweezers. *European Biophysics Journal*, 37(6):729–738, 2008.
13. Michelle D Wang, Hong Yin, Robert Landick, Jeff Gelles, and Steven M Block. Stretching dna with optical tweezers. *Biophysical Journal*, 72(3):1335, 1997.
14. Steven B Smith, Yujia Cui, and Carlos Bustamante. Overstretching b-dna: the elastic response of individual double-stranded and single-stranded dna molecules. *Science*, 271(5250):795–799, 1996.
15. Dmitri K Gramotnev and Sergey I Bozhevolnyi. Plasmonics beyond the diffraction limit. *Nature Photonics*, 4(2):83–91, 2010.
16. AN Grigorenko, NW Roberts, MR Dickinson, and Y Zhang. Nanometric optical tweezers based on nanostructured substrates. *Nature Photonics*, 2(6):365–370, 2008.
17. Yasuyuki Tsuboi, Tatsuya Shoji, Noboru Kitamura, Mai Takase, Kei Murakoshi, Yoshihiko Mizumoto, and Hajime Ishihara. Optical trapping of quantum dots based on gap-mode-excitation of localized surface plasmon. *The Journal of Physical Chemistry Letters*, 1(15):2327–2333, 2010.
18. Yoshito Tanaka and Keiji Sasaki. Optical trapping through the localized surface-plasmon resonance of engineered gold nanoblock pairs. *Optics Express*, 19(18):17462–17468, 2011.
19. Tatsuya Shoji, Junki Saitoh, Noboru Kitamura, Fumika Nagasawa, Kei Murakoshi, Hiroaki Yamauchi, Syoji Ito, Hiroshi Miyasaka, Hajime Ishihara, and Yasuyuki Tsuboi. Permanent fixing or reversible trapping and release of dna micropatterns on a gold nanostructure using continuous-wave or femtosecond-pulsed near-infrared laser light. *Journal of the American Chemical Society*, 135(17):6643–6648, 2013.
20. Kai Wang, Ethan Schonbrun, Paul Steinvurzel, and Kenneth B Crozier. Trapping and rotating nanoparticles using a plasmonic nano-tweezer with an integrated heat sink. *Nature Communications* 2, 469, 2011.
21. Yang Zhao, Amr AE Saleh, and Jennifer A Dionne. Enantioselective optical trapping of chiral nanoparticles with plasmonic tweezers. *Acs Photonics*, 3(3):304–309, 2016.
22. Preston R Huft, Joshua D Kolbow, Jonathan T Thweatt, and Nathan C Lindquist. Holographic plasmonic nanotweezers for dynamic trapping and manipulation. *Nano Letters*, 17(12):7920–7925, 2017.
23. Ali El Eter, Nyha M Hameed, Fadi I Baida, Roland Salut, Claudine Filiatre, Dusan Nedeljkovic, Elie Atie, Samuel Bole, and Thierry Grosjean. Fiber-integrated optical nano-tweezer based on a bowtie-aperture nano-antenna at the apex of a snom tip. *Optics Express*, 22(8):10072–10080, 2014.
24. Johann Berthelot, Srdjan S Aćimović, Mathieu L Juan, Mark P Kreuzer, Jan Renger, and Romain Quidant. Three-dimensional manipulation with scanning near-field optical nanotweezers. *Nature Nanotechnology*, 9(4):295–299, 2014.
25. Yuanjie Pang and Reuven Gordon. Optical trapping of a single protein. *Nano Letters*, 12(1):402–406, 2012.
26. Mathieu L Juan, Reuven Gordon, Yuanjie Pang, Fatima Eftekhari, and Romain Quidant. Self-induced back-action optical trapping of dielectric nanoparticles. *Nature Physics*, 5(12):915–919, 2009.
27. Steven Jones, Ahmed A Al Balushi, and Reuven Gordon. Raman spectroscopy of single nanoparticles in a double-nanohole optical tweezer system. *Journal of Optics*, 17(10):102001, 2015.

28. Allen HJ Yang, Sean D Moore, Bradley S Schmidt, Matthew Klug, Michal Lipson, and David Erickson. Optical manipulation of nanoparticles and biomolecules in sub-wavelength slot waveguides. *Nature*, 457(7225):71–75, 2009.
29. Sudeep Mandal, Xavier Serey, and David Erickson. Nanomanipulation using silicon photonic crystal resonators. *Nano Letters*, 10(1):99–104, 2010.
30. Yih-Fan Chen, Xavier Serey, Rupa Sarkar, Peng Chen, and David Erickson. Controlled photonic manipulation of proteins and other nanomaterials. *Nano Letters*, 12(3):1633–1637, 2012.
31. Miguel A Santiago-Cordoba, Murat Cetinkaya, Svetlana V Boriskina, Frank Vollmer, and Melik C Demirel. Ultrasensitive detection of a protein by optical trapping in a photonic-plasmonic microcavity. *Journal of Biophotonics*, 5(8-9):629–638, 2012.
32. Huibo Fan, Xiaodong Gu, Dawei Zhou, Huili Fan, Li Fan, and Changquan Xia. Confined whispering-gallery mode in silica double-toroid microcavities for optical sensing and trapping. *Optics Communications*, 434:97–103, 2019.
33. Adarsh Lalitha Ravindranath, Mirali Seyed Shariatdoust, Samuel Mathew, and Reuven Gordon. Colloidal lithography double-nanohole optical trapping of nanoparticles and proteins. *Optics Express*, 27(11):16184–16194, 2019.
34. Shih-Kun Liu, Wen-Kai Hsieh, Wei-Yi Sung, Yong-Jai Shen, Chia-Ching Tsai, and Fong-Min Hsu. Wavelength dependence of trapping efficiency in optical tweezers. In *2015 International Symposium on Next-Generation Electronics (ISNE)*, pages 1–3. IEEE, 2015.
35. Karel Svoboda and Steven M Block. Biological applications of optical forces. *Annual Review of Biophysics and Biomolecular Structure*, 23(1):247–285, 1994.
36. Ahmed A Al Balushi, Ana Zehtabi-Oskuie, and Reuven Gordon. Observing single protein binding by optical transmission through a double nanohole aperture in a metal film. *Biomedical Optics Express*, 4(9):1504–1511, 2013.

Chapter 8
Applications of Trapping to Protein Analysis and Interactions

Reuven Gordon

Abstract While optical tweezers have long allowed for studying proteins and their interactions, nanotechnology has allowed for fabrication of tweezers that are capable of trapping even single proteins. The use of apertures in sensing applications has been introduced in previous chapters, as well as a discussion of optical tweezers for single molecule analysis. The present chapter focuses on applications to protein analysis, showing different techniques that have been developed as well as challenges for the approach. This includes analysis based on intensity changes due to proteins and nanoparticles entering the optical trapping region, moving in the trapping region, and being modulated by high-frequency laser intensity beating (in the 1–1000 GHz regime).

8.1 Introduction

The advancement of nanotechnology has allowed for probing the nature of life at the single molecule level. Single molecule techniques show a diversity of interactions that are washed out by ensemble measurements: rare events, events that have a lower signal, events that operate at a different timescale [1]. They allow for seeing dynamics without synchronization: getting all the molecules to the same starting point requires some synchronization that perturbs the system away from its natural conditions. They allow for probing interactions at equilibrium, such as on-off binding, whereas most other kinetic measurements need to be perturbed away from equilibrium.

Fluorescence labeling is one of the primary ways of studying single molecules. While already established and continually providing significant scientific advances, fluorescence has the drawbacks of:

- limited time resolution due to the requirement for photon counting;
- blinking;
- bleaching;

R. Gordon (✉)
University of Victoria, Victoria, BC, Canada
e-mail: rgordon@uvic.ca

W. Bowen et al. (eds.), *Single Molecule Sensing Beyond Fluorescence*, Nanostructure Science and Technology,
https://doi.org/10.1007/978-3-030-90339-8_8

- tags altering biomolecule propertes; and
- indirect measurement of properties that are interpreted from changes in the fluorescence signal (e.g. Förster resonance energy transfer).

While the ideal fluorescent label will not change the properties of the protein, studies are showing that labelling changes the energy characteristics [2], and interactions [3]. To overcome these limitations, techniques that do not use fluorescence are desired.

Another common approach to understanding biomolecules is to use tethers. This has been applied to fluorescence studies, such as those that interrogate protein conformation changes [4], as well as to numerous works with optical tweezers [5]. Again, the tether immobilizes the biomolecule againsts its free motion, changes its natural properties and introduces steric hindrance. Ways of holding onto single molecules that do not require tethering are desired.

Optical tweezers use laser light to hold onto nanoparticles. By monitoring the scattering of this light, it is possible to gain information about the particle that is trapped in the tweezer. This can be done for extended periods, at the single particle level, without the need for tethers or labels, and at high speeds (potentially faster than nanoseconds). These features make optical tweezers, particularly nanoaperture tweezers capable of trapping single proteins, of interest to single molecule studies.

8.2 Optical Tweezers and Proteins

We consider a basic treatment of optical tweezers and how they hold onto nanoparticles. The aim is to show that the laser powers required for manipulating proteins are impractically large with conventional optical tweezers, and that focusing the optical power down to the nanoscale (extreme subwavelength) and using "self-induced back-action" (SIBA) are beneficial ways to make protein trapping practical. We also show the influence of the optical potential to change the shape of the trapped particle.

8.2.1 Stable Trapping Against Thermal Motion

We consider the problem of using an optical tweezer to manipulate a protein. The type of tweezer we consider is a non-resonant focused beam that scatters off a nominally neutral particle. Such a tightly focused beam was used to trap colloidal particles of 100 nm in size in 1986 [6]. This was then promptly applied to the trapping of bacteria and large viruses [7].

For trapping proteins, in the single digit nanometer range, impractically high intensities are required. The potential energy associated with a polarizable dipole (the model for the protein) in a electromagnetic field (the light) is given by:

$$U = -\frac{1}{2}\mathbf{p} \cdot \mathbf{E} \tag{8.1}$$

where **p** is the dipole moment and **E** is the electric field.

While the energy of the dipole itself increases, the total energy decreases by an equal amount and so the potential minimum occurs for location of the highest field intensity [8]. The optical potential energy should be significantly larger than the thermal energy $k_B T$ (where k_B is Boltzmann's constant and T is temperature in Kelvin) to hold the particle stably against thermal motion [6].

To give numerical examples, we estimate the intensity required to give the same potential energy as the thermal energy $k_B T$ at 300 K for a 5 nm diameter spherical particle (approximate size of a globular protein of 50 kDa), assuming the relative permittivity of the particle ϵ_d is 2.25 and the surrounding medium relative permittivity ϵ_b is 1.77.

Ignoring retardation and losses, the polarizability of a prolate spheroid (the approximate shape of the protein—say a globular protein) is:

$$\alpha = \frac{V\epsilon_0 (\epsilon_d - \epsilon_b)}{A(\epsilon_d - \epsilon_b) + \epsilon_b} \tag{8.2}$$

where V is the volume of the particle and ϵ_0 is the free-space permittivity and A is a geometric factor:

$$A = (\xi^2 - 1)\left(\frac{\xi}{2}\ln\left[\frac{\xi + 1}{\xi - 1}\right] - 1\right) \tag{8.3}$$

where $\xi = a/\sqrt{a^2 - b^2}$ and a and b are the long and short axes, with $A = 1/3$ for a sphere. Since $\mathbf{p} = \alpha\mathbf{E}$, using these values, we find that 10^{14} W/m^2 is the beam intensity required, which corresponds to approximately a 100 W laser beam focussed down to a micrometer squared, or a 10 mW beam focussed down to 10 nm squared. (Conventional lenses can focus down to a spot size of approximately the wavelength squared, so this is the reason for the micrometer squared diameter; whereas plasmonic structures allow for much tighter field confinement near the surface of a metal). It is clear that focusing down to subwavelength volumes allows for more practical beam powers. It also means that the particle is more confined. Even so, this is a substantial amount of tightly focussed power that is required to even reach the thermal energy threshold—even higher powers would be required for stable trapping.

8.2.2 Elongation and Orientation

We now consider the deformation of the nanoparticle/protein from the optical potential. Changes in the shape of the particle can also result in changes in **p**, and so the potential energy associated with those changes are also reflected in Eqs. 8.1 and 8.2. If the particle is extended by 10% along one axis while maintaining the volume, the polarizability changes by 1%. The amount of energy available to elongate the particle by 10% is about 1% of the trapping energy, so typically some (small) fraction of the thermal energy when stable trapping is achieved. This elongation is commonly referred to as electrostriction.

It is also noted that the polarizability along the long axis is the highest, and so the potential energy minimum occurs when the particle is oriented along the direction of the electric field of the light. This acts to rotate the nanoparticle/protein to be oriented with its long axis aligned with the electric field polarization.

8.2.3 SIBA

Additional trapping efficiency is gained if the nanoparticle/protein itself influences the field, **E**, in Eq. 8.1. One example is in a photonic crystal cavity: the nanoparticle/protein itself shifts the resonance of the cavity and thereby changes how much laser energy is coupled into the cavity [9]. Analysis in that work showed that stable trapping could be achieved for 100 nm particles that are about half the size of the photonic crystal holes using 100 mW of power inside the cavity. Actual trapping of mutant Wilson disease protein (estimated to be about 10 nm in diameter) in a photonic crystal cavity was published in 2012 [10]. In that work, the protein was fluorescently labelled to identify when it was trapped and around 10 mW of power was coupled into the cavity (although with a quality factor of several thousand, the intensity is expected to be 3–4 orders of magnitude larger inside the cavity, which is similar to the analysis above).

The nanoparticle did not have a large influence on the field distribution in the photonic crystal cavity case. Calculations of the force by energy methods (Maxwell Stress Tensor) or perturbative dipole analysis yielded the same results [9]. By contrast, in subwavelength circular apertures in metal films the nanoparticle has a large influence on the field distribution and so the force calculated with Maxwell Stress Tensor analysis was around 5 times larger [11]. Other works had previously investigated trapping with circular apertures prior to the investigation of SIBA [12, 13].

Trapping with metal apertures can be thought of as a resonance effect if plasmonic resonances are considered. Since the field distribution is changed substantially, it can also be thought of as effectively increasing the size of the aperture through "dielectric loading". Hans Bethe showed that apertures in thin metal films limit transmission as the fourth power of the aperture size [14], so that making the aperture bigger should give a substantially more transmission—for example, making the aperture 1% wider gives 4% more transmission. This is a primary benefit of using apertures—the transmission changes can be detected to monitor the trapping event and the amount of transmission is "amplified" by the effect of the aperture.

Greater field confinement is achieved by creating small gaps in the aperture about the size of the nanoparticle/protein of interest. The double-nanohole is one way to achieve this and trapping of a single protein with 3.4 nm hydrodynamic radius (Bovine Serum Albumin) was achieved for that aperture [15] (see Fig. 3 of Chap. 7). It has been shown that tapering in the double nanohole plays a significant role in enhancing the efficiency of trapping [16]. Other shapes such as coaxial structures [17–19], also allow for trapping of single proteins by confinement of the field to a narrow gap. Researchers have also studied rectangular [20] and bowtie [21, 22] apertures for trapping objects down to the single digit nanometer range.

8.3 Optical Tweezer Studies of Proteins with Tethering and/or Labelling

Conventional optical tweezers using a microscope objective to focus the beam cannot typically hold onto single proteins or other small biomolecules. For this reason, the proteins being studied are usually tethered to larger particles, lipid droplets, flat surfaces or micropipettes [5]. The optical tweezer allows for measuring forces and displacements in biomolecular interactions, for example: (i) in observing the transcription of DNA to RNA via the enzyme RNA polymerase [23], (ii) the motion of kinesin along a microtuble taking 8 nm steps [24, 25], (iii) the winding of DNA into a virus capsid by the molecular motor [26], and (iv) measuring the energy landscape of protein folding [27, 28]. While these approaches have been valuable in determining the properties of biomolecules and their interactions at the single molecule level, the impact of the tether on the natural function, hindrance for multiple small molecules interacting close to one another, and overall complexity all call for additional approaches to probe biomolecule interactions at the single molecule level.

8.4 Protein Analysis with Nanoapertures

Considering the interest in single protein and biomolecule studies that are label free and tether free, the aperture-based trapping, particularly with double-nanoholes, has been widely investigated for its analytical capability. In this section, we overview some of examples of the analysis that can be achieved.

8.4.1 Protein Mass and Mixtures

The polarizability of a nanoparticle varies with the size, approximately as the volume as in Eq. 8.2. When the nanoparticle/protein is trapped in an aperture, the thermal motion leads to intensity fluctuations in the transmission through the aperture that scale as the volume, or mass [29].

With this approach, resolution in the protein mass below a kDa is possible, allowing for measuring the mass of a small protein fragment removed during tryptic cleaving [29].

It is also possible to gain information from the autocorrelation of the intensity fluctuations. Considering that the Stokes drag, γ, on a particle in a nanofluidic environment scales as the radius of the particle, and the stiffness, k of the trap scales as the volume (from the polarizability above), we consider a Langevin analysis:

$$m\frac{\partial^2 x}{\partial t^2} + \gamma\frac{\partial x}{\partial t} - kx = F_l \tag{8.4}$$

where x is the particle position and F_l is a Langevin white noise term. The first inertial term is usually neglected in a highly damped environment, so the characteristic time constant for the autocorrelation of the trapping intensity fluctuations is [29]:

$$\tau = \gamma/k \propto m^{-2/3} \tag{8.5}$$

where m is the particle mass. In cases where intensity fluctuations arise from slower conformational changes of the protein, a two exponential fit to the autocorrelation data can be used to achieve time constants with the −2/3 scaling [30].

Since the tweezer is a single molecule approach, repeated or parallel measurements can be used to sample a heterogeneous mixture. In this way, the distribution of protein sizes can be found from an unprocessed (dirty) solution [30]. Figure 8.1 shows the comparison in the autocorrelation functions for pure protein solutions compared with buffer solution diluted (but otherwise unprocessed) egg white. This demonstrates the use of single protein analysis to determine the components of an unprocessed "dirty" solution.

8.4.2 *Conformational Changes*

Several works have studied conformational changes (including folding) of proteins using fluorescence [4] and tweezers [28]. Equations 8.1and 8.2 show that changes in the polarizability of the protein from conformational changes will be transduced into changes in energy of the optical tweezer and also changes in the transmission for the aperture optical tweezer. There is as driving force on this conformational change from the optical tweezer, as described above.

For example, Bovine Serum Albumin, a protein with conformation changes from a heart-shaped F isoform to and extended N isoform was trapped in a double-nanohole optical tweezer and spontaneous jumps in the trapping intensity were seen after the initial trapping event [15] (see Fig. 3b of Chap. 7). These jumps were attributed to the transition between the two isoforms, which was confirmed by forcing the protein into the N state by reducing the pH (at which point only the higher intensity level was seen from the trapping laser transmission through the aperture).

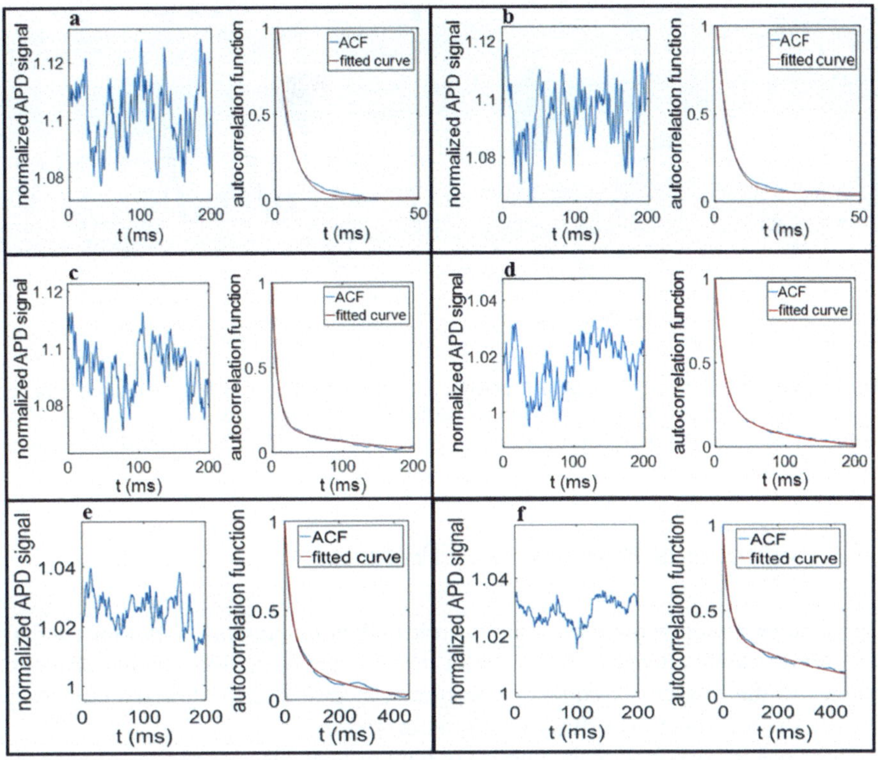

Fig. 8.1 Autocorrelation of fluctuations of laser signal through double-nanohole when trapping a single protein. For pure protein samples (**a**, **c**, **e**) and various trapping events on buffer-diluted (but otherwise unprocessed) egg white (**b**, **d**, **f**). **a** Pure ovotransferrin sample. **b** Protein trapped from egg-white with similar time-constants to ovotransferrin. **c** Pure ovalbumin sample. **d** Protein trapped from egg-white with similar time-constants to ovalbumin. **e** Pure ovomucoid sample. **f** Protein trapped from egg-white with similar time-constants to ovomucoid. Reproduced with permission from Ref. [30], ©American Chemical Society 2018

It is important to stress that this conformation change is observed without any fluorescent markers and without tethering, so it represents the unhindered motion of the natural protein.

Since photodiode detection can readily extend into the GHz regime, this approach is envisioned to be useful for testing fundamental questions like: how fast do proteins fold? What is the speed of conformational changes? Simulations have pointed to timescales in excess of tens of nanoseconds to milliseconds [31, 32].

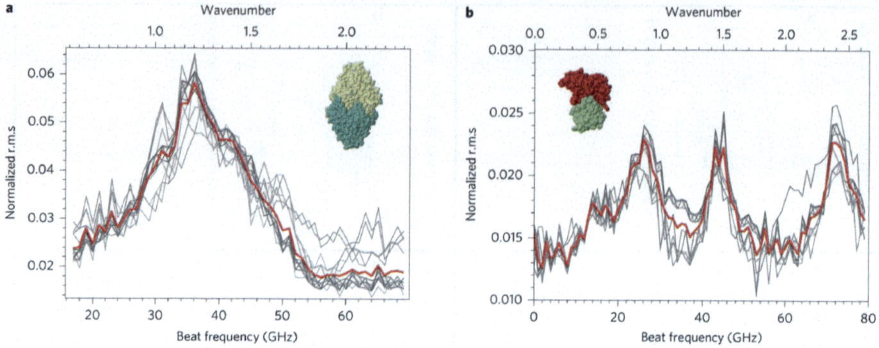

Fig. 8.2 Vibrational resonances of **a** Carbonic Anhydrase protein and **b** Conalbumin protein, probed with electrostriction excited increased noise fluctuations in an optical tweezer setup. Multiple sweeps are shown in grey. Reproduced with permission from Ref. [36], ©Nature Publishing Group 2015

8.4.3 Vibrational Resonances of Proteins

The vibrational motion of proteins plays a role in their functioning. Vibrations underlie conformational changes [33]. Even in the absence of conformational changes, changes in the vibration frequency spectrum can give rise to allostery (signaling between distinct functional sites of the protein) [34].

These mechanical modes are coupled to light through electrostriction (as changes in the shape lead to changes in the energy of the protein within an optical field, as per Eq. 8.2), which has been used to probe ensemble dynamics in an optical Kerr effect setup [35].

For single protein analysis, it has been realized that interference beating of two lasers of different frequency to produce GHz to THz vibrations allows for probing protein dynamics in a double-nanohole optical tweezer setup [36]. This approach has also been used to probe the resonances of single stranded DNA [37]. The approach has spectral resolution below the GHz range, which allows for probing slightly non-degenerate vibrational modes [36]. Figure 8.2 shows the vibration spectra of two different proteins probed via electrostriction in the optical tweezer setup.

Normal mode analysis has been used to compare the vibration spectra of the Raman active modes in the optical tweezer setup with their expected theoretical values, showing good agreement [38]. Other theory works showed qualitatively similar results [39], which may lead to improved modeling and understanding of protein dynamics [40].

For the DNA studies, it was possible to not only distinguish different lengths of DNA strands, but also different base compositions, since the bases have slightly different masses and this shifts the vibration resonances [37]. This has potential for sequencing applications.

Higher frequency vibrations may be probed with conventional Raman methods, and the Raman spectra of individual trapped polystyrene and titania nanoparticles has been achieved using the double-nanohole optical tweezer setup [41], as well as a rectangular aperture tweezer setup to probe polystyrene particles [42]. In the latter work, the onset of a shifted Raman peak was found with increasing power and it was speculated that this was the result of the nanoparticle interacting with the surface.

8.5 Analysis of Protein Interactions at Single Molecule Level

The function of proteins comes from their interactions, and therefore, it is useful to study these interactions at the single molecule level. The techniques using optical tweezers to probe protein–small molecule, protein–protein and protein–DNA interactions are described in this section.

8.5.1 Protein–Small Molecule Interactions

It was recognized that binding to a protein could alter its fluctuations within the double nanohole optical tweezer [43]. This can be explained by changes in the flexibility of the protein being translated into changes in the thermally driven motion. A simple example of this was demonstrated first for the well-studied biotin-streptavidin interaction. Not only does binding of biotin to streptavidin produce a change in the fluctuations, but a mutant version of streptavidin was also investigated with only a single binding site, showing distinct autocorrelation of the intensity fluctations. As a practical example for drug research, the binding of salicylic acid (commonly known as aspirin) to the protein cyclooxygenase-2 was measured, showing distinct autocorrelation for the bound and unbound cases.

Of course, these binding event examples were practically irreversible, and so extended observation periods could be taken for the bound and unbound cases, but not the transition between these two cases. For lower binding affinities, it is possible to measure the binding by changes in the amplitude of the trapping laser power transmitted through the double nanohole [44]. It was demonstrated that the dissociation constants for the small molecules tolubutamide and phenytoin to protein human serum albumin matched to literature reported values. The case of tolubutamide is shown in Fig. 8.3.

Unlike other methods to determine binding affinities, these studies were done at a single concentration, at the single molecule level and at equilibrium. These favourable characteristics are unique to such single molecule studies. Compared to other single molecule approaches, these studies were done without labels or tethers that are known to influence binding kinetics [45].

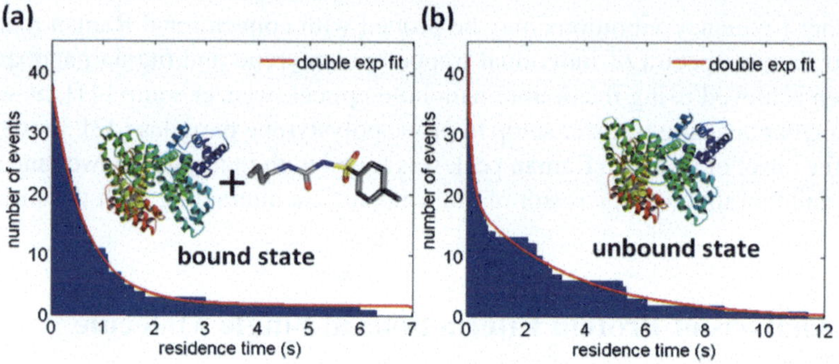

Fig. 8.3 On and off residence time histogram for tolbutamide and human serum albumin. From this single concentration single molecule measurement, the binding affinity was found to be within the range reported in the literature. From Ref. [44], ©American Chemical Society 2014

It is also possible to measure binding of proteins to large nanospheres with a protein binding target coating [46]; however, the use of the nanosphere makes the approach similar to other techniques and so this is less desirable.

8.5.2 Protein–Protein Interactions

Protein-protein interactions are diverse and they can be monitored at the single molecule level by making use of sizing and transmission methods outlined in the previous sections. For example, protein—antibody interactions have been demonstrated using trapping of bovine serum albumin protein followed by an antibody to that protein [47]. The initial trapping event showed a jump in transmission, as did the subsequent event. The second jump was confirmed not to be co-trapping by later experiments which showed that co-trapping was unlikely without binding [46].

The tryptic cleaving of ovalbumin was measured in a protein sizing experiment by comparing the measured molecular weight of the trypsin—ovalbumin complex with the sum of their individual molecular weights [29]. The complex was 4.4 kDa lower as the result of tryptic cleaving of part of the ovalbumin [48].

Many other interactions may be studied with this approach including generalized antibody development studies and understanding how protein complexes function.

8.5.3 Protein–DNA Interactions

The interaction between p53 (wildtype and a mutant) and DNA was studied by the optical tweezer technique [49]. A 20 base hairpin was trapped in the double nanohole tweezer setup, and showed a characteristic unzipping time of ~100 ms. This

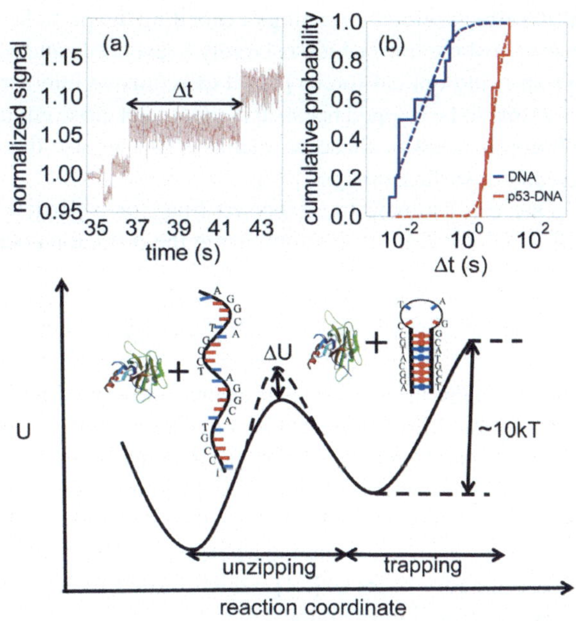

Fig. 8.4 **a** Shows double nanohole trapping jump for p53 bound to hair-pin DNA. The intermediate level is the hairpin state and the final level is the unzipped hairpin. **b** The time to unzip increases by over an order of magnitude when p53 is added. A schematic of the reaction co-ordinate for unzipping is shown at the bottom where the addition of p53 increases the unzipping barrier. From Ref. [49], ©Optical Society of America 2014

unzipping was likely assisted by the small amount of electrostriction described above. With the addition of p53, the unzipping time increased by 1–2 orders of magnitude, which corresponds to an increase in the potential energy barrier introduced by the p53 protein binding to the DNA of around 2×10^{-20} J. This is assuming an Arrhenius behavior. Mutant p53 had no effect on the unzipping time, even though it is known to bind to DNA still [50]. It is possible to use this approach as an assay for detecting if the mutant p53 could have the wildtype function restored by adding small molecules [51].

Many other proteins that interact with DNA and other biomolecules can be studied in this manner to determine how their behavior is modified and uncover the underlying biophysics of the interaction (Fig. 8.4).

8.6 Challenges to Trapping with Nanostructures

While many of the applications above focused on double-nanohole experiments, trapping proteins and biomolecules using nanostructures presents several challenges.

8.6.1 Surface Interactions

Since the trapping using nanostructures usually occurs close to the surface of the nanostructure, this requires the particle to get close to the surface region. The surface can become charged, which is common at an aqueous boundary. This can have one of two effects, either attracting or repelling the particle. For attraction, this can create an electrostatic trap, which is largely irreversible and not desirable if the intention is to trap using optical forces [52]. For repulsive interactions, the particle may take a long time to be trapped because it cannot get close enough to the surface. A common solution to each of these problems is to increase the ionic strength in solution to screen out surface charges [53].

One can estimate the expected time for a particle to get to the trap based on diffusion in a viscous medium, using the diffusion coefficient:

$$D = \frac{k_B T}{6\pi \eta a} = l^2/\tau \tag{8.6}$$

where η is the viscosity, a is the particle radius, l is the average particle separation and τ is the diffusion time. Considering a 2 nm diameter particle in water with an average separation of 2 microns between particles, τ is around 4 ms. In practice, the time to trap in aqueous solution has been observed to be much greater than this—of the order of seconds, which is attributed to surface interactions. The time to trap in hexane for inorganic nanoparticles was in the millisecond range, consistent with the estimate for τ above [54], which further underscores the importance of surface charge mediated through the solvent.

Proteins can also stick to surfaces. A typical approach to avoid this is to use a mPEG coating, attaching with mPEG-thiol to a gold surface, for example. This was used to prevent sticking of BSA in initial experiments, but found not to be crucial since the proteins were not overly sticky [15]. The trapping usually shows increased noise in the laser fluctuations, indicating that the protein is diffusing freely in the trap and not stuck. Some have used surfactant in solution to limit sticking [10], but this can denature proteins or otherwise change their characteristics.

Surfaces change the natural diffusion of the particle due to hydrodynamic interactions. A simplification to form proposed by Brenner [55] gives an approximate correction to the perpendicular diffusion as [56]:

$$D_{\perp} = D(1 - \gamma) \tag{8.7}$$

where $\gamma = a/h$ and h is the height from the surface measured from the centre whereas the approximate correction to the parallel diffusion from Faxén is:

$$D_{\parallel} = D\left(1 - \frac{9}{16}\gamma + \frac{1}{8}\gamma^3 - \frac{45}{256}\gamma^4 - \frac{1}{16}\gamma^5\right) \tag{8.8}$$

which approaches roughly a third of the free diffusion at small distances. These effects should be taken into account when calculating the properties of the particle in the trap, but also when calibrating the stiffness of the trap itself [57].

8.6.2 *Thermal Effects*

Nanostructured metals can concentrate the light intensity to extremely small volumes, which can lead to local heating. Excessive heating can damage the sample being studied (e.g., denaturing proteins). Moderate local heating can have benefits, such as raising the local temperature to physiological levels, or providing thermophoretic forces to attract particles to the trapping aperture. Apertures have a metal film surrounding them, so this helps to remove heat due to the thermal conductivity of the metal. Comparison between isolated metal nanoparticles and apertures with similar field enhancement showed that the temperature increase was 3 orders of magnitude lower for the aperture [58].

Heating at bowtie and double nanohole apertures has been measured and calculated to be in the range of 0.5–5 K/mW focused on a micron squared area, depending on the geometry, the wavelength of operation and the materials used. An order of magnitude lower heating was achieved by exciting on the low-loss side of the aperture (opposite to the adhesion layer which is a metal with higher losses), by using a more conductive substrate and by using a lower loss adhesion layer [59, 60].

Heating can also lead to convective or thermophoretic effects that can either assist or hinder in trapping [61–63].

8.7 Technological Advances

8.7.1 *Trapping Characterization*

Autocorrelation of the noise in the trapping laser gives an indication of the time constant for diffusion and can be used to find the trapping efficiency [57]. In addition, the transient to the trapped state has a similar time dependence. From these measurements, an efficiency was estimated of ~0.1 fN/(nm mW) [57], which is among the highest values. Fano resonances have been employed to achieve higher efficiencies [64].

8.7.2 *Aperture Fabrication*

Ion beam milling is a direct approach to create apertures in metal films with a variety of shapes [65–67]. Cusps, like those found in bowtie and double-nanoholes, help to enhance the local field [68].

Electron beam/UV lithography processes have also been used in an approach akin to those used in near-field aperture probes [20]. Anisotropic etching of silicon creates a taper to an aperture, which can be coated with a metal.

Nanosphere lithography allows for low cost fabrication of circular [13] and double-nanoholes [69]. Randomly deposited colloids serve as a mask for metal evapouration—the random distribution makes finding the apertures of interest challenging. By plasma etching the spheres prior to deposition of the metal, the gap size at the cusps of double-nanoholes, as well as the size of nanoholes could be tuned. Gaps as small as 10 nm were made with this approach [70].

A pattern can be transferred from a silicon master by making use of the poor adhesion of a metal overlayer [71]. This *template stripping* method has been used to make double-nanoholes with gaps below 10 nm [72].

Atomic layer deposition allows for making coaxial structures with gaps in the nanometer range [73]. Protein trapping was demonstrated with these coaxial structures [19].

8.7.3 Microfluidic Integration

Fluidic delivery can allow for delivering proteins to the trap as well as introducing particles for them to interact with [10, 47, 74]. Having a well-defined flow also allows for characterizing the strength of the trap against the drag force.

8.7.4 Fiber Based Methods

By integrating the tweezer on the end of a translatable probe, it is possible to simplify the setup (removing the microscope, for example) and translate the trapped particle [21, 75, 76]. Cleaved fibers can be used as well, which is simpler and less fragile than tapered ones and may be fabricated using template stripping [77, 78].

8.7.5 Potential for Complementary Analysis

Aperture optical trapping has been combined with Raman spectroscopy, using the trapping laser as the excitation source, allowing for identifying the material of the trapped particle [41, 42].

There are several works that have started combining nanoaperture and plasmonic tweezers with nanopores that monitor particles by translocation in an ionic liquid. Nanopores are used widely in nanoscale analysis, as outlined in this book in Chaps. 10, 12, 11 and 13 and elsewhere [79–93]. There has been interest in slowing nanopore translocation using a tweezer force [94].

The use of plasmonic apertures with flow through a membrane has been considered for over a decade [95–99]. Pressure and electrically driven flow can be used to concentrate particles of interest in the vicinity of the apertures.

This has been adapted for a single nanopore and a single nanoaperture for optical trapping or sensing of particles delivered to and through the pore, such as DNA and proteins, as well as monitoring their interactions [100–104]. One potential avenue of investigation is for membrane proteins over apertures, since membranes have been fabricated with nanopores [105]. Several works have looked at apertures for fluorescence correlation spectroscopy [106], and there is potential to combine this with tweezer approaches.

8.8 Conclusions and Outlook

Nanoapertures in metal films allow for trapping and studying particles below 100 nm, and down to the single digit nanometer size. This enables trapping of biomolecules including proteins and DNA, as well as studying molecular interactions. The technique benefits from working at the single molecule level, without the need for tethers or labels. This allows for processing in dirty samples (e.g., raw blood, or egg white) and gets rid of the need for processing and tagging. The molecular weight of proteins can be found with this method, their binding affinities to small molecules and even the vibrational modes of proteins and DNA. This may lead to better understanding of protein vibrations and the role they play in biological function. It may also allow for molecular identification and potentially even sequencing.

Due to the early stage of its development, the technique is still being refined and adapted by different research groups worldwide. It is unlikely that nanoaperture optical tweezers will replace embedded methods for drug discovery and protein analysis in the near future. Nevertheless, the potential to obtain direct measurements about the structure and dynamics of biomolecules is expected to be immediately useful in verifying the biophysical processes involved in new drug candidates. For example, the study of p53-DNA interactions demonstrated using the nanoaperture trapping is a powerful technique to validate the role of small molecules in p53 function restoration [107]. Significant advances have been made in nanofabrication, and it is envisioned that "nanopipettes" may soon be available where multiple nanoapertures are integrated on fibers to enable parallel well-plate interrogation in an automated fashion.

Some other challenges remain in the widescale adoption of this technique. Reliability in time to trap needs to be addressed—presently trapping can take a few seconds to a few hours depending on sample preparation. The influence of surface charging should be addressed, and it has been demonstrated that ionic strength tuning will affect binding events (e.g., see Ref. [53]). The problem may be solved by combining the approach with existing methods from the nanopore community. Alternatively, a deeper understanding of the nanofluidics may provide the insight to make the onset of trapping more predictable.

References

1. WE Moerner and David P Fromm. Methods of single-molecule fluorescence spectroscopy and microscopy. *Review of Scientific instruments*, 74(8):3597–3619, 2003.
2. Kapil Dave, Hannah Gelman, Chu Thi Hien Thu, Drishti Guin, and Martin Gruebele. The effect of fluorescent protein tags on phosphoglycerate kinase stability is nonadditive. *The Journal of Physical Chemistry B*, 120(11):2878–2885, 2016.
3. Feng Liang, Yuzheng Guo, Shaocong Hou, and Qimin Quan. Photonic-plasmonic hybrid single-molecule nanosensor measures the effect of fluorescent labels on dna-protein dynamics. *Science advances*, 3(5):e1602991, 2017.
4. Hoi Sung Chung, Kevin McHale, John M Louis, and William A Eaton. Single-molecule fluorescence experiments determine protein folding transition path times. *Science*, 335(6071):981–984, 2012.
5. Hongbao Xin, Yuchao Li, Yong-Chun Liu, Yao Zhang, Yun-Feng Xiao, and Baojun Li. Optical forces: from fundamental to biological applications. *Advanced Materials*, 32(37):2001994, 2020.
6. Arthur Ashkin, James M Dziedzic, JE Bjorkholm, and Steven Chu. Observation of a single-beam gradient force optical trap for dielectric particles. *Optics Letters*, 11(5):288–290, 1986.
7. Arthur Ashkin and James M Dziedzic. Optical trapping and manipulation of viruses and bacteria. *Science*, 235(4795):1517–1520, 1987.
8. John David Jackson. *Classical electrodynamics*. John Wiley & Sons, 2007.
9. Michael Barth and Oliver Benson. Manipulation of dielectric particles using photonic crystal cavities. *Applied Physics Letters*, 89(25):253114, 2006.
10. Yih-Fan Chen, Xavier Serey, Rupa Sarkar, Peng Chen, and David Erickson. Controlled photonic manipulation of proteins and other nanomaterials. *Nano Letters*, 12(3):1633–1637, 2012.
11. Mathieu L Juan, Reuven Gordon, Yuanjie Pang, Fatima Eftekhari, and Romain Quidant. Self-induced back-action optical trapping of dielectric nanoparticles. *Nature Physics*, 5(12):915, 2009.
12. K Okamoto and S Kawata. Radiation force exerted on subwavelength particles near a nanoaperture. *Physical Review Letters*, 83(22):4534, 1999.
13. Eun-Soo Kwak, Tiberiu-Dan Onuta, Dragos Amarie, Radislav Potyrailo, Barry Stein, Stephen C Jacobson, WL Schaich, and Bogdan Dragnea. Optical trapping with integrated near-field apertures. *The Journal of Physical Chemistry B*, 108(36):13607–13612, 2004.
14. Hans Albrecht Bethe. Theory of diffraction by small holes. *Physical Review*, 66(7-8):163, 1944.
15. Yuanjie Pang and Reuven Gordon. Optical trapping of a single protein. *Nano Letters*, 12(1):402–406, 2012.
16. Mostafa Ghorbanzadeh, Steven Jones, Mohammad Kazem Moravvej-Farshi, and Reuven Gordon. Improvement of sensing and trapping efficiency of double nanohole apertures via enhancing the wedge plasmon polariton modes with tapered cusps. *ACS Photonics*, 4(5):1108–1113, 2017.
17. Amr AE Saleh and Jennifer A Dionne. Toward efficient optical trapping of sub-10-nm particles with coaxial plasmonic apertures. *Nano Letters*, 12(11):5581–5586, 2012.
18. Amr AE Saleh, Sassan Sheikhoelislami, Steven Gastelum, and Jennifer A Dionne. Grating-flanked plasmonic coaxial apertures for efficient fiber optical tweezers. *Optics Express*, 24(18):20593–20603, 2016.
19. Daehan Yoo, Gurunatha Kargal, Daniel Mohr, Han-Kyu Choi, Reuven Gordon, and Sang-Hyun Oh. Low-power optical trapping of nanoparticles and proteins with resonant coaxial nanoapertures using 8 nm gaps. *Nano Letters*, 2018.
20. Chang Chen, Mathieu L Juan, Yi Li, Guido Maes, Gustaaf Borghs, Pol Van Dorpe, and Romain Quidant. Enhanced optical trapping and arrangement of nano-objects in a plasmonic nanocavity. *Nano Letters*, 12(1):125–132, 2012.

21. J Berthelot, SS Aćimović, ML Juan, MP Kreuzer, J Renger, and R Quidant. Three-dimensional manipulation with scanning near-field optical nanotweezers. *Nature Nanotechnology*, 9(4):295–299, 2014.
22. Russell A Jensen, I-Chun Huang, Ou Chen, Jennifer T Choy, Thomas S Bischof, Marko Lončar, and Moungi G Bawendi. Optical trapping and two-photon excitation of colloidal quantum dots using bowtie apertures. *ACS Photonics*, 3(3):423–427, 2016.
23. Hong Yin, Michelle D Wang, Karel Svoboda, Robert Landick, Steven M Block, and Jeff Gelles. Transcription against an applied force. *Science*, 270(5242):1653–1657, 1995.
24. Steven M Block, Lawrence SB Goldstein, and Bruce J Schnapp. Bead movement by single kinesin molecules studied with optical tweezers. *Nature*, 348(6299):348–352, 1990.
25. Karel Svoboda, Christoph F Schmidt, Bruce J Schnapp, and Steven M Block. Direct observation of kinesin stepping by optical trapping interferometry. *Nature*, 365(6448):721–727, 1993.
26. Douglas E Smith, Sander J Tans, Steven B Smith, Shelley Grimes, Dwight L Anderson, and Carlos Bustamante. The bacteriophage φ29 portal motor can package dna against a large internal force. *Nature*, 413(6857):748–752, 2001.
27. Bharat Jagannathan, Phillip J Elms, Carlos Bustamante, and Susan Marqusee. Direct observation of a force-induced switch in the anisotropic mechanical unfolding pathway of a protein. *Proceedings of the National Academy of Sciences*, 109(44):17820–17825, 2012.
28. Krishna Neupane, Ajay P Manuel, and Michael T Woodside. Protein folding trajectories can be described quantitatively by one-dimensional diffusion over measured energy landscapes. *Nature Physics*, 12(7):700–703, 2016.
29. Skyler Wheaton and Reuven Gordon. Molecular weight characterization of single globular proteins using optical nanotweezers. *Analyst*, 140(14):4799–4803, 2015.
30. Noa Hacohen, Candice J X Ip, and REuven Gordon. Analysis of egg white protein composition with double nanohole optical tweezers. *ACS Omega*, page to appear, 2018.
31. Kresten Lindorff-Larsen, Stefano Piana, Ron O Dror, and David E Shaw. How fast-folding proteins fold. *Science*, 334(6055):517–520, 2011.
32. Sudha Chakrapani and Anthony Auerbach. A speed limit for conformational change of an allosteric membrane protein. *Proceedings of the National Academy of Sciences*, 102(1):87–92, 2005.
33. Aline Thomas, Konrad Hinsen, Martin J Field, and David Perahia. Tertiary and quaternary conformational changes in aspartate transcarbamylase: a normal mode study. *Proteins: Structure, Function, and Bioinformatics*, 34(1):96–112, 1999.
34. A Cooper and DTF Dryden. Allostery without conformational change. *European Biophysics Journal*, 11(2):103–109, 1984.
35. David A Turton, Hans Martin Senn, Thomas Harwood, Adrian J Lapthorn, Elizabeth M Ellis, and Klaas Wynne. Terahertz underdamped vibrational motion governs protein-ligand binding in solution. *Nature Communications*, 5:3999, 2014.
36. Skyler Wheaton, Ryan M Gelfand, and Reuven Gordon. Probing the raman-active acoustic vibrations of nanoparticles with extraordinary spectral resolution. *Nature Photonics*, 9(1):68–72, 2015.
37. Abhay Kotnala, Skyler Wheaton, and Reuven Gordon. Playing the notes of dna with light: extremely high frequency nanomechanical oscillations. *Nanoscale*, 7(6):2295–2300, 2015.
38. Timothy DeWolf and Reuven Gordon. Theory of acoustic raman modes in proteins. *Physical Review Letters*, 117(13):138101, 2016.
39. Adrien Nicolaï, Patrice Delarue, and Patrick Senet. Theoretical insights into sub-terahertz acoustic vibrations of proteins measured in single-molecule experiments. *The Journal of Physical Chemistry Letters*, 7(24):5128–5136, 2016.
40. Adrien Nicolaï, Fatima Barakat, Patrice Delarue, and Patrick Senet. Fingerprints of conformational states of human hsp70 at sub-thz frequencies. *ACS Omega*, 1(6):1067–1074, 2016.
41. Steven Jones, Ahmed A Al Balushi, and Reuven Gordon. Raman spectroscopy of single nanoparticles in a double-nanohole optical tweezer system. *Journal of Optics*, 17(10):102001, 2015.

42. Sarp Kerman, Chang Chen, Yi Li, Wim Van Roy, Liesbet Lagae, and Pol Van Dorpe. Raman fingerprinting of single dielectric nanoparticles in plasmonic nanopores. *Nanoscale*, 7(44):18612–18618, 2015.
43. Ahmed A Al Balushi and Reuven Gordon. Label-free free-solution single-molecule protein–small molecule interaction observed by double-nanohole plasmonic trapping. *ACS Photonics*, 1(5):389–393, 2014.
44. Ahmed A Al Balushi and Reuven Gordon. A label-free untethered approach to single-molecule protein binding kinetics. *Nano Letters*, 14(10):5787–5791, 2014.
45. Ahmed A Al Balushi, Abhay Kotnala, Skyler Wheaton, Ryan M Gelfand, Yashaswini Rajashekara, and Reuven Gordon. Label-free free-solution nanoaperture optical tweezers for single molecule protein studies. *Analyst*, 140(14):4760–4778, 2015.
46. Ahmed A Al Balushi, Ana Zehtabi-Oskuie, and Reuven Gordon. Observing single protein binding by optical transmission through a double nanohole aperture in a metal film. *Biomedical Optics Express*, 4(9):1504–1511, 2013.
47. Ana Zehtabi-Oskuie, Hao Jiang, Bryce R Cyr, Douglas W Rennehan, Ahmed A Al-Balushi, and Reuven Gordon. Double nanohole optical trapping: dynamics and protein-antibody co-trapping. *Lab on a Chip*, 13(13):2563–2568, 2013.
48. Débora Bonenfant, Tobias Schmelzle, Estela Jacinto, José L Crespo, Thierry Mini, Michael N Hall, and Paul Jenoe. Quantitation of changes in protein phosphorylation: a simple method based on stable isotope labeling and mass spectrometry. *Proceedings of the National Academy of Sciences*, 100(3):880–885, 2003.
49. Abhay Kotnala and Reuven Gordon. Double nanohole optical tweezers visualize protein p53 suppressing unzipping of single dna-hairpins. *Biomedical Optics Express*, 5(6):1886–1894, 2014.
50. Dmitry B Veprintsev and Alan R Fersht. Algorithm for prediction of tumour suppressor p53 affinity for binding sites in dna. *Nucleic Acids Research*, 36(5):1589–1598, 2008.
51. Xin Yu, Adam R Blanden, Sumana Narayanan, Lalithapriya Jayakumar, David Lubin, David Augeri, S David Kimball, Stewart N Loh, and Darren R Carpizo. Small molecule restoration of wildtype structure and function of mutant p53 using a novel zinc-metallochaperone based mechanism. *Oncotarget*, 5(19):8879, 2014.
52. Madhavi Krishnan, Nassiredin Mojarad, Philipp Kukura, and Vahid Sandoghdar. Geometry-induced electrostatic trapping of nanometric objects in a fluid. *Nature*, 467(7316):692, 2010.
53. Martin D Baaske and Frank Vollmer. Optical observation of single atomic ions interacting with plasmonic nanorods in aqueous solution. *Nature Photonics*, 10(11):733, 2016.
54. Amirhossein Alizadehkhaledi, Adriaan L Frencken, Mohsen Kamandar Dezfouli, Stephen Hughes, Frank CJM van Veggel, and Reuven Gordon. Cascaded plasmon-enhanced emission from a single upconverting nanocrystal. *ACS Photonics*, 6(5):1125–1131, 2019.
55. Howard Brenner. The slow motion of a sphere through a viscous fluid towards a plane surface. *Chemical engineering science*, 16(3-4):242–251, 1961.
56. MIM Feitosa and ON Mesquita. Wall-drag effect on diffusion of colloidal particles near surfaces: a photon correlation study. *Physical Review A*, 44(10):6677, 1991.
57. Abhay Kotnala and Reuven Gordon. Quantification of high-efficiency trapping of nanoparticles in a double nanohole optical tweezer. *Nano Letters*, 14(2):853–856, 2014.
58. Pavel N Melentiev, Anton E Afanasiev, Artur A Kuzin, Andrey S Baturin, and Victor I Balykin. Giant optical nonlinearity of a single plasmonic nanostructure. *Optics Express*, 21(12):13896–13905, 2013.
59. Quanbo Jiang, Benoit Rogez, Jean-Benoît Claude, Guillaume Baffou, and Jérôme Wenger. Temperature measurement in plasmonic nanoapertures used for optical trapping. *ACS Photonics*, 6(7):1763–1773, 2019.
60. Quanbo Jiang, Benoît Rogez, Jean-Benoît Claude, Antonin Moreau, Julien Lumeau, Guillaume Baffou, and Jérôme Wenger. Adhesion layer influence on controlling the local temperature in plasmonic gold nanoholes. *Nanoscale*, 12(4):2524–2531, 2020.

61. Justus C Ndukaife, Alexander V Kildishev, Agbai George Agwu Nnanna, Vladimir M Shalaev, Steven T Wereley, and Alexandra Boltasseva. Long-range and rapid transport of individual nano-objects by a hybrid electrothermoplasmonic nanotweezer. *Nature nanotechnology*, 11(1):53, 2016.
62. Abhay Kotnala, Pavana Siddhartha Kollipara, Jingang Li, and Yuebing Zheng. Overcoming diffusion-limited trapping in nanoaperture tweezers using opto-thermal-induced flow. *Nano letters*, 20(1):768–779, 2019.
63. Quanbo Jiang, Benoît Rogez, Jean-Benoît Claude, Guillaume Baffou, and Jérôme Wenger. Quantifying the role of the surfactant and the thermophoretic force in plasmonic nano-optical trapping. *Nano Letters*, 2020.
64. Domna G Kotsifaki, Viet Giang Truong, and Síle Nic Chormaic. Fano-resonant, asymmetric, metamaterial-assisted tweezers for single nanoparticle trapping. *Nano Letters*, 20(5):3388–3395, 2020.
65. Thomas W Ebbesen, HJ Lezec, HF Ghaemi, Tineke Thio, and PA Wolff. Extraordinary optical transmission through sub-wavelength hole arrays. *Nature*, 391(6668):667, 1998.
66. JA Matteo, DP Fromm, Y Yuen, PJ Schuck, WE Moerner, and L Hesselink. Spectral analysis of strongly enhanced visible light transmission through single c-shaped nanoapertures. *Applied Physics Letters*, 85(4):648–650, 2004.
67. R Gordon, AG Brolo, A McKinnon, A Rajora, B Leathem, and KL Kavanagh. Strong polarization in the optical transmission through elliptical nanohole arrays. *Physical Review Letters*, 92(3):037401, 2004.
68. Antoine Lesuffleur, L Kiran Swaroop Kumar, and Reuven Gordon. Enhanced second harmonic generation from nanoscale double-hole arrays in a gold film. *Applied Physics Letters*, 88(26):261104, 2006.
69. Tiberiu-Dan Onuta, Matthias Waegele, Christopher C DuFort, William L Schaich, and Bogdan Dragnea. Optical field enhancement at cusps between adjacent nanoapertures. *Nano Letters*, 7(3):557–564, 2007.
70. Adarsh Lalitha Ravindranath, Mirali Seyed Shariatdoust, Samuel Mathew, and Reuven Gordon. Colloidal lithography double-nanohole optical trapping of nanoparticles and proteins. *Optics Express*, 27(11):16184–16194, 2019.
71. Prashant Nagpal, Nathan C Lindquist, Sang-Hyun Oh, and David J Norris. Ultrasmooth patterned metals for plasmonics and metamaterials. *Science*, 325(5940):594–597, 2009.
72. Ana Zehtabi-Oskuie, Aurora A Zinck, Ryan M Gelfand, and Reuven Gordon. Template stripped double nanohole in a gold film for nano-optical tweezers. *Nanotechnology*, 25(49):495301, 2014.
73. Hyungsoon Im, Kyle C Bantz, Nathan C Lindquist, Christy L Haynes, and Sang-Hyun Oh. Vertically oriented sub-10-nm plasmonic nanogap arrays. *Nano Letters*, 10(6):2231–2236, 2010.
74. Ana Zehtabi-Oskuie, Jarrah Gerald Bergeron, and Reuven Gordon. Flow-dependent double-nanohole optical trapping of 20 nm polystyrene nanospheres. *Scientific Reports*, 2:966, 2012.
75. Ali El Eter, Nyha M Hameed, Fadi I Baida, Roland Salut, Claudine Filiatre, Dusan Nedeljkovic, Elie Atie, Samuel Bole, and Thierry Grosjean. Fiber-integrated optical nano-tweezer based on a bowtie-aperture nano-antenna at the apex of a snom tip. *Optics express*, 22(8):10072–10080, 2014.
76. Nyha M Hameed, Ali El Eter, Thierry Grosjean, and Fadi I Baida. Stand-alone three-dimensional optical tweezers based on fibred bowtie nanoaperture. *IEEE Photonics Journal*, 6(4):1–10, 2014.
77. Ryan M Gelfand, Skylar Wheaton, and Reuven Gordon. Cleaved fiber optic double nanohole optical tweezers for trapping nanoparticles. *Optics Letters*, 39(22):6415–6417, 2014.
78. Jamal M Ehtaiba and Reuven Gordon. Template-stripped nanoaperture tweezer integrated with optical fiber. *Optics Express*, 26(8):9607–9613, 2018.
79. Stefan Howorka, Stephen Cheley, and Hagan Bayley. Sequence-specific detection of individual dna strands using engineered nanopores. *Nature Biotechnology*, 19(7):636, 2001.

80. Grégory F Schneider, Stefan W Kowalczyk, Victor E Calado, Grégory Pandraud, Henny W Zandbergen, Lieven MK Vandersypen, and Cees Dekker. Dna translocation through graphene nanopores. *Nano Letters*, 10(8):3163–3167, 2010.
81. Peng Chen, Jiajun Gu, Eric Brandin, Young-Rok Kim, Qiao Wang, and Daniel Branton. Probing single dna molecule transport using fabricated nanopores. *Nano Letters*, 4(11):2293–2298, 2004.
82. Stijn Van Dorp, Ulrich F Keyser, Nynke H Dekker, Cees Dekker, and Serge G Lemay. Origin of the electrophoretic force on dna in solid-state nanopores. *Nature Physics*, 5(5):347, 2009.
83. Grégory F Schneider and Cees Dekker. Dna sequencing with nanopores. *Nature Biotechnology*, 30(4):326, 2012.
84. Erik C Yusko, Jay M Johnson, Sheereen Majd, Panchika Prangkio, Ryan C Rollings, Jiali Li, Jerry Yang, and Michael Mayer. Controlling protein translocation through nanopores with bio-inspired fluid walls. *Nature Nanotechnology*, 6(4):253, 2011.
85. Erik C Yusko, Brandon R Bruhn, Olivia M Eggenberger, Jared Houghtaling, Ryan C Rollings, Nathan C Walsh, Santoshi Nandivada, Mariya Pindrus, Adam R Hall, David Sept, et al. Real-time shape approximation and fingerprinting of single proteins using a nanopore. *Nature Nanotechnology*, 12(4):360, 2017.
86. Breton Hornblower, Amy Coombs, Richard D Whitaker, Anatoly Kolomeisky, Stephen J Picone, Amit Meller, and Mark Akeson. Single-molecule analysis of dna-protein complexes using nanopores. *Nature Methods*, 4(4):315, 2007.
87. Stefan W Kowalczyk, Adam R Hall, and Cees Dekker. Detection of local protein structures along dna using solid-state nanopores. *Nano Letters*, 10(1):324–328, 2009.
88. Wenhong Li, Nicholas AW Bell, Silvia Hernández-Ainsa, Vivek V Thacker, Alana M Thackray, Raymond Bujdoso, and Ulrich F Keyser. Single protein molecule detection by glass nanopores. *ACS Nano*, 7(5):4129–4134, 2013.
89. Kyle Briggs, Harold Kwok, and Vincent Tabard-Cossa. Automated fabrication of 2-nm solid-state nanopores for nucleic acid analysis. *Small*, 10(10):2077–2086, 2014.
90. Zachary Roelen, José A Bustamante, Autumn Carlsen, Aidan Baker-Murray, and Vincent Tabard-Cossa. Instrumentation for low noise nanopore-based ionic current recording under laser illumination. *Review of Scientific Instruments*, 89(1):015007, 2018.
91. Paolo Cadinu, Giulia Campolo, Sergii Pud, Wayne Yang, Joshua B Edel, Cees Dekker, and Aleksandar P Ivanov. Double barrel nanopores as a new tool for controlling single-molecule transport. *Nano Letters*, 18(4):2738–2745, 2018.
92. Benjamin N Miles, Aleksandar P Ivanov, Kerry A Wilson, Fatma Doğan, Deanpen Japrung, and Joshua B Edel. Single molecule sensing with solid-state nanopores: novel materials, methods, and applications. *Chemical Society Reviews*, 42(1):15–28, 2013.
93. Jeffrey D Uram, Kevin Ke, Alan J Hunt, and Michael Mayer. Submicrometer pore-based characterization and quantification of antibody–virus interactions. *Small*, 2(8-9):967–972, 2006.
94. Chang Chen, Pol Van Dorpe, Kai Cheng, Tim Stakenborg, and Liesbet Lagae. Plasmonic force manipulation in nanostructures, January 9 2018. US Patent 9,862,601.
95. Fatemeh Eftekhari, Carlos Escobedo, Jacqueline Ferreira, Xiaobo Duan, Emerson M Girotto, Alexandre G Brolo, Reuven Gordon, and David Sinton. Nanoholes as nanochannels: flow-through plasmonic sensing. *Analytical Chemistry*, 81(11):4308–4311, 2009.
96. Ahmet Ali Yanik, Min Huang, Alp Artar, Tsung-Yao Chang, and Hatice Altug. Integrated nanoplasmonic-nanofluidic biosensors with targeted delivery of analytes. *Applied Physics Letters*, 96(2):021101, 2010.
97. Ahmet A Yanik, Min Huang, Osami Kamohara, Alp Artar, Thomas W Geisbert, John H Connor, and Hatice Altug. An optofluidic nanoplasmonic biosensor for direct detection of live viruses from biological media. *Nano letters*, 10(12):4962–4969, 2010.
98. Carlos Escobedo, Alexandre G Brolo, Reuven Gordon, and David Sinton. Optofluidic concentration: plasmonic nanostructure as concentrator and sensor. *Nano letters*, 12(3):1592–1596, 2012.

99. Ana ZehtabiOskuie, Jarrah Bergeron, Yuanjie Pang, Matthew Moffitt, and Reuven Gordon. Optical trapping of an encapsulated quantum dot using a double nanohole aperture in a metal film. In *Optical Trapping and Optical Micromanipulation IX*, volume 8458, page 84581P. International Society for Optics and Photonics, 2012.
100. Daniel V Verschueren, Sergii Pud, Xin Shi, Lorenzo De Angelis, L Kuipers, and Cees Dekker. Label-free optical detection of dna translocations through plasmonic nanopores. *ACS Nano*, 13(1):61–70, 2018.
101. Daniel Verschueren, Xin Shi, and Cees Dekker. Nano-optical tweezing of single proteins in plasmonic nanopores. *Small Methods*, 3(5):1800465, 2019.
102. Muhammad Usman Raza, Sai Santosh Sasank Peri, Liang-Chieh Ma, Samir M Iqbal, and George Alexandrakis. Self-induced back action actuated nanopore electrophoresis (sane). *Nanotechnology*, 29(43):435501, 2018.
103. Sai Santosh Sasank Peri, Manoj K Sabnani, Muhammad Usman Raza, Soroush Ghaffari, Susanne Gimlin, Debra D Wawro, Jung Soo Lee, Min Jun Kim, Jon Weidanz, and George Alexandrakis. Detection of specific antibody-ligand interactions with a self-induced back-action actuated nanopore electrophoresis sensor. *Nanotechnology*, 31(8):085502, 2019.
104. Sai Santosh Sasank Peri, Manoj Kumar Sabnani, Muhammad Usman Raza, Elizabeth L Urquhart, Soroush Ghaffari, Jung Soo Lee, Min Jun Kim, Jon Weidanz, and George Alexandrakis. Quantification of low affinity binding interactions between natural killer cell inhibitory receptors and targeting ligands with a self-induced back-action actuated nanopore electrophoresis (sane) sensor. *Nanotechnology*, 32(4):045501, 2020.
105. Hyungsoon Im, Nathan J Wittenberg, Antoine Lesuffleur, Nathan C Lindquist, and Sang-Hyun Oh. Membrane protein biosensing with plasmonic nanopore arrays and pore-spanning lipid membranes. *Chemical Science*, 1(6):688–696, 2010.
106. Raju Regmi, Ahmed A Al Balushi, Hervé Rigneault, Reuven Gordon, and Jérôme Wenger. Nanoscale volume confinement and fluorescence enhancement with double nanohole aperture. *Scientific Reports*, 5:15852, 2015.
107. Vladimir JN Bykov, Sofi E Eriksson, Julie Bianchi, and Klas G Wiman. Targeting mutant p53 for efficient cancer therapy. *Nature Reviews Cancer*, 18(2):89, 2018.

Chapter 9
Towards Single-Molecule Chiral Sensing and Separation

Riley Sanders, Yaoran Liu, and Yuebing Zheng

Abstract Molecular chirality refers to molecules with the same composition but with different three-dimensional orientation. In biological systems, almost all molecules have chiral structures. Since molecules with different chirality may vary differently in their biochemical reaction, it is important to detect and separate these molecules in the biomedical and pharmaceutical industry. Despite significant progress made toward single-molecule sensing, it is still challenging to differentiate and detect chiral molecules at single-molecule resolution in racemic mixtures. Herein, we discuss the existing techniques towards single chiral molecule sensing and separation. We start with traditional methods, specifically chiral chromatographic methods, which label the chiral molecules with surfactants or other molecules in order to separate and detect them. New techniques using electromagnetic fields for label-free chiral sorting will also be explored. We then review the use of nanophotonic platforms to increase chiro-optical responses for chiral sensing with high sensitivity down to picogram quantities. We finalize with our perspective on opportunities and challenges for future development.

R. Sanders
Walker Department of Mechanical Engineering, The University of Texas at Austin, 204 E. Dean Keeton St., Austin, USA
e-mail: therealrileysanders@utexas.edu

Y. Liu
Department of Electrical and Computer Engineering, The University of Texas at Austin, 2501 Speedway, Austin, USA
e-mail: yaoran@utexas.edu

Y. Zheng (✉)
Walker Department of Mechanical Engineering, Texas Materials Institute, Department of Electrical and Computer Engineering, and Department of Biomedical Engineering, The University of Texas at Austin, 204 E. Dean Keeton St., Austin, USA
e-mail: zheng@austin.utexas.edu

W. Bowen et al. (eds.), *Single Molecule Sensing Beyond Fluorescence*,
Nanostructure Science and Technology,
https://doi.org/10.1007/978-3-030-90339-8_9

9.1 Introduction

So complex are protein structures and biomolecular interactions that two molecules with the exact same chemical formula can have completely different effects on a given organism. In 1848, French biochemist Louis Pasteur observed a mirror image molecule of tartaric acid, paratartaric acid, when looking at wine under a microscope [1]. Now, we call tartaric and paratartaric acid *enantiomers*, or non-superimposable mirror image molecules.

Since Pasteur's discovery, the field of molecular *chirality*, or asymmetry such that a molecule cannot be superimposed over its mirror image, has uncovered that simple molecular asymmetry has many practical applications. For example, ibuprofen, the active ingredient in Advil, is a chiral molecule; that is, there are two enantiomers of ibuprofen [2]: ibuprofen R− (rectus, or right-handed), and ibuprofen S+ (sinister, or left-handed). These molecules are both considered ibuprofen, and they both have the same chemical formula, but inside the human body, their different shapes play a role in how they interact with intricately shaped proteins and bioreceptors. The Advil in our cabinets includes both ibuprofen enantiomers, making it a racemic mixture, but only ibuprofen S+ works as an anti-inflammatory. The other enantiomer is largely inactive [2].

Ibuprofen is not the only example of a chiral biomolecule; there are hundreds of examples, from medicines to agrochemicals to the amino acids that make up the proteins in our bodies. In fact, it has been shown that the left to right-handed (L to D) *racemization*, or conversion from one enantiomer to another, of amino acids in biological tissues can be a predictor of diseases like Parkinson's, Alzheimer's, and more [3]. The enantiomeric ratio of amino acids is also a fairly reliable tool in geological dating [4].

This chapter will review developments in the field of chiral sensing and separation, as well as discuss recent strides in the field towards separation and sensing down to single biomolecules by use of circularly polarized light (CPL) and nanoplasmonics.

9.2 Chromatographic and Related Methods

Early attempts towards enantioseparation began in the 1960s and utilized chromatographic methods with chiral selective stationary phases. Specifically, gas chromatography and gas-liquid chromatography emerged as prominent methods in the field with high throughput and lower retention times [4]. Figure 9.1a shows a typical chiral gas chromatography sensor. Gas flow carries the molecule to the column and the column interacts differently between enantiomers inducing different retention times before the detection. Early experiments focused on finding stationary phases that could selectively bind to chiral molecules due to their specific geometries [5, 6]. For example, a 1966 Israeli experiment used an optically active stationary phase in order to preferentially bond to diasteriomeric derivatives of enantiomers of N-TFA-alanine

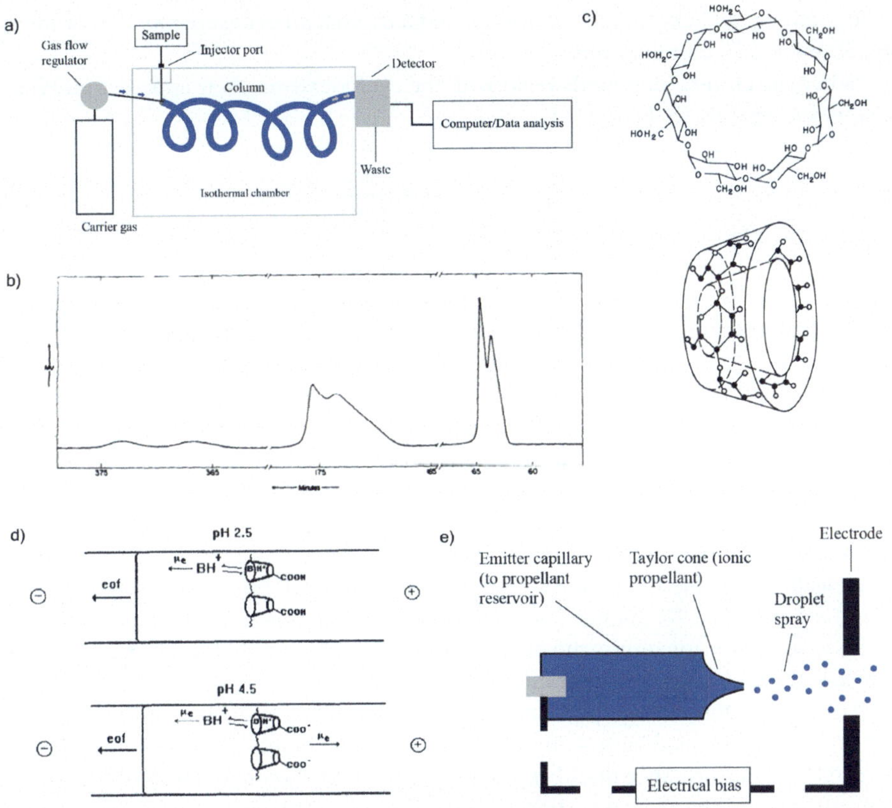

Fig. 9.1 **a** Schematic of typical gas chromatography setup. **b** Chromatogram of derivatized enantiomers of isopropanol, n-butanol, and cyclopentanol. Horizontal axis is "Minutes" and vertical axis is "MV" for millivolts. **c** Illustration of an example of a cyclodextrin with opening (top) view and side (bottom) view. **d** Illustration of electrophoretic process with β-cyclodextrin as a chiral additive to the background electrolyte. **e** Schematic of electrospray ionization setup. *Source* **a, e** Provided by Naftal Mat Mautia. Copyright (2020). **b** Reprinted with permission from [16]. Copyright (1966) (Elsevier). **c** Reprinted with permission from [7]. Copyright (1996) (Wiley). **d** Reprinted with permission from [8]. Copyright (1989) (Wiley)

to make chromatographic separation possible as shown in Fig. 9.1b. However, the enantiomers had to be derivatized and labelled, destroying them and rendering them useless after the analysis.

Studies towards better separation techniques were made after the implementation of cyclodextrins (Fig. 9.1c) [7]. Cyclodextrins are conical surfactant molecules with hydrophilic exteriors and hydrophobic interiors that stereoselectively bond with enantiomers of certain large chiral molecules. This results in enantioselective labelling of the molecules for chromatographic separation. Cyclodextrins are also used in capillary electrophoresis, a separation process that utilizes an electric field gradient, a background electrolyte, and a cyclodextrin to separate two or more enantiomers in a capillary tube.

When the chiral compounds bond with the cyclodextrins, their apparent mobility changes according to the Eq. 9.1:

$$\mu_{app} = \frac{lL}{t_x V} + \frac{lL}{t_0 V} \tag{9.1}$$

where μ_{app} is the apparent mobility of the compounds, l is the distance from the capillary inlet to the detector, L is the total capillary length, and t_x and t_0 are the times elapsed for the compound and electroosmotic marker to travel to the detector, respectively. The bound compounds are mobilized towards or away from a cathode depending on their mobilities.

An experiment shown in Fig. 9.1d depicts high resolving power using capillary electrophoresis for all but two enantiomers tested, as well as much faster resolution time than traditional chromatographic methods [5]. Differently from flow driven chiral chromatography, separation is achieved by applying high voltage, initiating the migration of the sample from the anode to the cathode through the chiral capillary channel. By this method, the size of the column can be significantly reduced. However, the need for higher concentrations of enantiomers as well as cyclodextrins leaves room for improvement from this method towards the ultimate goal of single-molecule, labelless separation.

The field of chromatographic separation has continued to evolve, utilizing different chiral selectors and varied, more tailored chromatographic systems. Sørensen et al. used high performance liquid chromatography (HPLC) coupled with electrospray ionization-mass spectrometry (ESI-MS) shown in Fig. 9.1e in order to separate and measure L and D aspartic acid from human teeth [9]. The use of electrospray ionization resulted in increased sensing resolution in the subsequent mass spectrometry step. This ground-breaking experiment leveraged addition of formic acid to optimize eluent pH for maximum resolution and tested different spray-needle voltages aiming for forensic diagnostics and age determination [9]. Rather than derivatizing the aspartic acid or using cyclodextrins, this process used teicoplanin, a chiral commercially available antibiotic biomolecule, as a chirally selective stationary phase in the separation column. The retention times were 4 min and 6 min for L- and D-aspartate respectively, which is fast compared to most other chiral selective chromatography. Sørensen et al. found that applying a zero-voltage to the spray needle as opposed to a 4 kV voltage increased resolution and signal to noise ratio, and proposes that a zero voltage may also be advantageous in the enantio-differential detection of other polar chiral biomolecules [9]. The experiment ultimately obtained a minimum detection resolution of 0.2 nanograms, which is excellent compared to other forms of chromatography.

Another labelled chiral separation method involves using enantioselective lipase enzymes to react with one racemate in order to separate chiral molecules [10]. The particular lipase used, an enzyme from *Candida antarctica* yeast called *Candida antarctica* Lipase B (CALB), selectively binds to R enantiomers in a certain temperature and pH range [10]. The CALB was used in conjunction with an enzymatic membrane reactor (EMR) which binds to the CALB after the enantioselective (R)-ketoprofen-EMR binding step. These two binding mechanisms were able to facilitate chiral selection. The experiment achieved a maximum (R)-ketoprofen conversion rate of up to 73% with all parameters optimized [10]. However, the immobilized CALB had to be left to react with the racemates for 24 h [10], which is not competitive with some chromatographic methods that take only minutes.

More recently, research has improved the chiral selective properties of chromatographic separation methods by focusing on the stationary phase. Molecular imprinting, a method in which the stationary phase is manufactured using the target molecule for separation, has shown promise when used in conjunction with HPLC as well as supercritical fluid chromatography (SFC) [11].

One of the most effective documented use cases of molecular imprinting was demonstrated by Gutierrez-Climente et al. with the chiral drug citalopram and its derivatives [12]. This experiment overcame some limitations of traditional chiral chromatography by decreasing the particle size of the chiral selective stage and by covalently immobilizing molecularly imprinted nanoparticles onto silica beads [12]. This technique provided sufficient access to chiral selective binding sites and allowed for the use of various solvents that would otherwise cause the dissolution of the chiral selective stage [13]. Percent recovery varied from 91.5 to 103.7% across a wide range of concentrations (0.1 to 5 μg of analyte per milliliter) [12], which is acceptably high for commercial use. However, the resolution (R) and separation (α) factors were not impressive compared to some more specialized liquid chromatography methods using proteins or polysaccharides for chiral selective stationary phases. For example, chromatographic columns using amylose have shown separation factors of up to 4.60 and resolutions of up to 10.60 under highly controlled and optimized conditions [14], while many of the samples tested in the silica-MIN case had resolution factors of less than 1, meaning resolution was not achieved [12]. Equations 9.2 and 9.3 show the separation factor, which is a measure of relative retention time, and the resolution factor, which quantifies how well the analytes are separated with respect to one another:

$$\alpha = \frac{t_{r,2} - t_m}{t_{r,1} - t_m} \tag{9.2}$$

$$R = \frac{t_{r,2} - t_{r,1}}{\frac{1}{2}(W_1 + W_2)} \tag{9.3}$$

where t_r is the resolution time, W is the width of the resolution peak of an enantiomer on a resolution versus time graph, t_m is the time required for the eluent to pass through the column, and the 1 and 2 subscripts refer to the first and second enantiomers.

Chiral chromatography offers a variety of different methods for chiral separation, each having its advantages and disadvantages. Table 9.1 summarizes a few of the aforementioned examples. While chiral chromatography is extremely useful for bulk commercial enantioseparation, efforts to increase sensitivity and extend its application to the molecular scale are still ongoing.

Table 9.1 Select chromatographic methods compared

Method	Cyclodextrin-Based	Electrophoretic	Molecular Imprinting
Derivation	Necessary [6]	Not Necessary [5]	Not Necessary [13]
Resolving Time	10–20 mins [6]	<5 mins [5]	5–30 mins [14]
Resolution		2–4 (ideal) [6]	<1 (most amino acids) [13]
Separation Factor	1–1.3 [6]		1–1.5 [12]
Analyte Concentration		10^{-5} M [5]	10^{-4} M [15]

9.3 Optical Foundations for Chiral Selection

The mechanism of traditional chiral chromatography relies on the recognition of chiral molecules coupled by another material, inducing different retention times in a separation medium. Therefore, chromatographic methods require a chiral selector or chiral column, which is time-consuming to set up, complex, and often must be adapted for different racemic compounds [13]. To overcome the limitations of chromatographic methods, several methods have been proposed to achieve label-free chiral sorting through circularly polarized light (CPL). CPL is light which has a constant-magnitude electric field vector that rotates in the plane perpendicular to the propagating direction of the light. CPL can be either left-handed or right-handed, meaning that the electric field vector can rotate counterclockwise or clockwise in the rotational plane, respectively.

For molecular sorting with chiral light, two metrics are of primary concern: optical chirality (C) and optical helicity (H) [28]. The optical helicity of an electromagnetic field is the volume integral of the optical helicity density, h; likewise, the chirality of an electromagnetic field is the volume integral of its optical chirality density, χ. Helicity density results from the projection of an electromagnetic (EM) wave's angular momentum vector onto its linear momentum vector, and is analogous to the wave's potential to impart a chiral-selective force, which is proportional to its chirality density. Optical helicity density and chirality density are described analytically with Eqs. 9.4 and 9.5:

$$h = \frac{1}{2}\left[\sqrt{\frac{\varepsilon_0}{\mu_0}}\mathbf{A}(\nabla \times \mathbf{A}) + \sqrt{\frac{\mu_0}{\varepsilon_0}}\mathbf{C}(\nabla \times \mathbf{C})\right] \tag{9.4}$$

$$\chi = \frac{1}{2}\left[\varepsilon_0\mathbf{E}(\nabla \times \mathbf{E}) + \frac{1}{\mu_0}\mathbf{B}(\nabla \times \mathbf{B})\right] \tag{9.5}$$

where A is the magnetic vector potential, C is the electric vector potential, and E and B are the electric and magnetic fields, respectively. ϵ_0 and μ_0 are the permittivity and permeability of free space, respectively.

In 2014, Tkachenko and Brasselet first demonstrated using CPL to sort chiral microdroplets (diameter 20 microns) made of chiral liquid crystals [17]. The nature of the chirality of the droplets was cholesteric, meaning that the individual molecules making the liquid crystals were achiral while the crystals themselves were chiral. As shown in Fig. 9.2a, left and right-handed CPL lasers with wavelength of 532 nm illuminated the samples from left and right of the capillary containing the droplets, respectively, and the light interacted chiral-selectively with the droplets depending on their liquid crystal chirality. As the droplets travelled down the capillary tube at a constant velocity due to gravity, they also travelled laterally in the + or − z direction depending on their chirality. Achiral droplets did not travel laterally because the chiral light forces cancelled each other out.

As previously discussed, light with a non-zero helicity (and therefore a non-zero chirality), can impart a force proportional to the chirality of the light, which is ±1 for perfectly CPL. Equation 9.6 governs the optical force on a microdroplet.

$$F_z(x) = \frac{2n_{ext}PR^2(1-\cos^4\theta_{B,ext})}{w_0^2 c}\exp\left(-\frac{2x^2}{w_0^2}\right) \tag{9.6}$$

where R is the radius of the spherical droplets, x is the coordinate in the direction of gravity, c is the speed of light, and θ_B is the mathematical term that comes from the assumption that a certain area of the sphere, defined by the projection of a circular cross-section on the concave surface of the droplet, behaves as a perfect spherical mirror. n_{ext} is the extinction coefficient, and P and w_0 are parameters associated with the opposing and conversely-rotational chiral laser beams (beam power and beam waist, or diameter, respectively).

The mechanism is based on the opposite sorting force due to the different preferential absorption of opposite CPL between chiral liquid crystals. As shown in Fig. 9.2b, when illuminating chiral particles (from microscale chiral structures to individual molecules) with CPL, preferential absorption and/or refraction occurs. Therefore, after passing through a chiral medium with circular differential absorption or refraction, CPL experiences either a differential reduction in the magnitude of the left and right components (circular dichroism), or a phase offset between the left and right components (circular birefringence), respectively.

However, Tkachenko and Brasselet were only able to harness these principles to sort microscale particles. To further improve the sorting efficiency, several methods have been demonstrated. Zhang et al. were able to theoretically determine ideal optical field conditions for chiral trapping and sorting by illuminating chiral particles in the interference field of two evanescent laser beams [18]. This method achieves chiral particle separation from around 100 nm to micro-scale sizes, which shows significant improvement in size compared with the previous work [18]. This method, shown in Fig. 9.2c, takes advantage of evanescent waves and reflected waves by placing the chiral molecules at the interface between media of different refractive indices. It was found that the combination of one linearly polarized beam and one circularly polarized beam worked the best to separate particles based on their chirality, creating a non-chiral-selective force in the direction of propagation of the waves and

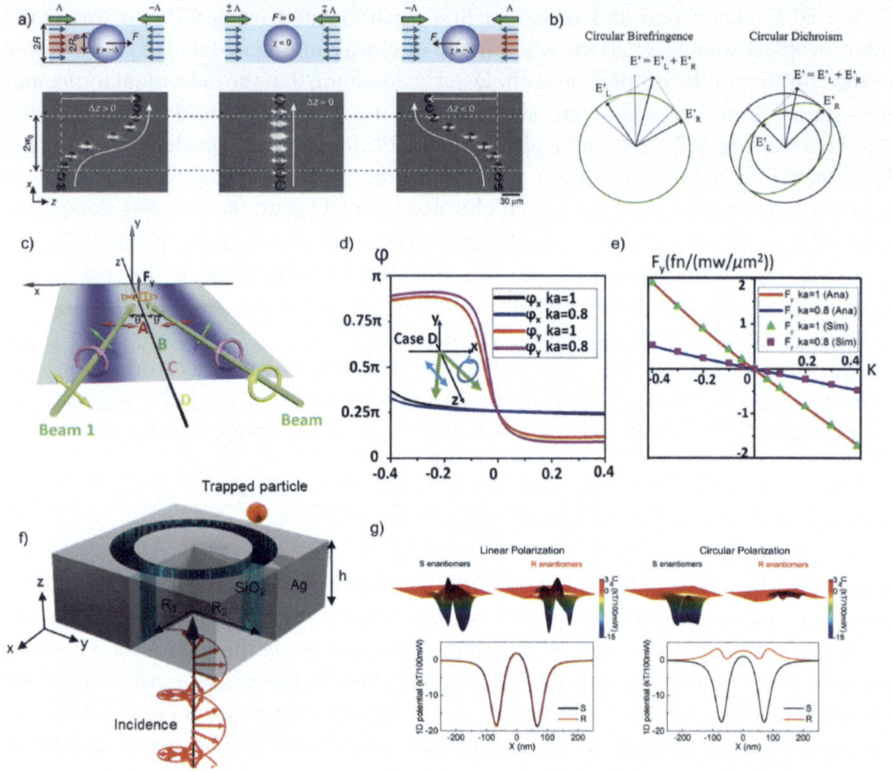

Fig. 9.2 **a** Schematic showing the light-matter interaction geometries with different light beam circular polarizations and dielectric droplet chiralities (top) and the corresponding superimposed snapshots demonstrating the droplet movements (below). Positive Λ denotes a counter-clockwise polarization and negative Λ denotes a clockwise polarization; the material chirality is denoted by χ. **b** Depiction of electric field vectors for light undergoing circular birefringence (left) and circular dichroism (right). **c** Schematic of two interference fields to achieve chiral separation. **d** The phase of the trapping force (φ_x) and the phase of the lateral force (φ_y) with the different chiral parameters for case D as shown in Fig. 9.2c. **e** The lateral (F_y) trapping forces with different chiral parameters for case D as shown in Fig. 9.2c. **f** Schematic showing chiral nanoplasmonic tweezers with illustration of CPL. **g** The calculated trapping potentials at 20 nm above the nanoaperture for R and S enantiomers when illuminating with linearly polarized and left-handed circularly polarized light. *Source* **a** Reprinted with permission from [17]. Copyright (2014) (Nature). **b** Provided by Naftal Mat Mautia. Copyright (2020). **c–e** Reprinted with permission from [18]. Copyright (2017) (ACS Publications). **f–g** Reprinted with permission from [19]. Copyright (2016) (ACS Publications)

a chirally selective perpendicular force shown in Fig. 9.2d, e [18]. Furthermore, the method has been proven to apply to a variety of particle chiralities and sizes, from the nanometer to micron scale.

Other chiral sorting methods have been demonstrated by coupling CPL with nanostructures to achieve low concentration chiral sorting with nanoscale particles. In another theoretical paper, nanoscale optofluidic sorting was proposed by leveraging nanoplasmonic "tweezers" with a coaxial design [19]. These nanoplasmonic tweezers and their applications are discussed at length in the Chap. 7. First, it was demonstrated that the coaxial plasmonic nanostructure can enantioselectively trap particles under 10 nm as shown in Fig. 9.2f. The mechanism relies on a dielectric molecule's electromagnetic (EM) polarizability, a metric of its susceptibility to optical forces. The EM polarizability of a molecule is given by Eq. 9.7:

$$\alpha_{EM} = -12\pi r^3 \frac{j\kappa\sqrt{\mu_0\varepsilon_0}}{(\varepsilon_r + 2\varepsilon_{rm})(\mu_r + 2\mu) - \kappa^2} \tag{9.7}$$

where r is the effective radius between the particle's EM dipoles, κ, is the molecular chirality parameter, and ϵ_0 and μ_0 are the permittivity and permeability, respectively. The subscripts 0, r, and rm are "free space", "relative", and "relative of the medium" respectively.

The optical tweezers are designed to trap particles at near-IR wavelengths, which have negligible circular dichroism [20]. The equation for the optical force acting transverse to the coaxial opening of the plasmonic tweezers are defined in Eq. 9.8 as:

$$F_{tr} \approx \frac{\Re(\alpha_{ee})}{4}\nabla|\mathbf{E}|^2 + \Im(\alpha_{em})\frac{1}{2}\nabla\Im(\mathbf{E}\cdot\mathbf{H}) \tag{9.8}$$

where the ee subscript denotes the electric polarizability and the em subscript denotes the magnetic polarizability. The first term in Eq. 9.8 depends on the electric field magnitude and the second term depends on the optical chirality density as well as the chirality of the enantiomer [19]. As a result, part of the force depends on the handedness of the molecule, which gives the optical tweezers their sorting capability.

In order to match the transverse force field to the shape of the coaxial tweezers and effectively trap enantiomers 20 nm above the coaxial opening, the tweezers were illuminated with left-handed CPL with incidence in the positive z direction. Figure 9.2g shows the corresponding force gradients. R enantiomers are repelled while S enantiomers are attracted towards the nanoaperture under the left-handed CPL while both S and R enantiomers are trapped using linearly polarized light.

The chiral sorting described in the plasmonic tweezers article is limited to chiral nanoparticles with extremely high chiral parameters [19]. Therefore, further research needs to be done in order to achieve sorting with natural chiral biomolecules. Nonetheless, this research is a promising stride towards single-molecule sorting.

9.4 Harnessing Enantiospecific Magnetic Forces for Chiral Sorting

Another recent chiral sorting method relies on enantiospecific interactions with a ferromagnetic substrate [21]. This method exploits the different electron transport patterns between chiral molecules, which results in differential spin polarizations for

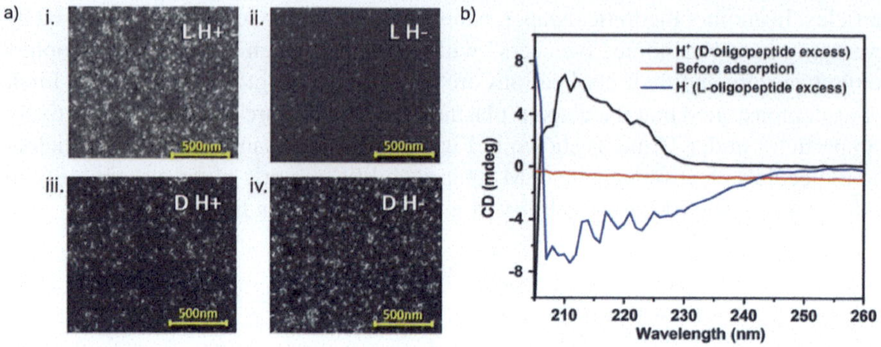

Fig. 9.3 **a** SEM image of the chiral PAL oligopeptide adsorbed on ferromagnetic samples magnetized with the magnetic dipole pointing up (H+) or down (H−) relative to the substrate surface. Panels (i) and (ii) L-PAL and (iii) and (iv) D-PAL were adsorbed for 2 s on a substrate magnetized up or down. **b** CD spectra of a racemic oligopeptide mixture before (red) and after (black and blue) the magnetic chiral separation. *Source* Reprinted with permission from [21]. Copyright (2018) (American Association for the Advancement of Science)

the two enantiomers [22]. The magnetized substrate induces electric dipole polarization in a racemic solution flowing over it, which causes an enantiospecific spin orientation at the poles of the molecules. Subsequently, the enantiomers are adsorbed onto the substrate at different rates as shown in Fig. 9.3a [21].

Banergee-Ghosh et al. harnessed this magnetic enantiospecific force by using a column coated with an externally magnetized ferromagnetic metal alloy and allowing a racemic oligopeptide mixture to flow through it. Figure 9.3b shows the CD spectra at the inlet and outlet of the column, which is clear evidence of high-resolution chiral sorting. The results indicate that the spin polarizations for L and D enantiomers are attracted to "up" and "down" polarized magnetic surfaces, respectively [21].

It is fascinating to achieve selector free chiral separation using light and a magnetic field. By tailoring the EM field with a nanoaperture, it is possible to achieve single moleclue chiral separation. However, most studies are still at theoretical levels. Meanwhile, a limitation of current selector free chiral separation is that current methods require extremely close (nanometer scale) distances between molecule and substrate. Therefore, quickly bringing the molecule of interest closer to the substrate is necessary for enantioseparation of chiral molecules with low concentrations.

9.5 Using Achiral Nanoparticles to Increase Chiro-Optical Signaling

Plasmonic nanostructures can be used not only for chiral sorting but also for chiral sensing. Most chiral mediums display both circular birefringence and circular dichroism. Cohen and Tang first discovered the link between the optical chirality of an electromagnetic field and the differential excitation of enantiomers in 2010 [23]. Furthermore, it was proven analytically that at the nodes of a superchiral field, chiral signals from excited enantiomers were enhanced 400-fold. Therefore, enhancing local optical field chirality gives an extremely large advantage with high sensitivity.

On the quest to enhance the inherently weak chiro-optical signals of individual molecules, we concern ourselves with the optical chirality parameter, κ, ascribed to a chiral molecular medium and representing the differential refraction of left and right CPL, leading to a circular birefringent chiro-optical signal [24]. The optical chirality parameter is complex, where the real part of the parameter describes the circular birefringent signal and the imaginary part describes the circular dichroism signal of the molecule [25]. Since circular birefringence results from a differential effect of a medium on CPL, the chirality parameter of a chiral medium describes how the wave-vector, and by extension, wavelength, are modulated by the medium according to Eqs. 9.9 and 9.10:

$$k_{L,R} = (n \pm \kappa)k_0 \tag{9.9}$$

$$k_0 = \frac{2\pi}{\lambda} \tag{9.10}$$

where k is the wave vector, λ is the wavelength, and n is the refractive index of the medium. Yoo and Park's research on signal enhancement using nanoparticles in the Mie regime (radius of the nanoparticle on the same order as the wavelength of the light) demonstrates that high refractive index plasmonic and dielectric nanoparticles increase light scattering, leading to chiral differential signal enhancement [24].

Yoo and Park focused on particles in the Mie regime, but their findings translate to the Rayleigh regime as well. They first experimentally and theoretically determine that different frequencies of light can lead to different scattering cross-sections by illuminating particles in the Mie regime with different wavelengths of light. They show that the degree of non-differential scattering affected by both plasmonic and dielectric achiral nanoparticles (Ag and Si, respectively) is linearly proportional to the measured chirality parameter of the medium in the near field of the nanoparticles, as shown in Fig. 9.4a.

Optical chirality, as previously mentioned in the context of electromagnetic fields, is a measure of the enantioselective potential of an electric field [26], which can be defined in (9.11):

$$C = \frac{\epsilon_0}{2\omega}\Im[\mathbf{E} \cdot \mathbf{B}] \tag{9.11}$$

where ϵ_0 is the permittivity of free space, ω is the angular frequency of the incident light, and **E** and **B** are the electric and magnetic fields.

Yoo and Park showed that scattering by dielectric and plasmonic nanoparticles can affect optical chirality conservation, but an analysis of scattering effects on chirality enhancement is not complete without a parallel discussion of absorption, the

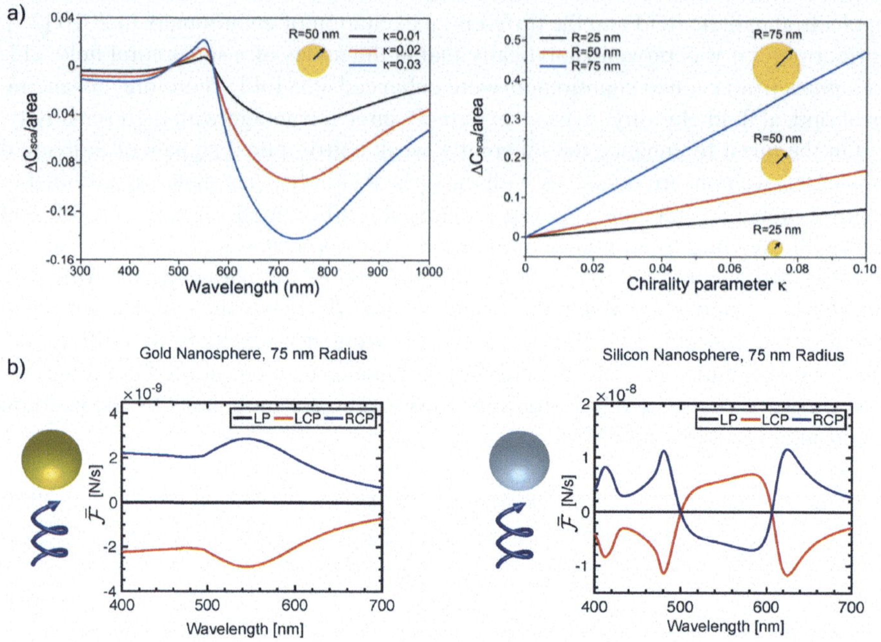

Fig. 9.4 **a** Left: circular differential scattering cross-section for a spherical gold nanoparticle of radius 50 nm in a medium of refractive index = 1.5 with different chirality parameters. Right: The maximum value of circular differential scattering cross-section for a spherical gold nanoparticle in a medium of refractive index = 1.5 with different chirality parameters. **b** Total volume-integrated chirality flux calculated for plasmonic (left) and dielectric (right) nanoparticles of 75 nm radii versus wavelength. *Source* **a** Reprinted with permission from [24]. Copyright (2015) (Nature). **b** Reprinted with permission from [28]. Copyright (2019) (Multidisciplinary Digital Publishing Institute)

mechanism responsible for circular dichroic signals. Together, differential scattering and absorption of light constitute the light's circular differential extinction [27]. A difference in the measured extinction coefficient for left-handed and right-handed CPL constitutes a CD signal.

The authors find that the differential extinction coefficient is a function of the plasmonic response (present only in metallic particles) as well as the electric and magnetic dipoles, quadrupoles, and even octupoles for both metallic and dielectric particles [24, 27]. Figure 9.4b shows the wavelength dependent resonant response of metallic and dielectric nanoparticles, indicating that the plasmonic resonant response dominates and acts in opposition to the electric and magnetic polar resonant responses. Therefore, optical chirality flux, or the change in the control volume "flow" of the chirality parameter has a much larger response when using dielectric particles versus plasmonic particles, interestingly enough.

Further research into the optical enhancement of dielectric and plasmonic nanostructures by Raziman et al. in 2019 parameterizes certain desired optical effects and analytically determines how nanodisk arrays made of silver and silicon alter these effects in the near field [29]. Logically, inherently achiral nanophotonic platforms have no circular differential effect; therefore, Raziman et al. illuminated the array with left CPL in order to enhance the chirality parameter and the degree of circular polarization (DOCP).

Raziman's team was interested in the effect of the nanophotonic platforms on the near field intensity of the electric field as well as the near field chirality. The ideal nanoparticles need to enhance the intensity of the electric field in order to increase sensitivity to chiral molecules in the near field, while preserving the original circular polarization of the incident light so as to not create a signal that obfuscates the signal of the chiral molecules [29]. The parameters used to describe these changes are presented in Eqs. (9.12, 9.13, 9.14):

Degree of Circular Polarization:

$$DOCP = \frac{V}{I} \tag{9.12}$$

where I is the Intensity of the electric field and V is the difference in left and right CPL intensities.

Electric energy density:

$$U_e = \frac{\epsilon_0}{4}|\mathbf{E}|^2 \tag{9.13}$$

where ϵ_0 is the permittivity of free space and **E** is the electric field vector.

Circular dichroism enhancement is given by:

$$CDE = \frac{C}{U_e} \tag{9.14}$$

where C is the optical chirality and U_e is the electric energy density.

The ideal nanophotonic platform, when illuminated by left CPL, as discussed above, should show a high near field increase in both intensity and electric energy density, while providing a spatially invariant degree of DOCP equal to one (or greater than one in the case of superchiral fields) and a high CDE. In the metallic nanodisks, the plasmonic response at the resonant wavelength creates strong electric field around the disk while the stokes parameter shows a lower value as shown in Fig. 9.5a. Meanwhile, the CD enhancement is suppressed and the abrupt change of DOCP from positive to negative values leads to its lower value, as also shown in Fig. 9.5a. Therefore, metallic nanodisks show poor performance to enhance the chiral interaction in the near field.

Another study conducted by Lee, Yoo, and Park clarified surface enhanced CD from the microscopic perspective. Briefly, the near field circular dichroism effect can be measured by a medium's differential reflectance, transmittance, and absorption on the two circularly polarized superpositions of linearly polarized light, which is incident on the medium [25]. Herein, reflection, transmission, and absorption make up the collectively exhaustive set of what happens to the light incident on the medium;

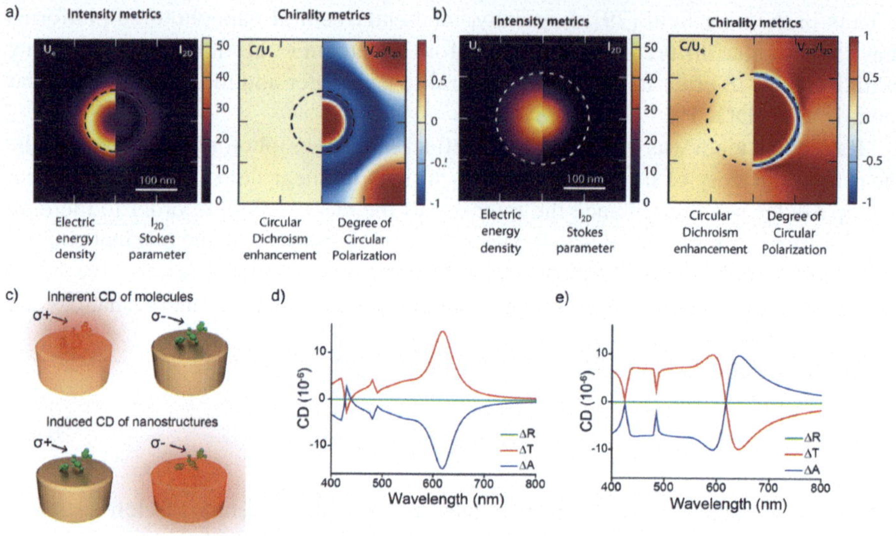

Fig. 9.5 **a** Intensity metrics and chirality metrics for gold nanodisk with dimension of H = 40 nm, R = 75 nm, P = 400 nm. **b** Intensity metrics and chirality metrics for dielectric nanodisk with dimension of H = 40 nm, R = 110 nm, P = 400 nm. **c** The inherent CD of chiral molecules comes from the enhanced near fields on the nanostrucure. The induced CD comes from the induced chirality due to the coupling between nanostructure and chiral molecule. Circular dichroism measured by the differential reflectance Δ R (green lines), transmittance Δ T (red lines), and absorptance Δ A (blue lines) for chiral molecule with **d** chirality parameter = 0.001 and **e** chirality parameter = 0.001i. *Source* **a–b** Reprinted with permission from [29]. Copyright (2019) (ACS Publishing). **c–e** Reprinted with permission from [25]. Copyright (2017) (ACS Publishing)

therefore, absorption and circular differential absorption (i.e. circular dichroism) can be calculated by measuring the differential reflection and transmission:

$$A + R + T = 1 \tag{9.15}$$

where A, R, and T represent the fraction of light absorbed, reflected, and transmitted, respectively.

Once transmission is measured, the circular dichroism, with units in degrees, is calculated by Eq. 9.16:

$$CD = \arctan\left(\frac{T_L - T_R}{T_L + T_R}\right) \tag{9.16}$$

where T_L and T_R are the fraction of left and right-handed CPL transmitted, respectively.

Lee, Yoo, and Park examine the origins of nanoparticle enhanced circular dichroism by splitting up circular dichroism signals into two distinct causes: inherent and induced CD as shown in Fig. 9.5c. Specifically, the inherent CD is the molecular CD enhanced by the near field of the nanoparticles. The induced CD is the asymmetric excitation and absorption of the chiral EM field due to the chiral molecules [25].

To study the contribution from the inherent CD and induced CD for chiral molecular sensing, the authors compare circular dichroism signals using either a purely real or purely imaginary chirality parameter as shown in Fig. 9.5d, e. The chiral molecule with the purely real chirality parameter can introduce neither differential absorption nor differential transmittance. However, from the simulation results, we observe the differential absorption is same as differential transmittance with the peak near plasmonic resonance of the nanoparticles indicating the induced CD. The chiral molecule with pure imaginary chirality parameters shows totally different trends. Herein, both the inherent CD and induced CD can contribute to the total CD signal. Therefore, the CD spectra is enhanced both in UV and visible range [25].

Achiral plasmonic nanostructures have been shown to have an ability to increase the magnitude of the electric field, optical helicity, and optical chirality in the near field, which make them useful for chiral molecular sensing in small molecular quantities. Another advantage of achiral plasmonic and dielectric nanostructures, as previously stated, is that they have no background CD spectra to skew or obfuscate the CD emission spectra of the chiral molecules subject to sensing. This gives achiral nanostructures an advantage over chiral nanostructures for sensing purposes.

9.6 Chiral Sensing with Inherently Chiral Metasurfaces

Apart from the achiral nanoparticles for chiral molecule sensing, alternative structures such as nanoparticles or metasurfaces with inherent chirality are also shown to be promising for chiral sensing. Similarly to chiral molecules, these structures can be either left handed or right handed chiral structure depending on the optical response to CPL.

In particular, Fig. 9.6a shows an example of the two chiral metasurfaces fabricated using stacked gold nanorods with $\pm 60°$ angles between them [26]. The structural chirality of the resulting metasurface exhibits a distinct optical response of the metasurface with adsorbed chiral molecules, termed an output CD response (CD_O) under different CPL where the response is calculated in Eq. 9.17:

$$CD_O = CD_i + 4kw\Im[\kappa_m]\frac{|T_{LR}T_{RL}|^2 - |T_{LL}T_{RR}|^2}{(|T_{LR}|^2 + |T_{RR}|^2)^2 + (|T_{LL}|^2 + |T_{RL}|^2)^2} \tag{9.17}$$

where CD_i is the background CD signal of the metasurface, k is the wave vector, w is the width of the metasurface, and T is the transmission coefficient, with first and second subscripts denoting the circular handedness of the coefficient and the circular polarization of incident light, respectively [26].

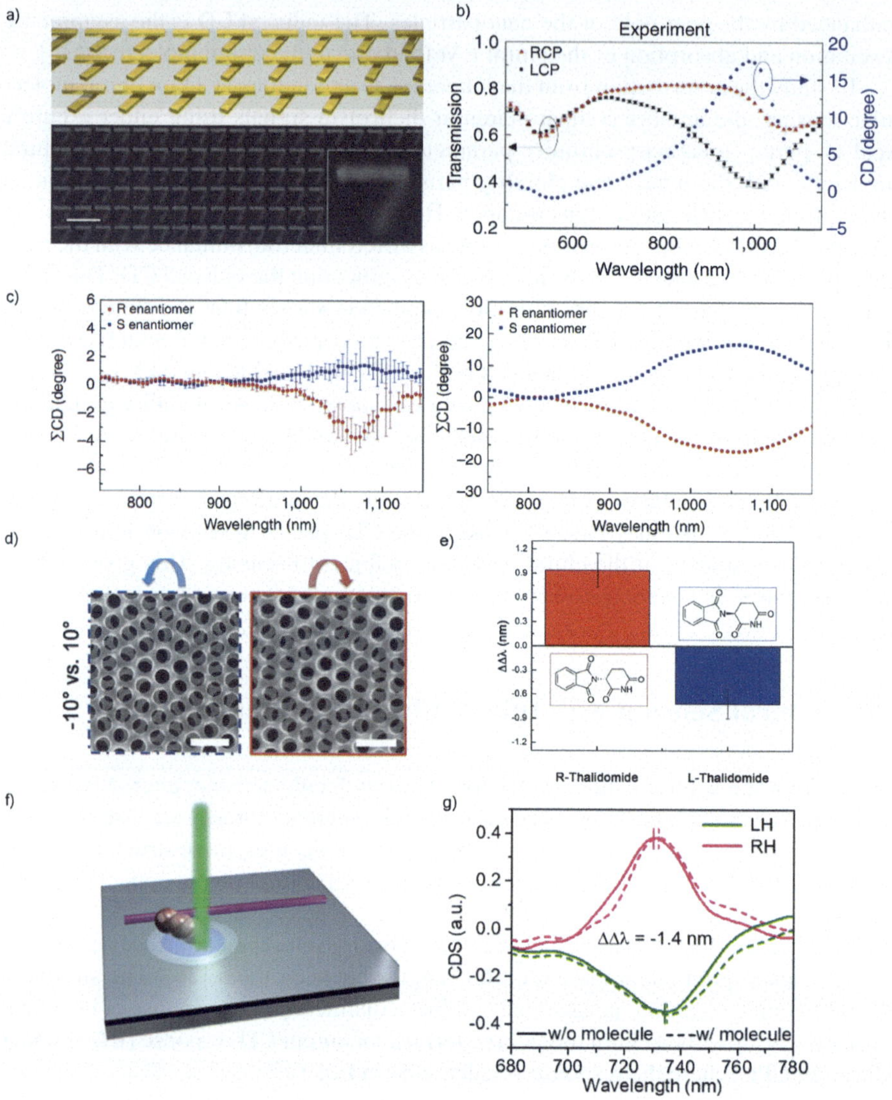

Fig. 9.6 **a** Schematic and scanning electron microscope images of +60° gold rod chiral metamaterial. Scale bar 500 nm. **b** with RCP (red curve) and LCP (black curve) excitation. **c** Experimentally measured transmission and CD of the +606o chiral metamaterial with RCP (red curve) and LCP (black curve) excitation. **d** Simulated CD summation of ±60° chiral metamaterial. **e** Scanning electron microscope image of nanohole MCM with layers turned ±10°. Scale bar 500 nm. **f** ΔΔλ values for R and S thalidomide. **f** Schematic drawing of OPN laser manipulation of colloidal particles. **g** CDS spectra of left-handed and right-handed structures with (solid line) and without (dashed line) adsorbed L-phenylalanine *Source* **a–c** Reprinted with permission from [26]. Copyright (2017) (Nature). **d–e** Reprinted with permission from [31]. Copyright (2017) (Wiley). **f–g** Reprinted with permission from [32]. Copyright (2019) (ACS Publishing)

Zhao et al. first conducted calibration to isolate background CD signals originating from the chiral nanostructures themselves [25] as well as to control for discrepancies caused by fabrication errors.

The experimental CD results with $+60^{\circ}$ angles chiral metasurfaces shows opposite response as shown in Fig. 9.6b. The chiral metasurface shows different transmission under RCP and LCP inducing CD spectra peak near 1000 nm. Then enantiopure solutions of R and S Propanediol were added, and the team was able to isolate the background CD signals and successfully sense the molecules down to zeptomole (10^{-21}) quantities as shown in Fig. 9.6c. They also conducted a numerical simulation (Fig. 9.6c) and the result shows a similar trend as the experimental results.

Such picogram level sensitivity is a step towards single-molecule sensing. The authors further propose that these chiral metasurfaces, if used in conjunction with nanophotonic optofluidic platforms [19, 30], may even be able to sense enantiopure clusters of just a few biomolecules. The biomolecules sensed can also be a wide range of molecular weights and can be sensed extremely quickly (in fractions of a second), which may suit this method to high throughput operations requiring high precision and versatility.

Wu et al. recently developed alternative chiral metamaterials called Moiré chiral metamaterials (MCMs) [31], as shown in Fig. 9.6e. The metamaterial consists of two monolayers of gold nanohole arrays stacked with an in-plane rotation. Depending on the relative orientation of the two nanohole arrays, MCMs can be either left-handed or right-handed structures, inducing strongly enhanced localized optical chirality. Wu et al. further demonstrated the capability of MCMs for chiral sensing. The adsorption of chiral proteins on the MCMs caused asymmetric spectral shifts of measured CD for the MCMs with opposite handedness [31]. The team also used the bilayer metamaterial to distinguish small enantiomers, specifically R-thalidomide and its and its "evil twin" as shown in Fig. 9.6f. The reconfigurable and high sensitivity MCMs can achieve detecting variable chiral molecules with a variety of molecular weights.

Emerging research through optical photon nudging (OPN) can be used to fabricate chiral structure for chiral sensing [32]. The mechanism of OPN is based on the optical scattering forces [33, 34] created by the illumination of gold nanoparticles with a laser. The heat generated laser mobilizes particles, allowing them to release from a surfactant substrate. Once the particles are mobilized, they can be pushed by the laser beam to the desired location as shown in Fig. 9.6f. Li et al. further demonstrated the application of OPN to assemble both metal and silicon nanoparticles as a tunable nanostructure [32]. Specifically, they fabricated the dielectric chiral metasurface using a single silicon nanowire and a single silicon nanoparticle for chiral molecule sensing. As shown in Fig. 9.6g, left-handed and right-handed chiral structures can successfully detect the handedness of L-phenylalanine at low quantities.

Further research has been done into chiral plasmonic nanostructures [35], with varying degrees of complexity in shape. For example, a bilayer chiral metamaterial substrate proves that by using two mirror image chiral arrays, the chirality parameter and circular dichroism can be enhanced in the near field with chiral nanoplasmonics while controlling for background circular dichroism [26]. After this proof of concept, it was also shown that a racemic array of plasmonic nanoparticles can mitigate background circular dichroism signals by cancelling them [36]. The advantage of using a racemic metasurface instead of a chiral metasurface is that it can cancel structural CD signals without any decrease in near field chirality or electric field magnitude compared to purely handed arrays [36].

Shown in Fig. 9.7a is an example of a racemic array made using gammadion nanostructures. After deposition of D, L, and DL phenylalanine, the racemic plasmonic gammadion array displayed distinct residual circular dichroism signals for each coating as shown in Fig. 9.7b. These signals arose from selective plasmonic coupling between the enantiomers and specific left or right-handed gammadions in the array, which increased both the local optical chirality and electric field enhancement in the near field of the gammadion array.

García et al. observed mirror image circular dichroism enhancement of the D and L enantiomers, as well as the aforementioned electric field and its chiral dissymmetry. Note that the sign of C determines which enantiomer will interact with which gammadion pair. The chiral dissymmetry, Δ C, is a measure of the platform's capacity for differential chiral signalling [36]. This signal enhancement resulted in opposite plasmonic coupling between D and L enantiomers, enabling clear chiral differentiation.

Along with the gammadion structure, other chiral plasmonic structures such as shuriken-shaped structures have also been studied. These structures produced superchiral nearfield chirality with the chirality parameter reaching magnitudes of up to 100 in some near field areas [37] as shown in Fig. 9.7c. These chiral fields were shown to interact asymmetrically with the nanostructure's plasmonic dipolar and quadrupolar modes, a phenomenon referred to as "dichroic coupling" [37]. Kelly et al. propose that dichroic coupling can be used to manipulate a number of optical properties, which has practical applications in the chiral sensing field [37].

Figure 9.7d shows how shuriken structures submerged in water, an isotropic dielectric medium, are used for chiral sensing of an enantiopure solution containing the amino acid alpha-pipene. When the handedness of the structure and the chiral biomolecules agree, this causes the distance between the reflectivity minima to be closer to that of the pure dielectric (in this case, water) than when they do not agree [37]. This method of analysis, when using pure left-handed CPL, can be used to determine the handedness of an enantiomeric coating 20 nm thick [37]. While this concentration is a far cry from single-molecule, it is an improvement from the 150 nm thickness required for the gammadion structures.

9.7 Cavity-Based Nanophotonic Platforms for Chiral Sensing and Sorting

Chiral nanostructures interact selectively with chiral light, and as a result, couple selectively to chiral molecules [36, 37]. Recently, experimentation with chiral nanocavities has taken advantage of the inverse of this effect, exploiting the chiral

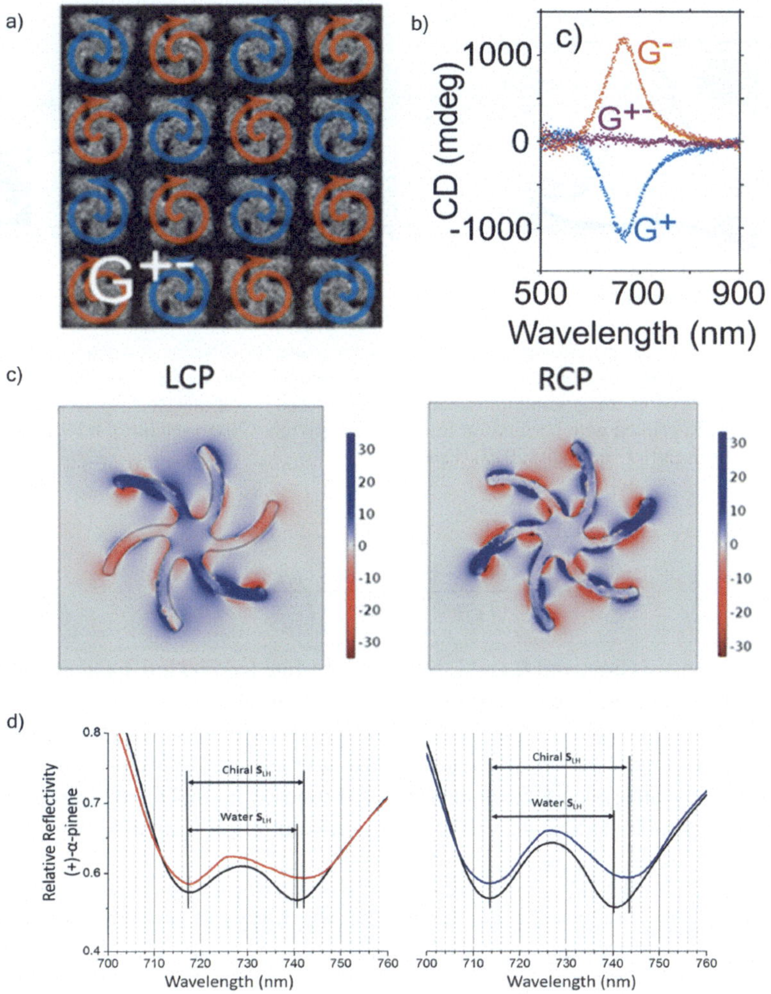

Fig. 9.7 **a** SEM images of racemic array made using gammadion nanostructures. **b** CD spectrum for enantiomer detection. The pink curve is racemic mixture. The blue curve is L-phenylalanine. The red curve is D-phenylalanine. **c** Simulated chiral near field on the top surface of left-handed shuriken chiral nanostructures under LCP and RCP illumination. **d** Relative reflectivity measurements for left-handed (left) and right-handed (right) shuriken nanostructures in water with + αpipine in the near field. *Source* **a–b** Reprinted with permission from [36]. Copyright (2018) (ACS Publishing). **c–d** Reprinted with permission from [37]. Copyright (2017) (ACS Publishing)

selective properties of helical nanoapertures [38] (Fig. 9.8a) and sawtooth shaped handed metasurfaces [39] (Fig. 9.8b). When studying nanoapertures, a metric of interest is the circular dichroism transmission (CDT), which is a measure of the differential transmission of the two circularly polarized superpositions of incident light, and is defined analytically in Eq. 9.18:

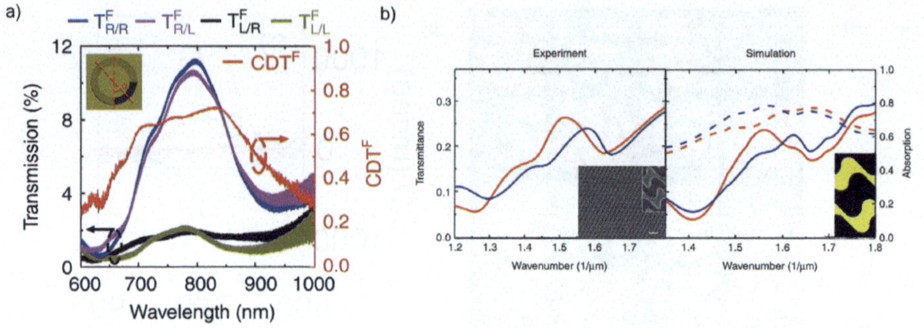

Fig. 9.8 **a** CDT and transmission from helical nanoaperture for different incidence and handedness combinations. **b** Experimental and simulated transmittance spectra for sawtooth nanosurface. The insert shows SEM images and the sawtooth cavity-based chiral metasurface used for simulation. *Source* **a** Reprinted with permission from [38]. Copyright (2019) (Nature). **b** Reprinted with permission from [39]. Copyright (2019) (De Gruyter)

$$CDT = \frac{(T_{R/R} + T_{R/L}) - (T_{L/R} + T_{L/L})}{(T_{R/R} + T_{R/L}) + (T_{L/R} + T_{L/L})} \tag{9.18}$$

where T is transmission, and R and L refer to right-handed and left-handed CPL. The first subscript indicates the type of incident light, and the second subscript indicates the type of light transmitted through the medium or metamaterial. While research has not been conducted on the actual chiral sensing capabilities of cavity-based nanophotonic platforms, electromagnetic field metrics and CDT measured in their near field suggest that they are viable as high resolution, low analyte concentration biosensors [35].

The strong chiral field using cavity-based nanophotonics can significantly facilitate chiral sensing. Similarly to optimizing the localized CD in a chiral metasurface for enantiomeric differentiation, a nanophotonic cavity can be engineered to enhance localized CD for chiral sensing.

One cavity-based platform for CD enhancement uses two mirror-symmetric types of helical cavities etched into a gold film to enable selective transmission of handed light [38]. Figure 9.8a shows the transmission and CD spectra of these helical cavities. The enhancement of the electric field outside the nanocavity is stronger when the light polarization matches the direction of the helical cavity indicating strong selective plasmonic coupling and wave propagation inside the nanoaperture [38]. In addition, the structures exhibit strong CDT on a broadband spectrum, which is preferable for enantiomeric sensing [40]. Further research is required to optimize this structure for that purpose.

Another example of cavity based chiral nanostructures is the sawtooth surface shown in Fig. 9.8b. These substrates are etched out of 30 nm gold film and placed between two dielectric glass substrates. The cavity leverages the principles of Fabry-Perot interferometry in order to selectively couple to CPL [39]. It is worth mentioning that Δ T peaks in a very narrow range of incident wavelengths around 650 nm [39], which makes this metasurface promising for extremely low analyte volume chiral sensing [40].

9.8 Conclusion

We have discussed several methods toward single chiral molecule sensing and detection. Traditional chromatographic methods can achieve high resolution chiral separation in biological samples. However, these methods are bulky and require chiral selectors to achieve chiral separation. Other techniques using optical or magnetic fields have been recently proposed to overcome this challenge. Of these techniques, the plasmonic nanoaperature tweezers are particularly promising. Based on theoretical analysis, it is possible to leverage this technique to achieve chiral molecular separation with a few molecules, or possibly even a single molecule. Meanwhile, plasmonic structures can also significantly enhance the local optical chirality and therefore achieve high resolution chiral molecule detection with small quantities. However, it is still challenging to achieve chiral separation using plasmonic structures at the molecular levels due to the small scale. With new developing technology constantly supplementing and replacing existing methods, we believe there is still much room for improvement left towards single chiral molecule detection and separation.

Acknowledgements The authors acknowledge the financial support of the National Science Foundation (NSF–ECCS–2001650), the National Aeronautics and Space Administration Early Career Faculty Award (80NSSC17K0520), and the National Institute of General Medical Sciences of the National Institutes of Health (DP2GM128446). The authors also acknowledge the generous contribution of three illustrations by Naftal Mat Mautia.

References

1. J. Gal, "Molecular chirality in chemistry and biology: Historical milestones," *Helvetica Chimica Acta*, vol. 96, no. 9, pp. 1617–1657, 2013.
2. A. M. Evans, "Comparative pharmacology of s (+)-ibuprofen and (rs)-ibuprofen," *Clinical rheumatology*, vol. 20, no. 1, pp. 9–14, 2001.
3. J. Kumar, H. Eraña, E. López-Martínez, N. Claes, V. F. Martín, D. M. Solís, S. Bals, A. L. Cortajarena, J. Castilla, and L. M. Liz-Marzán, "Detection of amyloid fibrils in parkinson,Äôs disease using plasmonic chirality," *Proceedings of the National Academy of Sciences*, vol. 115, no. 13, pp. 3225–3230, 2018.
4. K. Kalíková, T. Šlechtová, and E. Tesařová, "Enantiomeric ratio of amino acids as a tool for determination of aging and disease diagnostics by chromatographic measurement," *Separations*, vol. 3, no. 4, p. 30, 2016.
5. W. König, R. Krebber, and P. Mischnick, "Cyclodextrins as chiral stationary phases in capillary gas chromatography. part v: Octakis (3-o-butyryl-2, 6-di-o-pentyl)-γ-cyclodextrin," *Journal of High Resolution Chromatography*, vol. 12, no. 11, pp. 732–738, 1989.

6. S. Fox, H. Strasdeit, S. Haasmann, and H. Brückner, "Gas chromatographic separation of stereoisomers of non-protein amino acids on modified γ-cyclodextrin stationary phase," *Journal of Chromatography A*, vol. 1411, pp. 101–109, 2015.
7. T. Loftsson and M. E. Brewster, "Pharmaceutical applications of cyclodextrins. 1. drug solubilization and stabilization," *Journal of pharmaceutical sciences*, vol. 85, no. 10, pp. 1017–1025, 1996.
8. Z. Aturki, "Use of β-cyclodextrin polymer as a chiral selector in capillary electrophoresis," *Journal of Chromatography A*, vol. 680, no. 1, pp. 137–146, 1994.
9. M. B. Sørensen, P. Aaslo, H. Egsgaard, and T. Lund, "Determination of d/l-amino acids by zero needle voltage electrospray ionisation," *Rapid Communications in Mass Spectrometry: An International Journal Devoted to the Rapid Dissemination of Up-to-the-Minute Research in Mass Spectrometry*, vol. 22, no. 4, pp. 455–461, 2008.
10. A. L. Ong, A. H. Kamaruddin, S. Bhatia, and H. Y. Aboul-Enein, "Enantioseparation of (r, s)-ketoprofen using candida antarctica lipase b in an enzymatic membrane reactor," *Journal of separation science*, vol. 31, no. 13, pp. 2476–2485, 2008.
11. B. Sellergren, "Imprinted chiral stationary phases in high-performance liquid chromatography," *Journal of Chromatography A*, vol. 906, no. 1-2, pp. 227–252, 2001.
12. R. Gutierrez-Climente, A. Gomez-Caballero, A. Guerreiro, D. Garcia-Mutio, N. Unceta, M. A. Goicolea, and R. J. Barrio, "Molecularly imprinted nanoparticles grafted to porous silica as chiral selectors in liquid chromatography," *Journal of Chromatography A*, vol. 1508, pp. 53–64, 2017.
13. J. Teixeira, M. E. Tiritan, M. M. Pinto, and C. Fernandes, "Chiral stationary phases for liquid chromatography: Recent developments," *Molecules*, vol. 24, no. 5, p. 865, 2019.
14. R. Sardella, F. Ianni, A. Lisanti, S. Scorzoni, F. Marini, S. Sternativo, and B. Natalini, "Direct chromatographic enantioresolution of fully constrained β-amino acids: exploring the use of high-molecular weight chiral selectors," *Amino acids*, vol. 46, no. 5, pp. 1235–1242, 2014.
15. C.-C. Hwang and W.-C. Lee, "Chromatographic resolution of the enantiomers of phenylpropanolamine by using molecularly imprinted polymer as the stationary phase," *Journal of Chromatography B: Biomedical Sciences and Applications*, vol. 765, no. 1, pp. 45–53, 2001.
16. E. Gil-Av, B. Feibush, and R. Charles-Sigler, "Separation of enantiomers by gas liquid chromatography with an optically active stationary phase," *Tetrahedron Letters*, vol. 7, no. 10, pp. 1009–1015, 1966.
17. G. Tkachenko and E. Brasselet, "Optofluidic sorting of material chirality by chiral light," *Nature communications*, vol. 5, no. 1, pp. 1–7, 2014.
18. T. Zhang, M. R. C. Mahdy, Y. Liu, J. H. Teng, C. T. Lim, Z. Wang, and C.-W. Qiu, "All-optical chirality-sensitive sorting via reversible lateral forces in interference fields," *ACS nano*, vol. 11, no. 4, pp. 4292–4300, 2017.
19. Y. Zhao, A. Saleh, and J. Dionne, "Enantioselective optical trapping of chiral nanoparticles with plasmonic tweezers."
20. I. Liberal, I. Ederra, R. Gonzalo, and R. W. Ziolkowski, "Near-field electromagnetic trapping through curl-spin forces," *Physical Review A*, vol. 87, no. 6, p. 063807, 2013.
21. K. Banerjee-Ghosh, O. B. Dor, F. Tassinari, E. Capua, S. Yochelis, A. Capua, S.-H. Yang, S. S. Parkin, S. Sarkar, L. Kronik, *et al.*, "Separation of enantiomers by their enantiospecific interaction with achiral magnetic substrates," *Science*, vol. 360, no. 6395, pp. 1331–1334, 2018.
22. A. Kumar, E. Capua, M. K. Kesharwani, J. M. Martin, E. Sitbon, D. H. Waldeck, and R. Naaman, "Chirality-induced spin polarization places symmetry constraints on biomolecular interactions," *Proceedings of the National Academy of Sciences*, vol. 114, no. 10, pp. 2474–2478, 2017.
23. Y. Tang and A. E. Cohen, "Optical chirality and its interaction with matter," *Physical review letters*, vol. 104, no. 16, p. 163901, 2010.
24. S. Yoo and Q.-H. Park, "Enhancement of chiroptical signals by circular differential mie scattering of nanoparticles," *Scientific reports*, vol. 5, no. 1, pp. 1–8, 2015.
25. S. Lee, S. Yoo, and Q.-H. Park, "Microscopic origin of surface-enhanced circular dichroism," *ACS Photonics*, vol. 4, no. 8, pp. 2047–2052, 2017.

26. Y. Zhao, "Chirality detection of enantiomers using twisted optical metamaterials," *Nature Communications*, vol. 8, no. 1, pp. 1–8, 2017.
27. P. Gutsche and M. Nieto-Vesperinas, "Optical chirality of time-harmonic wavefields for classification of scatterers," *Scientific reports*, vol. 8, no. 1, pp. 1–13, 2018.
28. L. V. Poulikakos, J. A. Dionne, and A. García-Etxarri, "Optical helicity and optical chirality in free space and in the presence of matter," *Symmetry*, vol. 11, no. 9, p. 1113, 2019.
29. T. Raziman, R. H. Godiksen, M. A. Müller, and A. G. Curto, "Conditions for enhancing chiral nanophotonics near achiral nanoparticles," *ACS Photonics*, vol. 6, no. 10, pp. 2583–2589, 2019.
30. M. Soltani, J. Lin, R. A. Forties, J. T. Inman, S. N. Saraf, R. M. Fulbright, M. Lipson, and M. D. Wang, "Nanophotonic trapping for precise manipulation of biomolecular arrays," *Nature nanotechnology*, vol. 9, no. 6, p. 448, 2014.
31. Y. Liu, Z. Wu, and Y. Zheng, "Moiré chiral metamaterials," *Communication*, vol. 5, no. 16, 2017.
32. J. Li, M. Wang, Z. Wu, H. Li, G. Hu, T. Jiang, J. Guo, Y. Liu, K. Yao, Z. Chen, *et al.*, "Tunable chiral optics in all-solid-phase reconfigurable dielectric nanostructures," *Nano Letters*, 2020.
33. M. Selmke, U. Khadka, A. P. Bregulla, F. Cichos, and H. Yang, "Theory for controlling individual self-propelled micro-swimmers by photon nudging i: directed transport," *Physical Chemistry Chemical Physics*, vol. 20, no. 15, pp. 10502–10520, 2018.
34. B. Qian, D. Montiel, A. Bregulla, F. Cichos, and H. Yang, "Harnessing thermal fluctuations for purposeful activities: the manipulation of single micro-swimmers by adaptive photon nudging," *Chemical Science*, vol. 4, no. 4, pp. 1420–1429, 2013.
35. W. A. Paiva-Marques, F. Reyes Gómez, O. N. Oliveira, and J. R. Mejía-Salazar, "Chiral plasmonics and their potential for point-of-care biosensing applications," *Sensors*, vol. 20, no. 3, p. 944, 2020.
36. J. García-Guirado, M. Svedendahl, J. Puigdollers, and R. Quidant, "Enantiomer-selective molecular sensing using racemic nanoplasmonic arrays," *Nano letters*, vol. 18, no. 10, pp. 6279–6285, 2018.
37. C. Kelly, L. Khosravi Khorashad, N. Gadegaard, L. D. Barron, A. O. Govorov, A. S. Karimullah, and M. Kadodwala, "Controlling metamaterial transparency with superchiral fields," *ACS Photonics*, vol. 5, no. 2, pp. 535–543, 2018.
38. Y. Chen, X. Yang, and J. Gao, "3d janus plasmonic helical nanoapertures for polarization-encrypted data storage," *Light: Science & Applications*, vol. 8, no. 1, pp. 1–9, 2019.
39. R. Zhang, Q. Zhao, X. Wang, W. Gao, J. Li, and W. Y. Tam, "Measuring circular phase-dichroism of chiral metasurface," *Nanophotonics*, vol. 8, no. 5, pp. 909–920, 2019.
40. M. Schäferling, X. Yin, N. Engheta, and H. Giessen, "Helical plasmonic nanostructures as prototypical chiral near-field sources," *Acs Photonics*, vol. 1, no. 6, pp. 530–537, 2014.

Part IV
Nanopores

While apertures in metal films are conduits for light, pores are conduits for mass and charge. The last part of this book outlined how the passage of light can be modified by biomolecules; in this Part, it is shown how nanopores be used to monitor biomolecules by changes in the charge transport. Nanotechnology again plays a vital role in fabrication and performance, as detailed in Chap. 10. Chapter 11 considers two dimensional films as a special class of nanopore. It is possible to combine the optical methods of the last Part with nanopores. This has the advantage of trapping or slowing down the biomolecules in the pore. It also provides an optical method to detect them, in addition to the usual ionic current. The principle is outlined in Chaps. 12 and 13, and as well as applications to proteins and their interactions. Overall, this part of the book demonstrates that nanopores provide a powerful tool for analysing biomolecules, assisted in some cases by optical detection and manipulation.

Chapter 10
Experimental Approaches to Solid-State Nanopores

Y. M. Nuwan D. Y. Bandara, Jugal Saharia, Min Jun Kim, Scott Renkes, and George Alexandrakis

Abstract Solid-state nanopores have seen widespread adoption as a promising single-molecule device class with applications spanning a plethora of bio- and synthetic molecules and particles that have biochemical, biomedical, and pharmaceutical relevance. The operational principle is ostensibly simple where an analyte added to one side (*cis*) of a membrane is electrically transported to the other side (*trans*) through a nanopore, stamping analyte-specific information through perturbations to the open-pore current (i.e., events characterized by their width and depth). The transport is fundamentally through electrophoresis or electroosmosis (and sometimes diffusion) which could be opposing or reinforcing depending on the surface charge of the nanopore and the analyte. Mechanistically, the translocations could either be diffusion-limited or barrier-limited with the former characterized by a linear increase in capture rate with applied voltage and the latter with an exponential capture rate. The open-pore noise plays a vital role in the signal-to-noise ratio of the measurement (and by extension, the detectability of an event) and several chemical, physical, electronic, and architectural measures are available for noise reduction and improved event-detectability and characterization. This chapter is dedicated to solid-state nanopores with insight into transport phenomena and their modeling, experiment planning, electrical signal measurements, nanopore characterization, noise and, improvements in measurements.

Y. M. N. D. Y. Bandara
Research School of Physics, Australian National University, Canberra, Australia

J. Saharia · M. J. Kim
Department of Mechanical Engineerings, Southern Methodist University, Dallas, TX, USA

S. Renkes · G. Alexandrakis (✉)
Department of Bioengineering, University of Texas at Arlington, Arlington, TX, USA
e-mail: galex@uta.edu

W. Bowen et al. (eds.), *Single Molecule Sensing Beyond Fluorescence*,
Nanostructure Science and Technology,
https://doi.org/10.1007/978-3-030-90339-8_10

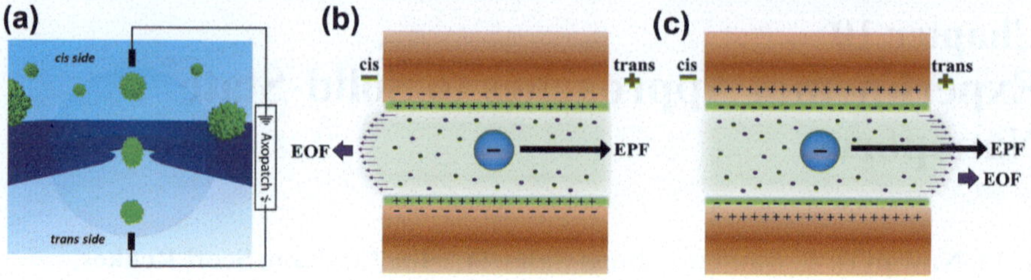

Fig. 10.1 (**a**) Typical representation of a nanopore setup where the analyte (adeno associated virus in this case) is added to the *cis* side and a voltage is applied to the *trans* side: *from Reference* [23] *published in 2020 and reproduced by permission of The Royal Society of Chemistry (RSC) on behalf of the Centre National de la Recherche Scientifique (CNRS) and the RSC*. Electrophoretic force (EPF, orange arrow) on a negatively charged particle in response to a positive applied voltage and the electroosmotic force (EOF, red arrow) originating from (**b**) negatively and (**c**) positively charged nanopore surface. EPF and EOF are opposing in **b** whereas reinforcing in **c**

10.1 Introduction

A nanoscale aperture spanning an impervious membrane (natural or solid-state) which functions as the sole fluidic pathway between two electrolyte reservoirs defines a nanopore in the most elementary manner. The operational principle is ostensibly simple where the analyte is added to one side (typically the *cis*/grounded side) of the membrane and then a voltage bias is applied to the other side (*trans*) to drive the molecules (preferably one at a time) across the nanopore fundamentally through electrophoresis and/or electroosmosis as seen in Fig. 10.1. This perturbs the open-pore ionic current (i.e., events) stamping analyte characteristic information. Each event is characterized by its depth and width (see Sect. 10.5 for more details).

This technology has expanded to characterize, for example, DNA/RNA [1–4], proteins [5–8], polysaccharides [9, 10], liposomes [11, 12], viruses [13–15] and synthetic nanoparticles [16] and polymers [17] using appropriate biological, solid-state and hybrid nanopore platforms. Note that the narrowest constriction of the most ubiquitous biological nanopores (e.g., α-hemolysin and MspA) are not wide enough to permit the transit of most of the analytes mentioned above and are typically used for the sensing of single-stranded DNA and RNA [18], some synthetic polymers [17], unfolded proteins [19], digested polysaccharides [20] and, extracted genomic content of viruses [21]. Despite these limitations, low noise, size reproducibility, and bio-engineering capability are among key factors that have propelled the advancement of biological nanopores compared to their solid-state counterparts. For example, biological nanopores have evolved into commercial devices (e.g., Oxford Nanopore Technologies), while solid-state nanopore platforms have not gained the same level of traction in commercialization. However, solid-state nanopores are size-tunable (sub-nanometer to a few hundred nanometers), available in a range of materials (e.g., silicon-based, polymer-based, graphene, etc.), resistant to extreme chemical and physical conditions, and surface tunable. The size tunability is a key advantage solid-state nanopore possess that allows these platforms to characterize a wide range of proteins, polysaccharides, liposomes and, viruses absent any digestion or denaturing steps. Until 2014, solid-state nanopores were primarily fabricated using transmission electron microscope (TEM), focused ion beam (FIB), helium ion microscope (HIM), and track etching methods. Most of these methods require access to an electron/ion microscope facility and the throughput of nanopore fabrication is limited by the capabilities of the electron/ion microscope and/or user. The nanopore technology has seen widespread adoption and democratization with the introduction of the controlled dielectric breakdown (CDB) method which is low-cost, time-efficient, and solution-based [22]. Conveniently, there are now companies that sell membrane chips (both intact membranes and membranes with nanopores), which eliminates the fabrication requirement from a research-group point-of-view. The fabrication method is inextricably linked with the surface chemistry of the nanopore and thus on the final sensing throughput, noise, and, signal-to-noise ratio (SNR) of the device. Sensing throughput, noise, SNR, analyte transport mechanism would also depend on the solution chemistry and electronic settings as well. This chapter is dedicated to solid-state nanopores with insight into transport phenomena, experiment planning, measurements, models, characterization, noise and, improvements in somewhat detail.

10.2 Transport Phenomena in Nanopores

10.2.1 Analyte Capture

The translocation of an analyte through a nanopore takes place fundamentally in three steps. When an electric field is applied, an analyte (i) undergoes free diffusion from the bulk to the capture zone, (ii) changes its transport mechanism from diffusive to drift-dominated as it gets funneled to the pore entrance in response to the applied voltage, and (iii) translocates through the pore provided applied voltage is sufficient to overcome the entropic penalty (in case of longer coiled biopolymers such as DNA) and/or electrostatic barriers. The first step is characterized by free diffusion of the analyte that is initiated far away from the pore. Nanopore experiments typically involve sub-10 nM concentrations (and sometimes in pM and fM range) of analytes. Assuming negligible intramolecular interactions, the capture rate, R_c (molecules/s) can be expressed as:

$$R_c = J/C \tag{10.2.1}$$

where J is the analyte flux rate $(1/m^2 s)$ and C is the analyte concentration $(molecules/m^3)$. At the vicinity of the pore, however, the analyte must overcome entropic (in the case of long-chain polymers such as DNA) as well as electrostatic barriers to finally translocate through the pore (rather than colliding with the pore opening). Therefore, analyte transport can be explained by two limiting cases: diffusion-limited, and barrier-limited regimes. Diffusion-limited transport could be explained using classical Smoluchowski theory, whereas barrier-limited transport could be elucidated with classical Kramer theory [24].

a **Diffusion-limited regime:** This is typically applicable to the capture of long-chain molecules (e.g., DNA with $> \sim 10^4$ base pairs [24]) under a high electric field. If this is the limiting case, a linear increase of the capture rate with the applied voltage is seen. The capture rate in this regime (R_c^{diff}) can be obtained from the Smoluchowski's rate equation; $J = 2\pi Dr^*C$. Therefore,

$$R_c^{\text{diff}} = 2\pi Dr^* \tag{10.2.2}$$

where D is the analyte diffusion coefficient and r^* is the capture radius—defined in such a way that free diffusion of the analyte takes place beyond the capture radius in the bulk solution while the analyte is captured irreversibly within the capture radius, and drift-dominated motion prevails. *Grosberg et al.* obtained the following expression for r^*[26]:

$$r^* = \frac{r^2\mu}{2lD}V \tag{10.2.3}$$

Equations (10.2.2) and (10.2.3) induce

$$R_c^{\text{diff}} = \frac{\pi r^2\mu}{l}V \tag{10.2.4}$$

where, V, μ, l and r are the applied voltage, electrophoretic mobility, pore length, and pore radius, respectively.

b **Barrier-limited regime:** The barrier-limited regime is typically for short molecules (e.g., proteins, short DNA strands) under weak electric fields. In this regime, the capture rate (R_c^{bar}) increases exponentially with the applied voltage. The exponential relationship could be understood from the fact that, as the voltage is increased, the barrier against the translocation decreases, leading to an exponential increase in the event frequency. *Pelta et al.* used the following Van't Hoff Arrhenius formalism to describe the capture rate in this regime [27, 28]:

$$R_c^{\text{bar}} = R_0 \exp(|V|/V_0) \tag{10.2.5}$$

where R_0 is the zero-voltage capture rate given by $R_0 \propto f^* \exp(-U^*/k_BT)$, k_B is the Boltzmann constant and T is the temperature. The activation energy U^*($f^*$is a frequency factor) originates from entropic and electrostatic barriers and controls R_0. The barrier at the pore entrance dominates the capture rate and the height of this barrier decreases with increasing the applied voltage. The barrier reduction factor due to the applied voltage is given by the ratio $|V|/V_0 = (zeV/k_BT)$, where z is the magnitude of the effective total number of elementary charges of the analyte and e is the magnitude of the elementary charge. V_0 represents the applied potential required by the charged analyte to overcome the Brownian motion. With DNA as the canonical test molecule, it has been shown that the capture rate increases nonlinearly with its length (barrier-limited) and plateaus at higher molecular lengths (diffusion-limited) [23]. If one examines the event rate across a wide voltage range (e.g., 50–1000 mV), a transition from barrier-limited transport to diffusion-limited transport could be encountered. This is because, as the voltage increases, sufficient energy to compensate for the barrier against threading is provided making diffusion the rate-limiting step.

10.2.2 Electrokinetic Flows in Nanopores

At the nanoscale, due to the high surface-to-volume ratio, the surface charge density (σ) becomes an important parameter in transport phenomena. Electrokinetic effects arising from the surface charges can be utilized to tune and drive nanoscale flows. The solid surface may be positive, negative, or net neutral depending on the type of surface head groups (i.e., acidic or basic or both) and the solution pH. For a surface rich in acidic head groups, following (or similar) equilibrium is applicable:

$$AH \rightleftharpoons A^- + H^+ \qquad \text{(Equilibrium 2.1)}$$

where AH and A^- are the protonated and deprotonated surface head groups governed by an equilibrium constant K_a. At acidic pH values, the equilibrium would favor the left side (neutral) and at basic pH values, it would favor the right side (negatively charged). Similarly, for a surface rich in basic head groups the equilibrium can be expressed as:

$$BH + H^+ \rightleftharpoons BH_2^+ \qquad \text{(Equilibrium 2.2)}$$

where BH and BH_2^+ deprotonated and protonated headgroups with an equilibrium constant K_b with acidic pH values favoring the right side (positively charged surface) and basic pH values favoring the left side (neutral side). The behavior of Equilibrium 2.1 and 2.2 was demonstrated by *Bandara et al.* by modifying the silicon nitride (Si_xN_y) surface with acidic (hydroxyl and carboxyl termination) and basic (amine termination) head groups [29]. To explain these equilibria more qualitatively, we need to study how the surface parameters—σ, pK and pH—are related to each other. From Grahame equation we have:

$$\sigma = \frac{2\epsilon_r\epsilon_0\kappa}{\beta e}\sinh\left(\frac{\beta e\phi_{\text{pore}}}{2}\right) \qquad (10.2.6)$$

where $\epsilon_r\epsilon_0$, ϕ_{pore}, β and, κ^{-1} are the permittivity of the solution, the diffuse layer potential of the pore, inverse of thermal energy ($\beta = 1/k_B T$), and Debye screening length, respectively [30–32]. The κ^{-1} can be calculated from:

$$\kappa = \sqrt{\frac{\beta e^2 n_{MX}}{\epsilon_r \epsilon_0}} \tag{10.2.7}$$

where n_{MX} is the numerical concentration of the electrolyte MX. Due to the presence of surface charge, oppositely charged ions or counterions from the solution are attracted towards the surface while similarly charged ions are repelled from it. The layer of counterions immediately next to the wall is immobilized due to the strong electrostatic forces. This layer is known as the Stern layer and its thickness is typically one ionic diameter [33, 34]. The loosely bound mobile layer of counterions, immediately next to the Stern layer, is known as the diffuse layer. The Stern layer and the diffuse layer together constitute the electrical double layer (EDL). The electric potential at the shear plane between the Stern layer and the diffuse layer is called the zeta potential (ζ) [35]. The thickness of the EDL can be estimated by the Debye length. For a negatively charged surface, ϕ_{pore} can be written as [36]:

$$\phi_{\text{pore}} = \frac{1}{\beta e} \ln\left[\frac{-\sigma}{e\Gamma + \sigma}\right] - \frac{\ln(10)}{\beta e}(pH - pK_a) - \frac{\sigma}{C_s} \tag{10.2.8}$$

where Γ is the number of surface chargeable groups and C_s is Stern layer capacitance. Equations (10.2.6) and (10.2.8) must be solved self consistently to solve for σ. Circumventing this requirement, another approximation for σ has been developed [28, 36]:

$$|\sigma| \cong \frac{C_{\text{eff}}}{\beta e} W\left(\frac{\beta e}{C_{\text{eff}}} \exp\left((pH - pK_a)\ln(10) + \ln(e\Gamma)\right)\right) \tag{10.2.9}$$

where C_{eff}, and W are effective Stern layer capacitance and Lambert W function respectively. For a surface rich in basic head groups, the $(pH - pK_a)$ in Eq. (10.2.9) has to be substituted with $(-pH + pK_b)$. Given the complexity associated with the surface chemistry of non-stoichiometric Si_xN_y—the material most commonly used in nanopores—it is somewhat customary to explain its nature using silanol (Si-OH) and $Si\text{-}NH_2$ models:

$$Si - OH(\sigma = 0) \rightleftharpoons Si - O^-(\sigma < 0) + H^+ \qquad \text{(Equilibrium 2.3)}$$

$$Si - NH_2(\sigma = 0) + H^+ \rightleftharpoons Si - NH_2^+(\sigma > 0) \qquad \text{(Equilibrium 2.4)}$$

Natively, Si_xN_y would have both these chemistries. Therefore, like amino acids, it would have an isoelectric point (pI) and when pI=pH, it would be net-neutral (usually at pH $\sim$ 4.1–4.3 [29, 37, 38]). The native surface chemistry can be changed during fabrication as demonstrated by our recent work with chemically-tuned controlled dielectric breakdown (CT-CDB) through Si_xN_y membranes [36], where the surface chemistry of the resulting nanopores resembled acid-rich head groups rather than being natively amphiphilic [36]. Consequently, σ can also be tuned through post-fabrication modifications [29, 39, 40].

10.2.2.1 Electroosmosis (EO)

Since hydrodynamic resistance is extremely high in nano-sized conduits, a macroscopic gradient (for example, an external electric field) is required for the generation of flow. Under the influence of an external electric field, the charged ions or particles in the diffuse layer are electrophoretically driven along the electric field [41]. Due to viscous interactions, the ions drag the liquid, producing an effective slip velocity. This movement of the bulk solution against the charged surface is called EO. The thickness of the EDL plays a significant role in the electroosmotic force (EOF). This thickness is primarily dependent on the concentration of the electrolyte used. Thicker EDL results from lower electrolyte concentrations (Eq. (10.2.7)). For smaller nanopores and/or at low electrolyte concentrations, the thicker EDL could overlap and prevent the formation of bulk electrolyte at the pore center [42]. The conditions in typical nanopore experiments produce thinner EDLs so that the ion concentration far from the pore wall can be approximated to be equal to that of the bulk solution.

The Helmholtz-Smoluchowski equation gives the magnitude of the electroosmotic mobility for an infinitesimally thin EDL [43]:

$$\mu_{eo} = \frac{\epsilon_r \epsilon_0 \zeta_w}{\eta} \tag{10.2.10}$$

where η is the dynamic viscosity of the electrolyte solution and ζ_w is the zeta potential of the pore wall. The electroosmotic migration velocity, given by $v_{eo} = \mu_{eo} E$, can therefore be expressed as:

$$v_{eo} = \frac{\epsilon_r \epsilon_0 \zeta_w}{\eta} E \tag{10.2.11}$$

where E is the externally applied electric field (V/l). Although EO is not the most common dominant transport mechanism in solid-state nanopore experiments, there are some notable examples of analyte classes where EO dominates, and we note these here briefly:

a **DNA:** *Yang et al.* demonstrated the transport of single nucleotides through a $\sim$1.8 nm diameter Si_xN_y pore by EO dominant transport [44]. In most cases, EOF is used to slow down transport by opposing the electrophoretic force (EPF) [24] and tune the conformation of the translocating molecule [45].

b **Proteins:** *Zhang et al.* showed that streptavidin can be transported through nanopores using EO with a diameter slightly larger than their length, producing capture rates $\sim$17$\times$ greater compared to a previous study [46]. *Huang et al.* used

FragaceatoxinC(FraC) nanopores to discriminate peptide and protein biomarkers [47]. EO was used as the transport mechanism of unfolded polypeptides.

c **Polysaccharides:** *Karawdeniya et al.* demonstrated the EO dominant transport of alginate through Si_xN_y nanopores. Not all polysaccharides are transported through EO as evident by their work [9]. Later we showed that maltodextrin—a charge-neutral polysaccharide—also translocates through EO using Si_xN_y nanopores fabricated through CT-CDB method with event frequency increasing with increasing solution pH (attributed to the increase in σ with increasing pH, which leads to an increase in EOF) [37].

10.2.2.2 Electrophoresis (EP)

Electrophoresis is the migration of charged particles in a stationary conducting liquid under the influence of an applied electric field. Electrophoretic transport of macromolecules is the most common mode of analyte transport in nanopore experiments and has been observed in both biological and synthetic nanopores. A charged analyte placed in an electrolyte will be screened by an EDL composed of counterions. When an electric field is applied, the oppositely charged mobile counterions in the EDL will also experience an electrophoretic force. However, these counterions will experience a force in the opposite direction to that experienced by the charged analyte (F_{bare}). The resulting hydrodynamic drag on the analyte (F_{drag}) counters the electrical force and slows down the translocation process [48, 49]. The net EPF acting on the analyte driving its translocation is, therefore, given by $F_{\text{elec}} = F_{\text{bare}} - F_{\text{drag}}$. *Schoch et al.* report that the thickness of the EDL influences the electrophoretic mobility of the analyte. Henry's formulation for electrophoretic motion is given by [50]:

$$\mu_{ep} = \frac{2\epsilon_r \epsilon_0 \zeta_p}{3\eta} f(\kappa a) \tag{10.2.12}$$

where a is the radius of the analyte, ζ_p is the zeta potetial of the particle/analyte and $f(\kappa a)$ is known as Henry's function [30], $f(\kappa a)$ depends on the thickness of the EDL. For a thick EDL where $\kappa a \ll 1, f(\kappa a)$ can be approximated as 1, leading to the Hückel-Onsager limit:

$$\mu_{ep} = \frac{2\epsilon_r \epsilon_0 \zeta_p}{3\eta} \tag{10.2.13}$$

In this case, the Stokes drag counters the Coulomb forces and analyte motion is controlled by the resultant of these two forces. Thin EDLs ($\kappa a \gg 1), f(\kappa a) \to \frac{3}{2}$, lead to the Helmholtz-Smoluchowski limit:

$$\mu_{ep} = \frac{\epsilon_r \epsilon_0 \zeta_p}{\eta} \tag{10.2.14}$$

Thin EDLs are commonly encountered in nanopore experiments, which justifies the use of Eq. (10.2.14) in most cases.

The translocation of an analyte through a nanopore is eventually dependent on the interplay between the EPF and the EOF. Consider the translocation of a DNA molecule through a nanopore drilled through a Si_xN_y membrane whose surface charge is negative at the physiological pH. When a positive bias voltage is applied across the nanopore, the negatively charged DNA molecule will experience an EPF pulling it towards the positive electrode. At the same time, the positive counterions in the EDL screening the pore wall will be repelled away from the positive electrode. As a result, the EOF will counter the EPF, slowing down DNA transport. If, on the other hand, the nanopore wall is positively charged, the EO flow of the negatively charged counterions will reinforce the EPF acting on the molecule. The translocation direction and speed are determined by the magnitude of the EOF and EPF, which are dependent on the applied voltage, molecule charge, and the surface charge of the pore walls. In some instances, translocation of an analyte can be observed at both voltage polarities. For example, *Saharia et al.* [6] and *Rant et al.* [51] showed protein translocations at pH $\sim$ 4 at both voltage polarities and attributed this phenomenon to simple diffusion. The event frequency in such cases is generally lower than those due to EP or EO.

10.3 Planning a Nanopore Experiment

Before proceeding to a nanopore experiment, one has to be mindful of the chemical (e.g. electrolyte pH, electrolyte type and concentration), physical (e.g. temperature), electronic (e.g. applied voltage, low-pass setting, acquisition frequency) and membrane (e.g. type, fabrication method, thickness and pore diameter) parameters/properties as they would ultimately govern translocation and open-pore characteristics. We discuss these here briefly to provide some insight into choosing the appropriate conditions for an experiment.

10.3.1 Chemical Conditions

10.3.1.1 pH

The pH is especially important as it governs the net charge of the analyte and the nanopore surface (see Equilibrium 2.1–2.4). The analyte could have a pK_a, pK_b or pI depending on its chemical nature.

a **Proteins:** They are amphiphilic and have a pI. Operating at pH values closer to the pI would render the net EPF imparted on the molecule to be meager as the molecule would be net-neutral at this pH. The native protein structure depends on the pH as well. For example, holo-human serum transferrin protein—a glycoprotein with an average molecular weight of ~80 kDa [52], which delivers otherwise insoluble iron to cells via receptor-mediated endocytosis—has a pI of ~ 5.2–5.6 and is natively folded when $pH > pI$ and natively unfolded when $pH < pI$. Proteins are also susceptible to extreme pH values. Thus, before proceeding with any experiments, the pH conditions where proteins denature, or undergo conformational changes should be found through a literature search.

b **Polysaccharides:** Although a pH of ~7 would render Si_xN_y fabricated through CDB or TEM to be net negatively charged, CT-CDB fabricated Si_xN_y pores would be near-neutral. The consequences are evident through translocation of maltodextrin—a charge-neutral glycan translocating through the EO mechanism—where no translocations were seen at pH ~ 7 when using CT-CDB fabricated Si_xN_y nanopores. However, when pH was increased to ~9, events were observed as the surface was now net-negative in charge, activating the EO mechanism. Thus, for charge-neutral molecules, the pH must be tuned such that the nanopore is appropriately charged to facilitate the transport through the EO mechanism.

c **DNA:** Considered as a gold standard in nanopore science, DNA is generally regarded to have a $pI \sim 5$ (the value depends on the sequence). Thus, operating at pH values where EP is meager would render the capture rate impractically low. The capture rate of 1 kb ds-DNA in 4 M LiCl and at an applied voltage of 200 mV in experiments with CT-CDB fabricated Si_xN_y nanopores was found to be <3 events/s for pH ~ 5 and ~6 whereas at pH ~ 7 and ~8, it was at least 9.5× greater [53].

d **Pore Surface:** Pores fabricated through Si_xN_y using TEM and CDB are amphoteric with a pI of ~4.1–4.3 [29, 38], whereas those fabricated through CT-CDB resemble a surface rich in acid head groups with a pK_a of ~9.7 [37].

10.3.1.2 Electrolyte Chemistry

Generally, the chlorides of group IA cations are used in nanopore sensing. Group IIA cations such as Ca^{2+} and Mg^{2+} are not desirable as they can lead to crosslinking of biopolymers and adhering them to the nanopore surface [54, 55]. The more ubiquitous electrolytes are LiCl and KCl (NaCl is also seldom used). LiCl is known to slow down the translocation of DNA [31]. Although this would reduce the capture probability, it would increase the dwell time of the DNA molecule. Tuning the dwell time is important to be above the lowpass filter response (discussed later in this chapter). The choice of electrolyte, according to the Hofmeister series, would have a salting-in (chaotropic—destabilizing effects) or salting-out (kosmotropic—stabilizing effects) effects on protein solubility [56]. The work of *Green et al.* [57] and *Medda et al.* [58] provides more insight into the concentration ranges where these effects are active, with the former work claiming Hofmeister series effects are prominent at higher salt concentrations (0.5–3.0 M) whereas the latter work showed that these effects can be present at physiological concentrations as well. Although DNA and polysaccharide experiments are present across a range of electrolyte concentrations (up to 4 M KCl and LiCl), such high salt concentrations could denature proteins. However, it is not uncommon to see protein experiments done up to about 2 M salt concentration [6, 7]. A high electrolyte concentration would essentially diminish the contribution of EO to the overall transport mechanism. At lower concentrations, the

EO mechanism becomes active (Debye length increases inversely with electrolyte concentration), which has been shown to reserve the resistive-pulse direction [31] and promote single file translocation of DNA [45].

10.3.2 Electronics

10.3.2.1 Applied Voltage

The applied voltage can govern the translocation speed, transport mechanism (e.g., diffusion or barrier-limited [25, 59]), conformation (e.g., folded, unfolded [60]) and, structural properties (e.g., electro-deformation [12, 13]) of the analyte. Analyte translocation can occur only when the applied voltage is sufficient to overcome the entropic penalty and/or the electrostatic barriers. If the applied voltage is not sufficient to overcome the energetic barriers, the analyte may simply collide with the pore entrance rather than penetrating it. Although the applied voltage would increase the capture probability and accommodates sufficient energy to overcome energetic barriers, it would increase the translocation speed as well. If the analyte is moving too fast it could completely go undetected or its signal metrics could be attenuated by the electronic response time (i.e., rather than the actual metrics, what the user would see is the response time of the electronics—discussed later in this chapter [61, 62]). The magnitude of the applied voltage would also determine whether the transport is diffusion or barrier-limited as discussed in Sect. 10.2.1. Unlike DNA, which can withstand an appreciable voltage range, proteins could undergo voltage-driven unfolding [6, 26]. Above a certain voltage, the protein could be fully stretched or may reach its limit of unfolding [27]. Although it is not uncommon to perform protein experiments at high voltages, they could force the protein to unfold and information on the native state may not be achievable. The applied voltage also deforms soft particles which depend on, for example, the particle membrane mechanical properties and on cargo content [12, 13, 23, 63]. Voltage-induced deformation of viruses were shown to depend on virus maturity [13] and cargo content [23], both of which are of clinical relevance. Thus, voltage should be carefully chosen to inhibit or promote deformation depending on the study.

10.3.2.2 Lowpass Filter Setting

Although this filter removes high-frequency noise features from the measurement, if not used properly, it could also attenuate the signal and a distorted picture of translocation characteristics may arise. The rise time (T_r) of the lowpass filter is given by [61, 64]:

$$T_r = 0.3321/f_c \tag{10.3.1}$$

where f_c is the cutoff frequency of the filter. Dwell times $< 2T_r$ would be attenuated—the minimum falling edge to overcome the filter response is T_r and similarly, the minimum rising edge must be T_r as well. Thus, the lowpass filter requires $2T_r$ of the event dwell time to respond properly. If the event is shorter than $2T_r$, the recording would be limited by the lowpass filter response and the data would correlate to the filter response rather than to the actual transit time. Several methods have been proposed to correct the dwell time in such instances including, for example, the integrated area [61], adaptive time series [65], falling-edge [60], and full width half maximum (FWHM) methods [62]. One option to overcome such attenuation is to opt for a higher lowpass filter setting with the caveat of higher background noise. Recently, complementary metal-oxide-semiconductor (CMOS) integrated nanopore platforms have been used to reach MHz level bandwidths due to the low parasitic capacitance such designs possess [66]. With a 1 MHz lowpass filter, the $2T_r$ would theoretically be $\sim$0.6 μs. Another option is to slow down the translocation such that dwell times become $> 2T_r$ (see Sect. 10.7.1 for more details). Unfortunately, increased dwell times could negatively impact the event frequency and a balance should be achieved between throughput and the lowpass filter setting. If the Axopatch 200A (Molecular Devices, LLC) is used, the two most common settings are the 10 and 50 kHz whereas, in the case of the Axopatch 200B, they are 10 and 100 kHz. The sampling rate (discussed below) should also be sufficient for the filter setting. If no useful/extra information is gained by operating at 50 or 100 kHz compared to the 10 kHz lowpass setting, it (10 kHz) is typically preferred as this setting would have a lower r.m.s. (root mean square) open-pore noise compared to the highest lowpass setting. Additionally, for current-voltage (I-V) curves, the 1 kHz low pass setting is more than adequate.

10.3.2.3 Acquisition/Sampling Frequency

This depends on whether the data is analyzed in the frequency domain (e.g. noise analysis) or in the time domain (current traces). For the former, twice the bandwidth is acceptable provided the anti-aliasing filter is ideal. Since ideal anti-aliasing filters are somewhat theoretical, it is typical to sample $\sim 2.5\times$ the bandwidth. Time-domain analysis requires ideal anti-aliasing and reconstruction filters. For such, $5\times$ the bandwidth is typical, whereas $10\times$ the bandwidth is considered optimal. For example, a 10 kHz lowpass filtering may require 25 kHz sampling in the frequency domain and 50–100 kHz sampling in the time domain. Thus, high filter settings require higher sampling rates. Therefore, the digitizer must be able to cope with the filter setting of the amplifier for the proper (re-)construction of the analog signal from digitized samples, without distortion.

10.3.3 Membrane Properties

10.3.3.1 Membrane Thickness

The ubiquitous membrane architectures are 2D, silicon-based, high-κ dielectric, and polymer-based. The adoption of each type depends on the application. For example, sequencing is best suited with ultra-thin membranes (i.e., 2D material) whereas volume estimation is better suited with relatively thick membranes. Thicker membranes offer low capture rates while also increasing the risk of (irreversible) analyte sticking. Therefore, depending on the analyte and the application of interest, the appropriate thickness should be chosen. The pore fabrication would also depend on the membrane type and thickness. For example, CDB is more suited for $l < 50$ nm (although, 30 nm could ideally be an upper limit based on our experience) with fabrication time increasing with increasing l. Thus, choosing a thin enough membrane could save time in fabrication. The controlled breakdown is also used in 2D materials such as graphene and MoS_2 [67, 68]. However, the more conventional way to make pores through 2D material is by using a TEM, which requires expert user input.

10.3.3.2 Membrane Type

The more commonly used membrane types include:

a **Silicon-based:** e.g. low-stress Si_xN_y and silicon oxide (SiO_2). Si_xN_y membranes are typically deposited on Si supports using LPCVD, PECVD, or PVD. Although the ubiquitous material of choice, this architecture suffers somewhat from high dielectric and $1/f$ noise. Consequently,Si_xN_y membranes fabricated with a SiO_2 underlayer (between Si_xN_y and the Si support) have lower capacitance noise compared to SiO_2-free conventional ones [66] (see Sect. 10.6 for more information on noise).

b **2D material:** e.g. Graphene, MoS_2, hexagonal boron nitride (h-BN). These are usually transferred on to silicon-based substrates. Although these materials carry the promise for next-generation solid-state nanopore-based sequencing, they suffer from high (DNA) transport velocity, pinhole formation and membrane stability issues.

c **High-κ dielectric material:** e.g. HfO_2 [69], Al_2O_3 [68, 70], Al_2O_3 stacked with graphene [69, 71]. Although not commonly used, these are typically known for low-noise and enhanced mechanical properties.

d **Polymer-based:** Typical examples include poly-ethylene terephthalate (PET) [70] and, polycarbonate (PC) [71]. Thickness is usually in the micron scale. Pore fabrication is done by etching the tracks created by bombarding the membrane with heavy ions, where fluence of the ion source must be controlled to control the pore density. Single swift heavy ion sources enable the fabrication of single pores, e.g. GSI (Darmstadt, Germany) [70].

Table 10.1 Concise overview of ubiquitous solid-state nanopore fabrication methods

Fabrication Method	References	Reported Material[a]	Pore Diameter	Fabrication Medium	Source
Chemical Etching[b]	[72–74]	PET,PC, Si_xN_y	Lower diameter pores are challenging	Aqueous	e.g. KOH, hot H_3PO_4
TEM	[1, 5, 75, 76]	Si_xN_y, SiO_2, MoS_2, h-BN, Graphene, etc.	Larger diameter pores take more time	Vacuum	Electron
FIB	[77]	Si_xN_y	Typically, >10 nm diameter pores	Vacuum	Heavy-ion (e.g. Gallium)
HIM	[78]	Si_xN_y	Permits <10 nm diameter pores[c]	Vacuum	Helium ion
CDB	[79, 80]	Si_xN_y, MoS_2 Graphene	<~40 nm[d]	Electrolyte	Voltage
LACB	[81]	Si_xN_y	Reported on ~20–50 nm pores	Electrolyte	Laser+ Voltage
MPVI	[82]	Si_xN_y	Reported on $sub1 - 3$ nm[e]	Electrolyte	Voltage
CT-CDB	[37]	Si_xN_y	<~50 nm[d]	Electrolyte	Voltage
TCAM	[83]	Si_xN_y	No Limitations noted	Ultra-Pure Water	Voltage

The choice of membrane depends on the application of interest and the tools available for fabrication as outlined in Table 10.1.

10.3.3.3 Fabrication Method

A multitude of fabrication methods exists. Ubiquitously used ones are (i) chemical etching of track-damaged membranes, (ii) electron beam-based fabrication (e.g. TEM), (iii) ion based fabrication (e.g. HIM, FIB) (iv) controlled breakdown (CDB, CT-CDB, multi-level pulsed voltage injection (MPVI), tesla-coil assisted method (TCAM)) and, (v) laser-assisted (laser-assisted controlled breakdown (LACB)). The choice of method fundamentally depends on the membrane type, pore-size requirements, and availability of resources. Note that voltage-assisted methods (e.g. CDB, MPVI, CT-CDB, LACB) are typically used for <30–50 nm thick membranes—the thinner the better. Table 10.1 provides a general comparison of the methods mentioned above.

a: Only the more commonly reported materials are included. There could be other examples.
b: The number of pores depends on the fluence of the ion source and window size.
c: Most ion-beam methods are not conducive for <10 nm diameter pore fabrication and HIM permits such and even the coveted <5 nm regime with SNR comparable to TEM [84].
d: Not recommended to go beyond $\sim$30 nm.
e: Multiple pore formation if not optimized properly [85]

The table above is a general guide that can be used to determine the appropriate method for pore fabrication depending on the membrane properties and targeted pore properties.

10.3.3.4 Pore Diameter

Fine control is needed to allow the entrance of the molecule, minimize sticking to the nanopore surface, maintain an appreciable capture rate and maximize the signal to noise ratio (SNR). For example, if the pore diameter is too large SNR will be low and without adequate confinement. In that case, molecules may transport ballistically without been detected. In contrast, if the pore diameter is too small analytes may collide with the pore opening rather than translocate, clogging may become prominent and the capture rate would be too low to attain practically useful throughput rates.

10.4 Nanopore Characterization

After fabricating a pore, it is important to characterize it. This short section is intended to provide some insight into pertinent pore characterization steps:

a Size estimation
b Shape characterization
c Surface characterization
d Noise characterization (see Sect. 10.6)
e Analyte responsiveness/sensitivity
f Pore quality and resilience (e.g., sticking probability, rectification, open-pore current stability)

Note that, other than (a), the rest are generally known for a given material, fabrication method, electrolyte chemistry, experimental conditions (e.g. applied voltage, lowpass filtering), and analyte. However, properties such as noise, rectification, and open-pore stability are important parameters to check for each fabricated pore.

10.4.1 Size and Shape Characterization

Microscope-based methods (e.g. TEM, FIB, HIM) have the added advantage of being able to image the pore, from which the size could be estimated. TEM imaging was used by *Goto et al.* to observe multiple pores that were formed by an unoptimized MPVI method [85]. Pores fabricated from TEM and HIM, especially, could close somewhat after fabrication, the effect being more profound with smaller diameter pores. Thus, it is important to check pore size after mounting it to the flow cell using a proper conductance-based model. The two most common equations used in literature are:

$$G = \left(\frac{1}{G_{\text{bulk}} + G_{\text{surface}}} + \frac{1}{G_{\text{access}}}\right)^{-1} \tag{10.4.1}$$

$$G = \left(\frac{1}{G_{\text{bulk}}} + \frac{1}{G_{\text{access}}}\right)^{-1} \tag{10.4.2}$$

where G, G_{bulk}, G_{surface} and G_{access} are open-pore conductance, bulk conductance, surface conductance, and access conductance. G can be found from the slope of the I-V curve. Equation (10.4.1) simplifies to Eq. (10.4.2) at high electrolyte concentrations and at pH values where σ is negligible because under such conditions $G_{\text{surface}} \sim 0$. It is a common practice to use a cylindrical bulk model for both equations. For cylindrical pore shape, $G_{bulk} = K\frac{\pi r^2}{l}$, $G_{\text{surface}} = \mu|\sigma|\frac{2\pi r}{l}$ where K and μ, are electrolyte conductivity, and mobility of counter-ions, respectively. $G_{\text{access}} = (\frac{2}{K\alpha_a 2r + \beta_a \mu|\sigma|})^{-1}$ where α_a and β_a are model-dependent parameters (usually set to 2) [37, 86]. However, the nanopore shape is dependent on the fabrication method. An interested reader could explore existing publications [1, 32, 87–89] to find G_{bulk} and G_{surface} expressions for cylindrical, conical cylindrical, double conical, hyperbolic, truncated double cone and hourglass-shaped nanopores. Thus, before using either one of Eq. (10.4.1) or Eq. (10.4.2), its essential to know the shape of the fabricated nanopore which can be validated by TEM tomography, FIB/SEM tomography, cross-sectional SEM, or small-angle X-ray scattering [1, 90–92].

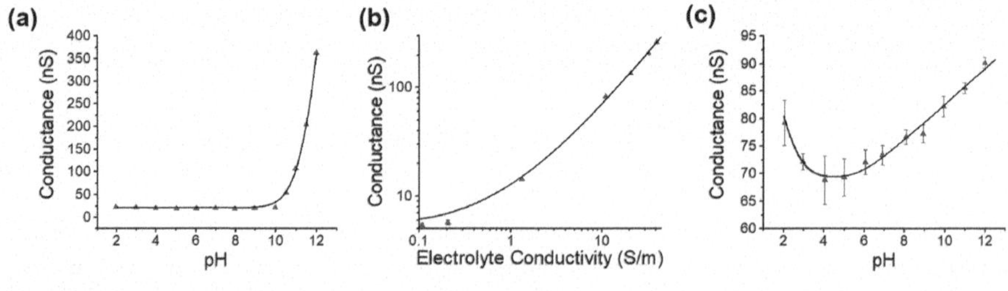

Fig. 10.2 (**a**) pH and (**b**) conductivity response of the open-pore conductance (G) of a ~11 nm diameter silicon nitride nanopores fabricated by the CT-CDB process. (**c**) pH response of G of a ~13 nm diameter silicon nitride nanopore fabricated by the CDB process. *Figure* 10.2c *was Reprinted with permission from ACS Appl. Mater. Interfaces 2019, 11, 33. Copyright (2019) American Chemical Society*

10.4.2 Surface Characterization

Both σ and ζ_w can be calculated from streaming potential measurements as outlined in [93]. These parameters are inextricably linked with transient interactions, EOF, capture rate, etc. One convenient way to estimate σ is to survey the pH dependent open-pore conductance (Fig. 10.2a), and then fit the data with Eq. (10.2.9). Given the approximations associated with Eq. (10.2.9), one could also plot electrolyte conductivity with open-pore conductance and fit the data with Eq. (10.4.1) [94] (Fig. 10.2). Small Angle X-ray Scattering (SAXS) has been recently used to characterize axial and radial ion track etch rates with unprecedented precision. This study showed a strong correlation between the ion energy and the morphology of the etched channels in thin films of amorphous SiO_2 [95].

10.4.3 Analyte Responsiveness/Sensitivity ($\Delta_{analyte}$)

DNA is the gold standard to gauge analyte responsiveness given its uniform charge, known length, and ruggedness. One can measure the rate of translocation as a function of (i) pore diameter, (ii) pH, (iii) electrolyte type and concentration, (iv) applied voltage, and then compare the results with a comparative (or competitive) fabrication method. The most common methods for event-rate quantification include (i) slope of cumulative events versus experiment time [25], (ii) survival probability method [25], (iii) exponential fit to arrival time distribution [96], (iv) the average number of events across the experiment time and, (v) average of the inter-translocation time across the experimental time. These methods could suffer from inter- and intra-pore variations and could be improved through controlled counting [25]. The slope of a calibration curve constructed with the desired event counting method against analyte concentration would yield the sensitivity of the pore to the analyte ($\Delta_{analyte}$). When the CT-CDB fabrication method was developed, its performance was gauged against the conventional CDB process using double-stranded DNA (dsDNA) from which we found Si_xN_y nanopores fabricated through the CT-CDB process were $\sim$6.5$\times$ more sensitive to dsDNA compared to their CDB counterpart [37].

10.4.4 Pore Quality and Resilience

a **Ionic Current Rectification (R_{ICR}):** Although the I-V curve of some pores are inherently asymmetric due to geometric and surface properties, it is desirable to have $R_{ICR} \sim 1$ in cylindrical pores. We would define R_{ICR} = G+/G-, where G+ and G- are the slope of the I-V curve at positive and negative voltage regimes. We have observed, pores (CDB, CT-CDB, and TEM fabricated Si_xN_y) that significantly deviate from $R_{ICR} \sim 1$ are noisy, less responsive to analytes, and prone to clogging.

b **Sticking Probability (S_p):** Transient perturbations due to analyte sticking on the nanopore walls that are much longer than regular translocations would change both chemical and physical properties with the potential to render the pore futile for further experiments. An intervention may be necessary (e.g., zapping, electrolyte exchange) unless sticking is self-corrected. A lower S_p and higher open-pore stability are preferred.

c **Open-Pore Current Stability:** One could measure this parameter by fitting a current trace with a function in the form $I_0 = I_{t=0} + \beta_1 t$, where $I_{t=0}$, β_1 and t are initial open-pore current, slope-factor and time, respectively. For an ultra-stable pore with no current drifts, $\beta_1 = 0$.

d **Open-Pore Lifetime (t_{open}):** This is the time duration during which a pore remains open before clogging irreversibly with the analyte, or undergoes a drastic change in current due to enlargement or pore closure. Higher t_{open} values are desirable.

10.5 Features Defining a Resistive Pulse

Although an event looks trivial, determining its bounds and depth is not always simple. A brief account of the common methods used for calculation of translocation time (t_d) and current drop (ΔI_B) are provided here. Note that most methods are calibrated by feeding pulses (using a pulse generator) of known height (h_{user}) and width (w_{user}) to the amplifier (i.e., the nanopore is replaced by a pulse-generator). That way the ideal ΔI_B and t_{d} would be known and based on the readout of the system, correlations between ΔI_B and h_{user} and t_d and w_{user} could be developed to better estimate the event characteristics.

10.5.1 Translocation Time (t_{d})

Since t_{d} is greatly affected by the rise-time of the lowpass filter (especially when $t_d < 2T_r$), the common approaches to estimate t_d include (i) FWHM of the peak, (ii) two sides of the blockage (analogous to B_1 and S_2 points of Fig. 10.3c), (iii) modified stop point [61, 62], and (iv) fitting and algorithmic approaches [65]. A given event would have three key regions: a falling edge, a plateau region, and a rising edge. These can be prominently seen in events with $t_d > 2T_r$ whereas in shorter events, the plateaued region can be absent, and the event would be represented solely by the falling and rising edge responses of the filter. The finite rise time of the filter would account for most of the rising edge width, which leads to an overestimation of the true pulse width with approaches that take two sides of the blockage. The modified stop-point method circumvents this by taking the last local minima before the signal starts to rise as the endpoint (Fig. 10.3c). The FWHM approach has been found to have performance comparable to this approach. Thus, one could define the pulse width as the distance between the starting point of the falling edge and the beginning of the rising edge [60]. However, for such approaches, the event would need to reach a plateau and the stop point should not coincide with the falling edge [62]. *Gu et al.* used a second order-differential-based calibration (DBC) to estimate t_d [61]. In this

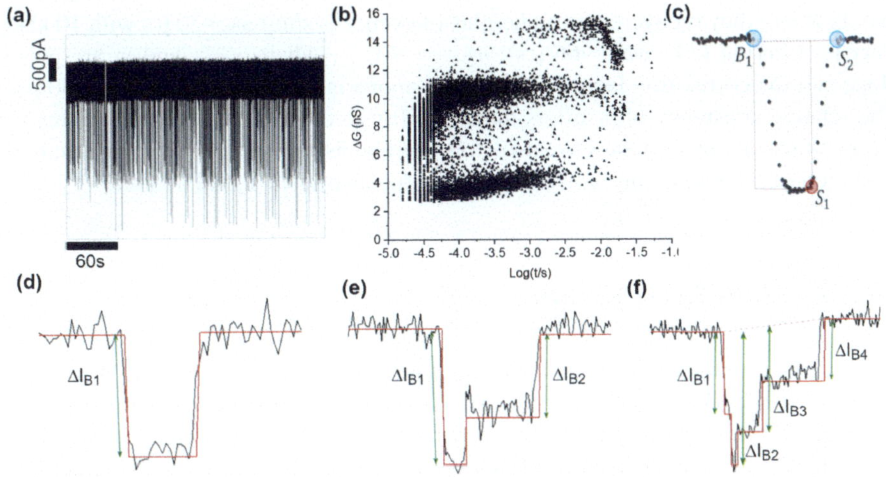

Fig. 10.3 **a** Current trace and **b** scatter plot of the log of translocation time versus change in conductance corresponding to 1 kb dsDNA translocating through a ~3.4 nm diameter pore (4 M LiCl, pH ~ 7, +200 mV). **c** A single level event with $t_d < 2T_r$. B_1 and S_2 correspond to the start and endpoint that take two sides of the blockage and the S_1 is the stop point corresponding to modified stop point methods. **d** Single level event with $t_d > 2T_r$, **e** folded-over event and **f** multi-level event; **d–f** were collected from a ~12 nm diameter pore. Red lines are fits made with a custom program

method, events were fitted with a fourth order Fourier series (although higher orders are possible) as a smoothing step followed by the DBC method, where inflection points at the two edges defined the primitive event boundaries before refining the end point by moving it to the local minima to account for the rising response of the filter to prevent overestimation of the t_d.

10.5.2 Current Drop (ΔI_B)

The local minima, or sometimes the mean current of the event, are most commonly used as ΔI_B. The former makes the method susceptible to sudden point-variations and the latter could under-estimate event amplitudes, especially those with no appreciable plateau. Consequently, like t_d characterization, some key approaches exist in literature to define ΔI_B. One approach uses a python-based MOSAIC software with Adaptive Time-Series Analysis (ADEPT) [65]. For attenuate-prone events, this method fits data to an electric circuit model to estimate the current drop. During the DBC implementation by *Gu et al.*, they have noted that the area of the event (e.g., one can calculate this by the *trapz* function of MATLAB) is seldom affected by the filter response [61]. If this is the case, after estimation of the event area and t_d by DBC, one can back-calculate ΔI_B (i.e., $\Delta I_B \times t_d$ = event area). *Pedone et al.* used a method based on the slope of the falling edge to estimate the height of truncated events where they recovered pulse heights of events as short as $\sim$30 μs with 10 kHz lowpass filtering ($2T_r$ was $\sim$66 μs): $h_{\text{user}} = m x_{\text{event}}$ where x_{event} and m are pulse slope and the correlation factor with m been appreciably constant for $t_d >\sim 30$ μs. The relative constancy of m offers the capability to correct ΔI_B without having to worry much about t_d. One caveat of this method is the somewhat high standard deviation of the histograms corresponding to reconstructed signal height.

10.5.3 Multi-Level Events

This is where things start to get somewhat tricky. For example, if the DNA translocates as a single-file it would produce a single-level event. However, DNA can also be captured as symmetrically folded-over, partially folded, circular, and knotted (e.g. prime-L knot, Fused knot, prime-H knot, prime-L knot, factored knot) structures [97, 98]. Although circular and symmetrically folded-over conformations would produce a single level ($\sim 2\times$ deeper than single-file) event structure, the rest would produce stepwise and sharp drop event structures. Thus, a simple single-level analysis could critically mask information that sub-events carry and undermine the final analysis. More notable multi-level event-analysis methods include CUSUM-based multilevel fitting [99–101], the MOSAIC algorithm [65] and hidden Markov model approaches [102, 103]. The *OpenNanopore* application (CUSUM-based) developed by the Radenovic group [99] has, to some extent, paved the way for a standardized analysis approach. Thus, one should develop or use an existing platform for event analysis that can detect and fit multi-level events. However, one should be cautious of how a given software estimates ΔI_B and t_d for reasons mentioned previously.

10.5.4 Event Analysis

Now that the experiment is done and the salient features are extracted, what is the next step? In an ideal world, if we use one molecule type, one would optimistically expect a single ΔI_B and t_d. However, this is far from reality and we would see a distribution of ΔI_B and t_d. This is because not every molecule translocates the same way due to, for example, the spatial trajectory through the pore [104], analyte-analyte and/or analyte-surface interactions (e.g. adsorption, electrostatic, specific binding, chelation/cross-linking [105]), analyte conformation (e.g. voltage induced structural changes [6, 8], DNA knots [98]). The result is a scattered population for ΔI_B and t_d.

10.5.5 Analysis of ΔI_B

The first step is to histogram the data with a suitable bin size. For DNA, one would typically see three peaks with the first corresponding to collisions, the second corresponding to single-file translocations, and the third corresponding to folded-over translocations, among other possibilities discussed above that give deeper blockages (see Supporting Information of [37] for a more detailed discussion). The histogram can be fitted with a summation of Gaussian functions, each in the form of $A_i \exp(-(\Delta G - \mu_i)^2/s_i^2)$ where A_i, μ_i and, s_i are the amplitude, mean, and standard deviation of the ith population. While this could be done in MATLAB and other programs, we have found Mathematica to be convenient for symbolic operations such as this. For dsDNA, the division of μ_i for $i = 3$ by $i = 2$ would ideally be $\sim$2: an internal check to make sure that data is from DNA but not from contamination, or other sources. Let the radius of dsDNA be r_{dsDNA} and the charge of dsDNA be Q_{dsDNA}. Assuming a cylindrical nanopore profile, the effective radius of the open-pore during dsDNA confinement would be $r_{\text{with DNA}} = \sqrt{r_0^2 - r_{\text{dsDNA}}^2}$. G can be evaluated using Eq. (10.4.1). Then, one can write, change in conductance due to dsDNA translocation (ΔG_{dsDNA}) as:

$$\Delta G_{\text{dsDNA}} = G - K\left(\frac{1}{\frac{\pi r_{\text{with DNA}}^2}{l} + \frac{\mu|\sigma|}{K}\frac{2\pi r_0}{l} + \frac{\mu}{K}\frac{q_{\lambda-\text{DNA}}}{l}} + \frac{2}{\alpha_a 2r_{\text{with DNA}} + \beta_a \frac{\mu|\sigma|}{K}}\right)^{-1} \tag{10.5.1}$$

The r_{dsDNA} and $Q_{(\text{dsDNA})}$ can be approximated to $\sim$1.1 nm and ~ -0.96 nC/m. Then, the theoretical $\Delta G_{(\text{dsDNA})}$ can be calculated using Eq. (10.5.1). However, one might find that the experimental and the theoretical values may differ because of, for example, variations in (i) l (typically ± 1–± 2), (ii) Q_{dsDNA} (can be changed by the electrolyte-DNA interactions among other possibilities) and (iii) r_{dsDNA} (i.e. hydrated radius ranging from 1.1–1.3 nm, [89, 106, 107]). Thus, one can use the extreme boundaries in each case and calculate a range for the theoretical ΔG_{dsDNA} to see whether the experimental value falls within that range. Disagreements may indicate flaws in, for example, data collection, data analysis, or contamination.

10.5.6 Volume Exclusion Models for ΔI_B

The formula widely used for spherical nanoparticles and globular proteins [5, 6, 108, 109] is:

$$\Delta I_B = K\frac{\gamma V \Lambda}{(l + 1.6r)^2} S_{r,d} \tag{10.5.2}$$

where γ, Λ, and $S_{r,d}$ are shape factor, transiently excluded electrolyte volume and the correction factor (typically assumed to be 1), respectively. It must be noted that the value of $\gamma = 1.5$ (a numerical constant) arises from the assumption that the translocating particle is a sphere [110] and is somewhat questionable to be treated as a variable, as seen from its derivation dating back to works of DeBlois [111] and Maxwell [112] in the early 1900s and later 1800s, respectively. Another key assumption is that the particle is completely confined within the nanopore (i.e., particle diameter $2a < l$). If that is not the case (i.e., $2a > l$), the following could be used instead [5, 113]:

$$\Delta I_B = I_0 \left(1 - \left(\frac{1+\alpha}{1+\alpha_b}\right)\left(\frac{1}{1+\frac{\Delta R}{R_p}}\right)\right) \tag{10.5.3}$$

where I_0 is the open pore current, $\alpha (= R_{\text{access}}/R_{\text{geometric}})$ is the ratio of access resistance (R_{access}) and geometric resistance ($R_{\text{geometric}}$), $\Delta R/R_p$ is the fractional resistance change inside the nanopore, and $\alpha_b = \alpha(1 - a^2/r^2)^{\beta_{corr}}$ where β_{corr} is the correction factor (assumed to be 0.8 for $2a > l$). We would like to refer interested readers to reference [113] for a detailed discussion of this topic.

10.5.7 Other Models for ΔI_B

a $\frac{\Delta I_B}{I_0} = \frac{a^2}{r^2}$. Although a simple formulation, it neglects (i) access resistance of the pore with and without the analyte, (ii) surface charge of the nanopore, and (iii) charge of the analyte.

b $\Delta I_B = KV\frac{A_{\text{analyte}}}{l_{\text{eff}}}$ where A_{analyte} and l_{eff} are the hydrodynamic cross-section of the analyte, and the effective membrane thickness [114].

c $\frac{\Delta I_B}{V} = \frac{6\pi k \epsilon_r \epsilon_0 \mu E}{(1+\kappa a)} f(\kappa a) + \frac{Q}{l^2}\mu - K\frac{\gamma\Lambda}{(l+1.6r)^2}$ where $k = k_1 k_2^2$ with k_1 being a fitting parameter and $k_2 = a/l$, Q is the initial effective charge of the particle [108]. Since the nanopore is an ionic conductor, the question arises whether resistive components can be added linearly as shown in this formula. Thus, there are reservations about the validity of this formula and we urge readers to exercise caution while adopting models where multiple resistive components are added linearly.

d $\frac{\Delta I_B}{V} = \frac{1}{l}\left(-\frac{\pi}{4}d_{\text{pol}}^2(\mu_{\text{cation}} + \mu_{\text{anion}})n_{\text{electrolyte}}e + \mu^* Q_{\text{pol}}^*\right)$ where d_{pol}, μ_{cation}, μ_{anion}, $n_{\text{electrolyte}}$, e, μ^* and Q_{pol}^* are the cross-sectional diameter of the polymer chain, mobility of cation of the electrolyte, mobility of the anion of the electrolyte, number density of the electrolyte, elementary charge, effective electrophoretic mobility of the counter ion (with respect to the polymer charge), and effective charge of the polymer chain, respectively [31]. This equation is used for polymer chains such as DNA and does not take access resistance into account.

e $\frac{\Delta I_B}{V} = K\left[\left(\frac{4l_{\text{eff}}}{\pi d_{\text{pore}}^2} + \frac{1}{d_{\text{pore}}}\right)^{-1} - \left(\frac{4l_{\text{eff}}}{\pi d_{\text{pore,analyte}}^2} + \frac{1}{d_{\text{pore,analyte}}}\right)^{-1}\right]$ where d_{pore} and $d_{\text{pore,analyte}}$ are open-pore diameter and the effective diameter of the pore containing the analyte respectively. Here, for a polymer chain such as DNA (assuming the analyte is a cylindrical rod), $d_{\text{pore,analyte}} = \sqrt{d_{\text{pore}}^2 - d_{\text{pol}}^2}$ [115]. This formula does not take surface charges into account.

10.5.8 Analysis of t_d

Literature suggests a couple of methods that are widely adopted in the nanopore community with the first passage time model and exponential decay function being the two prominent methods. We note that there is some underlying controversy in the adoption of the drift-diffusion model with *Ling et al.* pointing some inconsistencies in the literature [116]. We would leave the reader of this chapter to explore these inconsistencies and would only focus on the essence they have developed to improve the first passage time model based on the 1D biased diffusion model:

$$f(t) = \frac{l}{\sqrt{4\pi Dt^3}} \exp\left(-\frac{(l - v_{\text{analyte}}t)^2}{4Dt}\right) \tag{10.5.4}$$

where v_{analyte} is the drift velocity of the analyte. The peak of the fit obtained from Eq. (10.5.4) gives the translocation time, t_d. The l is also known as the trajectory length of the molecule and can be estimated using $l = l_{\text{eff}} + l_{\text{cont,analyte}}$ where $l_{\text{cont,analyte}}$ is the contour length of analyte. For DNA, $l_{\text{cont,analyte}} = 0.34N$, where N is the number of base pairs [117]. The nature of nonlinear fits is such that, depending on the initial parameter guesses and the boundaries, one can fit a host of data sets with a good regression coefficient even if the data set would not necessarily satisfy the conditions under which the equation was derived. Thus, one should pay attention to the output values of the parameters and experimental conditions as one can end up with unrealistic, or nonphysical values that would distort conclusions and outcomes. Transport is barrier-limited at low voltages and diffusion-limited at higher voltages (Sect. 10.2.1). *Ling et al.* claim that, at low voltages, the entropic barrier associated with the translocation of long polymer chains undermines the validity of Eq. (10.5.4). To check the validity of Eq. (10.5.4) at these two voltage regimes, the v_{analyte} was checked against Smoluchowski's linear electrophoresis and intrinsic diffusion constant (D_0) against Stokes—Einstein theory of diffusion. The v_{analyte} and the electrophoretic mobility of the analyte ($\overline{\mu}_{\text{analyte}}$), according to Smoluchowski's theory are linearly related by $v_{\text{analyte}} = E\,\overline{\mu}_{\text{analyte}}$. Thus, a linear relationship between v_{analyte} (from Eq. (10.5.4)) and V is expected, which was seen at higher applied voltages. A deviation from this expected linear behavior was seen at lower voltages. Similarly, the D (Eq. (10.5.4)) and V showed no clear relationship at lower voltages whereas at high voltages they shared a quadratic relationship. This behavior was explored with regards to the Taylor-dispersion effect due to EO, where an increase in the effective diffusion constant (from D to D_0) was seen due to a combination of radial diffusion and parabolic flow profiles [118, 119]. The D_0 calculated at high voltages was found to be in excellent agreement with the value obtained through the Stokes-Einstein theory of diffusion. Thus, we would urge readers to be careful when implementing Eq. 10.5.4 and check the validity of the resulting parameters by carrying out a vigilant study equivalent to *Ling et al.*

10.5.9 Other Models for t_d

a $t_d = C_p \frac{\eta L_m}{qV}$ where C_p, L_m and q are a proportionality constant, length of the molecule and the linear charge density of the molecule, respectively [114]. Since η depends on temperature, the above equation can be used to explain the behavior of t_d with temperature as well.

b $t_d = \frac{\eta l_{eff}^2}{(V+C_1)\zeta_p \epsilon_r \epsilon_0}$ where C_1 is the integration constant [30]. This is applicable for devices that satisfy $\kappa a \gg 1$ which allows the simplification of the Henry's function $(f(\kappa a))$ to derive the equation for t_d [30].

c $f(t) = \sum_{i=1}^{n} A_i \exp(-t/\tau_i)$ where A_i and, τ_i, are the pre-exponential constant and the time constant, respectively. In this method, the tail of the histogram corresponding to translocation time is fitted [120].

d $f(t) = \exp(-t/2s^2)\left((\sum_{i=1}^{n} A_i \exp(-t/\tau_i))\delta\right)$ where s, and δ are standard deviation and step function to truncate values below the mode threshold defined with respect to the translocation time corresponding to the peak of the histogram of the translocation time distribution. The peak of this fit is then taken as the corresponding t_d of the population [29].

e $f(t) = \frac{A}{t} \exp\left[-\left(\frac{ln(t)-\mu_t}{\sqrt{2}s}\right)^2\right]$ where all parameters bear the same definitions as above with μ_t being the mean of the log translocation time distribution. In this case, the log of translocation time is fitted with this equation and the antilog of μ_t is treated as the corresponding t_d of the population [29, 121].

10.5.10 Machine Learning

Although it is customary to characterize an event using a single t_d and ΔI_B, this, in some sense, overlooks the underlying information of the event and potentially undermines the prowess of the technology. With advancements in computing power, signal processing, and software development, we see more and more studies that incorporate machine learning (ML) and deep neural networks (DNN). ML approaches would play a key role in sequencing efforts where accurate assignment of current modulations to the correct nucleotide base is imperative for minimizing false positive and false negatives (and increase true negatives and true positives). Solid-state nanopores are yet to reach the sequencing advancement their biological counterparts have attained. ML approaches with solid-state nanopores are mostly seen with virus and bacterial studies where the following references would provide the interested reader with an understanding of the advancement and progress in ML implementations in solid-state nanopore technology [122–125]. More recently, *Karawdeniya et al.* showed the possibility of using DNN based on ResNet50 and solid-state nanopore-based electrodeformation to distinguish and discriminate between adeno associated viruses (AAVs) based on their cargo content (i.e., empty, ssDNA, and dsDNA) [23].

10.6 Noise in Solid-State Nanopores

The noise associated with nanopores can be segmented into four regimes based on origin and frequency: (i) $1/f$ noise ($S_{1/f}$, typically <100 Hz), (ii) white noise (S_w, typically 0.1–10 kHz), (iii) dielectric noise (S_d) and (iv) amplifier noise (S_a) [126]. The noise is typically evaluated through a power spectral density (PSD) plot. The next question is, how can one construct a PSD plot? The open pore current at a chosen operational condition is acquired and a few seconds-long segment is chosen for Fast-Fourier-Transform, which converts the time-domain signal into a frequency-domain signal. In-built functions such as fft of MATLAB can be easily used for this purpose. However, one must be mindful of the trace length and computational power available as an unnecessarily long trace would cause the program to crash or take an extremely long time to process. Typically, we try to keep the trace length <10 seconds long and evaluate multiple traces for statistical significance. We would discuss $S_{1/f}$ in somewhat detail and provide a concise overview of the other noise sources.

10.6.1 1/f Noise

Although the exact origin of this noise type is unclear, several mechanisms including fabrication and characterization conditions [127, 128], and structural and chemical properties [115, 129, 130] have been proposed. In this section, we will focus on the noise properties of pores fabricated using CT-CDB through Si_xN_y membranes that are nominally ~12 nm thick, as a model system. CT-CDB method is a compelling new method for pore fabrication as it offers the typical advantages of CDB (e.g., cost and time efficiency) in addition to other features (e.g. high temporal stability of the open-pore current, resilience against irreversible pore-clogging) that improves and address legacy issues that were once acceptable work practices in nanopore technology. The low-frequency noise can be expressed as:

$$S_{\mathrm{low,f}} = S_w + \frac{V^2}{f R_{\mathrm{total}}^4}\left[\frac{\alpha_{\mathrm{b}} R_{\mathrm{access}}^2}{N_{\mathrm{access}}} + \frac{\alpha_{\mathrm{b}} R_{\mathrm{pore}}^4}{R_{\mathrm{cyl}}^2 N_{\mathrm{cyl}}} + \frac{\alpha_{\mathrm{s}} R_{\mathrm{pore}}^4}{R_{\mathrm{surface}}^2 N_{\mathrm{surface}}}\right] \tag{10.6.1}$$

where N_{x}, R_{x}, α_{b} and α_{s} are the number of charge carriers corresponding to each ionic resistor component, resistive components of the nanopore sensor (e.g., surface, geometric, and access), and Hooge parameter for the bulk and surface $1/f$ noise, respectively [131, 132]. The definition of N_x is as follows: $N_{\text{cyl}} = \pi C N_{\text{A}} l \frac{d_{\text{pore}}^2}{4}$ (N_{A} is Avagadro's Number); $N_{\text{access}} = \pi C N_{\text{A}} \frac{d_{\text{pore}}^3}{6}$ and $N_{\text{surface}} = \pi \sigma l \frac{d_{\text{pore}}}{e}$. To understand the definitions of R_{x}, one has to look at the conductance model comprised of geometric, surface and access resistance contributions to the overall nanopore resistance (Sect. 10.4.1):

$$G = K\left(\frac{1}{\frac{\pi r^2}{l} + \frac{\mu|\sigma|}{K}\frac{2\pi r}{l}} + \frac{2}{\alpha_a 2r + \beta_a \frac{\mu|\sigma|}{K}}\right)^{-1} = \left(\frac{1}{\frac{1}{R_{\text{cyl}}} + \frac{1}{R_{\text{surface}}}} + R_{\text{access}}\right)^{-1}$$

or,

$$G = \left(R_{\text{pore}} + R_{\text{access}}\right)^{-1} \tag{10.6.2}$$

where R_{pore} is the nanopore resistance. Thus, R_{pore} in Eq. (10.6.1) and Eq. (10.6.2) are defined as the parallel addition of R_{cyl} and R_{surface} (i.e., $R_{\text{pore}} = \left(\frac{1}{R_{\text{cyl}}} + \frac{1}{R_{\text{surface}}}\right)^{-1}$). Thus, $R_{\text{total}} = R_{\text{pore}} + R_{\text{access}}$. Alternative expressions for $S_{\text{low,f}}$ exist too. For example, *Fragasso et al.* used a Lorentzian shape noise component for fitting purposes although this was most prominent with smaller diameter nanopores [133]. A sample PSD plot is shown in Fig. 10.4a that segregates $S_{1/\text{f}}$ and S_{w} by fitting it with Eq. (10.6.1).

10.6.2 *Effect of Solution pH*

Looking at Eqs. (10.6.1) and (10.6.2), the noise should vary with solution pH since σ is a function of the pH (Eqs. (10.2.6), (10.2.8), and (10.2.9)). From the pH-G response of CT-CDB pores (Fig. 10.2a), it is evident that at pH values $< \sim 8$, the G plateaus ($\sigma \sim 0$). Equation (10.6.2) shows that when $\sigma \to 0$, $R_{\text{surface}} \to \infty$, it reduces to $G = (R_{\text{cyl}} + R_{\text{access}})^{-1}$. Similarly, Eq. (10.6.1) reduces to:

$$S_{\text{low,f},\sigma\to 0} = S_{\text{w}} + \frac{V^2}{f R_{\text{total}}^4}\left[\frac{\alpha_{\text{b}} R_{\text{access}}^2}{N_{\text{access}}} + \frac{\alpha_{\text{b}} R_{\text{pore}}^4}{R_{\text{cyl}}^2 N_{\text{cyl}}}\right] \tag{10.6.3}$$

Thus, noise should be independent of pH at values where G plateaus, as seen in Fig. 10.4b. When G increases with pH, the noise also starts to increase and follows Eq. (10.6.1).

10.6.3 Effect of Applied Voltage

In nanopore experiments, the R.M.S. value of open-pore current (I_{rms}) increases with increasing applied voltage. To understand this behavior, let us eliminate the contribution from surface charges and consider Eq. (10.6.3) (the simplified form of Eq. (10.6.1)). Assuming d_{pore}, l, N_{A}, and α_{b} are constant for a given nanopore, it can be shown that:

$$S_{\mathrm{low,f},\sigma\to 0} = S_{\mathrm{w}} + \frac{K^2V^2}{fC}\Phi_{\sigma\to 0} \tag{10.6.4}$$

where $\Phi_{\sigma\to 0} = \frac{1}{\left(\frac{1}{\frac{\pi d_{\mathrm{pore}}^2}{4l}} + \frac{2}{\alpha\, d_{\mathrm{pore}}}\right)^4}\left[\frac{\alpha_b\left(\frac{2}{\alpha d_{\mathrm{pore}}}\right)^2}{\pi N_{\mathrm{A}}\frac{d_{\mathrm{pore}}^3}{6}} + \frac{\alpha_{\mathrm{b}}\left(\frac{1}{\frac{\pi d_{\mathrm{pore}}^2}{4l}}\right)^2}{\pi N_{\mathrm{A}} l \frac{d_{\mathrm{pore}}^2}{4}}\right]^2$. Now we see that $S_{\mathrm{low,f},\sigma\to 0}$ should increase with the square of the applied voltage. This is essentially the case seen in Fig. 10.4c, with the fit line corresponding to Eq. (10.6.4).

10.6.4 Effect of Electrolyte Concentration

Like applied voltage, an increase in the open-pore noise is observed with increasing electrolyte concentration, C, with some electrolytes performing better than others [134]. Equation (10.6.4) provides some proof for this observation. As seen, the $S_{low,f,\sigma\to 0}$ is proportional to K^2/C. As the electrolyte concentration increases, its conductivity would increase as well. Although with electrolytes such as KCl, K, and C vary linearly with analyte concentration at concentrations relevant to nanopore experiments, the same is not the case with LiCl. However, irrespective of the relationship between K and C, K^2/C does increase with increasing C. Thus, with increasing C it is not surprising to see $S_{1/f,1Hz}$ increasing as well (Fig. 10.4d). If one looks at Fig. 10.4d, the $S_{1/f,1Hz}$ cannot be fitted well with Eq. (10.6.4). However, when K^2 was replaced with K^{eff} (similar to an effective conductivity), the raw data points could be fitted well with the exponent (eff) being 4.3 ± 0.5. Although it is not clear why this is the case, this could be due to, for example, slight pore-to-pore variations in α_b, need for the refinement of N_{x} parameter estimates (N_{cyl}, N_{surface}, N_{access}) to better suit CT-CDB pores and variations in l (the manufacturer typically notes a ± 1–± 2 nm variation in thickness).

10.6.5 Effect of Pore Diameter

The $1/f$ noise is expected to decrease with increasing pore diameter, which according to *Fragasso et al.*, is weaker in pores <10 nm in diameter whereas it is more prominent at larger diameter pores [133]. Such a decrease in $1/f$ noise was attributed to two reasons: (i) pore-diameter dependent redistribution of the voltage across the access and pore regions with the former dominating in larger pores and the latter at smaller diameter pores, and (ii) surface contributions accounting for much of the $1/f$ noise than its bulk counterpart (evident by the few magnitudes larger α_s than α_b) with surface $1/f$ noise dominating in smaller diameter pores. In larger pores, the surface contributions are smaller and are dominated, as mentioned earlier, by access contributions (i.e., $1/f$ noise decrease). However, these results pertain to Si_xN_y pores fabricated by TEM. Pores fabricated by CT-CDB pores (or even CDB) on the other hand, exhibit an increase in $1/f$ noise with increasing pore diameter (Fig. 10.4e). This is counterintuitive based on the reasoning and equations provided so far. However, the equations do not take Joule heating into account, that occurs during breakdown-type voltage induced fabrication methods. Larger diameter pores would require higher voltages to fabricate, which would lead to higher Joule heating that may in turn lead to an increase in $1/f$ noise. While trying to make larger pores, higher voltage requirements could lead to unstable open-pore currents and multiple pore-formation that would affect the performance of the fabricated nanopores. Thus, CDB (and CT-CDB) is most suitable for the fabrication of <30 nm diameter pores (although successful nanopore fabrication up to ~50 nm diameter has been reported in literature).

10.6.6 Reducing Noise

Various approaches have been pursued to reduce noise components associated with nanopore measurements through strategies and measures such as:

a **Amplifier noise (S_a):** Although elementary, it is important to remember that the front switch of the Axopatch 200B (±200 mV) has less R.M.S. noise compared to the rear-switch (±1V) setting. Unless one requires > ±200 mV, it is best to operate in the front switch mode. The complementary metal-oxide semiconductor (CMOS) devices offer low noise with more compact designs such as *Nanopore Reader* (developed by Elements SRL) having low instrument footprint and portability.

b **Digitizer noise:** The analog signal of the amplifier must be digitized before it can be read by the software. We have observed a reduction in the R.M.S. noise from Digidata 1440A to Digidata 1550B (both are digitizers developed by Molecular Device, LLC).

c **$S_{1/f}$ and S_d reduction:** e.g., Polydimethylsiloxane (PDMS) coating ($S_{1/f}$ and S_d reduction) [135], Piranha treatment ($S_{1/f}$ reduction) [135] (should be handled with care and not suitable for chips susceptible to such harsh chemical treatments) and, Low-loss factor material (e.g., quartz, S_d reduction) [136].

d **Surface passivation:** e.g., Atomic layer deposition [127].

e **Vibration and electrical isolation:** Attention should be paid on using, for example, active vibration isolation table, Faraday cage, insulation of wires, and ded-

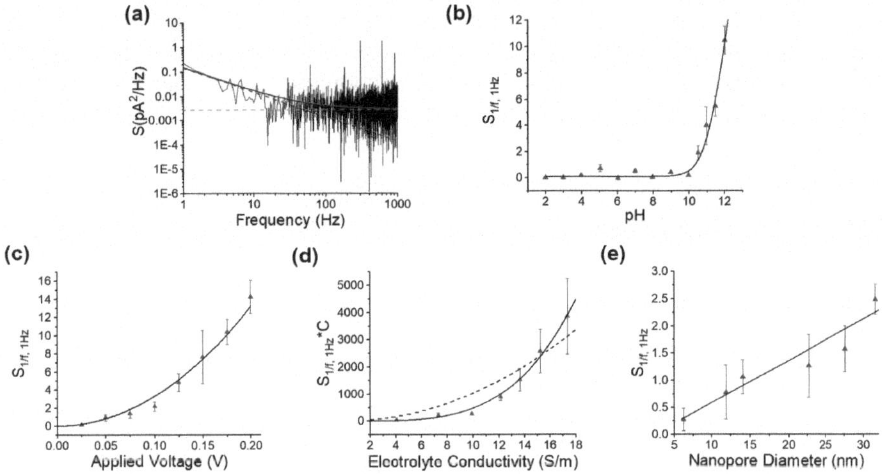

Fig. 10.4 (**a**) PSD with blue, magenta, and green fit lines corresponding to Eq. (10.6.1), $1/f$, and S_W components, respectively. $1/f$ noise at 1 Hz ($S_{1/f,1Hz}$) with (**b**) solution pH (the fit line from Eq. (10.6.1)), (**c**) applied voltage (the fit line from Eq. 10.6.4), (**d**) LiCl conductivity (solid and dashed fit lines from Eq. (10.6.4) and replacing K^2 of Eq. (10.6.4) with K^{eff} respectively) and (**e**) pore diameter (linear fit line as a guide to the eye). All traces were recorded at 10 kHz low pass filtering

icated power line. Although elementary facts, if not done properly, these could introduce noise features originating from surrounding vibrations, electrical noise, magnetic noise, electrical cross talk, and instrument cross talk. 60 Hz noise (50 Hz in some parts of the world) is easy to recognize—a tell-tale sign of poor electrical insulation—through the rhythmic wave-like noise pattern in the open-pore current. Something that is mostly overlooked is the high-frequency noise of computers—keep them as far away as possible or replace them with a low-noise modules.

The electrical performance of the most ubiquitous material used in solid-state nanopore technology, Si_xN_y, heavily hinges on the S_f and S_d noise characteristics. Unfortunately, the S_d of Si_xN_y is aggravated by its high loss factor. However, as explained above, methods are available to handle this issue without resorting to a change of material.

10.7 Improving Measurements

Ideally, one would require $t_d > 2T_r$ (see Sect. 10.5 for more details) and ΔI_B to be as deep as possible (higher SNR). Thus, it is desirable to slow down the translocation time and increase the blockage depth. Slowing down the translocation time typically comes at the expense of overall throughput (exceptions exist such as the salt-gradient method [24]). Thus, it is required to strike a balance between the throughput of the sensor and t_d. Ideally, we would recommend collecting 1000 events for statistical analysis (one should be mindful of subtle baseline variations that could sometimes give rise to false events). However, it is not uncommon to see analysis done with $<$1000 events or the number of events not been reported at all.

10.7.1 Slowing Down t_d

Generally used strategies are:

a Use of electrolytes that reduce the overall charge of the analyte, such as LiCl (and even NaCl to some extent) in DNA sensing experiments. Care must be exercised when using multivalent cations as they could crosslink analytes and/or attach analyte to the pore surface [54]. In case of proteins, as mentioned previously, the Hofmeister series positioning of the ions in the electrolyte must be inspected as well.

b Decreasing operational temperature [137]. There must be sufficient energy for diffusion and drift to overcome entropic (for long polymers) and other energetic barriers for translocation to be feasible.

c Increasing the viscosity of the electrolyte by adding additives such as glycerol [114].

d Tuning the electrolyte concentration. For example, in the case of DNA, increasing LiCl and, NaCl concentration is found to increase the translocation time with the added benefit of increasing ΔI_B [54].

e Ionic liquids: care must be taken to avoid any unnecessary analyte-electrolyte interactions in ionic liquids [138].

f Application of pressure gradient (typically configured to have opposing voltage and pressure forces) [139].

g Reducing the applied voltage (must be high enough to overcome the entropic and other energy barriers) [140].

h Salt gradient (typically decrease t_d while increasing the capture rate) [24].

i Organo-surface chemical coating to increase electrostatic interactions [29, 39].

j pH modulation to manipulate analyte and surface charge (care should be taken not to denature, or unfavorably cause pH-induced conformational changes) [39].

k Tuning of pore diameter and length—smaller diameters and longer lengths would typically increase t_d yet promote analyte clogging [141].

l Analyte tagging: Keyser's group have demonstrated the use of engineered DNA strands with protein binding sites which generate a knot-like signal with the deeper and sharper event structure corresponding to the protein-DNA complex [142, 143]. With such techniques, small proteins that would otherwise defy the electronic limitations of the amplifier could be detected.

m Optical trapping: care must be exercised to ensure that the laser power is low enough to minimize any structural changes of the molecule [144]. More details can be found in Chaps. 7 and 8.

n Topological modifications (e.g. electrospinning of nanofiber mesh [145], membrane metallization [146]).
o Pore material: *Larkin et al.* showed that strong interactions of the DNA phosphate backbone with HfO_2 nanopores yielded a reduction in the translocation speed compared to Si_xN_y nanopores, with the added advantage of admitting DNA at a lower operational voltage [69].

Although these strategies would decrease t_d, many of them would reduce the capture rate as well (lower throughput). One important factor to note is the distribution of t_d: the narrower the better. However, transport distribution depends on a host of factors including the initial conformation of the molecule (correlated to viscous-drag-induced velocity fluctuations) [147], non-transport collisions with the pore opening, access trajectory [148], transient conformational changes (e.g. protein unfolding), pore surface-analyte interactions (including kinetics and location [104]), interactions with extra-pore and intra-pore segments (e.g. DNA) [117], and thermal fluctuations. Since most nanopore technologies are developed with a focus on DNA sequencing, it is imperative to minimize the distribution width of t_d. For this, strategies such as entropic reduction through preconfinement [149], molecular stepper motors (biological nanopores) [150], and reducing the pore diameter for smooth transport [117] have been used.

10.7.2 Increasing ΔI_B

Unlike t_d, which could be manipulated through a multitude of approaches, tuning of ΔI_B is not associated with the same multitude of options. Guide inserted nanopores [151], smaller diameter pores [117], ultra-thin membranes [152], reducing R.M.S. noise of open-pore current [66, 136], increasing electrolyte concentration (increases $1/f$ noise) [31, 54], increasing the applied voltage (increases t_d and $1/f$ noise) [114] are some of the available options.

10.7.3 Increasing Analyte Detection Sensitivity/Throughput

Here we discuss some common approaches:

a **Salt Gradient:** By having a low electrolyte concentration in the *cis* side compared to the *trans* side, *Wanunu et al.* were able to reach sub-10 pM DNA concentrations while maintaining an appreciable capture rate (~1 event/s) [24]. The increase in capture rate was attributed to the effective polarization of the pore entrance due to transport of K^+ ions (electrolyte was KCl) along both the chemical and electrical potential gradients. Interestingly, the increase in capture did not induce fast DNA translocation, but the opposite. Similar experiments have been done with LiCl gradients as well [153, 154]. Note that I-V response, unlike with the usual symmetric electrolyte concentration case, would show a positive offset (current at zero applied voltage) for $C_{trans}/C_{cis} > 1$ and vise-versa for $C_{trans}/C_{cis} < 1$. The offset is also known as resting voltage (V_{rest}) and has a Nernstian relationship (slope ~60 mV) with $\log(C_{trans}/C_{cis})$.

b **Pore Fabrication Method:** Our work with Si_xN_y nanopores fabricated through CT-CDB showed a ~6.5× higher sensitivity to DNA (at pH ~ 7) compared to pores fabricated from the conventional CDB process [37]. This was attributed to the negligible EOF acting against the EPF in the CT-CDB compared to the CDB fabricated Si_xN_y pores. Our work with AAVs indicated negligible capture rates for Si_xN_y pores fabricated through FIB, whereas those fabricated from the TEM allowed statistically significant event capturing across five voltages (−50 mV to −125 mV) under ~30 min [23]. The capture rate difference was thought to be due to differences in surface chemistry during fabrication, as one used heavy ions (FIB) and the other used electrons (TEM) [27].

c **Dielectrophoretic Trapping:** One of the core problems in nanopore sensing is the disproportionate accessibility of molecules in the capture zone compared to the bulk stemming from the requirement of molecules to diffuse from the bulk to the capture zone of the nanopore. It has been shown that the capture zone has ~108× fewer molecules compared to the bulk. That is, for a 10 μL solution of ~1 nM concentration, there would merely be ~34 molecules inside a 3 μm capture radius [155]. The problem is further aggravated at pM concentrations where the average drops <1 molecule rendering the throughput to decrease drastically. As a solution, *Freedman et al.* showed, detection of DNA as low as 5 fM (at a capture rate of 315±147 molecules/minute) can be reached by applying an AC field (Au electrodes) for trapping followed by a DC field (Ag/AgCl electrodes) for sensing [155]. Other intricate details can be found through further reading of that paper. For more discussion on dielectrophoretic trapping, see Chap. 6.

d **Flow Rate:** This is applicable for nanopore devices integrated with a microfluidic flow channel. A good example of this is the very recent work of *Sohi et al.* where the application of a tangential flow over the nanopore increased the capture rate at moderate flow rates (~5×) although it subsequently dropped at higher flow rates [156].

e **Pressure-Electric Configuration:** With the finite temporal resolution of patch-clamp systems, it becomes challenging to detect small proteins (and even small DNA strands). *Li et al.* demonstrated, by applying an electric field that opposes the pressure-driven transport of proteins through a nanopore, proteins that may be invisible in electric-only configurations can eventually be detected [157]. One could argue that higher bandwidths could resolve this issue, but it comes at a high R.M.S. open-pore current noise.

f **Electric and Physical Tuning:** Increasing the applied voltage would increase the capture rate. This would increase t_d (can get bandwidth limited) and the open-pore noise. The pore radius can be enlarged to increase the capture rate. This would allow molecules to translocate with a multitude of configurations broadening both the t_d and ΔI_B distributions.

g **Surface Modifications**

(a) Increasing the hydrophilicity: Piranha treatment [158] (some pore types could etch or degrade), TiO_2 deposition (become super hydrophilic after UV treatment) [159, 160], and air/oxygen plasma treatment [161] are some of the common approaches.

(b) Chemical Attachments: If the molecules cannot be slowed down sufficiently, one way would be to modify the nanopore surface with a molecular layer that would bind specifically with the analyte. For a protein, this could be its receptor. Then, by measuring the open-pore current before and after equilibrating with the analyte and removing any non-specifically bound molecules (e.g., washing, subtle voltage/magnetic application), the ΔI_B would be characteristic of the analyte length and charge. *Chuah et al.* achieved a 0.8 fM detection limit for prostate-specific antigen using a similar method [162].

10.7.4 Multiple Recapturing

Conventionally, as discussed so far, translocations are single-capture events. That is, once they enter the *trans* side through the nanopore, they would eventually escape the sensing zone of the nanopore. However, multiple recapturing allows the same molecule to be analyzed multiple times through appropriate voltage reversal after detecting a translocation event. *Gershow* and *Golovchenko* demonstrated the recapture of 4–6 kbp dsDNA through a Si_xN_y nanopore [163]. With multiple recapturing, statistically significant data of a given particle can be obtained (something the conventional single-pass capture fails to attain), and by extending it to >1000 molecules, a statistically significant data pool representative of the sample can be obtained. In brief, after detecting a resistive pulse, the system automatically reverses the voltage polarity after a preset value of delay time, T_{delay} (Fig. 10.5). The T_{delay} is set so as to retain the analyte within the capture radius of the nanopore. To avoid data acquisition in the short-lived capacitive spike region as a result of voltage reversal, the pulse scanning is kept on hold for a preset period (T_{skip}) and then scanned for a period of $T_{\text{recapture}}$, until the analyte gets recaptured. The $T_{\text{recapture}}$ is not a preset value, but the time taken for the given analyte to be recaptured. The protocol can be operated at (i) symmetric voltage conditions (i.e. $V_{\text{Capture}} = V_{\text{Recapture}}$) or (ii) asymmetric voltage ($V_{\text{Capture}} \neq V_{\text{Recapture}}$) conditions. One must be mindful of the fact that if either V_{Capture} or $V_{\text{Recapture}}$ is too large, the analyte will escape the capture radius of the nanopore. A key advantage of asymmetric voltage is that it has the potential to measure the relaxation dynamics of analytes that may undergo electro-deformation or other voltage-induced structural changes by manipulating V_{Capture}, $V_{\text{Recapture}}$, and T_{delay}. More information on recapturing can be found elsewhere [12, 13].

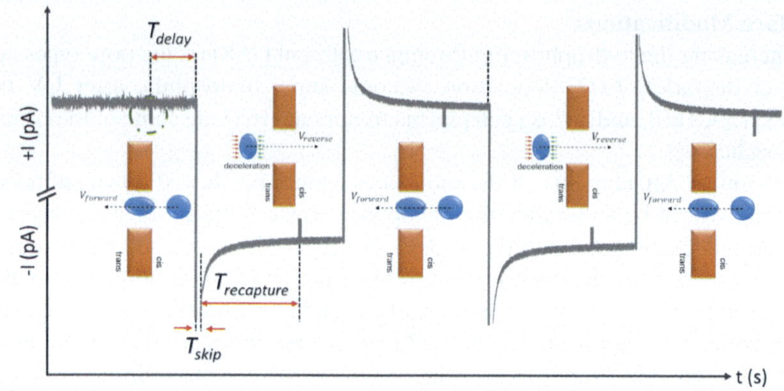

Fig. 10.5 Schematic of the multiple recapture protocol where upon the detection of a resistive pulse (green circle), the voltage is reversed after a preset time (T_{delay}). The data acquisition is kept hold for a preset time (T_{skip}) and then scanned for a period of $T_{\text{recapture}}$, until the protein gets recaptured

10.7.5 Electrode Maintenance

The most common electrode used is the Ag/AgCl electrode. It is an exhaustible electrode, yet it is reversible and has minimal polarization and predictable junction potential. The electrode reaction can be written as $AgCl + e \rightleftharpoons Ag + Cl^{-}$. However, when the AgCl is depleted, the bare Ag wire would be exposed which could cause silver ions to leak into the solution. The silver ions can react with biomolecules and alter their structure and function. If the bare Ag is exposed (electrode would look whitish), a freshly prepared Ag/AgCl electrode should be used. If the same bare Ag is required to be reused, it should be sanded down until it becomes glossy and then dipped in a bleach solution, preferably overnight. A proper Ag/AgCl electrode would be black in color. Additionally, one must be mindful of the composition of the buffer as well.

10.7.6 Analyte Concentration

Although it might seem a good idea to add as much analyte as possible because the capture rate is proportional to it, a higher capture rate would induce irreversible pore-clogging as well. As a rule of thumb, we aim to collect a minimum of 1000 events in 15 min. If a lower analyte concentration could achieve this, it would have the added benefit of minimal sample usage as well.

10.7.7 Optimizing Lowpass Filter Settings

See detailed discussed in Sects. 10.3 and 10.5 above.

10.8 Conclusions

In this chapter, we broadly discussed the transport phenomena and their modeling, experiment planning, electrical signal measurements, nanopore characterization, noise and, improvements in measurements corresponding to nanopore single molecule sensing. Moreover, the analyte translocation methods (e.g. EO,EP and diffusion), mechanisms (e.g. diffusion and barrier limited) and the impact of analyte chemistry, electrolyte chemistry, nanopore chemistry and electronic settings (e.g. applied voltage) on these are discussed in Sect. 10.2. The same conditions would also dictate the translocation characteristics such as ΔI_B and t_d and should be carefully optimized or chose as discussed in Sects. 10.3 and 10.5. Section 10.4 discuss the strategies ubiquitously used for physical and chemical nanopore characterization. For nanopore sensing to be successful, low noise is key and Sect. 10.6 discuss the noise properties associated with nanopores along with strategies to improve the inherent noise. Finally, further methods to improve sensing such as increasing t_d, ΔI_B (for higher SNR) and throughput along with vital maintenance tips are provided in Sect. 10.7. Broadly, this chapter discuss the key aspects of nanopore sensing essential to properly setup a nanopore experiment, characterization metrics, challenges associated with nanopores, strategies developed to overcome them.

10.9 Acronym Glossary

α_s: surface Hooge parameter
α_b: bulk Hooge parameter
β: inverse of thermal energy
Γ: number of surface chargeable groups
$\epsilon_r \epsilon_0$: permittivity of the solution
ζ_p: zeta potential of the analyte
ζ_w: zeta potential of the nanopore wall
η: dynamic viscosity of the electrolyte
κ^{-1}: Debye screening length
μ: electrophoretic mobility
σ: surface charge density
ϕ_{pore}: diffuse layer potential
C: analyte concentration
C_{eff}: effective stern layer capacitance
C_s: stern layer capacitance
D: analyte diffusion coefficient
e: elementary charge
E: electric field
f^*: frequency factor
f_c: cut-off frequency of the filter
G: open pore conductance
G_{bulk}: bulk conductance
G_{access}: access conductance
$G_{surface}$: surface conductance
I_{rms}: root mean square of open-pore current

J: analyte flux rate
K: electrolyte conductivity
K_a: acid dissociation constant
K_b: base dissociation constant
k_B: Boltzmann constant
l: pore length
r: pore radius
r^*: capture radius
R_c^{diff}: diffusion-limited capture rate
R_c^{bar}: barrier-limited capture rate
R_0: zero-voltage capture rate
$S_{1/f}$: $1/f$ noise
S_w: white noise
S_d: dielectric noise
S_a: amplifier noise
T: temperature
T_r: rise time
t_d: translocation/dwell time of the resistive pulse
U^*: activation energy
V: applied voltage
V_{cross}: cross-over voltage
W: lambert W function
pI: isoelectric point
ADEPT: adaptive time-series analysis
CDB: controlled dielectric breakdown
CMOS: complementary metal-oxide-semiconductor
CT-CDB: chemically tuned CDB
DBC: differential-based calibration
dsDNA: double-stranded DNA
EDL: electrical double layer
EO: electroosmosis
EOF: electrosomostic force
EP: electrophoresis
EPF: electrophoretic force
FIB: focused ion beam
FWHM: full width at the half maximum
h-BN: hexagonal boron nitride
HIM: helium ion microscope
ΔI_B: current drop of the resistive pulse
ICR: ion current rectification
LACB: laser-assisted controlled breakdown
LPCVD: low-pressure chemical vapor deposition
MPVI: multi-level pulsed voltage injection
PC: polycarbonate
PECVD: plasma-enhanced chemical vapor deposition
PET: polyethylene terephthalate
PSD: power spectral density
PVD: physical vapor deposition
R.M.S: root mean square
RNA: ribonucleic acid
SAXS: small angle X-ray scattering
SEM: scanning electron microscope
SNR: signal to noise ratio
Si_xN_y: silicon nitride
SiO_2: silicon dioxide
TCAM: tesla-coil assisted method
TEM: transmission electron microscope

References

1. Min Jun Kim, Meni Wanunu, David C Bell, and Amit Meller. Rapid fabrication of uniformly sized nanopores and nanopore arrays for parallel dna analysis. *Advanced Materials*, 18(23):3149–3153, 2006.
2. Kevin J Freedman, Chi Won Ahn, and Min Jun Kim. Detection of long and short dna using nanopores with graphitic polyhedral edges. *ACS Nano*, 7(6):5008–5016, 2013.
3. Amir Barati Farimani, Kyoungmin Min, and Narayana R Aluru. Dna base detection using a single-layer mos2. *ACS Nano*, 8(8):7914–7922, 2014.

4. Jiandong Feng, Ke Liu, Roman D Bulushev, Sergey Khlybov, Dumitru Dumcenco, Andras Kis, and Aleksandra Radenovic. Identification of single nucleotides in mos2 nanopores. *Nature Nanotechnology*, 10(12):1070, 2015.
5. Jugal Saharia, YM Nuwan DY Bandara, Jung Soo Lee, Qingxiao Wang, Moon J Kim, and Min Jun Kim. Fabrication of hexagonal boron nitride based 2d nanopore sensor for the assessment of electro-chemical responsiveness of human serum transferrin protein. *Electrophoresis*, 41(7-8):630–637, 2020.
6. Jugal Saharia, YM Nuwan DY Bandara, Gaurav Goyal, Jung Soo Lee, Buddini Iroshika Karawdeniya, and Min Jun Kim. Molecular-level profiling of human serum transferrin protein through assessment of nanopore-based electrical and chemical responsiveness. *ACS Nano*, 13(4):4246–4254, 2019.
7. Y. M. Nuwan D.Y. Bandara, Jiannan Tang, Jugal Saharia, Louis William Rogowski, Chi Won Ahn, and Min Jun Kim. Characterization of flagellar filaments and flagellin through optical microscopy and label-free nanopore responsiveness. *Analytical Chemistry*, 91(21):13665–13674, 2019.
8. Kevin J Freedman, Maike Jürgens, Anmiv Prabhu, Chi Won Ahn, Per Jemth, Joshua B Edel, and Min Jun Kim. Chemical, thermal, and electric field induced unfolding of single protein molecules studied using nanopores. *Analytical Chemistry*, 83(13):5137–5144, 2011.
9. Buddini Iroshika Karawdeniya, Y. M. Nuwan D. Y. Bandara, Jonathan W. Nichols, Robert B. Chevalier, and Jason R. Dwyer. Surveying silicon nitride nanopores for glycomics and heparin quality assurance. *Nature Communications*, 9(1):3278, 2018.
10. James T Hagan, Brian S Sheetz, YM Nuwan DY Bandara, Buddini I Karawdeniya, Melissa A Morris, Robert B Chevalier, and Jason R Dwyer. Chemically tailoring nanopores for single-molecule sensing and glycomics. *Analytical and Bioanalytical Chemistry*, 6:10.
11. Gaurav Goyal, Armin Darvish, and Min Jun Kim. Use of solid-state nanopores for sensing co-translocational deformation of nano-liposomes. *Analyst*, 140(14):4865–4873, 2015.
12. Jung Soo Lee, Jugal Saharia, YM Nuwan DY Bandara, Buddini Iroshika Karawdeniya, Gaurav Goyal, Armin Darvish, Qingxiao Wang, Moon J Kim, and Min Jun Kim. Stiffness measurement of nanosized liposomes using solid-state nanopore sensor with automated recapturing platform. *Electrophoresis*, 40(9):1337–1344, 2019.
13. Armin Darvish, Jung Soo Lee, Bin Peng, Jugal Saharia, Ramalingam VenkatKalyana Sundaram, Gaurav Goyal, Nuwan Bandara, Chi Won Ahn, Jungsuk Kim, and Prashanta Dutta. Mechanical characterization of hiv-1 with a solid-state nanopore sensor. *Electrophoresis*, 40(5):776–783, 2019.
14. Kaimeng Zhou, Lichun Li, Zhenning Tan, Adam Zlotnick, and Stephen C Jacobson. Characterization of hepatitis b virus capsids by resistive-pulse sensing. *Journal of the American Chemical Society*, 133(6):1618–1621, 2011.
15. Angus McMullen, Hendrick W De Haan, Jay X Tang, and Derek Stein. Stiff filamentous virus translocations through solid-state nanopores. *Nature Communications*, 5:4171, 2014.
16. Gaurav Goyal, Kevin J Freedman, and Min Jun Kim. Gold nanoparticle translocation dynamics and electrical detection of single particle diffusion using solid-state nanopores. *Analytical Chemistry*, 85(17):8180–8187, 2013.
17. Joseph WF Robertson, Claudio G Rodrigues, Vincent M Stanford, Kenneth A Rubinson, Oleg V Krasilnikov, and John J Kasianowicz. Single-molecule mass spectrometry in solution using a solitary nanopore. *Proceedings of the National Academy of Sciences*, 104(20):8207–8211, 2007.
18. John J Kasianowicz, Eric Brandin, Daniel Branton, and David W Deamer. Characterization of individual polynucleotide molecules using a membrane channel. *Proceedings of the National Academy of Sciences*, 93(24):13770–13773, 1996.
19. Jeff Nivala, Douglas B Marks, and Mark Akeson. Unfoldase-mediated protein translocation through an α-hemolysin nanopore. *Nature Biotechnology*, 31(3):247–250, mar 2013.
20. Aziz Fennouri, Régis Daniel, Manuela Pastoriza-Gallego, Loïc Auvray, Juan Pelta, and Laurent Bacri. Kinetics of Enzymatic Degradation of High Molecular Weight Polysaccharides through a Nanopore: Experiments and Data-Modeling. *Analytical Chemistry*, 85(18):8488–8492, sep 2013.

21. Thomas Hoenen, Allison Groseth, Kyle Rosenke, Robert J Fischer, Andreas Hoenen, Seth D Judson, Cynthia Martellaro, Darryl Falzarano, Andrea Marzi, and R Burke Squires. Nanopore sequencing as a rapidly deployable ebola outbreak tool. *Emerging Infectious Diseases*, 22(2):331, 2016.
22. Harold Kwok, Kyle Briggs, and Vincent Tabard-Cossa. Nanopore fabrication by controlled dielectric breakdown. *PloS one*, 9(3):e92880, 2014.
23. Buddini Iroshika Karawdeniya, Y. M. Nuwan D. Y. Bandara, Aminul Islam Khan, Wei Tong Chen, Hoang-Anh Vu, Adnan Morshed, Junghae Suh, Prashanta Dutta, and Min Jun Kim. Adeno-associated virus characterization for cargo discrimination through nanopore responsiveness. *Nanoscale*, 2020.
24. Meni Wanunu, Will Morrison, Yitzhak Rabin, Alexander Y Grosberg, and Amit Meller. Electrostatic focusing of unlabelled dna into nanoscale pores using a salt gradient. *Nature Nanotechnology*, 5(2):160, 2010.
25. Martin Charron, Kyle Briggs, Simon King, Matthew Waugh, and Vincent Tabard-Cossa. Precise dna concentration measurements with nanopores by controlled counting. *Analytical Chemistry*, 91(19):12228–12237, 2019.
26. Alexander Y Grosberg and Yitzhak Rabin. Dna capture into a nanopore: interplay of diffusion and electrohydrodynamics. *The Journal of chemical physics*, 133(16):10B617, 2010.
27. Benjamin Cressiot, Abdelghani Oukhaled, Gilles Patriarche, Manuela Pastoriza-Gallego, Jean-Michel Betton, Loïc Auvray, Murugappan Muthukumar, Laurent Bacri, and Juan Pelta. Protein transport through a narrow solid-state nanopore at high voltage: experiments and theory. *ACS Nano*, 6(7):6236–6243, 2012.
28. Abdelghani Oukhaled, Benjamin Cressiot, Laurent Bacri, Manuela Pastoriza-Gallego, Jean-Michel Betton, Eric Bourhis, Ralf Jede, Jacques Gierak, Loïc Auvray, and Juan Pelta. Dynamics of Completely Unfolded and Native Proteins through Solid-State Nanopores as a Function of Electric Driving Force. *ACS Nano*, 5(5):3628–3638, may 2011.
29. YM Nuwan DY Bandara, Buddini I Karawdeniya, James T Hagan, Robert B Chevalier, and Jason R Dwyer. Chemically functionalizing controlled dielectric breakdown silicon nitride nanopores by direct photohydrosilylation. *ACS Applied Materials Interfaces*, 11(33):30411–30420, 2019.
30. Nima Arjmandi, Willem Van Roy, Liesbet Lagae, and Gustaaf Borghs. Measuring the electric charge and zeta potential of nanometer-sized objects using pyramidal-shaped nanopores. *Analytical Chemistry*, 84(20):8490–8496, 2012.
31. Ralph MM Smeets, Ulrich F Keyser, Diego Krapf, Meng-Yue Wu, Nynke H Dekker, and Cees Dekker. Salt dependence of ion transport and dna translocation through solid-state nanopores. *Nano Letters*, 6(1):89–95, 2006.
32. Cameron M Frament, Nuwan Bandara, and Jason R Dwyer. Nanopore surface coating delivers nanopore size and shape through conductance-based sizing. *ACS Applied Materials & Interfaces*, 5(19):9330–9337, 2013.
33. George Karniadakis, Ali Beskok, and Narayan Aluru. *Microflows and nanoflows: fundamentals and simulation*, volume 29. Springer Science & Business Media, 2006.
34. J. Lyklema, S. Rovillard, and J. de Coninck. Electrokinetics: the properties of the stagnant layer unraveled. *Langmuir*, 14:5659–5663, 1998.
35. Patrick Abgrall and Nam Trung Nguyen. Nanofluidic devices and their applications. *Analytical Chemistry*, 80(7):2326–2341, 2008.
36. Sven H Behrens and David G Grier. The charge of glass and silica surfaces. *The Journal of Chemical Physics*, 115(14):6716–6721, 2001.
37. Y. M. Nuwan D. Y. Bandara, Jugal Saharia, Buddini I. Karawdeniya, James T. Hagan, Jason R. Dwyer, and Min Jun Kim. Beyond nanopore sizing: improving solid-state single-molecule sensing performance, lifetime, and analyte scope for omics by targeting surface chemistry during fabrication. *Nanotechnology*, 31(33):335707, 2020.
38. D. P. Hoogerheide, S. Garaj, and J. A. Golovchenko. Probing surface charge fluctuations with solid-state nanopores. *Physical Review Letters*, 102(25):256804, 2009.

39. Brett N Anderson, Murugappan Muthukumar, and Amit Meller. ph tuning of dna translocation time through organically functionalized nanopores. *ACS Nano*, 7(2):1408–1414, 2012.
40. Meni Wanunu and Amit Meller. Chemically modified solid-state nanopores. *Nano Letters*, 7(6):1580–1585, 2007.
41. Olivia M Eggenberger, Cuifeng Ying, and Michael Mayer. Surface coatings for solid-state nanopores. *Nanoscale*, 11(42):19636–19657, 2019.
42. Eszter Mádai, Bartłomiej Matejczyk, András Dallos, Mónika Valiskó, and Dezső Boda. Controlling ion transport through nanopores: modeling transistor behavior. *Physical Chemistry Chemical Physics*, 20(37):24156–24167, 2018.
43. Robert J Hunter. *Zeta potential in colloid science: principles and applications*, volume 2. Academic press, 2013.
44. Haojie Yang, Zhongwu Li, Wei Si, Kabin Lin, Jian Ma, Kun Li, Litao Sun, Jingjie Sha, and Yunfei Chen. Identification of single nucleotides by a tiny charged solid-state nanopore. *The Journal of Physical Chemistry B*, 122(32):7929–7935, 2018.
45. Niklas Ermann, Nikita Hanikel, Vivian Wang, Kaikai Chen, Nicole E Weckman, and Ulrich F Keyser. Promoting single-file dna translocations through nanopores using electro-osmotic flow. *The Journal of Chemical Physics*, 149(16):163311, 2018.
46. Yin Zhang, Jiabin Zhao, Wei Si, Yajing Kan, Zheng Xu, Jingjie Sha, and Yunfei Chen. Electroosmotic facilitated protein capture and transport through solid-state nanopores with diameter larger than length. *Small Methods*, page 1900893, 2020.
47. Gang Huang, Kherim Willems, Misha Soskine, Carsten Wloka, and Giovanni Maglia. Electro-osmotic capture and ionic discrimination of peptide and protein biomarkers with frac nanopores. *Nature Communications*, 8(1):1–11, 2017.
48. Stijn Van Dorp, Ulrich F Keyser, Nynke H Dekker, Cees Dekker, and Serge G Lemay. Origin of the electrophoretic force on dna in solid-state nanopores. *Nature Physics*, 5(5):347, 2009.
49. Reto B Schoch, Jongyoon Han, and Philippe Renaud. Transport phenomena in nanofluidics. *Reviews of Modern Physics*, 80(3):839, 2008.
50. James W Swan and Eric M Furst. A simpler expression for henry's function describing the electrophoretic mobility of spherical colloids. *Journal of Colloid and Interface Science*, 388(1):92–94, 2012.
51. M. Firnkes, D. Pedone, J. Knezevic, M. Doblinger, and U. Rant. Electrically facilitated translocations of proteins through silicon nitride nanopores: conjoint and competitive action of diffusion, electrophoresis, and electroosmosis. *Nano Letters*, 10(6):2162–2167, 2010.
52. Simon Welch. *Transferrin: the iron carrier*. CRC Press, 1992.
53. Jugal Saharia, YM Nuwan DY Bandara, Buddini I Karawdeniya, George Alexandrakis, and Min Jun Kim. Assessment of 1/f noise associated with nanopores fabricated through chemically-tuned controlled dielectric breakdown. *Electrophoresis*, 42:899–909, 2021.
54. S. W. Kowalczyk, D. B. Wells, A. Aksimentiev, and C. Dekker. Slowing down dna translocation through a nanopore in lithium chloride. *Nano Letters*, 12(2):1038–44, 2012.
55. R Russo, M Malinconico, and G Santagata. Effect of cross-linking with calcium ions on the physical properties of alginate films. *Biomacromolecules*, 8(10):3193–3197, 2007.
56. James M Broering and Andreas S Bommarius. Evaluation of hofmeister effects on the kinetic stability of proteins. *The Journal of Physical Chemistry B*, 109(43):20612–20619, 2005.
57. Arda Alden Green. Studies in the physical chemistry of the proteins x. the solubility of hemoglobin in solutions of chlorides and sulfates of varying concentration. *Journal of Biological Chemistry*, 95(1):47–66, 1932.
58. Luca Medda, Cristina Carucci, Drew F Parsons, Barry W Ninham, Maura Monduzzi, and Andrea Salis. Specific cation effects on hemoglobin aggregation below and at physiological salt concentration. *Langmuir*, 29(49):15350–15358, 2013.
59. Payam Rowghanian and Alexander Y Grosberg. Electrophoretic capture of a dna chain into a nanopore. *Physical Review E*, 87(4):042722, 2013.
60. Kevin J Freedman, S Raza Haq, Joshua B Edel, Per Jemth, and Min Jun Kim. Single molecule unfolding and stretching of protein domains inside a solid-state nanopore by electric field. *Scientific Reports*, 3:1638, 2013.

61. Zhen Gu, Yi-Lun Ying, Chan Cao, Pingang He, and Yi-Tao Long. Accurate data process for nanopore analysis. *Analytical Chemistry*, 87(2):907–913, 2015.
62. Daniel Pedone, Matthias Firnkes, and Ulrich Rant. Data analysis of translocation events in nanopore experiments. *Analytical Chemistry*, 81(23):9689–9694, 2009.
63. A. Morshed, B. I. Karawdeniya, Ymndy Bandara, M. J. Kim, and P. Dutta. Mechanical characterization of vesicles and cells: A review. *Electrophoresis*, 41(7-8):449–470, 2020.
64. C. Plesa, S. W. Kowalczyk, R. Zinsmeester, A. Y. Grosberg, Y. Rabin, and C. Dekker. Fast translocation of proteins through solid state nanopores. *Nano Letters*, 13(2):658–63, 2013.
65. Jacob H Forstater, Kyle Briggs, Joseph WF Robertson, Jessica Ettedgui, Olivier Marie-Rose, Canute Vaz, John J Kasianowicz, Vincent Tabard-Cossa, and Arvind Balijepalli. Mosaic: a modular single-molecule analysis interface for decoding multistate nanopore data. *Analytical Chemistry*, 88(23):11900–11907, 2016.
66. Jacob K Rosenstein, Meni Wanunu, Christopher A Merchant, Marija Drndić, and Kenneth L Shepard. Integrated nanopore sensing platform with sub-microsecond temporal resolution. *Nature Methods*, 9(5):487, 2012.
67. J Feng, Ke Liu, Michael Graf, Martina Lihter, Roman D Bulushev, Dumitru Dumcenco, Duncan TL Alexander, Daria Krasnozhon, Tomislav Vuletic, and Andras Kis. Electrochemical reaction in single layer mos2: nanopores opened atom by atom. *Nano Letters*, 15(5):3431–3438, 2015.
68. Aaron T Kuan, Bo Lu, Ping Xie, Tamas Szalay, and Jene A Golovchenko. Electrical pulse fabrication of graphene nanopores in electrolyte solution. *Applied Physics Letters*, 106(20):203109, 2015.
69. Joseph Larkin, Robert Henley, David C Bell, Tzahi Cohen-Karni, Jacob K Rosenstein, and Meni Wanunu. Slow dna transport through nanopores in hafnium oxide membranes. *ACS Nano*, 7(11):10121–10128, 2013.
70. Bala Murali Venkatesan, Brian Dorvel, Sukru Yemenicioglu, Nicholas Watkins, Ivan Petrov, and Rashid Bashir. Highly sensitive, mechanically stable nanopore sensors for dna analysis. *Advanced Materials*, 21(27):2771–2776, 2009.
71. Bala Murali Venkatesan, David Estrada, Shouvik Banerjee, Xiaozhong Jin, Vincent E Dorgan, Myung-Ho Bae, Narayana R Aluru, Eric Pop, and Rashid Bashir. Stacked graphene-al2o3 nanopore sensors for sensitive detection of dna and dna–protein complexes. *ACS Nano*, 6(1):441–450, 2012.
72. Youwen Zhang, Xiaohan Chen, Ceming Wang, Golbarg M Roozbahani, Hsueh-Chia Chang, and Xiyun Guan. Chemically functionalized conical pet nanopore for protein detection at the single-molecule level. *Biosensors and Bioelectronics*, page 112289, 2020.
73. Timothy Plett, Wenqing Shi, Yuhan Zeng, William Mann, Ivan Vlassiouk, Lane A Baker, and Zuzanna S Siwy. Rectification of nanopores in aprotic solvents–transport properties of nanopores with surface dipoles. *Nanoscale*, 7(45):19080–19091, 2015.
74. Ivan Vlassiouk, Pavel Y Apel, Sergey N Dmitriev, Ken Healy, and Zuzanna S Siwy. Versatile ultrathin nanoporous silicon nitride membranes. *Proceedings of the National Academy of Sciences*, 106(50):21039–21044, 2009.
75. Meni Wanunu, Tali Dadosh, Vishva Ray, Jingmin Jin, Larry McReynolds, and Marija Drndić. Rapid electronic detection of probe-specific micrornas using thin nanopore sensors. *Nature Nanotechnology*, 5(11):807–814, 2010.
76. Ke Liu, Jiandong Feng, Andras Kis, and Aleksandra Radenovic. Atomically thin molybdenum disulfide nanopores with high sensitivity for dna translocation. *ACS Nano*, 8(3):2504–2511, 2014.
77. Birgitta Schiedt, Loïc Auvray, Laurent Bacri, G Oukhaled, Ali Madouri, Eric Bourhis, Gilles Patriarche, Juan Pelta, Ralf Jede, and Jacques Gierak. Direct fib fabrication and integration of "single nanopore devices" for the manipulation of macromolecules. *Microelectronic Engineering*, 87(5-8):1300–1303, 2010.
78. Jijin Yang, David C Ferranti, Lewis A Stern, Colin A Sanford, Jason Huang, Zheng Ren, Lu-Chang Qin, and Adam R Hall. Rapid and precise scanning helium ion microscope milling of solid-state nanopores for biomolecule detection. *Nanotechnology*, 22(28):285310, 2011.

79. K. Briggs, H. Kwok, and V. Tabard-Cossa. Automated fabrication of 2-nm solid-state nanopores for nucleic acid analysis. *Small*, 10(10):2077–2086, 2014.
80. H. Kwok, K. Briggs, and V. Tabard-Cossa. Nanopore fabrication by controlled dielectric breakdown. *PLoS One*, 9(3):e92880, 2014.
81. Cuifeng Ying, Jared Houghtaling, Olivia M Eggenberger, Anirvan Guha, Peter Nirmalraj, Saurabh Awasthi, Jianguo Tian, and Michael Mayer. Formation of single nanopores with diameters of 20-50 nm in silicon nitride membranes using laser-assisted controlled breakdown. *ACS Nano*, 12(11):11458–11470, 2018.
82. I. Yanagi, R. Akahori, T. Hatano, and K. Takeda. Fabricating nanopores with diameters of sub-1 nm to 3 nm using multilevel pulse-voltage injection. *Scientific Reports*, 4:5000, 2014.
83. YM Nuwan DY Bandara, Buddini I Karawdeniya, and Jason R Dwyer. Push-button method to create nanopores using a tesla-coil lighter. *ACS Omega*, 4(1):226–230, 2019.
84. Jiri Kudr, Sylvie Skalickova, Lukas Nejdl, Amitava Moulick, Branislav Ruttkay-Nedecky, Vojtech Adam, and Rene Kizek. Fabrication of solid-state nanopores and its perspectives. *ELECTROPHORESIS*, 36(19):2367–2379, oct 2015.
85. Yusuke Goto, Itaru Yanagi, Kazuma Matsui, Takahide Yokoi, and Ken-ichi Takeda. Integrated solid-state nanopore platform for nanopore fabrication via dielectric breakdown, dna-speed deceleration and noise reduction. *Scientific Reports*, 6:31324, 2016.
86. Choongyeop Lee, Laurent Joly, Alessandro Siria, Anne-Laure Biance, Rémy Fulcrand, and Lydéric Bocquet. Large apparent electric size of solid-state nanopores due to spatially extended surface conduction. *Nano Letters*, 12(8):4037–4044, 2012.
87. YM Nuwan DY Bandara, Jonathan W Nichols, Buddini Iroshika Karawdeniya, and Jason R Dwyer. Conductance–based profiling of nanopores: Accommodating fabrication irregularities. *Electrophoresis*, 39(4):626–634, 2018.
88. YM Nuwan DY Bandara, Buddini Iroshika Karawdeniya, and Jason R Dwyer. Real-time profiling of solid-state nanopores during solution-phase nanofabrication. *ACS Applied Materials & Interfaces*, 8(44):30583–30589, 2016.
89. Stefan W Kowalczyk, Alexander Y Grosberg, Yitzhak Rabin, and Cees Dekker. Modeling the conductance and dna blockade of solid-state nanopores. *Nanotechnology*, 22(31):315101, 2011.
90. Angelika Holzinger, Gregor Neusser, Benjamin J J Austen, Alonso Gamero-Quijano, Grégoire Herzog, Damien W M Arrigan, Andreas Ziegler, Paul Walther, and Christine Kranz. Investigation of modified nanopore arrays using FIB/SEM tomography. *Faraday Discussions*, 210:113–130, 2018.
91. Oun-Ho Park, Joy Y Cheng, M Hart, Teya Topuria, Philip M Rice, Leslie E Krupp, Robert D Miller, Hiroshi Ito, and Ho-Cheol Kim. High-aspect-ratio cylindrical nanopore arrays and their use for templating titania nanoposts. *Advanced Materials*, 20(4):738–742, 2008.
92. A Hadley, C Notthoff, P Mota-Santiago, S Dutt, S Mudie, MA Carrillo-Solano, ME Toimil-Molares, C Trautmann, and P Kluth. Analysis of nanometer-sized aligned conical pores using small-angle x-ray scattering. *Physical Review Materials*, 4(5):056003, 2020.
93. Pradeep Waduge, Rui Hu, Prasad Bandarkar, Hirohito Yamazaki, Benjamin Cressiot, Qing Zhao, Paul C Whitford, and Meni Wanunu. Nanopore-based measurements of protein size, fluctuations, and conformational changes. *ACS Nano*, 11(6):5706–5716, 2017.
94. Jiandong Feng, Michael Graf, Ke Liu, Dmitry Ovchinnikov, Dumitru Dumcenco, Mohammad Heiranian, Vishal Nandigana, Narayana R Aluru, Andras Kis, and Aleksandra Radenovic. Single-layer mos2 nanopores as nanopower generators. *Nature*, 536(7615):197–200, 2016.
95. Andrea Hadley, Christian Notthoff, Pablo Mota-Santiago, UH Hossain, N Kirby, ME Toimil-Molares, C Trautmann, and P Kluth. Etched ion tracks in amorphous sio2 characterized by small angle x-ray scattering: influence of ion energy and etching conditions. *Nanotechnology*, 30(27):274001, 2019.
96. Amit Meller and Daniel Branton. Single molecule measurements of dna transport through a nanopore. *Electrophoresis*, 23(16):2583–2591, 2002.
97. Calin Plesa, Daniel Verschueren, Sergii Pud, Jaco van der Torre, Justus W. Ruitenberg, Menno J. Witteveen, Magnus P. Jonsson, Alexander Y. Grosberg, Yitzhak Rabin, and Cees

Dekker. Direct observation of dna knots using a solid-state nanopore. *Nature nanotechnology*, 11:1093, 2016.
98. Rajesh Kumar Sharma, Ishita Agrawal, Liang Dai, Patrick S Doyle, and Slaven Garaj. Complex dna knots detected with a nanopore sensor. *Nature Communications*, 10(1):1–9, 2019.
99. Camille Raillon, Pierre Granjon, Michael Graf, LJ Steinbock, and Aleksandra Radenovic. Fast and automatic processing of multi-level events in nanopore translocation experiments. *Nanoscale*, 4(16):4916–4924, 2012.
100. Ke Liu, Chao Pan, Alexandre Kuhn, Adrian Pascal Nievergelt, Georg E Fantner, Olgica Milenkovic, and Aleksandra Radenovic. Detecting topological variations of dna at single-molecule level. *Nature Communications*, 10(1):1–9, 2019.
101. Calin Plesa and Cees Dekker. Data analysis methods for solid-state nanopores. *Nanotechnology*, 26(8):084003, 2015.
102. Jianhua Zhang, Xiuling Liu, Yi-Lun Ying, Zhen Gu, Fu-Na Meng, and Yi-Tao Long. High-bandwidth nanopore data analysis by using a modified hidden markov model. *Nanoscale*, 9(10):3458–3465, 2017.
103. Jian-Hua Zhang, Xiu-Ling Liu, Zheng-Li Hu, Yi-Lun Ying, and Yi-Tao Long. Intelligent identification of multi-level nanopore signatures for accurate detection of cancer biomarkers. *Chemical Communications*, 53(73):10176–10179, 2017.
104. Lingzhi Wu, Hang Liu, Wenyuan Zhao, Lei Wang, Chuanrong Hou, Quanjun Liu, and Zuhong Lu. Electrically facilitated translocation of protein through solid nanopore. *Nanoscale Research Letters*, 9(1):140, 2014.
105. Yin Zhang, Lei Liu, Jingjie Sha, Zhonghua Ni, Hong Yi, and Yunfei Chen. Nanopore detection of dna molecules in magnesium chloride solutions. *Nanoscale Research Letters*, 8(1):1–8, 2013.
106. M Mercedes Tirado, Carmen López Martínez, and José García de la Torre. Comparison of theories for the translational and rotational diffusion coefficients of rod-like macromolecules. application to short dna fragments. *The Journal of chemical physics*, 81(4):2047–2052, 1984.
107. Shawn M Douglas, Hendrik Dietz, Tim Liedl, Björn Högberg, Franziska Graf, and William M Shih. Self-assembly of dna into nanoscale three-dimensional shapes. *Nature*, 459(7245):414–418, 2009.
108. Hongwen Wu, Hang Liu, Shengwei Tan, Jingjing Yu, Wenyuan Zhao, Lei Wang, and Quanjun Liu. The estimation of field-dependent conductance change of nanopore by field-induced charge in the translocations of aunps-dna conjugates. *The Journal of Physical Chemistry C*, 118(46):26825–26835, 2014.
109. E. C. Yusko, J. M. Johnson, S. Majd, P. Prangkio, R. C. Rollings, J. Li, J. Yang, and M. Mayer. Controlling protein translocation through nanopores with bio-inspired fluid walls. *Nature Nanotechnolagy*, 6(4):253–260, 2011.
110. NB Grover, J Naaman, S Ben-Sasson, and F Doljanski. Electrical sizing of particles in suspensions. *Biophysical Journal*, 9(11):1398–1414, 1969.
111. RW DeBlois and CP Bean. Counting and sizing of submicron particles by the resistive pulse technique. *Scientific Instruments*, 41(7):909–916, 1970.
112. James Clerk Maxwell. *A treatise on electricity and magnetism*, volume 1. Oxford: Clarendon Press, 1873.
113. Junrong Wang, Jian Ma, Zhonghua Ni, Li Zhang, and Guoqing Hu. Effects of access resistance on the resistive-pulse caused by translocating of a nanoparticle through a nanopore. *RSC Advances*, 4(15):7601–7610, 2014.
114. Daniel Fologea, James Uplinger, Brian Thomas, David S McNabb, and Jiali Li. Slowing dna translocation in a solid-state nanopore. *Nano Letters*, 5(9):1734–1737, 2005.
115. Eric Beamish, Harold Kwok, Vincent Tabard-Cossa, and Michel Godin. Precise control of the size and noise of solid-state nanopores using high electric fields. *Nanotechnology*, 23(40):405301, 2012.
116. Daniel Y Ling and Xinsheng Sean Ling. On the distribution of dna translocation times in solid-state nanopores: an analysis using schrödinger's first-passage-time theory. *Journal of Physics: Condensed Matter*, 25(37):375102, 2013.

117. Spencer Carson, James Wilson, Aleksei Aksimentiev, and Meni Wanunu. Smooth dna transport through a narrowed pore geometry. *Biophysical Journal*, 107(10):2381–2393, 2014.
118. Geoffrey Ingram Taylor. Dispersion of soluble matter in solvent flowing slowly through a tube. *Proceedings of the Royal Society of London. Series A. Mathematical and Physical Sciences*, 219(1137):186–203, 1953.
119. Rutherford Aris. On the dispersion of a solute in a fluid flowing through a tube. *Proceedings of the Royal Society of London. Series A. Mathematical and Physical Sciences*, 235(1200):67–77, 1956.
120. Christopher A. Merchant, Ken Healy, Meni Wanunu, Vishva Ray, Neil Peterman, John Bartel, Michael D. Fischbein, Kimberly Venta, Zhengtang Luo, A. T. Charlie Johnson, and Marija Drndić. Dna translocation through graphene nanopores. *Nano Letters*, 10(8):2915–2921, 2010.
121. RMM Smeets, Stefan W Kowalczyk, AR Hall, NH Dekker, and C Dekker. Translocation of reca-coated double-stranded dna through solid-state nanopores. *Nano Letters*, 9(9):3089–3095, 2009.
122. Makusu Tsutsui, Takeshi Yoshida, Kazumichi Yokota, Hirotoshi Yasaki, Takao Yasui, Akihide Arima, Wataru Tonomura, Kazuki Nagashima, Takeshi Yanagida, Noritada Kaji, Masateru Taniguchi, Takashi Washio, Yoshinobu Baba, and Tomoji Kawai. Discriminating single-bacterial shape using low-aspect-ratio pores. *Scientific Reports*, 7(1):1–9, 2017.
123. Makusu Tsutsui, Masayoshi Tanaka, Takahiro Marui, Kazumichi Yokota, Takeshi Yoshida, Akihide Arima, Wataru Tonomura, Masateru Taniguchi, Takashi Washio, Mina Okochi, and Tomoji Kawai. Identification of individual bacterial cells through the intermolecular interactions with peptide-functionalized solid-state pores. *Analytical Chemistry*, 90(3):1511–1515, 2018.
124. Masateru Taniguchi. Combination of single-molecule electrical measurements and machine learning for the identification of single biomolecules. *ACS Omega*, 5(2):959–964, 2020.
125. Akihide Arima, Ilva Hanun Harlisa, Takeshi Yoshida, Makusu Tsutsui, Masayoshi Tanaka, Kazumichi Yokota, Wataru Tonomura, Jiro Yasuda, Masateru Taniguchi, Takashi Washio, Mina Okochi, and Tomoji Kawai. Identifying single viruses using biorecognition solid-state nanopores. *Journal of the American Chemical Society*, 140(48):16834–16841, 2018.
126. Shengfa Liang, Feibin Xiang, Zifan Tang, Reza Nouri, Xiaodong He, Ming Dong, and Weihua Guan. Noise in nanopore sensors: sources, models, reduction, and benchmarking. *Nanotechnology Precision Engineering*, 3(1):9–17, 2020.
127. Peng Chen, Toshiyuki Mitsui, Damon B Farmer, Jene Golovchenko, Roy G Gordon, and Daniel Branton. Atomic layer deposition to fine-tune the surface properties and diameters of fabricated nanopores. *Nano Letters*, 4(7):1333–1337, 2004.
128. Aleksandra Radenovic, Eliane Trepagnier, Roseann Csencsits, Kenneth H Downing, and Jan Liphardt. Fabrication of 10 nm diameter hydrocarbon nanopores. *Applied Physics Letters*, 93(18):183101, 2008.
129. Sergey M Bezrukov and Mathias Winterhalter. Examining noise sources at the single-molecule level: 1/f noise of an open maltoporin channel. *Physical Review Letters*, 85(1):202, 2000.
130. Ashvani Kumar, Kyeong-Beom Park, Hyun-Mi Kim, and Ki-Bum Kim. Noise and its reduction in graphene based nanopore devices. *Nanotechnology*, 24(49):495503, 2013.
131. Chenyu Wen, Shuangshuang Zeng, Kai Arstila, Timo Sajavaara, Yu Zhu, Zhen Zhang, and Shi-Li Zhang. Generalized noise study of solid-state nanopores at low frequencies. *ACS Sensors*, 2(2):300–307, 2017.
132. RMM Smeets, NH Dekker, and C Dekker. Low-frequency noise in solid-state nanopores. *Nanotechnology*, 20(9):095501, 2009.
133. Alessio Fragasso, Sergii Pud, and Cees Dekker. 1/f noise in solid-state nanopores is governed by access and surface regions. *Nanotechnology*, 30(39):395202, 2019.
134. Kazuma Matsui, Yusuke Goto, Itaru Yanagi, Rena Akahori, Michiru Fujioka, Takeshi Ishida, Takahide Yokoi, Tatsuo Nakagawa, and Ken-ichi Takeda. Low-frequency noise induced by cation exchange fluctuation on the wall of silicon nitride nanopore. *Scientific Reports*, 10(1):1–8, 2020.

135. Vincent Tabard-Cossa, Dhruti Trivedi, Matthew Wiggin, Nahid N Jetha, and Andre Marziali. Noise analysis and reduction in solid-state nanopores. *Nanotechnology*, 18(30):305505, 2007.
136. Min-Hyun Lee, Ashvani Kumar, Kyeong-Beom Park, Seong-Yong Cho, Hyun-Mi Kim, Min-Cheol Lim, Young-Rok Kim, and Ki-Bum Kim. A low-noise solid-state nanopore platform based on a highly insulating substrate. *Scientific Reports*, 4:7448, 2014.
137. Li-Hsien Yeh, Mingkan Zhang, Sang W Joo, and Shizhi Qian. Slowing down DNA translocation through a nanopore by lowering fluid temperature. *ELECTROPHORESIS*, 33(23):3458–3465, dec 2012.
138. Ranulu Samanthi S de Zoysa, Dilani A Jayawardhana, Qitao Zhao, Deqiang Wang, Daniel W Armstrong, and Xiyun Guan. Slowing dna translocation through nanopores using a solution containing organic salts. *The Journal of Physical Chemistry B*, 113(40):13332–13336, 2009.
139. Hengbin Zhang, Qing Zhao, Zhipeng Tang, Song Liu, Qingtao Li, Zhongchao Fan, Fuhua Yang, Liping You, Xuemei Li, Jingmin Zhang, and Dapeng Yu. Slowing Down DNA Translocation Through Solid-State Nanopores by Pressure. *Small*, 9(24):4112–4117, dec 2013.
140. Stefan W Kowalczyk, Adam R Hall, and Cees Dekker. Detection of local protein structures along dna using solid-state nanopores. *Nano Letters*, 10(1):324–328, 2010.
141. Rena Akahori, Takanobu Haga, Toshiyuki Hatano, Itaru Yanagi, Takeshi Ohura, Hirotaka Hamamura, Tomio Iwasaki, Takahide Yokoi, and Takashi Anazawa. Slowing single-stranded dna translocation through a solid-state nanopore by decreasing the nanopore diameter. *Nanotechnology*, 25(27):275501, 2014.
142. Nicholas AW Bell and Ulrich F Keyser. Specific protein detection using designed dna carriers and nanopores. *Journal of the American Chemical Society*, 137(5):2035–2041, 2015.
143. Ulrich F Keyser. Enhancing nanopore sensing with dna nanotechnology. *Nature Nanotechnology*, 11(2):106–108, 2016.
144. Muhammad Usman Raza, Sai Santosh Sasank Peri, Liang-Chieh Ma, Samir M Iqbal, and George Alexandrakis. Self-induced back action actuated nanopore electrophoresis (sane). *Nanotechnology*, 29(43):435501, 2018.
145. Allison H Squires, Joseph S Hersey, Mark W Grinstaff, and Amit Meller. A nanopore-nanofiber mesh biosensor to control dna translocation. *Journal of the American Chemical Society*, 135(44):16304–16307, 2013.
146. Harold Kwok, Matthew Waugh, José Bustamante, Kyle Briggs, and Vincent Tabard-Cossa. Long Passage Times of Short ssDNA Molecules through Metallized Nanopores Fabricated by Controlled Breakdown. *Advanced Functional Materials*, 24(48):7745–7753, dec 2014.
147. Bo Lu, Fernando Albertorio, David P Hoogerheide, and Jene A Golovchenko. Origins and consequences of velocity fluctuations during dna passage through a nanopore. *Biophysical Journal*, 101(1):70–79, 2011.
148. Hongwen Wu, Yuhao Chen, Qizhao Zhou, Rongliang Wang, Baicheng Xia, Dejun Ma, Kaifu Luo, and Quanjun Liu. Translocation of rigid rod-shaped virus through various solid-state nanopores. *Analytical Chemistry*, 88(4):2502–2510, 2016.
149. Kyle Briggs, Gregory Madejski, Martin Magill, Konstantinos Kastritis, Hendrick W de Haan, James L McGrath, and Vincent Tabard-Cossa. Dna translocations through nanopores under nanoscale preconfinement. *Nano Letters*, 18(2):660–668, 2018.
150. Andrew H Laszlo, Ian M Derrington, Henry Brinkerhoff, Kyle W Langford, Ian C Nova, Jenny Mae Samson, Joshua J Bartlett, Mikhail Pavlenok, and Jens H Gundlach. Detection and mapping of 5-methylcytosine and 5-hydroxymethylcytosine with nanopore mspa. *Proceedings of the National Academy of Sciences*, 110(47):18904–18909, 2013.
151. Kidan Lee, Hyomin Lee, Seung-Hyun Lee, Hyun-Mi Kim, Ki-Bum Kim, and Sung Jae Kim. Enhancing the sensitivity of dna detection by structurally modified solid-state nanopore. *Nanoscale*, 9(45):18012–18021, 2017.
152. Julio A Rodriguez-Manzo, Matthew Puster, Adrien Nicolai, Vincent Meunier, and Marija Drndić. Dna translocation in nanometer thick silicon nanopores. *ACS Nano*, 9(6):6555–6564, 2015.
153. Trang Vu, Julia Borgesi, Joanna Soyring, Melissa D'Alia, Shanna-Leigh Davidson, and Jiwook Shim. Employing licl salt gradient in the wild-type α-hemolysin nanopore to slow

down dna translocation and detect methylated cytosine. *Nanoscale*, 11(21):10536–10545, 2019.

154. Julian Bello, Maksudul Mowla, Nicholas Troise, Joanna Soyring, Julia Borgesi, and Jiwook Shim. Increased dwell time and occurrence of dsdna translocation events through solid state nanopores by licl concentration gradients. *Electrophoresis*, 40(7):1082–1090, 2019.
155. Kevin J Freedman, Lauren M Otto, Aleksandar P Ivanov, Avijit Barik, Sang-Hyun Oh, and Joshua B Edel. Nanopore sensing at ultra-low concentrations using single-molecule dielectrophoretic trapping. *Nature Communications*, 7:10217, 2016.
156. Ali Najafi Sohi, Eric Beamish, Vincent Tabard-Cossa, and Michel Godin. Dna capture by nanopore sensors under flow. *Analytical Chemistry*, 2020.
157. Ji Li, Rui Hu, Xiaoqing Li, Xin Tong, Dapeng Yu, and Qing Zhao. Tiny protein detection using pressure through solid-state nanopores. *ELECTROPHORESIS*, 38(8):1130–1138, apr 2017.
158. Yann Astier, Lucien Datas, Randy Carney, Francesco Stellacci, Francesco Gentile, and Enzo DiFabrizio. Artificial surface-modified si3n4 nanopores for single surface-modified gold nanoparticle scanning. *Small*, 7(4):455–459, 2011.
159. Christopher A Merchant, Ken Healy, Meni Wanunu, Vishva Ray, Neil Peterman, John Bartel, Michael D Fischbein, Kimberly Venta, Zhengtang Luo, and AT Charlie Johnson. Dna translocation through graphene nanopores. *Nano Letters*, 10(8):2915–2921, 2010.
160. Rong Wang, Kazuhito Hashimoto, Akira Fujishima, Makoto Chikuni, Eiichi Kojima, Atsushi Kitamura, Mitsuhide Shimohigoshi, and Toshiya Watanabe. Photogeneration of highly amphiphilic tio2 surfaces. *Advanced Materials*, 10(2):135–138, 1998.
161. Dae-Sik Lee, Seokhan Park, Yong Duk Han, Jae Eun Lee, Hu Young Jeong, Hyun C Yoon, Moon Yeon Jung, Sang Ouk Kim, and Sung-Yool Choi. Selective protein transport through ultra-thin suspended reduced graphene oxide nanopores. *Nanoscale*, 9(36):13457–13464, 2017.
162. Kyloon Chuah, Yanfang Wu, SRC Vivekchand, Katharina Gaus, Peter J Reece, Adam P Micolich, and J Justin Gooding. Nanopore blockade sensors for ultrasensitive detection of proteins in complex biological samples. *Nature Communications*, 10(1):1–9, 2019.
163. Marc Gershow and Jene Andrew Golovchenko. Recapturing and trapping single molecules with a solid-state nanopore. *Nature nanotechnology*, 2(12):775, 2007.

Chapter 11
Challenges in Protein Sequencing Using 2-D MoS_2 Nanopores

Adrien Nicolaï and Patrick Senet

Abstract Solid-state nanopores (SSN) made of 2-D materials such as molybdenum disulfide (MoS_2) have emerged as one of the most versatile sensors for single-molecule detection. One of the most promising applications of SSN is DNA and protein sequencing, at a low cost and faster than the current standard methods. The detection principle relies on measuring the relatively small variations of ionic current as charged biomolecules immersed in an electrolyte traverse the nanopore, in response to an external voltage applied across the membrane. The passage of a biomolecule through the pore yields information about its structure and chemical properties, as demonstrated experimentally particularly for DNA molecules. Indeed, protein sequencing using SSN remains highly challenging since the protein ensemble is far more complex than the DNA one. In the present chapter, we focus on challenges in protein sequencing using 2-D MoS_2 nanopores. Three challenges are highlighted using Molecular Dynamics to simulate protein translocation experiments through MoS_2 nanopores. First, the threading of the protein through the nanopore is discussed. Second, the modification of the nanopore dimensions in order to slow down the passage of the protein through the pore is detailed. Finally, the application of time series analysis tools in order to identify protein sequence motif from measured raw data is presented.

11.1 Introduction

After three decades of research [1], development of label-free, electrical-based nanodevice for DNA sequencing is a mature field [2]. The first sequence reads of DNA strands using nanopores was published 10 years ago [3] and the commercial sequencer MinION of Oxford Nanopore Technology was successfully applied to sequence a human genome [4]. Both biological nanopores [1–4] as well as solid-

A. Nicolaï · P. Senet (✉)
Laboratoire Interdisciplinaire Carnot de Bourgogne, UMR 6303 CNRS-Université de Bourgogne Franche-Comté, 9 Av. A. Savary, BP 47 870, 21078 Dijon Cedex, France
e-mail: psenet@u-bourgogne.fr

W. Bowen et al. (eds.), *Single Molecule Sensing Beyond Fluorescence*, Nanostructure Science and Technology,
https://doi.org/10.1007/978-3-030-90339-8_11

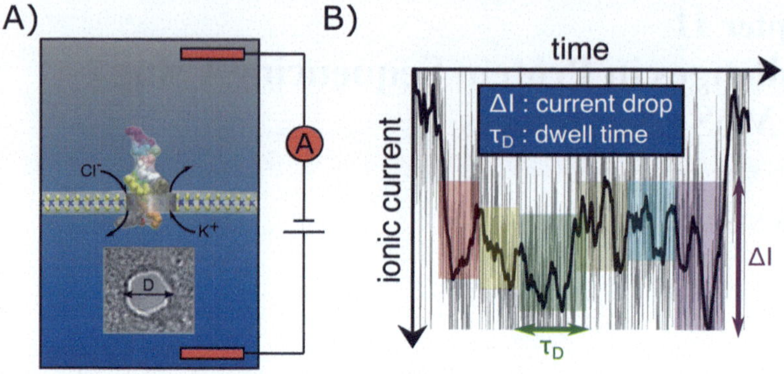

Fig. 11.1 Principle and detection method of solid-state nanopore technology. **A** Schematic representation of protein sensing using MoS_2 nanopore. Two reservoirs filled with a KCl electrolyte are separated by a SL-MoS_2 membrane and connected via a single pore of nanometer size (see TEM image as inset). Electrodes (in red) are immersed in the two compartments and a constant voltage bias is applied across the membrane causing an ionic-current flow through the pore. **B** Ionic current trace as a function of time during the passage of a protein through the pore. Usually, the ionic current decreases and the features of the current drop, such as its amplitude ΔI, duration $\equiv$ dwell time τ_D and fluctuations, contain important information about the protein, including its size, charge, shape or sequence, as represented with transparent color rectangles in the given trace

state nanopores (SSN) [5–8] were developed to sequence single DNA molecules. Protein sequencing using nanopores is less advanced [9, 10] and is the focus of the present chapter.

The general principle of electrical-based nanopore sequencing of a biopolymer by a SSN is illustrated in Fig. 11.1. A 2-D material, MoS_2 nanoporous membrane, discussed further below, is shown (panel A). The charged biopolymer to be detected and sequenced is driven by an electric field to a pore of dimension comparable to its molecular dimensions (typical values are 1–3 nm) and is immersed in an electrolyte (typically NaCl or KCl). The ionic current induced by the electric potential difference applied to the membrane is measured with a high temporal resolution [11] (up to 100 ns). Obstruction of the nanopore while the biopolymer is passing through the channel decreases the ionic conductance of the device. Consequently, the ionic current variations depend on the molecular fragment residing in the nanopore. The read of the current drops (ΔI) and their duration (τ_D) as a function of time provide fingerprints of the individual fragments translocating through the nanopore (Fig. 11.1, Panel **B**). At best, each monomer is expected to be revealed by a distinct fingerprint, allowing the identification of the biopolymer primary structure.

Invention of the electrical-based nanopore sequencing in 1998 featured a method for evaluating a polymer molecule which includes linearly (connected) sequential monomer residues and envisioned protein sequencing [1]. The two main advantages of the nanopore technology for protein sequencing compared to other technique of protein identification such as mass spectroscopy for instance are: (i) its sensitivity, as single-molecule can be detected and (ii) the possibility to develop a portable medical diagnostic device. Protein sequencing is essential for early disease diagnostic as DNA sequencing of cells does not fully characterize human diseases [12]. A particular application of protein sequencing is for example the identification of protein tumor biomarkers, which can assist in determining the presence, absence or the evolution of a cancer [13]. The protein ensemble is far more complex than the DNA ensemble. The proteosome includes proteins with post-translational modifications [14], as for example the phosphorylation which may alter the location, the function and even the folded state of a protein [15]. In contrast to the negatively uniformly charged double strands of nucleotides, which is the common shared structure of DNA molecules, proteins occur in many different folded structures with various heterogeneous charge states, and as already mentioned various post-translational modifications [14].

Identification of single amino acid connected to a charged tag was achieved using biological (aerolysin) nanopore [16], paving the way to polypeptide sequencing. Because biological nanopores are fragile, SSN for protein sequencing emerged as an alternative promising material because of their mechanical stability, chemical stability to extreme denaturant solvent conditions, their tunable physical properties (opto-electronic, semiconducting, ...), their ease to be chemically modified (functionalization) and their possible integration in a multi-probe device adding plasmonic or spectroscopic signals [8]. Moreover, SSN exhibit signal levels that are at least one order of magnitude higher than biological pores due to the thinner extent of the membrane. Common SSN materials are silicon nitride thin films (SiN_x, thickness $h \approx 10$ nm) in which a pore of sub-nm diameter (up to a few nm) is drilled. Potential identification of individual amino acids was reached by using a sub-nanometer SiN_x nanoporous membrane as SSN [17, 18]. Assuming that the ionic current fluctuations are linearly related to the amino acid volumes obstructing the nanopore, the ionic current signal of protein translocation could be decomposed in fingerprints of quadromer by using machine learning algorithms [18].

As the ionic conductance of the nanopore is inversely proportional to its membrane thickness [19], interest for DNA and protein sequencing using 2-D materials, of sub-nm to a few nm effective thickness, such as graphene [5, 20], atomically thin MoS_2 layers [6, 7, 19, 21–26] or graphene-MoS_2 heterostructures [27] is steadily growing [28]. In addition to MoS_2, other transition metal dichalcogenide (TMD) membranes such as WS_2 [29] are promising SSN, complementary to graphene, because of their visible bandgap, the absence of strong surface hydrophobic interactions with DNA bases and hydrophobic amino acids and their rich optoelectronic properties [30, 31]. They also present a larger SNR than graphene ($\approx$10 for MoS_2 compared to $\approx$3 for graphene for a pore of similar dimension) [6]. DNA identification using ionic current signal from MoS_2 SSN [6, 7] and correlated ionic and transverse electronic currents across the membrane [21] showed promising results, which could extend in principle to protein sequencing. Finally, detection of protein translocation through MoS_2 nanopores was achieved recently [22]. The large pore diameter (11 nm) does not allow to sequence the protein but to detect translocation events characterized by current drops up to 2.5 times larger than for the detection of the same protein using thicker SiN_x SSN of similar pore diameter [22]. In addition, ability to discriminate single amino acid with MoS_2 SSN was simulated by Molecular Dynamics [23–27].

In spite of recent progress, protein sequencing using nanopore technology remains challenging [9, 10] and no systematic current method is able to visually differentiate two polypeptides with different sequences at the single amino-acid level, although the difference in unfolded states [32] or between homopolymeric peptides [33] can be detected. Three challenges may be identified to achieve single amino-acid resolution using 2-D TMD materials, which are particularly promising to sequence proteins using SSN.

- The first challenge is threading the protein into the nanopore. It requires to unfold its compact tertiary structure for example by chemical [17, 34] or thermal denaturation [17, 35]. However, the unfolded state of a protein in solution remains compact; if the unfolded chain of N residues is represented by a self-avoiding random chain, its gyration radius can be approximated to $R_g \approx 0.059 \times (N - 1)^{0.6}$ nm, i.e. $\approx$14 nm for $N = 100$ amino acids. Therefore, the interaction of the unfolded protein with the surface of the nanopore is an important parameter which facilitates or slow down the protein diffusion towards the nanopore. Moreover, as the passage of monomers across the membrane is driven by an electric field, an appropriate control of the protein net charge using tags [12, 25] or chemical denaturant [17] is also necessary.
- The second challenge is to measure the translocation events within the limited temporal resolution of the ionic current recording experimentally, which is between 100 ns and 10–100 μs. Because of the small thickness of the nanoporous membrane, the residence time of molecular fragments may be much shorter than the bin of the ionic current time series. The geometrical parameters of the SSN and the solvent properties can be modified in order to slow down the passage of monomers across the membrane.
- The third challenge is to associate specific ionic current fingerprints from the measured current time series to each amino-acid sequence. In this context, time series analysis methods are applied to extract relevant information about the peptide from ionic current measurements.

These challenges are illustrated next from Molecular Dynamics (MD) simulations of peptide translocation through MoS_2 nanopores. In fact, polypeptide sequencing is the preliminary step to protein sequencing.

11.2 Methods

In this section, we briefly describe all-atom modeling of MoS_2 nanopore device and MD simulations of the translocation of biological peptides through MoS_2 nanopores. In addition, an analytical model of ionic conductance of MoS_2 nanopores is presented. Finally, ionic current measurements and statistical distributions extracted from MD are shown. More theoretical and technical details can be found in the References [19, 25, 26].

11.2.1 Modeling of MoS_2 Nanopore Device and Molecular Dynamics

MoS_2 nanopore device studied here is comprised of three distinct elements: a solid-state nanoporous membrane, a biological peptide, both immersed in a KCl electrolyte (Fig. 11.2A). The membrane is of dimension $\approx 10 \times 10$ nm^2 and the pore of cylindrical shape, which is drilled at the center of the membrane ($\rho_{pore} = 0$ and $z_{pore} = 0$), is defined by two geometrical parameters at the nanoscale: its diameter D and its thickness h. In the present work, we consider three different nanoporous membranes: a single-layer MoS_2 (SL-MoS_2, $h = 0.3$ nm) with pore diameter $D = 2.0$ nm, a bilayer-MoS_2 (BL-MoS_2, $h = 0.9$ nm) with pore diameter $D = 2.0$ nm and a single-layer MoS_2 (SL-MoS_2, $h = 0.3$ nm) with pore diameter $D = 1.5$ nm (Fig. 11.2B). Moreover, translocation experiments through MoS_2 nanopores for three different peptide sequences are presented (Fig. 11.2C): a Lysine dipeptide of sequence KK, a Met-Enkephalin pentapeptide of sequence YGGFM [36] with poly-Lysine tag of different lengths and an α-Synuclein hexapeptide of sequence KTKEGV [37]. The total charge of each peptide is $+2$, 0 and $+1$, respectively. Concentration of KCl electrolyte is taken as 1 M. In total, the simulation box is comprised of $\approx 200,000$ atoms and is globally neutral.

All-atom Molecular Dynamics using Periodic Boundary Conditions (PBC) were carried out using the LAMMPS software package [38]. In the present simulations, we do not consider MoS_2 to be rigid since the dynamics of the nanoporous membrane play a role in the diffusion of the biomolecule on the surface and in the threading of the biomolecule through the pore [25]. Consequently, a Stillinger-Weber (SW) potential is used [39] to simulate the dynamics of Mo-S bonded interactions. The biological peptide is modeled using the Amber ff99SB-ILDN force-field [40]. The water model used in the present work is the TIP3P model [41] and parameters for K^+ and Cl^- ionic species were taken from reference [42]. Finally, non-bonded interactions between each subsystem, i.e. MoS_2 atoms, peptide atoms, water molecules and ions are described using a Lennard-Jones (LJ) plus Coulomb potential. Originally, the biological peptide is placed above the nanoporous membrane at a normal distance $z = 2.0$ nm (the membrane is placed at $z = 0$ in the simulation box). By doing that, we avoid biased threading when the peptide is originally placed into the pore. Peptide translocation during MD production runs was enforced by imposing a uniform electric field, directed normal to the nanoporous membrane, to all atomic partial charges in the simulation box. The corresponding applied voltage simulated is $V = -EL_z$, where L_z is the length of the simulation box in the z-direction ($L_z \approx 20$ nm), with $V = 1$ V for all MD runs presented hereafter. Before running MD production runs, energy minimization followed by a two-stage equili-

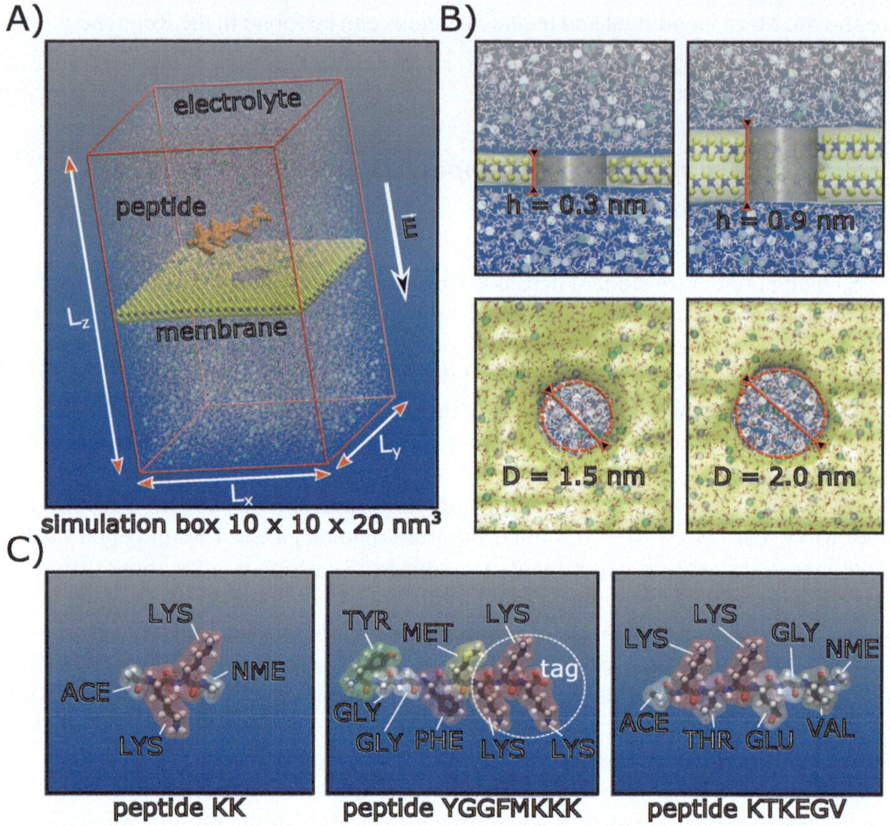

Fig. 11.2 Atomic representation of MoS_2 nanopore device introduced in the present chapter. **A** Simulation box of dimension $\approx 10 \times 10 \times 20$ nm^3. MoS_2 membrane is shown in yellow, the pore being represented by a gray cylinder, the biological peptide is shown in orange and the the electrolyte is shown with transparent balls and sticks. The direction of the applied electric field **E** is also given. **B** Geometrical characteristics of MoS_2 nanoporous membranes described in the present chapter: SL- and BL-MoS_2 (top panels) of thickness $h = 0.3$ and $h = 0.9$ nm, respectively. Pore diameters $D = 1.5$ nm and $D = 2.0$ nm (bottom panels). The color code is the same as in panel A. **C** Sequences and atomic representations of biological peptides studied in the present chapter. Lysine residues (K) are shown in red, Glycine (G) in white, Tyrosine (Y) in green, Phenylalanine (F) in purple, Methionine (M) in yellow, Threonine (T) in mauve and Glutamic acid (E) in pink. Capping groups (ACE and NME) are shown in gray

bration without any applied electric field, first in NVT for 50 ps ($T = 300$ K) and second in NPT for 100 ps ($T = 300$ K and $P = 1$ bar), were performed. During equilibration process, the peptide is maintained at its original position using restraints. Finally, several independent MD production runs using different initial conditions were carried out for each peptide. The duration goes from 50 ns for the shortest peptide (KK) to 500 ns for the longer ones (YGGFM and KTKEGV), with atom coordinates saved every 1 ps.

11.2.2 Ionic Conductance of MoS_2 Nanopores

Physically, nanoporous membranes used as single-molecule sensors are characterized by their ionic conductance. Ionic conductance is an intrinsic property of the nanopore and can be estimated from the knowledge of its geometrical dimensions [43]. The continuum model used to predict the ionic conductance G_0 of a channel of effective diameter D^* and thickness h^* is given by:

$$G_0 = \sigma_{\text{bulk}} \left(\frac{4h^* + \pi D^*}{\pi D^{*2}} \right)^{-1}, \tag{11.1}$$

where σ_{bulk} is the ionic conductivity of the electrolyte (in S m^{-1}). Effective pore diameter D^* and membrane thickness h^* correspond to the effective dimensions of the ionic conducting cylindrical channel of the nanoporous membrane experienced by solvent molecules and were extracted from the probability distributions of solvent molecules inside the pore [44]. Effective pore diameter is given by $D^* = D - 0.3$ nm and effective thickness for SL-MoS_2 is $h^* = 0.96$ nm ($h^* = 1.44$ nm for BL-MoS_2).

However, such an analytical model cannot be applied at the nanoscale since ions are confined in volumes whose dimensions are of similar sizes to that of the ionic radii. Therefore, ion concentration, mobilities and hydration are very different than their bulk counterparts. Therefore, ionic conductivity of the electrolyte in nanopores, i.e. σ_{pore}, is expected to deviate drastically from its bulk value, i.e. σ_{bulk}, and the ionic conductance of nanopores predicted by Eq. 11.1 is likely to be inaccurate. Thanks to all-atom MD simulations for SL-MoS_2 membranes with sub-5 nm pores, we derived a correction of the original model of the electrolyte conductivity at room temperature, as a function of the pore effective diameter D^*. As shown in Fig. 11.3A, the ion conductivity inside the pore deviates significantly from the bulk electrolyte conductivity for the range of diameters studied here (1–3 nm), which corresponds to those used for biomolecule sensing. For instance, the deviation is around 50% for a nanopore of diameter $D^* = 1.5$ nm.

The corrected continuum model of ionic conductance $\widetilde{G}_0$ is given by Eq. 11.2. It has been demonstrated that the model is in very good agreement with ionic conductance of MoS_2 nanopores measured experimentally [19] (Fig. 11.3B).

$$\widetilde{G}_0 = \sigma_{\text{bulk}} \left[\frac{1}{2} \sum_i \exp\left(\frac{-4\varphi^i}{\pi D^{*2}} \right) \frac{D^*}{\delta^i + \epsilon^i D^*} \right] \left(\frac{4h^* + \pi D^*}{\pi D^{*2}} \right)^{-1} \tag{11.2}$$

where the index i represents the ionic species. In addition, φ, δ and ϵ are constant values associated to concentrations (φ in nm^2) and mobilities of ions (δ in nm and ϵ dimensionless). These constant values are available for K^+ and Cl^- ions in reference [19]. Finally, this model can be used by experimentalists to extract dimensions of a nanopore from conductance measurements and vice-versa, as done very recently for 0-D Angstrom-size defect MoS_2 nanopores [45, 46]. With the abacus

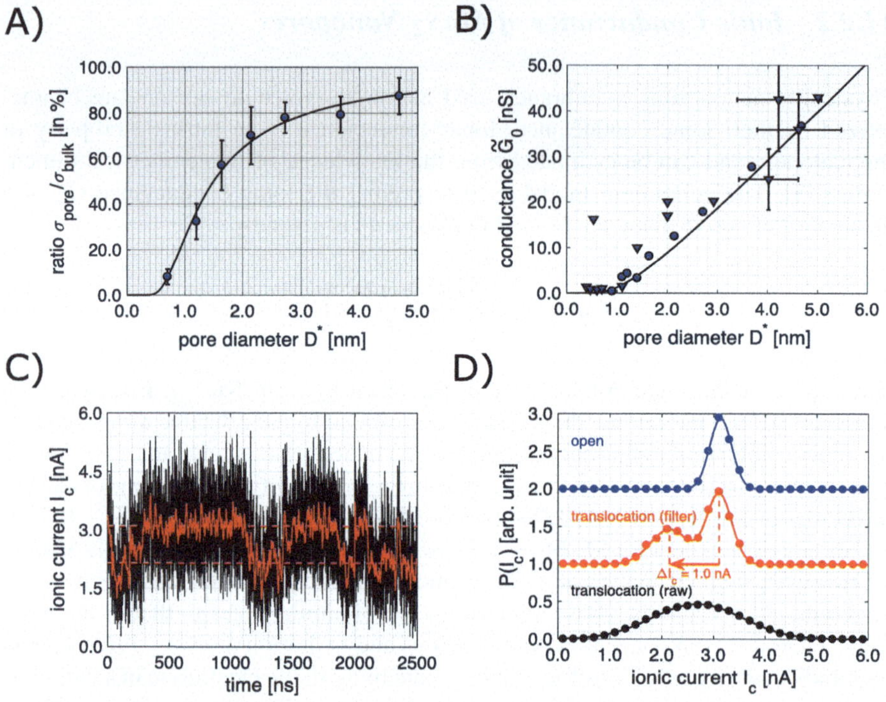

Fig. 11.3 Ionic conductance of MoS_2 nanopores and ionic current measurements extracted from MD translocation experiments. **A** Abacus graph of the correction to apply to the conductance model [in %] as a function of effective diameter D^* of the nanopore [in nm]. **B** Ionic conductance $\widetilde{G}_0$ [in nS] of MoS_2 nanopores as a function of effective pore diameter D^* [in nm]. Circles and triangles represent conductance extracted from MD and experimental data, respectively. **C** Ionic current measurements I_c [in nA] as a function of time [in ns] extracted from MD simulations of the translocation of KTKEGV peptide through SL-MoS_2 nanopores of diameter $D = 1.5$ nm. Black and red lines represents raw and filtered data, respectively. **D** Probability distributions of ionic current $P(I_c)$ extracted from MD simulations in the open pore scenario (blue) and during translocation of KTKEGV peptide (black and red). Black and red curves correspond to ionic current time series shown in panel C (raw and filtered data, respectively)

graph presented in Fig. 11.3A, the correction to apply to experimental conductance measurements can be easily extracted and use.

11.2.3 Ionic Current Measurements from Translocation Experiments

During nanopore translocation experiments, charged biomolecules immersed in an electrolyte are driven by a transverse applied voltage V through MoS_2 nanopore. In the meantime, the ionic current $I_c(t) = V \times G(t)$ is monitored, in order to detect the passage of biomolecule at sub-microsecond temporal resolution [11] (up to 100 ns from experiments and up to 1 ns from simulations). From MD trajectories, the ionic current $I_c(t)$ is computed as follows:

$$I_c(t) = \frac{1}{\Delta t L_z} \sum_{i=1}^{N} q_i \left[z_i(t + \Delta t) - z_i(t) \right], \tag{11.3}$$

where Δt is the time between MD snapshots chosen for the calculations (1 ns), L_z is the dimension of the simulation box in the z-direction, which is the direction of the applied electric field, N is the total number of ions, q_i is the charge of the ion i and $z_i(t)$ is the z-coordinate of the ion i at time t.

Figure 11.3C shows ionic current time series recorded during the translocation of KTKEGV peptide through SL-MoS_2 nanopore ($D = 1.5$ nm). First, fluctuations of the signal I_c^{RMS} are very large and noisy, the corresponding probability distribution $P(I_c)$ being very broad and unimodal (black line, Fig. 11.3D). It means that drops of current associated with the passage of biomolecule through the pore would not be distinguishable in the raw signal. From this observation, we decided to filter the data extracted from MD in order to remove high frequency fluctuations (low-pass filter). To do so, we computed the filtered signal as the moving mean of the raw signal over $T = 10,000$ samples. As shown in Fig. 11.3D (red line), current drops visually appear in the filtered signal, which is confirmed by the bimodal characteristics of the probability distribution. The maximum peak of $P(I_c)$ is centered around the mean value of open pore current $\overline{I}_0 = 3.1$ nA, with the same width as the one computed from the open pore current simulation (blue line, Fig. 11.3D). This clearly demonstrates that this part of the translocation signal is associated to an open pore situation. In addition, the second peak of the distribution, which is centered around 2.1 nA, corresponds to a drop associated with the passage of the biomolecule through the nanopore. Finally, the signal-to-noise ratio (SNR) calculated from MD simulations is around 5.0 with $I_c^{RMS} = 0.2$ nA and $\Delta I_c = 1.0$ nA. This value is very close to the experimental SNR of SL-MoS_2 nanopore of comparable dimension ($D = 1.4$ nm, $V = 300$ mV, $\overline{I}_0 = 2.6$ nA, $I_c^{RMS} = 0.1$ nA and $\Delta I_c = 0.5$ nA) [47].

In the following sections, ionic current traces from MD translocation experiments through different membranes and for different sequences of biological peptide are presented, in order to highlight the three main challenges in protein sensing using 2-D MoS_2 nanopores described in the Introduction Section.

11.3 Threading of Proteins Through MoS_2 Nanopores (Challenge One)

In contrast to DNA strands which are highly negatively charged biomolecules, i.e. the total charge being proportional to the number of nucleotides, proteins can be globally neutral, independently of its number of residues, sequence or size. Therefore, driving such biomolecules into the nanopore for sequencing applications requires a different strategy than just applying an electric field as driving force. Indeed, a compromise has to be made between facilitating the threading of the protein through the pore and controlling the translocation speed of the biomolecule, which should allow the detection of discernible current drops associated to its passage.

SSN with very high peptide translocation speed may limit their usability as protein sequencing device, which is the application of interest in our research group. The threading of biomolecule through the pore and the translocation speed of SSN can be tuned by adjusting different parameters in experiments such as solvent properties [7, 48, 49] (ionic species, concentration, temperature, viscosity, etc.); tuning materials size and shape [50] or by mixing different types of materials using heterostructures [27]. More drastically, the physical technique used to drive the biomolecule through the pore can also be modified, such as the use of a water-flow in addition to the presence of an applied voltage to drag proteins through the pore [24]. But the fragility of the membrane under such a large hydrostatic pressure gradient may be problematic.

Compared to all the techniques mentioned above, our approach focus on modifying the properties of the peptide instead of modifying the properties of the membrane. Technically, this alternative consists in the use of tags made of positively charged amino acids such as Lysine to functionalize the termination of proteins in order to promote their entrance into the nanopore. In biochemistry, polyionic tags such as short Lysine tags are used as enhancers of protein solubility in recombinant protein production. Due to their small size and their repetitive amino-acid content, they do not necessarily have an ordered three-dimensional conformation. As a result, they can exert their solubility-enhancing without interfering with the structure of the protein of interest or compromising its activity [51].

To explore the feasibility of using poly-Lysine tags to thread and fully translocate proteins through MoS_2 nanopores, we performed MD simulations of the translocation of Met-Enkephalin protein of sequence YGGFM, which is a neutral pentapeptide, through SL-MoS_2 ($D = 2.0$ nm). Tags made of Lysine residues, from 0 to 5 amino-acid length, were attached to the C-terminal part of Met-Enkephalin (Fig. 11.2C). Figure 11.4 shows results extracted from MD translocation experiments. First, from time series $z(t)$, which correspond to the z-coordinate of the center of mass of YGGFM peptide, we estimate the effective free-energy profile for the peptide to pass through the nanoporous membrane, i.e. $F(z) = -kT \log[P(z)/P_{\text{MAX}}]$, where $P(z)$ is the probability distribution of time series $z(t)$. Figure 11.4A shows free-energy profile $F(z)$ for YGGFMKKK peptide (3K). In this profile, three wells are identified and correspond to: ① the peptide diffusing in bulk solvent above the membrane, ② the peptide diffusing at the top surface of the membrane and ③ the peptide diffusing at the bottom surface of the membrane after a complete translocation through the nanopore. From this profile, two free-energy barriers ΔF are estimated, i.e. $\Delta F_{1\to2} = 3.3$ kT, which corresponds to free-energy barrier for the peptide to adsorb into the MoS_2 surface and $\Delta F_{2\to3} = 2.4$ kT, which corresponds to the free-energy barrier for the peptide to go through the MoS_2 nanopore. Basically, free-energy barrier $\Delta F_{1\to2}$ corresponds to the relaxation of the peptide at the beginning of the MD production run. The relaxation is due to the fact that the peptide "feels" the electric field for the first time, since there is no electric field applied during the equilibration process. Moreover, the position of the peptide is restrained during equilibration, in order to preserve the position of the peptide in bulk solvent above the membrane. Therefore, new initial random velocities plus the presence of the electric field involve a relaxation period that varies from a few nanoseconds for shorter peptides to a few tens of nanoseconds for longer ones.

The same type of profile with the same three wells is observed for YGGFMKK (2K), YGGFMKKKK (4K) and YGGFMKKKKK (5K), as shown in Fig. 11.4B. In addition, similar free-energy barriers ①→② are estimated from each profile ($\Delta F_{1\to 2} \approx 3.0$ kT, Fig. 11.4C). However, for YGGFM (0K) and YGGFMK (1K), only wells 1 and 2 are detected. In fact, there is no translocation of the peptide through MoS_2 nanopores for these two peptides with 0K or 1K tag. Therefore, free-energy barriers $\Delta F_{2\to 3}$ are *infinite* within the time-scale of our simulations for these two cases (gray rectangles in Fig. 11.4B). It is no surprising that without tag, since YGGFM is a neutral peptide and all its residues are neutral, the passage of the peptide through the pore is not observed since the applied electric field has no real impact into the peptide transverse displacement. For YGGFMK (1K), the total charge of the peptide is +1 and it appears that, for the given duration of MD runs performed here (250 ns), it is not sufficient to thread the peptide into the pore. However, we compare free-energy barriers $\Delta F_{2\to 3}$ between the four other tags 2K, 3K, 4K and 5K that lead to translocation events. As shown in Fig. 11.4C, the corresponding free-energy barrier decreases for 3K tag ($\Delta F^{3K}_{2\to 3} = 2.4$ kT) compare to 2K ($\Delta F^{2K}_{2\to 3} = 3.6$ kT), 4K ($\Delta F^{4K}_{2\to 3} = 4.2$ kT) and 5K tag ($\Delta F^{5K}_{2\to 3} = 6.4$ kT). At first, it is quite a surprising results owing to the fact adding more Lysine to the tag increases gradually the total charge of the peptide and so does the electric force $q\mathbf{E}$ acting on each amino acid. Despite that, free-energy barrier $\Delta F_{2\to 3}$ keeps increasing for 4K and 5K tag. It comes from the fact that the peptide becomes longer and bigger and consequently, it costs more free-energy to pass through the pore.

To quantify this behavior, we computed the radius of gyration R_g of the peptide during MD simulations. Figure 11.4D represents probability distributions $P(R_g)$ computed from R_g time series for each of the six YGGFM peptides, from 0K to 5K tag. As expected, adding more and more Lysine increases the size of the peptide, from $R_g = 0.5$ nm (YGGFM and YGGFMK) to $R_g = 0.9$ nm (YGGFMKKKKK). Moreover, from the distributions $P(R_g)$, we now understand why 3K tag behaves differently than the others. It is due to the fact that $P(R_g)$ is characterized by two peaks, one centered around $R_g = 0.55$ nm and the second one centered around $R_g = 0.75$ nm. Therefore, before entering the pore, YGGFMKKK peptide (3K) has a relatively small size compare to YGGFMKKKK (4K) and YGGFMKKKKK (5K). Finally, we also computed ionic current drops ΔI_c related to the passage of Met-Enkephalin peptides through MoS_2 nanopores. The largest drop is observed for YGGFMKKK (3K), with $\Delta I_c = 1.6$ nA. The fluctuations of the ionic current I_c^{RMS} being similar for all Met-Enkephalin peptides, YGGFMKKK (3K) is characterized by the largest SNR experiment. Finally, by looking at probability distributions $P(I_c)$ (Fig. 11.4D), the two peaks of the distribution corresponding to the open pore and blocked pore

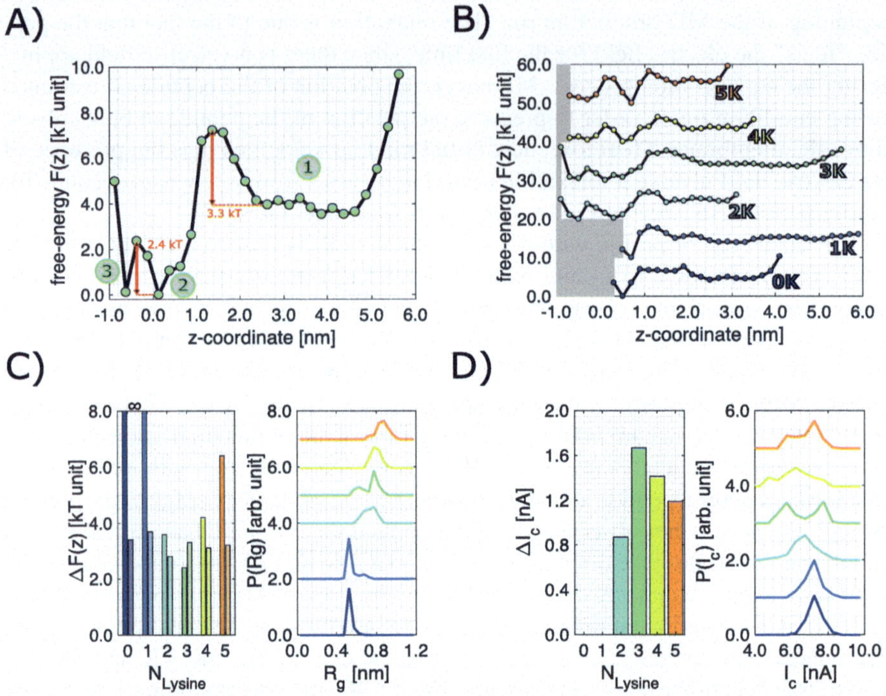

Fig. 11.4 Translocation of Met-Enkephalin through MoS_2 nanopores. **A** Effective free-energy profile F [in kT unit] as a function of z-coordinate [in nm] of the center of mass of the peptide computed from MD simulations of the translocation of YGGFMKKK peptide through SL-MoS_2 ($D = 2.0$ nm). Free-energy barriers are indicated with red labels. Well number 1 corresponds to the peptide diffusing in bulk solvent ($z > 2.0$ nm), well number 2 corresponds to the peptide attached to the top surface ($z > 0$ nm) and well number 3 corresponds to the peptide attached to the bottom surface ($z < 0$ nm). **B** Effective free-energy profiles F [in kT unit] as a function of z-coordinate [in nm] of the center of mass of the peptide computed from the translocation of YGGFM (0K, blue), YGGFMK (1K, lightblue), YGGFMKK (2K, cyan), YGGFMKKK (3K, green), YGGFMKKKK (4K, yellow) AND YGGFMKKKKK (5K, orange) peptide through SL-MoS_2 ($D = 2.0$ nm). Data have been shifted by 10.0 kT along the Y-axis for more clarity. **C** Effective free-energy barriers ΔF as a function of the length of Lysine tag attached to YGGFM (left panel, $\Delta F_{1 \to 2}$ is shown with transparent bar on the right and $\Delta F_{2 \to 3}$ is shown with plain bar on the left of the histogram) and probability distributions P of radius of gyration R_g (right panel). The color code is the same as in panel B. Data have been shifted by 2.0 along the Y-axis for more clarity. **D** Ionic current drops ΔI_c as a function of the length of Lysine tag attached to YGGFM (left panel) and probability distributions P of ionic current I_c (right panel). The color code is the same as in panel B. Data have been shifted by 1.0 along the Y-axis for more clarity

situation are well separated compared to other tags, which is crucial to sense single-biomolecule using nanopores.

Overall, we can conclude that free-energy landscape analysis reveals that a competition exists between the total charge of the peptide, which depends on the length of the poly-Lysine tag added to Met-Enkephalin protein, and the influence of the tag into the size and shape of the peptide in comparison with the dimension of the nanopore ($D = 2.0$ nm). Consequently, numerical testing can inform to determine the ideal length of the K tag to add to the peptide of interest. Moreover, one has to be careful that the addition of Lysine residues as tag to increase the total charge of the peptide must not decrease too drastically the SNR of the nanopore sensor. It is usually the case when the translocation happens too fast if the charge of the peptide increased synthetically becomes too large. In the case of a very fast translocation experiment, one approach used to to slow down the passage of charged peptides through MoS_2 nanopores consists to modify the pore dimensions D and h but, once again, SNR of the device must remain high, as well as its spatial and temporal resolution. This approach is fully discussed in the next section.

11.4 Slowing down Protein Translocation by Tuning Pore Dimensions (Challenge Two)

Here, we focus our interest to the modification of MoS_2 pore dimensions and its impact into ionic current traces recorded during MD simulations. To do so, a Lysine dipeptide (KK, Fig. 11.2C) is translocated through three different nanoporous membranes (Fig. 11.2B): (i) a SL-MoS_2 nanopore ($h = 0.3$) nm with a diameter $D = 2.0$ nm, which is considered here as the reference signal, (ii) a SL-MoS_2 nanopore ($h = 0.3$) nm with a diameter $D = 1.5$ nm and (iii) a BL-MoS_2 nanopore ($h = 0.9$) nm with a diameter $D = 2.0$ nm. From the reference ionic current signal $I_c(t)$, we quantify the differences observed in ionic current distributions $P(I_c)$ by first, reducing the diameter D of the pore and second, by increasing the thickness h of the membrane. The full description of MD runs presented here is available in reference [25].

The translocation of Lysine dipepetide through SL-MoS_2 of diameter $D = 2.0$ nm is characterized by a two-step process (Fig. 11.5A, top panel): the first Lysine amino acid (K1) enters the pore at $t \approx 34$ ns and is sensed for ≈ 1 ns and then, the second Lysine amino acid (K2) threads into the nanopore and the whole peptide leaves the ionic channel as one entity after $\tau_D \approx 1$ ns of dwell time. During this two-step sequence of event, the local ionic current drop detected is negligible with fluctuations of current being very large, which makes the drop visually indiscernible (Fig. 11.5A, top panel). From probability distributions $P(I_c)$ shown in Fig. 11.5B when the peptide is IN (red area) or OUT (blue area), we estimate the ionic current drop for this computer experiment to be $\Delta I_c = 1.3$ nA. Moreover, by looking at the whole distributions $P(I_c)$, we remark that there is a complete overlap between the values of ionic current recorded for both situations, IN and OUT. Finally, the distribution when the peptide is OUT is very wide (standard deviation $\sigma_{OUT} = 2.1$ nA), with values ranging from 4.0 to 10.0 nA. It means that without any information about the position of the peptide as it is the case in nanopore sensing MD experiments, we would not be able to detect the passage of a Lysine dipeptide through the pore and

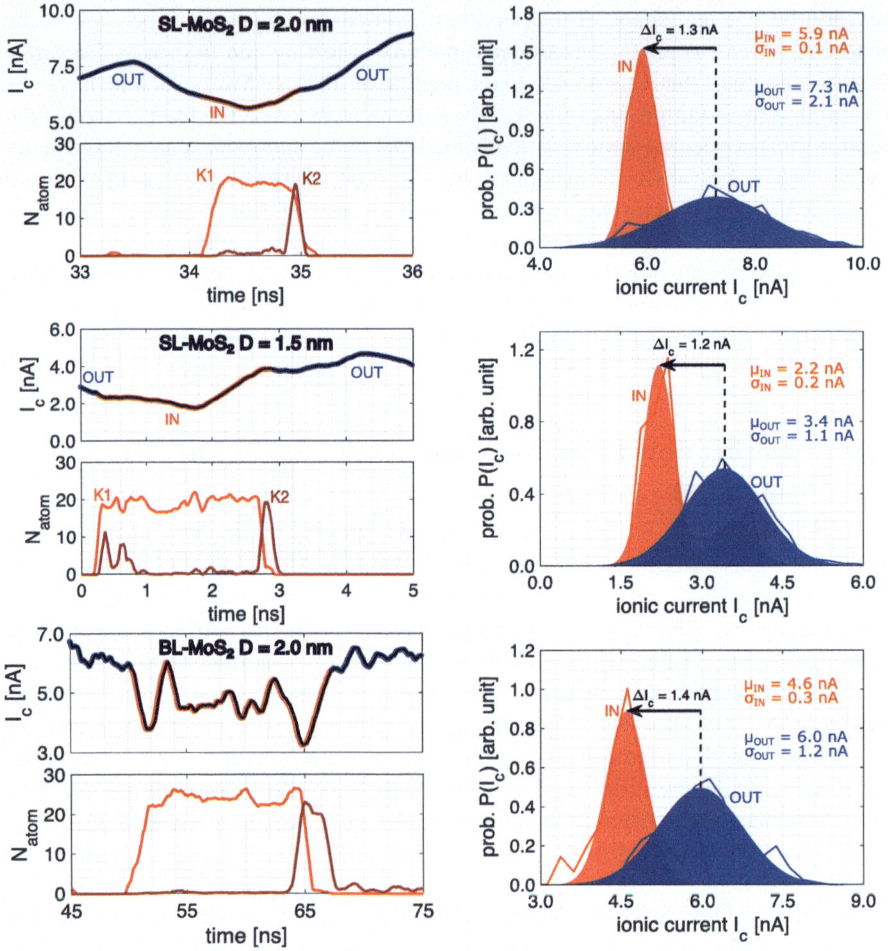

Fig. 11.5 Ionic current traces and probability distributions of ionic current recorded during the translocation of Lysine dipeptide through MoS_2 nanopores for different pore dimensions. **A** Ionic current I_c [in nA] (top panel) and number of atoms for each residue inside the pore N_{atom} (bottom panel) as a function of time [in ns]. Values of current which correspond to the presence (IN) or absence (OUT) of the peptide in the pore are indicated in red and blue, respectively. **B** Probability distributions $P(I_c)$ computed from I_c time series. The color code is the same as in panel A. Characteristics of Gaussian distributions, i.e. mean μ and standard deviation σ are indicated

more importantly we would not be able to statistically differentiate both amino acids K1 and K2. Therefore, reducing the translocation speed in order to get discernible ionic current drops out of the fluctuations of the time series is essential for the design of a nanopore sequencing device. Nevertheless, the fact that a two-step process, i.e. one residue translocating at a time is observed is an important preliminary result.

First, the diameter of SL-MoS_2 nanopore was reduced from $D = 2.0$ to $D = 1.5$ nm. This decrease of diameter leads to a pore conductance reduction from $\widetilde{G}_0 \approx 8$ to 4 nS, according to our model (Fig. 11.3B). As observed for SL-MoS_2 nanopore of diameter $D = 2.0$ nm, the translocation follows a two-step process, although an attempt of translocation of the peptide as a whole is observed at the beginning of the translocation process (Fig. 11.5A, middle panel). The N-terminal Lysine (K1) enters the pore first and after $\approx$3 ns, the C-terminal Lysine (K2) threads into the pore and the full translocation occurs. The dwell time τ_D is around 3 ns, which is three times longer than the one detected for the larger pore. From probability distributions of ionic current $P(I_c)$ (Fig. 11.5B, middle panel), we estimate the ionic current drop to be $\Delta I_c = 1.2$ nA, which is similar to the previous one detected for a larger pore. In addition, reducing diameter D of the nanopore involves smaller fluctuations of the ionic current in an open pore configuration, with $\sigma_{OUT} = 1.1$ nA. It means that SNR would be larger and the detection of the peptide translocation through the pore should be visually more clear, even though a second peak in the distribution when the peptide is IN is observed around 3.6 nA (corresponding to values of current for an open pore situation). Finally, as it is the case for diameter $D = 2.0$ nm, distinguishing each Lysine residue separately from ionic current traces recorded during MD translocation experiments remains unachievable.

Last but not least, the thickness of the MoS_2 membrane h was increased from single-layer ($h = 0.3$ nm) to bilayer ($h = 0.9$ nm), with a pore diameter of $D = 2.0$ nm. Compared to the decrease of the pore diameter, the conductance of the nanopore only drops from $\widetilde{G}_0 \approx 8.0$ to 6.0 nS, according to our model (Fig. 11.3B). The expected maximum conductance drop is then larger by increasing the thickness h than reducing the diameter D of the nanoporous membrane. Therefore, a larger SNR is expected (at constant I_c^{RMS}), leading to a good compromise in terms of sensor resolution. Figure 11.5A (bottom panel) shows ionic current trace recorded during the translocation of Lysine dipepetide through BL-MoS_2. As already observed for SL-MoS_2 membranes, the translocation process can de described by two steps: the N-terminal Lysine (K1) threading first into the pore followed by the C-terminal Lysine (K2), $\approx$15 ns later. In total, the dwell time $\tau_D \approx 18$ ns, reducing significantly the translocation speed compared to SL-MoS_2. Furthermore, ionic current trace recorded during the passage of Lysine dipeptide through the pore shows two distinct and discernible drops: the first one when K1 enters the pore at $t \approx 50$ ns and a second drop of the same magnitude when K2 threads into the pore at $t \approx 65$ ns. The corresponding current drop is $\Delta I_c = 1.2$ nA, estimated from probability distributions of ionic current shown in Fig. 11.5B (bottom panel). It corresponds to the same order of magnitude as the ones for the two other membranes (with similar fluctuations as SL-MoS_2 with $D = 1.5$ nm) but this time, two ionic current fingerprints of the signal are clearly and visually observed for the first time from MD.

This result is crucial for the design of protein sequencing devices. Indeed, for several years, experimentalists have been trying to reduce the thickness of the membrane, particularly using 2-D materials in order to get larger SNR and better spatial resolution. Here, we show that considering bilayer 2-D transition metal dichalcogenides materials instead of single-layer may lead to drops of ionic current directly related to the passage of individual amino acids through the nanopore. Consequences of improving the time resolution of the sensor by increasing the thickness of the nanoporous membrane is that the spatial resolution is downgraded. Therefore, visually reading the primary structure of proteins from ionic current measurements remains very challenging and time series algorithm are sometimes inevitable. This challenge is presented in the next section.

11.5 Identifying Protein Sequence Motifs from Ionic Current Measurements (Challenge Three)

The variability in ionic current traces associated with peptide translocation, as shown particularly in the previous section, requires a deep understanding and learning of the data. This can be done by applying time series analysis tools to ionic current measurements recorded during translocation experiments. Over the past few years, sophisticated algorithms have been developed to detect and statistically characterize ionic current drops and dwell time of biomolecule translocation through nanopore from experimental measurements [52, 53]. Indeed, it is necessary to *establish the non-linear relationship that exists between the presence of amino acids of the peptide inside the nanopore and the ionic current variations measured.* This can only be done if the position of the peptide is known at every single time step of the translocation experiment, which is what Molecular Dynamics is all about.

In order to understand the variability of peptide translocation fingerprints, a meaningful and reliable physical parameter is required to characterize the complexity of ionic current time series measured during nanopore sensing of proteins. In physics, the complexity of a time series is associated with the disorder degree, i.e. randomness and unpredictability. In this context, entropy is one of the most powerful metrics [54]. Specifically, Permutation Entropy (PE), which combined the concept of entropy and temporal order in a time series, measures information based on the occurrence or the absence of certain permutation patterns of the ranks of values in a time series [55]. In addition, PE can be calculated for arbitrary real-world time series [56], the method being extremely fast and robust. Here, we applied PE algorithm to ionic current time series recorded during the translocation of KTKEGV peptide through SL-MoS_2 of diameter $D = 1.5$ nm, the time series being shown in Fig. 11.3C (red line).

In nanopore experiments, single-molecule sensing events are extracted from ionic current time series using a current threshold value I_c^{thr}. A drop of ionic current ΔI_c is considered as a sensing event if values of current $I_c(t)$ measured during a certain amount of time, i.e. dwell time τ_D, are below the threshold. In order to mimic as close as possible the experimental investigation of sensing events from nanopore measurements, we first perform a blind detection of sensing events based on a current threshold, as in experiments. The current threshold was defined from the probability distribution of open pore ionic current as $I_c^{thr} = 5\sigma_{I_0}$, which corresponds to five times the standard deviation of the open pore ionic current I_0 probability distribution shown in Fig. 11.3D (blue line). In several experimental works, the current threshold used is $3\sigma_{I_0}$. However, in the case of MoS_2 nanopores, we demonstrated that a $3\sigma_{I_0}$ current threshold is not sufficient [26]. With the $5\sigma_{I_0}$ threshold, a total of 49 sensing events are detected (Fig. 11.6A), representing cumulatively around 20% of the total 2.5 µs ionic current time series. As observed experimentally and detailed recently for DNA in reference [57], there is a large variability of current versus time traces within sensing events. For instance, some events maintain fairly constant current drop and others show switching levels and bumps.

Therefore, it is impossible to visually associate each level of ionic current to the passage of specific amino acids through the nanopore. Nevertheless, information about the biomolecule may be hidden in the ionic current time series and the question is where is this information localized. In other words, which levels and bumps of current and its variations are relevant for protein sequencing and how much information do they contain. To answer these questions, we quantify the complexity of the ionic current traces extracted from MD using PE algorithm. PE measures the disorder of n successive values in a given time series by using the probability of different permutation patterns for a given time window of length T within the time series. Technically, PE is computed as:

$$PE = -\frac{1}{\log_2(n!)} \sum_{j=1}^{n!} p_j \log_2(p_j), \tag{11.4}$$

where p_j represents the relative frequencies of the possible permutation patterns ($n!$). More details about PE algorithm can be found in references [55, 56].

PE algorithm was applied to each ionic current trace shown in Fig. 11.6A with $n = 3$ and $T = 1{,}000$ samples, then sliding the time window by 500 samples [26]. Figure 11.6B shows PE as a function of time window for a specific sensing event (index 41, Fig. 11.6A), with a maximum current drop $\Delta I_c = 0.34$ nA and dwell time $\tau_D = 19.5$ ns. The corresponding ionic current trace is presented in Fig. 11.6B and is overall characterized by three different levels of ionic current, i.e. $\Delta I_c \approx 0.05$ nA for $\tau \in [0; 3]$ ns and $\tau \in [15; 19]$ ns, $\Delta I_c \approx 0.25$ nA for $\tau \in [6.5; 8.5]$ ns and $\Delta I_c \approx 0.30$ nA for $\tau \in [10; 13]$ ns. Moreover, several bumps are observed within levels of ionic current. Applying PE reveals four time windows for which PE is almost null, corresponding to regular linear drops of ionic current between each level and named hereafter sub-events [26].

Each sub-event is then depicted by the absolute value of its slope, i.e. $\Delta(\Delta I_c)/\Delta\tau$, which corresponds to an ionic current drop speed. This new parameter named $v_{\Delta I_c}$ is defined here to characterize ionic current traces variability from protein translocation experiments. Furthermore, each ionic current drop speed detected using PE algorithm can be associated to protein sequence motifs. Namely, an ionic current drop speed $v_{\Delta I_c}$ is associated to a protein sequence motif comprised of N amino acids if the N amino acids are inside the pore at the same time during its detection. Several examples about the detection of protein sequence motifs are described in details in Reference [26]. For the specific event shown in Fig. 11.6B, two distinct motifs are

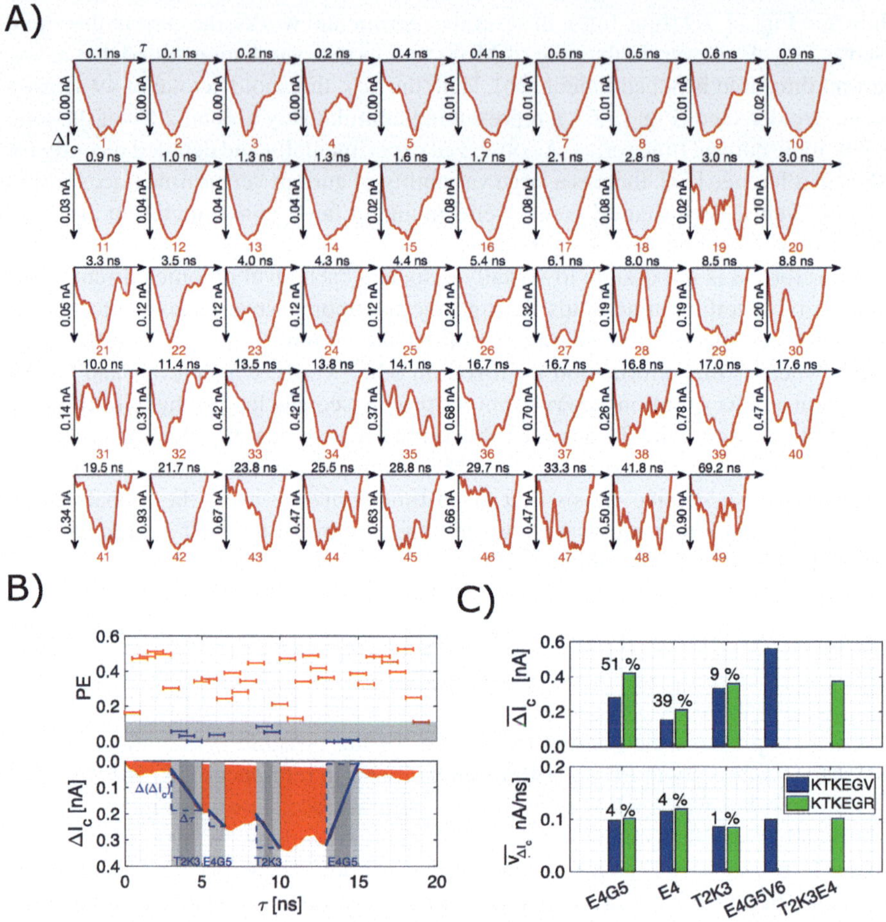

Fig. 11.6 Ionic current traces variability observed during the translocation of KTKEGV peptide through SL-MoS_2 nanopores ($D = 1.5$ nm). **A** Signatures of ionic current drop ΔI_c as a function of time τ for the 49 sensing events extracted from MD translocation experiments shown in Fig. 11.3C. **B** Permutation Entropy as a function of time [in ns] (top panel) for a specific sensing event (index 41, panel A) of 19.5 ns duration. Ionic current drop as a function of time [in ns] (bottom panel) is also shown. Blue lines correspond to parts of the signal for which PE <0.11, named hereafter sub-events. **C** Average current drop $\overline{\Delta I_c}$ [in nA] (top panel) and average ionic current drop speed $\overline{v_{\Delta I_c}}$ [in nA/ns] (bottom panel) per protein sequence motif

associated to ionic current drop speed, i.e. T2K3 and E4G5. Finally, we performed the same PE analysis for each of the 49 events shown in Fig. 11.6A. In total, eleven motifs are associated to ionic current drop speed $v_{\Delta I_c}$, the four most probable ones being: E4G5 (≈40%), E4 (≈28%), T2K3 (≈12%) and E4G5V6 (≈10%).

Lastly, we evaluate the impact of biological mutations onto ionic current traces and their characteristics $v_{\Delta I_c}$ described above for the translocation of KTKEGV peptide through SL-MoS_2 nanopores. The mutation selected here is replacing Valine V6 (hydrophobic neutral amino acid) by an Arginine R6 (positively charged amino acid). Consequently, the total charge of the peptide is modified from +1 to +2. Using the same procedure to detect sensing events, 26 events were detected, representing cumulatively 10% of the total ionic current time series. In addition, 11 protein sequence motifs were associated to ionic current traces of the mutant peptide KTKEGR. The four most probable motifs are given in Fig. 11.6C and were compared to the ones extracted from KTKEGV peptide. Among these four motifs, three are similar for both peptides: E4G5, E4, and T2K3. Moreover, these three motifs are characterized by similar average ionic current drop speed $\overline{v_{\Delta I_c}}$, which is a very important result for sequencing application. Indeed, from independent MD simulations, we demonstrate for the first time to the best of our knowledge that protein sequence motifs involve similar ionic current related parameters. A difference of maximum 4% is observed for $\overline{v_{\Delta I_c}}$ for identical motifs. For information, the comparison of average ionic current drop $\overline{\Delta I}_c$, usually usc to try to sequence biomolecule using SSN shows up to 51% difference for identical motifs, which is definitely not suitable for protein sequencing application. This result confirms that the application of PE algorithm to extract relevant characteristics of protein sequence motifs from the ionic current traces is appropriate and consistent.

11.6 Concluding Remarks

Identification of proteins by solid-state nanopores sequencing is very challenging. Most probably, it will require to cut into polypeptides pieces [58]. Understanding and control polypeptides sequencing using SSN is thus a prerequisite. Heterogeneity of amino-acid properties and of nanoporous membranes at the atomic scale, as well as the complicated conformational dynamics of a polypeptide chain, render difficult the prediction and the interpretation of translocation events. Experimentally, a detailed information at the atomic scale, as accessible for protein in their native state by NMR or XRD is not available. Therefore, MD simulations of protein sequencing using SSN may play a role in contributing to a better understanding of the complexity of translocation of polypeptides.

In the present chapter, all-atom MD simulations of translocation of model polypeptides through promising MoS_2 nanopores were analyzed. From them, several useful information were extracted:

1. "A larger net peptide charge is not necessarily better". A compromise must be found in the number of charged residues (Lysine or Arginine) added to improve the threading of the peptide through the nanopore. Increasing the length of the tag may increase the free-energy barrier to cross the membrane. Further studies are needed to quantify the relative contributions of entropy and enthalpy to the crossing barrier.

2. "A thinner membrane is not necessarily the best". Indeed, increasing the thickness decreases SNR of the nanopore sensor but also slows down the passage of the peptide to the other side of the membrane. As shown here, a comprise must be found between these two effects. For MoS_2 nanopores, a double-layer appears to be a working concept.
3. "Current drops are maybe not the best fingerprints". In fact, larger variations of current drops are observed in simulations for the translocation of the same protein sequence motif, whereas the current drop slope is much more discriminant.

Moreover, what MD can provide next? As demonstrated for DNA and poly-Lysine peptides [21], the combination of several signals characterizing the biopolymer translocation would improve the accuracy of biomolecule sequence reads. We envision that Extraordinary Acoustic Raman (EAR) spectroscopy [59], which characterize the acoustical modes of proteins ($<$100 GHz), might be integrated in principle to a SSN device. MD would be essential to interpret such possible 2-D signal (ionic current/spectroscopic data) as predictions of protein acoustical modes require all-atom description of their vibrational spectra [60, 61].

From a fundamental point of view, realistic simulations of polypeptide translocations including protein diffusion, state-of-the art atomic potentials and simulation of the current fluctuations, can contribute to decipher the nonlinear relations between the ionic current time series and its meaning, i.e. its information content. As shown here, application of time series analysis, as high frequency filtering and permutation entropy, better reveal information embedded in the noisy signal. Certainly, supervised machine learning techniques, already used in this context [18], may assist in the identification of motifs and ultimately in sequences of polypeptides.

What machine learning techniques coupled to MD could bring in the future is the relation between the polypeptide structures and the current measured. Indeed, every value in the time series corresponds to an ensemble of polypeptide configurations known by MD but invisible to experiments. Applying deep learning techniques to establish links between current and structures might be one of the next challenges.

Acknowledgements The simulations were performed using HPC resources from DSI-CCuB (Université de Bourgogne). The work was supported by a grant from the Air Force Office of Scientific Research (AFOSR), as part of a joint program with the Directorate for Engineering of the National Science Foundation (NSF), Emerging Frontiers and Multidisciplinary Office (Grant No. FA9550-17-1-0047). Moreover, this work is part of the project NANOSEQ (2018–2021) and NANO-NEURO-MED (2019–2022) supported by the EIPHI Graduate School (contract ANR-17- EURE-0002), the Conseil Régional de Bourgogne Franche-Comté and the European Union through the PO FEDER-FSE Bourgogne 2014/2020 programs.

References

1. Akeson Mark, Branton Daniel, Church Geroge, Deamer David W, and Church George. Characterization of individual polymer molecules based on monomer-interface interactions. December 2003.

2. David Deamer, Mark Akeson, and Daniel Branton. Three decades of nanopore sequencing. *Nat. Biotechnol.*, 34(5):518–524, May 2016.
3. Elizabeth A. Manrao, Ian M. Derrington, Andrew H. Laszlo, Kyle W. Langford, Matthew K. Hopper, Nathaniel Gillgren, Mikhail Pavlenok, Michael Niederweis, and Jens H. Gundlach. Reading DNA at single-nucleotide resolution with a mutant MspA nanopore and phi29 DNA polymerase. *Nat. Biotechnol.*, 30(4):349–353, March 2012.
4. Nicole Rusk. MinION takes center stage. *Nat. Methods*, 12(1):12–13, January 2015.
5. Christopher A. Merchant, Ken Healy, Meni Wanunu, Vishva Ray, Neil Peterman, John Bartel, Michael D. Fischbein, Kimberly Venta, Zhengtang Luo, A. T. Charlie Johnson, and Marija Drndić. DNA translocation through graphene nanopores. *Nano Lett.*, 10(8):2915–2921, August 2010.
6. Ke Liu, Jiandong Feng, Andras Kis, and Aleksandra Radenovic. Atomically Thin Molybdenum Disulfide Nanopores with High Sensitivity for DNA Translocation. *ACS Nano*, 8(3):2504–2511, March 2014.
7. Jiandong Feng, Ke Liu, Roman D. Bulushev, Sergey Khlybov, Dumitru Dumcenco, Andras Kis, and Aleksandra Radenovic. Identification of single nucleotides in MoS2 nanopores. *Nat. Nanotechnol.*, 10(12):1070–1076, December 2015.
8. Liang Xue, Hirohito Yamazaki, Ren Ren, Meni Wanunu, Aleksandar P. Ivanov, and Joshua B. Edel. Solid-state nanopore sensors. *Nature Reviews Materials*, 5(12):931–951, December 2020.
9. Laura Restrepo-Pérez, Chirlmin Joo, and Cees Dekker. Paving the way to single-molecule protein sequencing. *Nat. Nanotechnol.*, 13(9):786–796, September 2018.
10. Nitinun Varongchayakul, Jiaxi Song, Amit Meller, and Mark W. Grinstaff. Single-molecule protein sensing in a nanopore: a tutorial. *Chem. Soc. Rev.*, 47(23):8512–8524, November 2018.
11. Siddharth Shekar, David J. Niedzwiecki, Chen-Chi Chien, Peijie Ong, Daniel A. Fleischer, Jianxun Lin, Jacob K. Rosenstein, Marija Drndić, and Kenneth L. Shepard. Measurement of DNA Translocation Dynamics in a Solid-State Nanopore at 100 ns Temporal Resolution. *Nano Lett.*, 16(7):4483–4489, 2016.
12. Benjamin Cressiot, Laurent Bacri, and Juan Pelta. The Promise of Nanopore Technology: Advances in the Discrimination of Protein Sequences and Chemical Modifications. *Small Methods*, 4(11):2000090, 2020.
13. Carl A. K. Borrebaeck. Precision diagnostics: moving towards protein biomarker signatures of clinical utility in cancer. *Nat. Rev. Cancer*, 17(3):199–204, March 2017.
14. Robert S. B. Clark, Hülya Bayir, and Larry W. Jenkins. Posttranslational protein modifications. *Crit. Care Med.*, 33(12 Suppl):S407–409, December 2005.
15. Alaji Bah, Robert M. Vernon, Zeba Siddiqui, Mickaël Krzeminski, Ranjith Muhandiram, Charlie Zhao, Nahum Sonenberg, Lewis E. Kay, and Julie D. Forman-Kay. Folding of an intrinsically disordered protein by phosphorylation as a regulatory switch. *Nature*, 519(7541):106–109, March 2015.
16. Hadjer Ouldali, Kumar Sarthak, Tobias Ensslen, Fabien Piguet, Philippe Manivet, Juan Pelta, Jan C. Behrends, Aleksei Aksimentiev, and Abdelghani Oukhaled. Electrical recognition of the twenty proteinogenic amino acids using an aerolysin nanopore. *Nat. Biotechnol.*, 38(2):176–181, February 2020.
17. Eamonn Kennedy, Zhuxin Dong, Clare Tennant, and Gregory Timp. Reading the primary structure of a protein with 0.07 nm3 resolution using a subnanometre-diameter pore. *Nat. Nanotechnol.*, 11(11):968–976, November 2016.
18. Mikhail Kolmogorov, Eamonn Kennedy, Zhuxin Dong, Gregory Timp, and Pavel A. Pevzner. Single-molecule protein identification by sub-nanopore sensors. *PLOS Comput. Biol.*, 13(5):e1005356, 2017.
19. María Daniela Barrios Pérez, Adrien Nicolaï, Patrice Delarue, Vincent Meunier, Marija Drndić, and Patrick Senet. Improved model of ionic transport in 2-D MoS2 membranes with sub-5 nm pores. *Appl. Phys. Lett.*, 114(2):023107, January 2019.
20. Y. P. Shan, P. B. Tiwari, P. Krishnakumar, I. Vlassiouk, W. Z. Li, X. W. Wang, Y. Darici, S. M. Lindsay, H. D. Wang, S. Smirnov, and J. He. Surface modification of graphene nanopores for protein translocation. *Nanotechnology*, 24(49):495102, December 2013.

21. Michael Graf, Martina Lihter, Damir Altus, Sanjin Marion, and Aleksandra Radenovic. Transverse Detection of DNA Using a MoS2 Nanopore. *Nano Lett.*, 19(12):9075–9083, December 2019.
22. Chaoming Gu, Zhoubin Yu, Xiaojie Li, Xin Zhu, Zhen Cao, Zhi Ye, Chuanhong Jin, and Yang Liu. Experimental study of protein translocation through MoS2 nanopores. *Appl. Phys. Lett.*, 115(22):223702, November 2019.
23. Amir Barati Farimani, Mohammad Heiranian, and Narayana R. Aluru. Identification of amino acids with sensitive nanoporous MoS 2 : towards machine learning-based prediction. *npj 2D Materials and Applications*, 2(1):1–9, May 2018.
24. Huang Chen, Libo Li, Tao Zhang, Zhiwei Qiao, Jinhui Tang, and Jian Zhou. Protein Translocation through a MoS2 Nanopore:A Molecular Dynamics Study. *J. Phys. Chem. C*, 122(4):2070–2080, February 2018.
25. Adrien Nicolaï, Maria Daniela Barrios Pérez, Patrice Delarue, Vincent Meunier, Marija Drndić, and Patrick Senet. Molecular Dynamics Investigation of Polylysine Peptide Translocation through MoS2 Nanopores. *J. Phys. Chem. B*, 123(10):2342–2353, March 2019.
26. Adrien Nicolaï, Aniket Rath, Patrice Delarue, and Patrick Senet. Nanopore sensing of single-biomolecules: a new procedure to identify protein sequence motifs from molecular dynamics. *Nanoscale*, 12(44):22743–22753, November 2020.
27. Binquan Luan and Ruhong Zhou. Single-File Protein Translocations through Graphene-MoS2 Heterostructure Nanopores. *J. Phys. Chem. Lett.*, 9(12):3409–3415, June 2018.
28. Hadi Arjmandi-Tash, Liubov A. Belyaeva, and Grégory F. Schneider. Single molecule detection with graphene and other two-dimensional materials: nanopores and beyond. *Chem. Soc. Rev.*, 45(3):476–493, February 2016.
29. Gopinath Danda, Paul Masih Das, Yung-Chien Chou, Jerome T. Mlack, William M. Parkin, Carl H. Naylor, Kazunori Fujisawa, Tianyi Zhang, Laura Beth Fulton, Mauricio Terrones, Alan T. Charlie Johnson, and Marija Drndić. Monolayer WS2 Nanopores for DNA Translocation with Light-Adjustable Sizes. *ACS Nano*, 11(2):1937–1945, February 2017.
30. Humberto R. Gutiérrez, Nestor Perea-López, Ana Laura Elías, Ayse Berkdemir, Bei Wang, Ruitao Lv, Florentino López-Urías, Vincent H. Crespi, Humberto Terrones, and Mauricio Terrones. Extraordinary room-temperature photoluminescence in triangular WS2 monolayers. *Nano Lett.*, 13(8):3447–3454, August 2013.
31. Mike Tebyetekerwa, Jian Zhang, Zhen Xu, Thien N. Truong, Zongyou Yin, Yuerui Lu, Seeram Ramakrishna, Daniel Macdonald, and Hieu T. Nguyen. Mechanisms and Applications of Steady-State Photoluminescence Spectroscopy in Two-Dimensional Transition-Metal Dichalcogenides. *ACS Nano*, 14(11):14579–14604, November 2020.
32. Fangzhou Hu, Borislav Angelov, Shuang Li, Na Li, Xubo Lin, and Aihua Zou. Single-Molecule Study of Peptides with the Same Amino Acid Composition but Different Sequences by Using an Aerolysin Nanopore. *Chembiochem*, 21(17):2467–2473, September 2020.
33. Fabien Piguet, Hadjer Ouldali, Manuela Pastoriza-Gallego, Philippe Manivet, Juan Pelta, and Abdelghani Oukhaled. Identification of single amino acid differences in uniformly charged homopolymeric peptides with aerolysin nanopore. *Nature Communications*, 9(1):966, March 2018.
34. Linda Payet, Marlène Martinho, Manuela Pastoriza-Gallego, Jean-Michel Betton, Loïc Auvray, Juan Pelta, and Jérôme Mathé. Thermal unfolding of proteins probed at the single molecule level using nanopores. *Anal. Chem.*, 84(9):4071–4076, May 2012.
35. Kevin J. Freedman, Maike Jürgens, Anmiv Prabhu, Chi Won Ahn, Per Jemth, Joshua B. Edel, and Min Jun Kim. Chemical, thermal, and electric field induced unfolding of single protein molecules studied using nanopores. *Anal. Chem.*, 83(13):5137–5144, July 2011.
36. François Sicard and Patrick Senet. Reconstructing the free-energy landscape of Met-enkephalin using dihedral principal component analysis and well-tempered metadynamics. *J. Chem. Phys.*, 138(23):235101, June 2013.
37. Ulf Dettmer, Andrew J. Newman, Victoria E. von Saucken, Tim Bartels, and Dennis Selkoe. KTKEGV repeat motifs are key mediators of normal α-synuclein tetramerization: Their mutation causes excess monomers and neurotoxicity. *Proc. Natl. Acad. Sci. U.S.A.*, 112(31):9596–9601, August 2015.

38. Steve Plimpton. Fast Parallel Algorithms for Short-Range Molecular Dynamics. *J. Comput. Phys.*, 117(1):1–19, March 1995.
39. Jin-Wu Jiang. Parametrization of Stillinger–Weber potential based on valence force field model: application to single-layer MoS 2 and black phosphorus. *Nanotechnology*, 26(31):315706, 2015.
40. Kresten Lindorff-Larsen, Stefano Piana, Kim Palmo, Paul Maragakis, John L. Klepeis, Ron O. Dror, and David E. Shaw. Improved side-chain torsion potentials for the Amber ff99SB protein force field. *Proteins: Structure, Function, and Bioinformatics*, 78(8):1950–1958, 2010.
41. William L. Jorgensen, Jayaraman Chandrasekhar, Jeffry D. Madura, Roger W. Impey, and Michael L. Klein. Comparison of simple potential functions for simulating liquid water. *J. Chem. Phys.*, 79(2):926–935, July 1983.
42. In Suk Joung and Thomas E. Cheatham. Determination of alkali and halide monovalent ion parameters for use in explicitly solvated biomolecular simulations. *J. Phys. Chem. B*, 112(30):9020–9041, July 2008.
43. Stefan W. Kowalczyk, Alexander Y. Grosberg, Yitzhak Rabin, and Cees Dekker. Modeling the conductance and DNA blockade of solid-state nanopores. *Nanotechnology*, 22(31):315101, 2011.
44. Marıa Daniela Barrios Perez, Patrick Senet, Vincent Meunier, and Adrien Nicola. Computational Investigation of the Ionic Conductance through Molybdenum Disulfide (MoS2) Nanopores. *WSEAS Transactions On Circuits And Systems*, 16, 2017.
45. Jothi Priyanka Thiruraman, Kazunori Fujisawa, Gopinath Danda, Paul Masih Das, Tianyi Zhang, Adam Bolotsky, Néstor Perea-López, Adrien Nicolaï, Patrick Senet, Mauricio Terrones, and Marija Drndić. Angstrom-Size Defect Creation and Ionic Transport through Pores in Single-Layer MoS2. *Nano Lett.*, 18(3):1651–1659, March 2018.
46. Jothi Priyanka Thiruraman, Paul Masih Das, and Marija Drndić. Stochastic Ionic Transport in Single Atomic Zero-Dimensional Pores. *ACS Nano*, 14(9):11831–11845, September 2020.
47. Alessio Fragasso, Sonja Schmid, and Cees Dekker. Comparing Current Noise in Biological and Solid-State Nanopores. *ACS Nano*, 14(2):1338–1349, February 2020.
48. Daniel Fologea, James Uplinger, Brian Thomas, David S. McNabb, and Jiali Li. Slowing DNA Translocation in a Solid-State Nanopore. *Nano Lett.*, 5(9):1734–1737, September 2005.
49. Stefan W. Kowalczyk, David B. Wells, Aleksei Aksimentiev, and Cees Dekker. Slowing down DNA Translocation through a Nanopore in Lithium Chloride. *Nano Lett.*, 12(2):1038–1044, February 2012.
50. Ke Liu, Martina Lihter, Aditya Sarathy, Sabina Caneva, Hu Qiu, Davide Deiana, Vasiliki Tileli, Duncan T. L. Alexander, Stephan Hofmann, Dumitru Dumcenco, Andras Kis, Jean-Pierre Leburton, and Aleksandra Radenovic. Geometrical Effect in 2D Nanopores. *Nano Lett.*, 17(7):4223–4230, July 2017.
51. Vasiliki Paraskevopoulou and Franco H. Falcone. Polyionic Tags as Enhancers of Protein Solubility in Recombinant Protein Expression. *Microorganisms*, 6(2), May 2018.
52. C. Raillon, P. Granjon, M. Graf, L. J. Steinbock, and A. Radenovic. Fast and automatic processing of multi-level events in nanopore translocation experiments. *Nanoscale*, 4(16):4916–4924, July 2012.
53. Calin Plesa and Cees Dekker. Data analysis methods for solid-state nanopores. *Nanotechnology*, 26(8):084003, February 2015.
54. C. E. Shannon. A mathematical theory of communication. *Bell Syst. Tech. J.*, 27(3):379–423, July 1948.
55. Christoph Bandt and Bernd Pompe. Permutation Entropy: A Natural Complexity Measure for Time Series. *Phys. Rev. Lett.*, 88(17):174102, April 2002.
56. M. Riedl, A. Müller, and N. Wessel. Practical considerations of permutation entropy. *Eur. Phys. J. Spec. Top.*, 222(2):249–262, June 2013.
57. Chen-Chi Chien, Siddharth Shekar, David J. Niedzwiecki, Kenneth L. Shepard, and Marija Drndić. Single-Stranded DNA Translocation Recordings through Solid-State Nanopores on Glass Chips at 10 MHz Measurement Bandwidth. *ACS Nano*, 13(9):10545–10554, September 2019.

58. G. Sampath. Amino acid discrimination in a nanopore and the feasibility of sequencing peptides with a tandem cell and exopeptidase. *RSC Adv.*, 5(39):30694–30700, March 2015.
59. Skyler Wheaton, Ryan M. Gelfand, and Reuven Gordon. Probing the Raman-active acoustic vibrations of nanoparticles with extraordinary spectral resolution. *Nature Photonics*, 9(1):68–72, January 2015.
60. Adrien Nicolaï, Patrice Delarue, and Patrick Senet. Theoretical Insights into Sub-Terahertz Acoustic Vibrations of Proteins Measured in Single-Molecule Experiments. *J. Phys. Chem. Lett.*, 7(24):5128–5136, December 2016.
61. Adrien Nicolaï, Fatima Barakat, Patrice Delarue, and Patrick Senet. Fingerprints of Conformational States of Human Hsp70 at Sub-THz Frequencies. *ACS Omega*, 1(6):1067–1074, December 2016.

Chapter 12
Single-Molecule Ionic and Optical Sensing with Nanoapertures

Wayne Yang and Cees Dekker

Abstract Solid-state nanoapertures (nanometer sized holes within a membrane) allow for the interrogation of single molecules by probing them within their nanoscale volume through which molecules pass in a single-file manner. Molecules are probed using two main techniques: ionic sensing where a salt solution and applied voltage is used to produce an ionic current through the nanopore, and optical sensing through a shift in the resonance wavelength of the plasmonic nanoaperture. Here, we briefly review the basic principles, applications, and challenges in sensing with solid-state nanoapertures, as well as some strategies for further improvements. We compare the complimentary features of the two approaches and highlight recent attempts to combine them into new sensing platforms.

12.1 Introduction

All biosensors can be categorized by the size of their sensing region which is important for their detection sensitivity. By constraining a sensor to a small enough volume and by tuning the concentration of the analyte to low values, even single molecules can be sensed and studied [1–3]. Properties on the single-molecule level often differ from ensemble studies [4–6]. For example, transient intermittent conformational states of proteins, which are important for an understanding of their function, are often lost with ensemble techniques such as crystallography [7–9]. Similarly, detection of single-base mutations in the DNA sequence (known as single nucleotide polymorphisms) which are only present on some DNA strands, are quickly lost in the background signal when an ensemble of DNA molecules is sequenced [10, 11]. Indeed, the ability for sensing on the single-molecule level is important for both

W. Yang · C. Dekker (✉)
Kavli Institute of Nanoscience, Delft University of Technology, Delft, The Netherlands
e-mail: c.dekker@tudelft.nl

W. Yang
e-mail: w.w.w.yang@tudelft.nl

W. Bowen et al. (eds.), *Single Molecule Sensing Beyond Fluorescence*,
Nanostructure Science and Technology,
https://doi.org/10.1007/978-3-030-90339-8_12

the understanding of biological processes as well as in applications for diagnostics [12–15].

Solid-state nanopores are a common way of building such single-molecule sensors. Here one etches a tiny hole (aperture) into a solid-state membrane which allows single molecules to physically pass by, as they are small enough to pass, similar to a molecular tunnel [16]. The apertures need to be reliably fabricated on the nanometer scale, i.e. the size scale of single biomolecules—in order to physically restrict the passage of molecules to a single-file manner. During the passage of the molecule, its properties can be probed. Interestingly, in cells, such pores are formed naturally within the membranes of cells where they serve to control the passage of molecules like DNA, RNA, and proteins. However, such biological pores are hard to integrate into solid-state platforms as the fabrication of such systems requires a wide range of material science, engineering, and nanofabrication techniques [1, 17, 18]. Fortunately, the rapid development and the increasing availability of nanolithography techniques in the last few decades have enabled the top-down fabrication of similar solid-state nanoapertures [19]. Apertures, of various shapes on the scale of one to many tens of nanometers, can be reliably formed in a variety of different solid-state membranes (e.g. silicon nitride-SiN). The use of novel 2D materials, such as graphene or hexagonal boron nitride (hBN), has pushed the membrane thickness and therefore the volume of the sensing region down to the ultimate thickness of a single atomic layer [20–22].

We define a nanoaperture to be a nanometer-sized hole (of varying shape and with lateral dimensions of 1–100 nm) that is fabricated within a free-standing membrane that is of a similar thickness. In this chapter, we cover two common ways to probe the passage of the molecules of interest through such nanoapertures. First, ionic sensing which involves the detection of a temporary blockade in the ionic current when molecules translocates through the nanopore [23]. Second, optical sensing where a change is detected in the light transmission or reflection upon the molecular passage, which arises due to a change in the optical resonance of the nanoaperture (which will be further explained below) [24]. Notably, both of these methods do not require the biomolecule of interest to be labelled. Such label-free techniques are to be preferred over other single-molecule techniques such as electron microscopy, fluorescence microscopy, or force techniques such as AFM or optical/magnetic tweezers, that all require the use of labels such as heavy-atom stains in electron microscopy, fluorophores in fluorescence techniques, and attachment groups for force spectroscopy, which might alter the native behavior of the molecule of interest [25–27]. Below, we introduce the ionic and optical sensing techniques, describe a few applications, and lay out the limitations and challenges facing each approach.

12.2 Principle of Ionic Current Sensing with Nanopores

A long-standing idea in biosensing has been to monitor blockades in the ionic current in order to detect single particles. This idea was first proposed by Wallace Coulter in the 1940s for the screening and counting of blood cells [28]. In so-called ‘Coulter

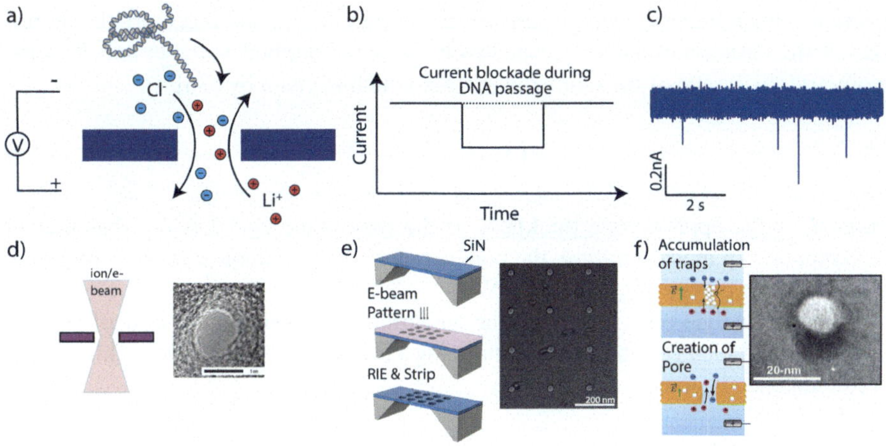

Fig. 12.1 Overview of ionic sensing and fabrication of nanopores a Schematic overview of ionic sensing. A nanoaperture is immersed in an ionic solution (pictured here with LiCl) and a voltage difference is applied. Ions in the solution are electrophoretically driven to the respective electrodes. Charged biomolecules such as DNA can also be driven through this nanoaperture. **b** Illustration of the current blockade during the DNA passage of the nanopore. When a molecule such as DNA enters the nanopore, it temporarily blocks the flow of ions leading to a measurable decrease in the current during the time of passage. **c** Example trace of the ionic current (100 mV, 1M LiCl) through an 15 nm nanopore, showing clear dips in the current each time a DNA molecule translocates (20 kbp, 1 ng/μl). **d** Illustration of ion/electron beam milling of SiN membranes for nanopores. Typically, the electron beam needs to be ~200 keV to be able to mill away the Silicon nitride membrane. Nanopore image on the right is from the first application of a TEM drilling of nanopores, reproduced with permission from [32], ©Springer Nature, 2003. **e** An array of nanopores produced with e-beam lithography and reactive ion etching. Reproduced from [33]. **f** Illustration of nanopores produced by dielectric breakdown of SiN membranes. A voltage is applied across an insulating membrane which causes charges to accumulate at a defect site. This leads the membrane to locally collapse and produce a nanopore. Reproduced from [34], Creative Commons CC-BY license, 2014

counters', a blood sample is diluted in an ionic solution such as NaCl, and an electric field is applied through a voltage difference between both reservoirs, driving ions through an aperture (which here was ~10 μm diameter). This narrow constriction provides a resistance in the ionic current because it constricts the flow of ions. When the particle of interest, here a blood cell, passes through the constriction, the flow of ions is temporarily disrupted, causing a measurable change in the ionic current. This sensing principle was later extended to single-molecule detection with nano-constrictions such as nanopores (~1–50 nm), where the passage of a single DNA causes a measurable change in the ionic current [1, 29–31] (Fig. 12.1a).

The current through such a nanopore is, to first approximation, given by [35]:

$$i_{\text{open pore current}(d)} = n(\mu_{\text{cation}} + \mu_{\text{anion}})eV * \frac{1}{R} \tag{12.1}$$

where n is the number of charge carriers in solution, μ_{cation} and μ_{anion} are the mobilities of the cation and anions, respectively, V is the applied voltage, and R is the geometric resistance of the nanoconstriction which is given by [36]:

$$R(d) = \frac{4L}{\pi d^2} + \frac{1}{d} \tag{12.2}$$

where L is the approximate thickness of the membrane and d is the diameter of the nanopore. In most literature, the conductance rather than the current is reported, which is given by $G = i_{\text{open pore current}}/V$.

When an analyte passes through the constriction, it disrupts the flow of ions by way of the excluded volume, thereby inducing a lower pore conductance. This change in conductance is given by:

$$\Delta G = G_{\text{open pore current}} - G_{\text{analyte}} = G(d) - G(d_{\text{reduced}}) \tag{12.3}$$

where G_{analyte} is the reduced ionic conductance of the nanopore when the analyte resides in the pore. For an extended object that spans the entire thickness of the pore (such as a long DNA molecule), this d_{reduced} is equal to the new effective diameter from the reduction of the open pore volume that is now occupied by the analyte:

$$d_{\text{reduced}} = \sqrt{d^2 - d_{\text{analyte}}^2} \tag{12.4}$$

where d_{analyte} is the diameter of the rod that approximates the analyte (2.2 nm for the case of DNA), see Fig. 12.1b. Figure 12.1c shows an example of ionic current trace for a buffer of 2M LiCl containing DNA. Clear dips can be seen which are caused by the passage of individual DNA molecules. Equation 12.3 highlights a few parameters that are crucial for ionic sensing: the salt concentration, the applied voltage which supplies the driving force to force the molecule through the constriction, and the size of the constriction that defines the geometric resistance [37, 38].

The nanoscale apertures for ionic sensing are usually fabricated in a thin membrane (such as 5–20 nm thick SiN membrane) through direct-milling techniques with an electron beam in a Transmission Electron Microscope (TEM), typically with electron energies over 200 KeV or with a Focused Ion Beam (FIB) [32, 39, 40] (Fig. 12.1d). In recent years, more accessible pore-fabrication techniques have been developed such as nanolithography combined with a dry etching (Fig. 12.1e) or dielectric breakdown by applying a large applied voltage (Fig. 12.1f) [33, 34, 41–43]. There also has been a drive to control the fabrication of such pores down to the precision of a single atoms with crystalline 2D materials and advanced milling techniques [21, 44].

12.3 Application of Ionic Sensing in Nanopore Experiments

Ionic sensing has been successfully applied to the sensing of single DNA molecules. Most applications with solid-state nanopores focused on a fundamental biophysical understanding of the polymer physics of the translocating DNA. Different polymer configurations are adopted by DNA as it translocates through the nanopore, as already observed in the earliest work on solid-state nanopores [45–47]. This spawned interest in nanopores as a tool to probe DNA polymer physics on the nanoscopic scale, an order of magnitude smaller than the typical sub-micrometer resolution of fluorescence-based techniques. More recently, the groups of Dekker and Garaj (Fig. 12.2a) used solid-state nanopores to study DNA folds and knots [48, 49] where they interrogated thousands of DNA at the single-molecule level to elucidate the behavior and physics of DNA knots. Nanopores have also been used to detect and identify single proteins [50, 51], for example, in work by Mayer et al. who used lipid-coated solid-state nanopores to determine the size, shape, and charge of single proteins based on differences in the ionic current [52] (Fig. 12.2b). Additionally, these hybrid membranes have allowed for bottom up construction of biomimetic nanopores, by tethering biological materials (such as nucleoporins and peptide strains) to an intermediate interface such as DNA origami structures that dock on top of solid state membranes. This approach has been successful in probing and reconstituting the transport behaviour of biological pores and channels such as nuclear pore complexes that mediate transport in and out of the cell's nucleus (NPCs) [53–55].

The success of solid-state nanopores to interrogate molecules has inspired nanopore-based diagnostic applications. As the nanopore sensing principle revolves around the sensitivity of the ionic current to the size of the object, people have started using locally bound proteins to fingerprint DNA, as such protein yield measurable spikes above the DNA-only blockade level (such as in RecA-coated DNA or streptavidin-biotin-tagged DNA [58–60]). Recent work has focused on CRISPR/dCas9 proteins that bind particular DNA sequences. Weckman et al. and Yang et al. showed that differences as small as a single base mutation can lead to detectable changes in the protein-binding rate to DNA, opening up applications in screening for single-nucleotide polymorphism [56, 61] (Fig. 12.2c). Alternatively, Edel et al. explored the use of aptamers, DNA sequences that are specially designed to bind specific proteins, to screen for the presence of certain proteins which are biomarkers or targets in human serum [57] (Fig. 12.2d). These efforts illustrate the application of nanopores to diagnostics applications beyond just biophysics experiments in a research setting.

12.4 Limitations and Challenges of Ionic Sensing

Though successful in such applications, the ionic-sensing approach suffers from several limitations. First, it requires the use of an ionic liquid as the signal scales with concentration of salt dissolved (typically 0.3–4 M) [62]. The higher the salt

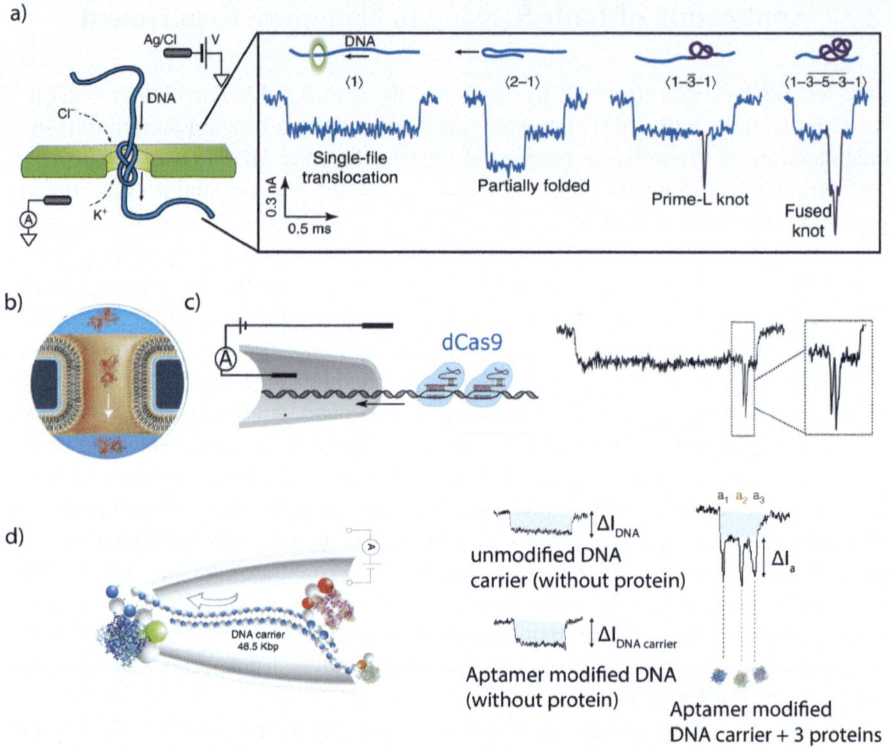

Fig. 12.2 Ionic-sensing application of nanopores a Sensing of DNA folds and knots in solid-state nanopore, which show up as distinct shapes in the ionic current blockades. The depth of the blockade is quantized according to the number of strands of the DNA that simultaneously resides in the pore. Reproduced from [48], Creative Commons CC-BY license 4.0, 2019. **b** Detection of proteins in a lipid-coated solid-state nanopore to discriminate between the size, shape, and charge of the proteins. Reproduced with permission from [52], ©Springer Nature, 2017. **c** Detection of CRISPR/dCas9 bound on DNA. Due to their larger physical size, show up as additional blockade on top of the DNA blockade current. Since the proteins are programmable to bind any target sequence, specific sequences along the DNA can be targeted for detection and fingerprinting of the DNA. Reproduced with permission from [56], ©American Chemical Society 2019. **d** Sensing of proteins with DNA aptamers. Specific DNA sequences allows for binding of certain proteins, thereby facilitating their detection and identification. Reproduced from [57], Creative Commons CC-BY license 4.0, 2017

concentration, the higher the conductivity of the liquid and therefore the higher the current obtained (Fig. 12.3a). However, high-salt solutions are usually detrimental for many biomolecules which require physiological conditions (0.1–0.25 M) as high salt interferes with the proper binding and functioning of proteins which are often mediated by electrostatic interactions [63]. Second, ionic current sensing with nanopores comes with a fair amount of noise. A wide range of electrical noise sources contributes across the frequency spectrum, as illustrated in Figure 12.3b. For an in depth look at each of these noise sources, we refer to the recent review by Fragasso et

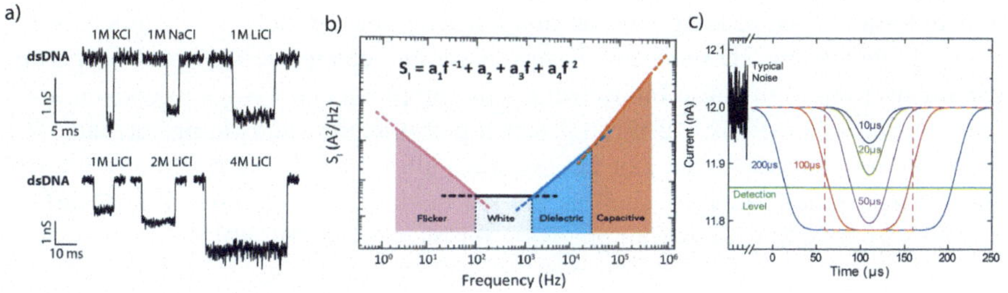

Fig. 12.3 Signals and noise in solid-state nanopores a DNA blockade signals in buffers with different concentrations and types of salt. Reproduced with permission from [62], ©American Chemical Society 2012. **b** Power spectrum density of the nanopore ionic current, with the main noise regions indicated. Reproduced with permission from [64], ©American Chemical Society 2020. **c** Depth of ionic blockade with different low-pass frequency filters. For the same input signal, the resultant signal is plotted after the different low pass filter is applied. The low pass filter distorts the maximum depth of the current blockade due to the different temporal resolution of the low pass filter. If a detection level is set (green line), the signal passed through the 10, 20 µs (100, 50 KHz) will not pass this cut off. Reproduced with permission from [68], ©American Chemical Society 2015

al. [64]. In brief, it is challenging to eliminate or control all these sources of noise, for example, those that arise from chemical processes happening on the membrane surface (e.g., transient binding and unbinding of ions to defect sites) which are very hard to prevent [64–67]. Third, there is a limit to the temporal resolution that can be achieved in ionic sensing. Molecules, driven by the electrophoretic force from the applied electric field, typically transverse very quickly across the nanosized apertures (<5µs for proteins and <0.5µs for each bp in a 10 kbp long DNA) [68, 69]. Many commercial amplifier have a bandwidth limitation of ~50 KHz (20 µs), making it impossible to resolve the full blockade of relevant analytes. Furthermore, low-pass filters are typically applied to eliminate high-frequency electrical noise, which further exacerbates the problem as it reduces the signal from the translocation event by clipping the full depth of the nanopore event thereby limiting event detection (See Fig. 12.3c).

Various solutions have been proposed to address these problems. Newly developed amplifiers and the miniaturization and integration of on-chip CMOS preamplifiers with solid-state nanopores have pushed the temporal resolution to the 10 MHz range (0.1 µs) [70]. Effective operation with such amplifiers, however, still need low-noise nanopores with specially designed membranes. New nanopore chips are being developed with glass support or additional insulating layers that lower the capacitance and hence lower the electrical noise [71, 72]. Other strategies revolves around slowing down the translocation of the molecule such as the NEOtrap, which uses an induced electro-osmotic current to trap the analyte of interest for long interrogation times [73, 74]. Furthermore, with the advent of 2D materials, membranes fabricated from a single or few layer graphene or hexagonal boron nitride have been explored [21, 75]. These membranes can be as thin as a single atomic layer which is the ultimate limit in terms of the sensing volume that a membrane can define. An added bonus feature is that many 2D materials have remarkable electronic transport properties which can be used to sense biomolecules in complementary ways to conventional ionic sensing schemes such as through sensing of tunneling and gating currents [76–78]. Moreover, the vast library of 2D materials allows for the fabrication of a novel class of nanofluidic devices, such as 2D slit devices, with geometric and surface tunability approaching angstrom precision for biosensing [79, 80].

12.5 Optical Sensing in Plasmonic Apertures

We now turn to a complimentary sensing technique, the monitoring of optical transmission through the nanoaperture. Classically, it was thought that nanoapertures with subwavelength dimensions (~100 nm) were thought to be too small for significant light transmission and hence for optical readout methods to be employed [81]. However, the report of extraordinary transmission through nanoapertures [82], where the magnitude of transmission was measured to be orders of magnitude higher than classically predicted, overturned this idea and reinvigorated the exploration of optical readout methods to probe the nanoscale volumes confined by a small aperture.

Central to the phenomenon of extraordinary transmission of light is the role of surface plasmons that are excited in plasmonic materials such as gold or silver films. These plasmonic oscillations of electrons in the metal can focus the incident electromagnetic (EM) field to the aperture, which gives rise to the enhancement of scattered and transmitted light [87]. Transmission of light though these apertures strongly depends on the wavelength of the incident light and the size as well as geometry of the aperture. The apertures are designed typically to show a resonance peak at a particular wavelength. As the resonant wavelength, i.e. where transmission is maximum, is very sensitive to the surrounding refractive index of the medium enclosing the nanoaperture, small modulations, such as a protein or molecule occupying the nanoaperture volume, can lead to a measurable redshift in the wavelength of the resonance peak. Figure 12.4a illustrates this principle of optical detection showing a redshift of the resonance wavelength when an object enters near the nanoaperture. Notably, this can lead to a decrease or an increase of the measured transmission signal: if the incident laser wavelength was lower than the resonant peak wavelength (i.e., on the left of the peak of the back curve in Fig. 12.4a), there will be a decrease in the transmission when the peak is redshifted, while there will be an increase in the transmission if the laser wavelength was higher than the resonant wavelength (right of the peak as drawn in Fig. 12.4a). The change in the transmitted light can be detected with a fast photodetector, typically an avalanche photodiode (APD). This has inspired a new class of sensors with optical readout methods that are based on detecting this resonance-wavelength shift in either the transmitted or reflected light.

Additionally, the tight focusing of the incident optical light produces strong optical gradient within the nanoaperture, yielding a trapping force on the particle in the aperture, similar to that of an optical tweezer. This can be employed as nanotweezers

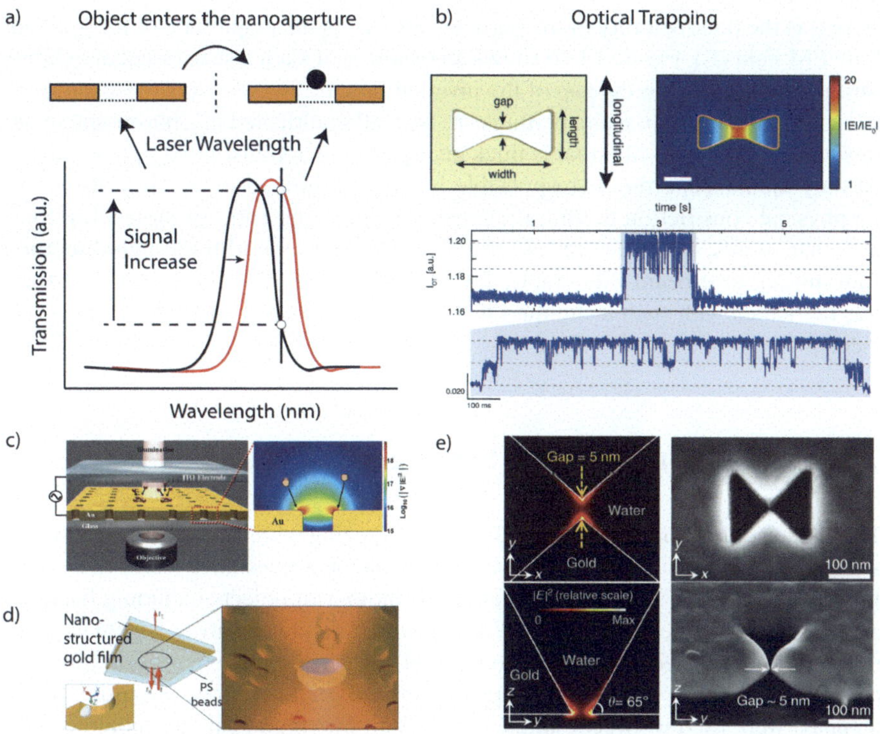

Fig. 12.4 Optical detection and trapping with nanoapertures a Illustration of the sensing principle. A nanoaperture has a peak wavelength where the transmission through the aperture is the highest (black curve). A laser is parked at a wavelength slightly detuned from the resonant wavelength peak. When an object of a higher refractive enters the aperture, the resonant wavelength is redshifted (red curve). This increases the transmission through the nanoaperture. **b** Optical trapping in a nanoaperture. A gold nanoaperture focuses the incident E-field to the center of the bowtie. The accompanying field gradient gives rise to an optical trapping force similar to optical tweezers. Small objects such as proteins can be trapped in the nanoaperture. Shown here is the trapping signal from a beta-amylase protein. Reproduced with permission from [83], ©John Wiley and Sons 2019. **c–e** Examples of various nanoapertures that produces an E-field gradient and trapping force. Shapes vary from round holes of 100 nm to inverted bowtie structures with features <5 nm in scale. Reproduced with permissions from [84–86], ©Springer Nature 2009, American Chemical Society 2014, Creative Commons CC-BY license 4.0, 2018

to hold the molecule in place for sensing [84, 88, 89]. Finite-difference time-domain method (FDTD) simulations of the EM field in the structures estimate these EM gradients for a variety of different structures (Fig. 12.4b–c). They produce a gradient force:

$$F = \frac{1}{2}\alpha \nabla E^2 \tag{12.5}$$

where α is the polarizability of the particle and ∇E^2 is the magnitude of the gradient of the EM field (E). Figure 12.4b shows an example of such a structure that exhibits a strong gradient within the gap of the inverted bowtie.

For optical detection, nanoapertures are typically fabricated in a plasmonic material such as a thin gold film that is thick enough to block transmitted light to ensure that only the nanoaperture is responsible for the transmitted light. They also serve as a physical constriction to limit the sensing volume and entry of molecules. Similar to nanopores, the structures are usually milled into the film through direct ion beam milling or e-beam lithography [90–92]. Figure 12.4c–e shows various structures fabricated with such techniques that were used to detect and tweeze nanoscale objects.

12.6 Application of Optical Sensing in Nanopores

Though the application of optical sensing and tweezing with nanoapertures is still in its infancy, early success with the tweezing of nanoparticles have led to the exploration of sensing and trapping of a variety of biological objects including bacterial cells, DNA, and single proteins [93–95]. Figure 12.5a shows an array of gold nanoantennas (here fabricated as simply two parallel Au wires that were 500 nm long and spaced ~30 nm apart, thereby generating an optical gradient in the proximity of the antennas) were used to tweeze single E-coli [96]. Moving to smaller biomolecules, gold nanoapertures were already employed in 2012 to optically tweeze and detect single proteins [97, 98] by the group of Gordon et al. (Fig. 12.5b). Bovine serum albumin (BSA) was used as a model proteins in their double nanohole traps. When BSA was trapped, two different intensity levels could be seen above the baseline which they attributed to the two different protein conformations [99]. The same traps were also used to study a variety of reactions including binding rates of protein-antibodies [99]. Finally, Verschueren et al. and Shi et al. were able to optically detect single DNA molecules that traversed a plasmonic nanopore, in both transmission and reflection mode [83, 100] (Fig. 12.5c). The DNA molecules were electrophoretically driven into the nanoapertures where they caused a shift in the resonance wavelength of the laser and a measurable change in the transmission (in this instance, a decrease). They verified their optical signal with conventional current blockade signal from ionic sensing. Notably, even the number of folds in the DNA during the passage through the nanoaperture led to an observable difference in the optical signal.

These studies show some of the many exciting features of label-free biosensing. There is also the ability for multiplex sensing of parallel apertures where a high density of these sensors can be packed into a small volume and simultaneously read out, because each nanoaperture can be independently measured. Notably, optical sensing is independent of buffer solutions unlike ionic sensing which requires a salt solution, allowing native physiological buffers to be used [101]. The noise sources for optical readout methods are very different from the noise sources affecting electrical readout methods. They are usually dictated by sources in the far-field such as the

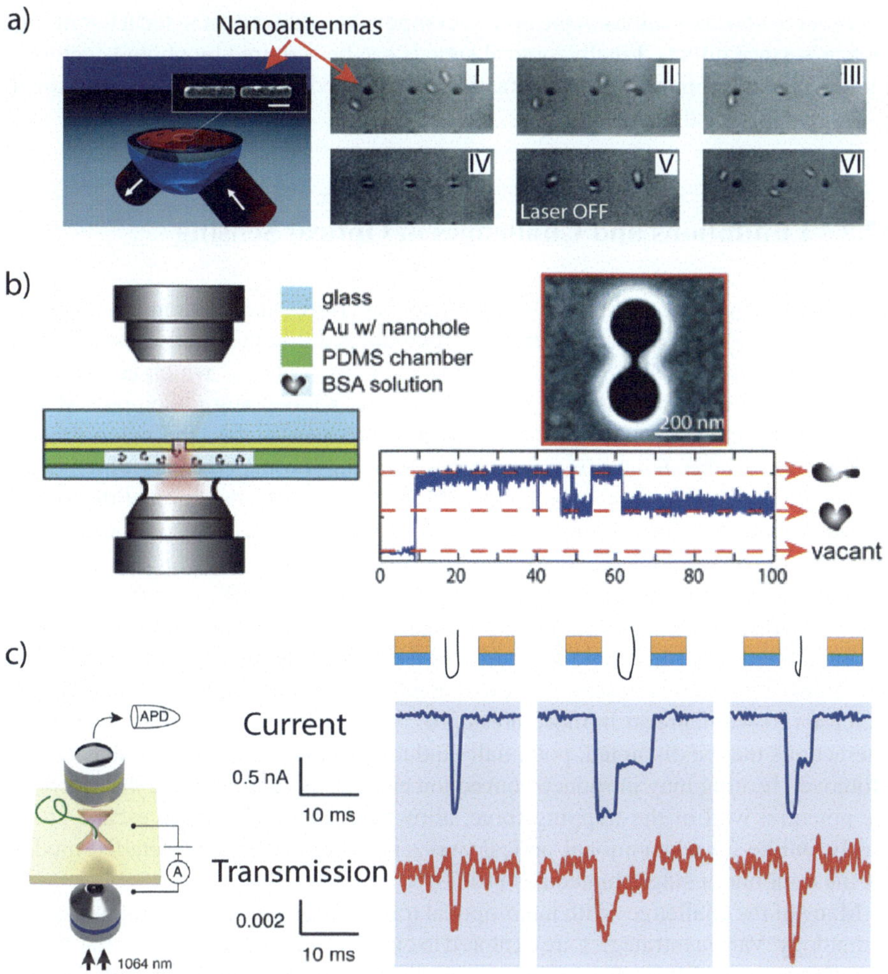

Fig. 12.5 Examples of optical detection and trapping of biomolecules in Au nanoapertures a Trapping of Escherichia coli bacteria with an array of Au nanoantenna. Scale bar is 200 nm. The Au nanoantenna was 500 nm with a 30 nm gap between them. Time series shows the E. coli (white) localising to the nanoantennas (the pair of nanoantennas (1030 nm) show up as black dots in the image). At time V, the incident laser was switched off and the E.coli was released from the vicinity of the nanoantennas indicating that they had been optically trapped. Reproduced with permission from [96], ©American Chemistry Society 2009. **b** Trapping of BSA proteins in a double hole nanoapertures. When a single protein is trapped, two different intensity levels are observed, associated with two different conformational shapes. Reproduced with permission from [97], ©American Chemistry Society 2012. **c** Detection of DNA in Au nanoapertures. When DNA is electrophoretically driven into an inverted-bowtie-shaped nanoaperture, the ionic and optical signals can be simultaneously detected. Reproduced with permission from [83], ©American Chemistry Society 2019

laser source and aberrations in the optical components, rather than in local conditions such as surface effects. Finally, optical signals can be acquired by photodetectors at a much higher temporal resolution (as fast as picoseconds or 10^{12} Hz), fast enough to observe protein dynamics.

12.7 Limitations and Challenges in Optical Sensing

Yet, as any technique, optical sensing with nanoapertures also has some limitations. First, the fabrication of nanoapertures is extremely difficult because the plasmon response are very sensitive to small (~nm) geometric changes which can shift the resonance and thus decrease the sensitivity of the nanoaperture. Complex shapes with a well-defined sensing region are hard to produce as the fabrication process quickly reduces the sharpness of the structure, giving it rounded edges which damp out the plasmons and the focusing of the EM field. Generally, the fabrication process is critical and difficult to control and consistently reproduce. Second, the use of the laser source for the readout method can cause local heating which may lead to denaturation of proteins. While typical laser powers (5–20 mW) over the laser spot size (~1 μm in diameter) are unlikely to cause an overall temperature increase in the bulk, the use of plasmonic nanostructures can concentrate and focus the energy into highly localised nanoscale hotspots thereby generating significant temperature increase (20–40 degrees) in those areas [102–106]. At these temperatures, peptide interactions maybe disrupted, potentially inducing protein unfolding or denaturing. Moreover, heating may introduce convection currents near the surface that decrease the potential well of the trapping force, allowing trapped objects to escape [107]. Lastly, unlike for conventional optical tweezers, there is a lack of detailed models for the trapping of single molecules [108].

Many of the challenges with nano-optical traps relate to the relative infancy of the technology. Various strategies are explored to circumvent the above listed challenges. The limits of nanofabrication techniques can be pushed for example, from the current effectively ~5 nm scale [90, 109, 110] (set by the resolution of the resist used) with e-beam lithography and laser interferometery, to the 1–2 nm scale using maskless direct-milling techniques [92]. An interesting approach to circumvent resolution issues is to simply fabricate a huge array of structures on the same gold film [91], with each slightly different geometries and laser scanning to find structures that are ideal for optical sensing and trapping. Second, local heating from the impinging laser beam can be reduced by the use of heat sinks surrounding the nanoapertures [111, 112]. Third, attempts are underway to improve modelling of nano-optical tweezers in order to better model probe and understand the nanoscale forces in the nanoaperture and guide future designs of such nanotraps [84, 89].

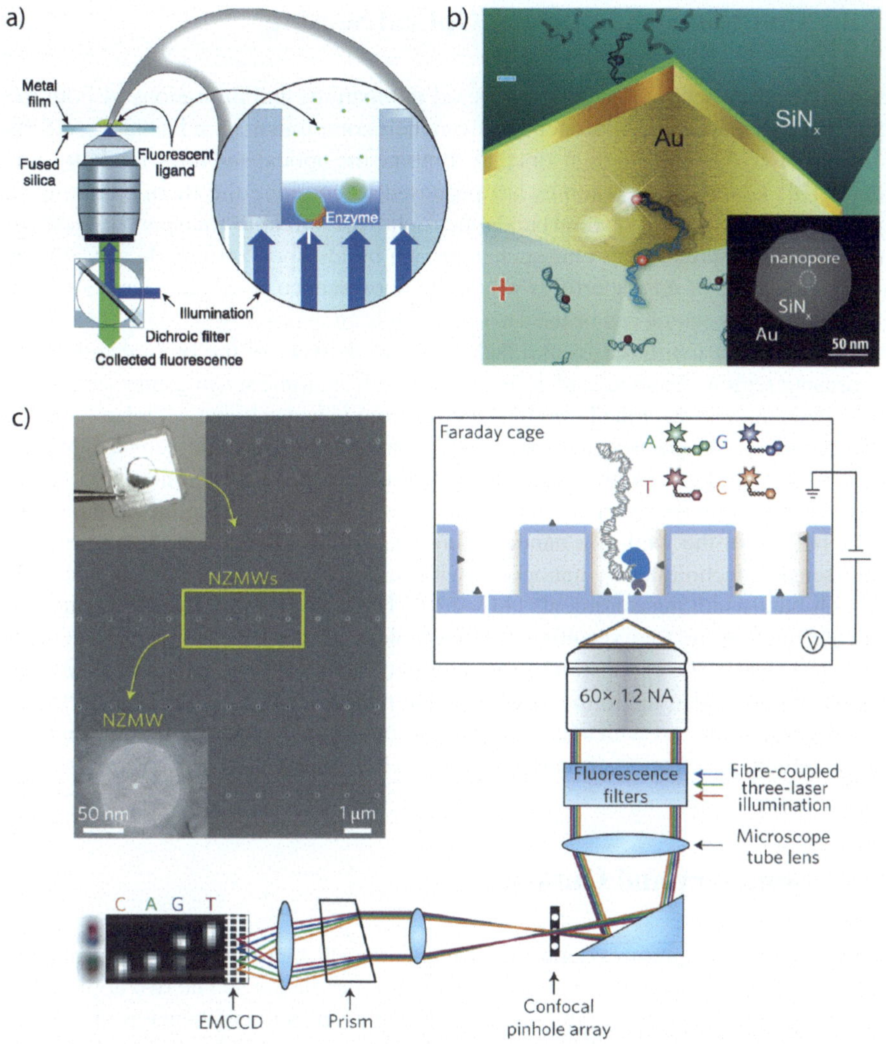

Fig. 12.6 Combined optical-ionic nanopores a Zero mode waveguides where an aperture is fabricated in a thin metal film to sense single enzymes. At the bottom of the well, a DNA polymerase complex is bound that produces a fluorescence signal that can be used to identify the DNA sequence of interest. Reproduced with permission from [113], ©The American Association for the Advancement of Science, 2003. **b** Illustration of a nanowell milled in a thin gold film with an integrated nanopore termed "opto-nanopore". The nanopore is used to electrophoretically drive the DNA into the sensing area where it is optically sensed. Reproduced with permission from [114], ©John Wiley and Sons 2017. **c** Integration of a zero-mode waveguide with a nanopore that drives DNA molecules to the sensing area. Reproduced with permission from [115], ©Springer Nature 2017

12.8 Simultaneous Ionic and Optical Sensing

In the preceding sections, we have covered the features and limitations of both the ionic and optical sensing approaches. Given their complimentary advantages, efforts are ongoing to develop new sensors that combine the optical and ionic readout readouts [116]. A variety of schemes are proposed to combine the thermophoretic (a byproduct of the optical readout) and electrophoretic effect to transport particles to targeted sensing area [117]. One such approach builds upon zero mode waveguides which are a class of nanoapertures where the fluorescence is enhanced in a strongly localised regions, which can be used to sense single enzymes (Fig. 12.6a) [113]. However, these devices suffered from a lack of active delivery of the biomolecules into the sensing region. Wanunu et al. took inspiration from ionic sensing approaches and integrated a nanopore at the bottom of the zero-mode waveguide well where a DNA polymerase was assembled, in order to deterministically drive and deliver DNA for sequencing (Fig. 12.6c) with a fluorescent readout method [115]. Similarly, Meller et al. developed what they called "opto-nanopores" where they integrated a nanopore at the bottom of the well of a nanoaperture (Fig. 12.6b), thus allowing to combine ionic sensing techniques in nanopores with a variety of optical readout methods for high-bandwidth measurements [118–120]. They, along with others, demonstrate optical detection and fingerprinting of DNA molecules through the use of fluorescent labels [83, 100]. The expansion of research in combining ionic and optical sensing in nanoapertures also led to the discovery that nanopores can be directly fabricated in thin SiN membranes with a laser beam. This opens up yet another scalable avenue for the reliable fabrication of nanopore arrays [121, 122].

12.9 Summary and Outlook

Summing up, nanoapertures define a sensing volume that can be probed with an ionic current or through optical sensing at the single-molecule level. The two approaches offer complimentary advantages and recent attempts to combine them have led to the advent of a new class of optical nanopores for detection and manipulation of single biomolecules. These nanosensors will increasingly be beneficial for a fundamental understanding of biology as well as for real-world applications such as the detection of biomarkers in point-of-care devices.

Acknowledgements We would like to thank Alessio Fragasso, Xin Shi and Sonja Schmid for fruitful discussions. C.D would like to acknowledge funding from ERC grant SynDiv 669598 and the Netherlands Organization for Scientific Research (NWO/OCW) as part of the Frontiers of Nanoscience and Basyc programs.

References

1. Daniel Branton, David W. Deamer, Andre Marziali, Hagan Bayley, Steven A. Benner, Thomas Butler, Massimiliano Di Ventra, Slaven Garaj, Andrew Hibbs, Xiaohua Huang, Stevan B. Jovanovich, Predrag S. Krstic, Stuart Lindsay, Xinsheng Sean Ling, Carlos H. Mastrangelo, Amit Meller, John S. Oliver, Yuriy V. Pershin, J. Michael Ramsey, Robert Riehn, Gautam V. Soni, Vincent Tabard-Cossa, Meni Wanunu, Matthew Wiggin, and Jeffery A. Schloss. The potential and challenges of nanopore sequencing. 26(10):1146–1153, 2008.
2. Laura Restrepo-Pérez, Chirlmin Joo, and Cees Dekker. Paving the way to single-molecule protein sequencing. *Nature nanotechnology*, 13(9):786–796, 2018.
3. John J Kasianowicz and Sergey M Bezrukov. On 'three decades of nanopore sequencing'. *Nature Biotechnology*, 34(5), 2016.
4. Jordanka Zlatanova and Kensal van Holde. Single-Molecule Biology: What Is It and How Does It Work? *Molecular cell*, 24(3):317–329, 2006.
5. Taekjip Ha, Alexander G. Kozlov, and Timothy M. Lohman. Single-molecule views of protein movement on single-stranded DNA. *Annual Review of Biophysics*, 2012.
6. Weichun Zhang, Martín Caldarola, Xuxing Lu, Biswajit Pradhan, and Michel Orrit. Single-molecule fluorescence enhancement of a near-infrared dye by gold nanorods using DNA transient binding. *Physical Chemistry Chemical Physics*, 20(31):20468–20475, 2018.
7. K Ravi Acharya and Matthew D Lloyd. The advantages and limitations of protein crystal structures. *Trends in pharmacological sciences*, 26(1):10–14, 2005.
8. Xavier Michalet, Shimon Weiss, and Marcus Jäger. Single-molecule fluorescence studies of protein folding and conformational dynamics. *Chemical reviews*, 106(5):1785–1813, 2006.
9. Eugene Kim, Martin D. Baaske, Isabel Schuldes, Peter S. Wilsch, and Frank Vollmer. Label-free optical detection of single enzyme-reactant reactions and associated conformational changes. *Science Advances*, 3(3):e1603044, 2017.
10. J Kong, J Zhu, and U F Keyser. Single molecule based SNP detection using designed DNA carriers and solid-state nanopores. *Chemical Communications*, 53(2):436–439, 2017.
11. Q. Zhao, G. Sigalov, V. Dimitrov, B. Dorvel, U. Mirsaidov, S. Sligar, A. Aksimentiev, and G. Timp. Detecting SNPs using a synthetic nanopore. *Nano Letters*, 7(6):1680–1685, 2007.
12. Andrea M. Armani, Rajan P. Kulkarni, Scott E. Fraser, Richard C. Flagan, and Kerry J. Vahala. Label-free, single-molecule detection with optical microcavities. *Science*, 317(5839):783–787, 2007.
13. D Cibran Perez-Gonzalez and J Carlos Penedo. *Single-molecule strategies for DNA and RNA diagnostics. RNA and DNA Diagnostics*, Springer: 297-312, 2015.
14. David R. Walt. Optical methods for single molecule detection and analysis. *Analytical Chemistry*, 85(3):1258–1263, 2013.
15. Eugene Kim, Martin D. Baaske, and Frank Vollmer. Towards next-generation label-free biosensors: recent advances in whispering gallery mode sensors. *Lab on a Chip* 17(7):1190–1205, 2017.
16. Murugappan Muthukumar, Calin Plesa, and Cees Dekker. Single-molecule sensing with nanopores. *Physics Today*, 68(8):40–46, 2015.
17. Nitinun Varongchayakul, Jiaxi Song, Amit Meller, and Mark W. Grinstaff. Single-molecule protein sensing in a nanopore: a tutorial. *Chemical Society reviews*, 47(23):8512–8524, 2018.
18. Yanxiao Feng, Yuechuan Zhang, Cuifeng Ying, Deqiang Wang, and Chunlei Du. Nanopore-based Fourth-generation DNA Sequencing Technology. *Genomics, Proteomics & Bioinformatics*, 13(1):4–16, 2015.
19. Qi Chen and Zewen Liu. Fabrication and applications of solid-state nanopores. *Sensors*, 19(8):1886, 2019.
20. S. Garaj, W. Hubbard, A. Reina, J. Kong, D. Branton, and J. A. Golovchenko. Graphene as a subnanometre trans-electrode membrane. *Nature*, 467(7312):190–193, 2010.
21. Ke Liu, Martina Lihter, Aditya Sarathy, Sabina Caneva, Hu Qiu, Davide Deiana, Vasiliki Tileli, Duncan T. L. Alexander, Stephan Hofmann, Dumitru Dumcenco, Andras Kis, Jean-Pierre

Leburton, and Aleksandra Radenovic. Geometrical effect in 2d nanopores. *Nano Letters*, 17(7):4223–4230, 07 2017.
22. Christopher A Merchant, Ken Healy, Meni Wanunu, Vishva Ray, Neil Peterman, John Bartel, Michael D Fischbein, Kimberly Venta, Zhengtang Luo, AT Charlie Johnson, et al. Dna translocation through graphene nanopores. *Nano letters*, 10(8):2915–2921, 2010.
23. John J. Kasianowicz, Eric Brandin, Daniel Branton, and David W. Deamer. Characterization of individual polynucleotide molecules using a membrane channel. *Proceedings of the National Academy of Sciences of the United States of America*, 93(24):13770–13773, 1996.
24. Yi Xu, Ping Bai, Xiaodong Zhou, Yuriy Akimov, Ching Eng Png, Lay-Kee Ang, Wolfgang Knoll, and Lin Wu. Optical Refractive Index Sensors with Plasmonic and Photonic Structures: Promising and Inconvenient Truth. *Advanced Optical Materials*, 7(9):1801433, 2019.
25. Helen Miller, Zhaokun Zhou, Jack Shepherd, Adam J.M. Wollman, and Mark C. Leake. Single-molecule techniques in biophysics: a review of the progress in methods and applications. *Reports on Progress in Physics*, 81(2):024601, 2017.
26. Björn Hellenkamp, Sonja Schmid, Olga Doroshenko, Oleg Opanasyuk, Ralf Kühnemuth, Soheila Rezaei Adariani, Benjamin Ambrose, Mikayel Aznauryan, Anders Barth, Victoria Birkedal, et al. Precision and accuracy of single-molecule fret measurements—a multi-laboratory benchmark study. *Nature methods*, 15(9):669–676, 2018.
27. M. Götz, P. Wortmann, S. Schmid, and T. Hugel. A Multicolor Single-Molecule FRET Approach to Study Protein Dynamics and Interactions Simultaneously. *Methods in enzymology* 581:487–516, 2016.
28. Wallace H Coulter. Means for counting particles suspended in a fluid. US Patent 2,656,508, 1953.
29. Hagan Bayley and Charles R Martin. Resistive-pulse sensing from microbes to molecules. *Chemical reviews*, 100(7):2575–2594, 2000.
30. Meni Wanunu. Nanopores: A journey towards DNA sequencing. *Physics of Life Reviews*, 9(2):125–158, 2012.
31. Tim Albrecht. Single-Molecule Analysis with Solid-State Nanopores. *Annual Review of Analytical Chemistry*, 12(1):371–387, 2019.
32. A. J. Storm, J. H. Chen, X. S. Ling, H. W. Zandbergen, and C. Dekker. Fabrication of solid-state nanopores with single-nanometre precision. *Nature Materials*, 2(8):537–540, 2003.
33. Daniel V. Verschueren, Wayne Yang, and Cees Dekker. Lithography-based fabrication of nanopore arrays in freestanding SiN and graphene membranes. *Nanotechnology*, 29(14):145302, 2018.
34. Harold Kwok, Kyle Briggs, and Vincent Tabard-Cossa. Nanopore fabrication by controlled dielectric breakdown. *PloS one*, 9(3):92880, 2014.
35. Benjamin N. Miles, Aleksandar P. Ivanov, Kerry A. Wilson, Fatma Dogan, Deanpen Japrung, and Joshua B. Edel. Single molecule sensing with solid-state nanopores: Novel materials, methods, and applications. *Chemical Society Reviews*, 42(1):15–28, 2013.
36. Stefan W Kowalczyk, Alexander Y Grosberg, Yitzhak Rabin, and Cees Dekker. Modeling the conductance and DNA blockade of solid-state nanopores. *Nanotechnology*, 22:315101–315106, 2011.
37. Ralph M.M. Smeets, Ulrich F. Keyser, Diego Krapf, Meng Yue Wu, Nynke H. Dekker, and Cees Dekker. Salt dependence of ion transport and DMA translocation through solid-state nanopores. *Nano Letters*, 6(1):89–95, 2006.
38. Daniel Fologea, James Uplinger, Brian Thomas, David S. McNabb, and Jiali Li. Slowing DNA translocation in a solid-state nanopore. *Nano Letters*, 5(9):1734-1737, 2005.
39. Jiali Li, Derek Stein, Ciaran McMullan, Daniel Branton, Michael J. Aziz, and Jene A. Golovchenko. Ion-beam sculpting at nanometre length scales. *Nature*, 412(6843):166–169, 2001.
40. Laurent D. Menard and J. Michael Ramsey. Fabrication of Sub-5 nm Nanochannels in Insulating Substrates Using Focused Ion Beam Milling. *Nano Letters*, 11(2):512–517, 2011.
41. Sergii Pud, Daniel Verschueren, Nikola Vukovic, Calin Plesa, Magnus P. Jonsson, and Cees Dekker. Self-Aligned Plasmonic Nanopores by Optically Controlled Dielectric Breakdown. *Nano Letters*, 15(10), 2015.

42. Kyle Briggs, Martin Charron, Harold Kwok, Timothea Le, Sanmeet Chahal, José Bustamante, Matthew Waugh, and Vincent Tabard-Cossa. Kinetics of nanopore fabrication during controlled breakdown of dielectric membranes in solution. *Nanotechnology*, 26(8), 2015.
43. Yuning Zhang, Yoichi Miyahara, Nassim Derriche, Wayne Yang, Khadija Yazda, Xavier Capaldi, Zezhou Liu, Peter Grutter, and Walter Reisner. Nanopore Formation via Tip-Controlled Local Breakdown Using an Atomic Force Microscope. *Small Methods*, 3(7):1900147, 2019.
44. Jothi Priyanka Thiruraman, Paul Masih Das, and Marija Drndic. Stochastic Ionic Transport in Single Atomic Zero-D Pores. *ACS Nano*, 14(9):11831-11845, 2020.
45. Jiali Li, Marc Gershow, Derek Stein, Eric Brandin, and Jene Andrew Golovchenko. Dna molecules and configurations in a solid-state nanopore microscope. *Nature materials*, 2(9):611–615, 2003.
46. Mirna Mihovilovic, Nicholas Hagerty, and Derek Stein. Statistics of dna capture by a solid-state nanopore. *Physical Review Letters*, 110(2):028102, 2013.
47. A. J. Storm, J. H. Chen, H. W. Zandbergen, and C. Dekker. Translocation of double-strand DNA through a silicon oxide nanopore. *Physical Review E - Statistical, Nonlinear, and Soft Matter Physics*, 71(5):051903, 2005.
48. Rajesh Kumar Sharma, Ishita Agrawal, Liang Dai, Patrick S. Doyle, and Slaven Garaj. Complex DNA knots detected with a nanopore sensor. *Nature Communications*, 10(1):4473, 2019.
49. C Plesa, D Verschueren, S Pud, J Van Der Torre, J W Ruitenberg, M J Witteveen, M P Jonsson, A Y Grosberg, Y Rabin, and C Dekker. Direct observation of DNA knots using a solid-state nanopore. *Nature Nanotechnology*, 11(12):1093–1097, 2016.
50. Calin Plesa, Stefan W Kowalczyk, Ruben Zinsmeester, Alexander Y Grosberg, Yitzhak Rabin, and Cees Dekker. Fast translocation of proteins through solid state nanopores. *Nano letters*, 13(2):658–663, 2013.
51. Roman D Bulushev, Sanjin Marion, Ekaterina Petrova, Sebastian J Davis, Sebastian J Maerkl, and Aleksandra Radenovic. Single molecule localization and discrimination of dna–protein complexes by controlled translocation through nanocapillaries. *Nano letters*, 16(12):7882–7890, 2016.
52. Erik C. Yusko, Brandon R. Bruhn, Olivia M. Eggenberger, Jared Houghtaling, Ryan C. Rollings, Nathan C. Walsh, Santoshi Nandivada, Mariya Pindrus, Adam R. Hall, David Sept, Jiali Li, Devendra S. Kalonia, and Michael Mayer. Real-time shape approximation and fingerprinting of single proteins using a nanopore. *Nature Nanotechnology*, 12(4), 2017.
53. Alessio Fragasso, Hendrik W De Vries, John Andersson, Eli O Van Der Sluis, Erik Van Der Giessen, Andreas Dahlin, Patrick R Onck, and Cees Dekker. A designer fg-nup that reconstitutes the selective transport barrier of the nuclear pore complex. *Nature communications*, 12(1):1–15, 2021.
54. Philip Ketterer, Adithya N Ananth, Diederik S Laman Trip, Ankur Mishra, Eva Bertosin, Mahipal Ganji, Jaco van der Torre, Patrick Onck, Hendrik Dietz, and Cees Dekker. Dna origami scaffold for studying intrinsically disordered proteins of the nuclear pore complex. *Nature communications*, 9(1):1–8, 2018.
55. Stefan W Kowalczyk, Timothy R Blosser, and Cees Dekker. Biomimetic nanopores: learning from and about nature. *Trends in biotechnology*, 29(12):607–614, 2011.
56. Nicole E. Weckman, Niklas Ermann, Richard Gutierrez, Kaikai Chen, James Graham, Ran Tivony, Andrew Heron, and Ulrich F. Keyser. Multiplexed DNA Identification Using Site Specific dCas9 Barcodes and Nanopore Sensing. *ACS Sensors*, 4(8):2065–2072, 2019.
57. J Y Y Sze, A P Ivanov, A E G Cass, and J B Edel. Single molecule multiplexed nanopore protein screening in human serum using aptamer modified DNA carriers. *Nature Communications*, 8(1):1-10, 2017.
58. RMM Smeets, Stefan W Kowalczyk, AR Hall, NH Dekker, and C Dekker. Translocation of reca-coated double-stranded dna through solid-state nanopores. *Nano letters*, 9(9):3089–3095, 2009.
59. Kaikai Chen, Matyas Juhasz, Felix Gularek, Elmar Weinhold, Yu Tian, Ulrich F Keyser, and Nicholas AW Bell. Ionic current-based mapping of short sequence motifs in single dna molecules using solid-state nanopores. *Nano letters*, 17(9):5199–5205, 2017.

60. Aleksandre Japaridze, Wayne Yang, Cees Dekker, William Nasser, and Georgi Muskhelishvili. Dna sequence-directed cooperation between nucleoid-associated proteins. *iScience*, 24(5):102408, 2021.
61. Wayne Yang, Laura Restrepo-Pérez, Michel Bengtson, Stephanie J. Heerema, Anthony Birnie, Jaco van der Torre, and Cees Dekker. Detection of CRISPR-dCas9 on DNA with Solid-State Nanopores. *Nano Letters*, 18(10):6469–6474, 2018.
62. Stefan W Kowalczyk, David B Wells, Aleksei Aksimentiev, and Cees Dekker. Slowing down dna translocation through a nanopore in lithium chloride. *Nano letters*, 12(2):1038–1044, 2012.
63. Stephen E Halford and John F Marko. How do site-specific dna-binding proteins find their targets? *Nucleic acids research*, 32(10):3040–3052, 2004.
64. Alessio Fragasso, Sonja Schmid, and Cees Dekker. Comparing current noise in biological and solid-state nanopores. *ACS nano*, 14(2):1338–1349, 2020.
65. Alessio Fragasso, Sergii Pud, and Cees Dekker. 1/f noise in solid-state nanopores is governed by access and surface regions. *Nanotechnology*, 30(39):395202, 2019.
66. Chenyu Wen, Shuangshuang Zeng, Kai Arstila, Timo Sajavaara, Yu Zhu, Zhen Zhang, and Shi Li Zhang. Generalized Noise Study of Solid-State Nanopores at Low Frequencies. *ACS Sensors*, 2(2):300–307, 2017.
67. Ralph MM Smeets, Ulrich F Keyser, Nynke H Dekker, and Cees Dekker. Noise in solid-state nanopores. *Proceedings of the National Academy of Sciences*, 105(2):417–421, 2008.
68. Calin Plesa, Justus W. Ruitenberg, Menno J. Witteveen, and Cees Dekker. Detection of individual proteins bound along DNA using solid-state nanopores. *Nano Letters*, 15(5):3153-3158, 2015.
69. Calin Plesa, Nick Van Loo, Philip Ketterer, Hendrik Dietz, and Cees Dekker. Velocity of dna during translocation through a solid-state nanopore. *Nano letters*, 15(1):732–737, 2015.
70. Jacob K Rosenstein, Meni Wanunu, Christopher A Merchant, Marija Drndic, and Kenneth L Shepard. Integrated nanopore sensing platform with sub-microsecond temporal resolution. *Nature methods*, 9(5):487, 2012.
71. Adrian Balan, Chen-Chi Chien, Rebecca Engelke, and Marija Drndić. Suspended solid-state membranes on glass chips with sub 1-pf capacitance for biomolecule sensing applications. *Scientific reports*, 5(1):1–8, 2015.
72. Min-Hyun Lee, Ashvani Kumar, Kyeong-Beom Park, Seong-Yong Cho, Hyun-Mi Kim, Min-Cheol Lim, Young-Rok Kim, and Ki-Bum Kim. A low-noise solid-state nanopore platform based on a highly insulating substrate. *Scientific reports*, 4(1):1–7, 2014.
73. Sonja Schmid, Pierre Stömmer, Hendrik Dietz, and Cees Dekker. Nanopore electro-osmotic trap for the label-free study of single proteins and their conformations. *bioRxiv*, 2021.
74. Sonja Schmid and Cees Dekker. Nanopores: a versatile tool to study protein dynamics. *Essays in Biochemistry*, 65(1):93–107, 2021.
75. Grégory F. Schneider, Qiang Xu, Susanne Hage, Stephanie Luik, Johannes N. H. Spoor, Sairam Malladi, Henny Zandbergen, and Cees Dekker. Tailoring the hydrophobicity of graphene for its use as nanopores for DNA translocation. *Nature Communications*, 4(1):2619, 2013.
76. Stephanie J. Heerema and Cees Dekker. Graphene nanodevices for DNA sequencing. *Nature Nanotechnology*, 11(2):127–136, 2016.
77. Michael Graf, Martina Lihter, Damir Altus, Sanjin Marion, and Aleksandra Radenovic. Transverse Detection of DNA Using a MoS2 Nanopore. *Nano Letters*, 19(12):9075–9083, 2019.
78. Mingye Xiong, Michael Graf, Nagendra Athreya, Aleksandra Radenovic, and Jean-Pierre Leburton. Microscopic detection analysis of single molecules in mos2 membrane nanopores. *ACS nano*, 14(11):16131–16139, 2020.
79. Wayne Yang, Boya Radha, Adnan Choudhary, Yi You, Gangaiah Mettela, Andre K Geim, Aleksei Aksimentiev, Ashok Keerthi, and Cees Dekker. Translocation of dna through ultrathin nanoslits. *Advanced Materials*, 33(11):2007682, 2021.
80. B Radha, Ali Esfandiar, FC Wang, AP Rooney, K Gopinadhan, Ashok Keerthi, Artem Mishchenko, Amritha Janardanan, Peter Blake, Laura Fumagalli, et al. Molecular transport through capillaries made with atomic-scale precision. *Nature*, 538(7624):222–225, 2016.

81. C. Genet and T. W. Ebbesen. Light in tiny holes. *Nature*, 445(7123):39–46, 2007.
82. Thomas W Ebbesen, Henri J Lezec, HF Ghaemi, Tineke Thio, and Peter A Wolff. Extraordinary optical transmission through sub-wavelength hole arrays. *Nature*, 391(6668):667–669, 1998.
83. Daniel V. Verschueren, Sergii Pud, Xin Shi, Lorenzo De Angelis, L. Kuipers, and Cees Dekker. Label-Free Optical Detection of DNA Translocations through Plasmonic Nanopores. *ACS Nano*, 13(1):61–70, 2019.
84. Mathieu L. Juan, Reuven Gordon, Yuanjie Pang, Fatima Eftekhari, and Romain Quidant. Self-induced back-action optical trapping of dielectric nanoparticles. *Nature Physics*, 5(12):915–919, 2009.
85. Avijit Barik, Lauren M Otto, Daehan Yoo, Jincy Jose, Timothy W Johnson, and Sang-Hyun Oh. Dielectrophoresis-Enhanced Plasmonic Sensing with Gold Nanohole Arrays. *Nano Letters*, 14:2020, 2014.
86. Seung Ju Yoon, Jungmin Lee, Sangyoon Han, Chang Kyu Kim, Chi Won Ahn, Myung Ki Kim, and Yong Hee Lee. Non-fluorescent nanoscopic monitoring of a single trapped nanoparticle via nonlinear point sources. *Nature Communications*, 9(1):341401, 2018.
87. James V Coe, Joseph M Heer, Shannon Teeters-Kennedy, Hong Tian, and Kenneth R Rodriguez. Extraordinary transmission of metal films with arrays of subwavelength holes. *Annu. Rev. Phys. Chem.*, 59:179–202, 2008.
88. Reuven Gordon. Biosensing with nanoaperture optical tweezers. *Optics & Laser Technology*, 109:328–335, 2019.
89. Pau Mestres, Johann Berthelot, Srdjan S Aćimović, and Romain Quidant. Unraveling the optomechanical nature of plasmonic trapping. *Light: Science & Applications*, 5(7):16092–16092, 2016.
90. Serap Aksu, Ahmet A. Yanik, Ronen Adato, Alp Artar, Min Huang, and Hatice Altug. High-throughput nanofabrication of infrared plasmonic nanoantenna arrays for vibrational nanospectroscopy. *Nano Letters*, 10(7):2511–2518, 2010.
91. Xin Shi, Daniel Verschueren, Sergii Pud, and Cees Dekker. Integrating Sub-3 nm Plasmonic Gaps into Solid-State Nanopores. *Small*, 14(18):1703307, 2018.
92. Myung Ki Kim, Hongchul Sim, Seung Ju Yoon, Su Hyun Gong, Chi Won Ahn, Yong Hoon Cho, and Yong Hee Lee. Squeezing Photons into a Point-Like Space. *Nano Letters*, 15(6):4102–4107, 2015.
93. Alemayehu Nana Koya, Joao Cunha, Tian-Long Guo, Andrea Toma, Denis Garoli, Tao Wang, Saulius Juodkazis, Dan Cojoc, and Remo Proietti Zaccaria. Novel plasmonic nanocavities for optical trapping-assisted biosensing applications. *Advanced Optical Materials*, 8(7):1901481, 2020.
94. Mathieu L Juan, Maurizio Righini, and Romain Quidant. Plasmon nano-optical tweezers. *Nature photonics*, 5(6):349, 2011.
95. Abhay Kotnala and Reuven Gordon. Double nanohole optical tweezers visualize protein p53 suppressing unzipping of single dna-hairpins. *Biomedical optics express*, 5(6):1886–1894, 2014.
96. M Righini, P Ghenuche, S Cherukulappurath, Viktor Myroshnychenko, Francisco Javier Garcia de Abajo, and Romain Quidant. Nano-optical trapping of rayleigh particles and escherichia coli bacteria with resonant optical antennas. *Nano letters*, 9(10):3387–3391, 2009.
97. Yuanjie Pang and Reuven Gordon. Optical trapping of a single protein. *Nano Letters*, 12(1):402–406, 2012.
98. Daniel Verschueren, Xin Shi, and Cees Dekker. Nano-Optical Tweezing of Single Proteins in Plasmonic Nanopores. *Small Methods*, 3(5):1800465, 2019.
99. Ana Zehtabi-Oskuie, Hao Jiang, Bryce R. Cyr, Douglas W. Rennehan, Ahmed A. Al-Balushi, and Reuven Gordon. Double nanohole optical trapping: Dynamics and protein-antibody co-trapping. *Lab on a Chip*, 13(13):2563–2568, 2013.
100. Xin Shi, Daniel V. Verschueren, and Cees Dekker. Active Delivery of Single DNA Molecules into a Plasmonic Nanopore for Label-Free Optical Sensing. *Nano Letters*, 18(12):8003–8010, 2018.

101. Noa Hacohen, Candice JX Ip, and Reuven Gordon. Analysis of egg white protein composition with double nanohole optical tweezers. *ACS omega*, 3(5):5266–5272, 2018.
102. Quanbo Jiang, Benoit Rogez, Jean-Benoît Claude, Guillaume Baffou, and Jérôme Wenger. Temperature measurement in plasmonic nanoapertures used for optical trapping. *ACS photonics*, 6(7):1763–1773, 2019.
103. Liselotte Jauffred, Akbar Samadi, Henrik Klingberg, Poul Martin Bendix, and Lene B Oddershede. Plasmonic heating of nanostructures. *Chemical reviews*, 119(13):8087–8130, 2019.
104. Colin R Crick, Pablo Albella, Hyung-Jun Kim, Aleksandar P Ivanov, Ki-Bum Kim, Stefan A Maier, and Joshua B Edel. Low-noise plasmonic nanopore biosensors for single molecule detection at elevated temperatures. *ACS Photonics*, 4(11):2835–2842, 2017.
105. Ulrich F Keyser, Diego Krapf, Bernard N Koeleman, Ralph MM Smeets, Nynke H Dekker, and Cees Dekker. Nanopore tomography of a laser focus. *Nano letters*, 5(11):2253–2256, 2005.
106. Francesca Nicoli, Daniel Verschueren, Misha Klein, Cees Dekker, and Magnus P Jonsson. Dna translocations through solid-state plasmonic nanopores. *Nano letters*, 14(12):6917–6925, 2014.
107. Brian J Roxworthy, Abdul M Bhuiya, Surya P Vanka, and Kimani C Toussaint. Understanding and controlling plasmon-induced convection. *Nature communications*, 5(1):1–8, 2014.
108. Erik Schäffer, Simon F Nørrelykke, and Jonathon Howard. Surface forces and drag coefficients of microspheres near a plane surface measured with optical tweezers. *Langmuir*, 23(7):3654–3665, 2007.
109. Bita Malekian, Kunli Xiong, Evan SH Kang, John Andersson, Gustav Emilsson, Marcus Rommel, Takumi Sannomiya, Magnus P Jonsson, and Andreas Dahlin. Optical properties of plasmonic nanopore arrays prepared by electron beam and colloidal lithography. *Nanoscale Advances*, 1(11):4282–4289, 2019.
110. Liang Pan, Yongshik Park, Yi Xiong, Erick Ulin-Avila, Yuan Wang, Li Zeng, Shaomin Xiong, Junsuk Rho, Cheng Sun, David B Bogy, and Xiang Zhang. Maskless plasmonic lithography at 22 nm resolution. *Scientific reports*, 1(1):1–6, 2011.
111. Quanbo Jiang, Benoît Rogez, Jean-Benoît Claude, Antonin Moreau, Julien Lumeau, Guillaume Baffou, and Jérôme Wenger. Adhesion layer influence on controlling the local temperature in plasmonic gold nanoholes. *Nanoscale*, 12(4):2524–2531, 2020.
112. Kai Wang, Ethan Schonbrun, Paul Steinvurzel, and Kenneth B. Crozier. Trapping and rotating nanoparticles using a plasmonic nano-tweezer with an integrated heat sink. *Nature Communications*, 2(1):1–6, 2011.
113. H. J. Levene, J. Korlach, S. W. Turner, M. Foquet, H. G. Craighead, and W. W. Webb. Zero-mode waveguides for single-molecule analysis at high concentrations. *Science*, 299(5607):682–686, 2003.
114. Ossama N. Assad, Tal Gilboa, Joshua Spitzberg, Matyas Juhasz, Elmar Weinhold, and Amit Meller. Light-Enhancing Plasmonic-Nanopore Biosensor for Superior Single-Molecule Detection. *Advanced Materials*, 29(9):1605442, 2017.
115. Joseph Larkin, Robert Y Henley, Vivek Jadhav, Jonas Korlach, and Meni Wanunu. Length-independent dna packing into nanopore zero-mode waveguides for low-input dna sequencing. *Nature nanotechnology*, 12(12):1169–1175, 2017.
116. Joshua D Spitzberg, Adam Zrehen, Xander F van Kooten, and Amit Meller. Plasmonic-nanopore biosensors for superior single-molecule detection. *Advanced Materials*, 31(23):1900422, 2019.
117. Justus C Ndukaife, Alexander V Kildishev, Agbai George, Agwu Nnanna, Vladimir M Shalaev, Steven T Wereley, and Alexandra Boltasseva. Long-range and rapid transport of individual nano-objects by a hybrid electrothermoplasmonic nanotweezer. *Natue Nanotechnology*, 11, 2016.
118. Ben McNally, Alon Singer, Zhiliang Yu, Yingjie Sun, Zhiping Weng, and Amit Meller. Optical recognition of converted dna nucleotides for single-molecule dna sequencing using nanopore arrays. *Nano letters*, 10(6):2237–2244, 2010.

119. Brett N Anderson, Ossama N Assad, Tal Gilboa, Allison H Squires, Daniel Bar, and Amit Meller. Probing solid-state nanopores with light for the detection of unlabeled analytes. *ACS nano*, 8(11):11836–11845, 2014.
120. Tal Gilboa and Amit Meller. Optical sensing and analyte manipulation in solid-state nanopores. *Analyst*, 140(14):4733–4747, 2015.
121. Tal Gilboa, Adam Zrehen, Arik Girsault, and Amit Meller. Optically-monitored nanopore fabrication using a focused laser beam. *Scientific reports*, 8(1):1–10, 2018.
122. Tal Gilboa, Eran Zvuloni, Adam Zrehen, Allison H. Squires, and Amit Meller. Automated, Ultra-Fast Laser-Drilling of Nanometer Scale Pores and Nanopore Arrays in Aqueous Solutions. *Advanced Functional Materials*, 30(18):1900642, 2020.

Chapter 13
Self-induced Back-Action Actuated Nanopore Electrophoresis (SANE) Sensing

Scott Renkes, Sai Santosh Sasank Peri, Muhammad Usman Raza, Jon Weidanz, Min Jun Kim, and George Alexandrakis

Abstract We present a method to trap nanoscale analytes in double nanohole (DNH) nanoapertures integrated on top of solid-state nanopores (ssNPs). The analytes are propelled by an electrophoretic force from the *cis* towards the *trans* side of the nanopore but are trapped in the process when they have reached the vicinity of the DNH-ssNP interface. The self-induced back action (SIBA) force, created by the plasmonic field between the tips of the DNH, opposes the electrophoretic force and enables simultaneous optical and electrical sensing of a single nanoparticle for many seconds. The SIBA actuated nanopore electrophoresis (SANE) sensor was fabricated using two-beam gas field ion source (GFIS) focused ion beam (FIB). Firstly, Ne FIB milling was used to create the DNH features and was combined with end pointing to stop milling at the metal-dielectric interface. Subsequently, He FIB was used to drill a 25 nm nanopore through the center of the DNH. The capabilities of the device are demonstrated using a series of three experiments involving nanoparticles, high-affinity protein ligand interactions and low-affinity protein-ligand interactions. The presence of optical trapping in the SANE sensor extended electrical sensing and translocation times by up to four orders of magnitude over classical nanopores. In addition, SANE sensor measurements enabled quantification of bimodal optical-electrical parameters that were quantified concurrently for each trapping event, which enabled distinguishing analytes from each other, specific from non-specific binding

S. Renkes · G. Alexandrakis (✉)
Department of Bioengineering, University of Texas at Arlington, Arlington, TX, USA
e-mail: galex@uta.edu

S. Renkes
e-mail: sar9179@mavs.uta.edu

S. S. S. Peri · M. U. Raza
Department of Electrical Engineering, University of Texas at Arlington, Arlington, TX, USA

J. Weidanz
Department of Biology, University of Texas at Arlington, Arlington, TX, USA
e-mail: weidanz@uta.edu

M. J. Kim
Department of Mechanical Engineering, Southern Methodist University, Dallas, TX, USA
e-mail: mjkim@lyle.smu.edu

W. Bowen et al. (eds.), *Single Molecule Sensing Beyond Fluorescence*,
Nanostructure Science and Technology,
https://doi.org/10.1007/978-3-030-90339-8_13

events, and protein complex formation. Importantly, the SANE sensor enabled ultra-high sensitivity in protein-ligand interaction detection. Electrically driven focusing of reactants into the sensor's nanoscopic optical trap volume enabled formation of significant bound fractions (30–50% range) at reactant concentrations up to three orders of magnitude lower than the free solution equilibrium binding constant. Furthermore, the SANE sensor could measure the off-binding rate of low-affinity (micromolar) protein-ligand interactions that are challenging to measure with current label-free commercial assays.

13.1 Introduction

The self-induced back action nanopore electrophoresis (SANE) sensor utilizes a self-induced back action (SIBA) force [1], created by a plasmonic trap, to oppose an electrophoretic force trying to push molecules past the trap and through an underlying nanopore. As a result, the SIBA effect provides an opportunity to observe molecules, or molecular complexes, label-free and tether-free for many seconds before they translocate through the sensor. In addition, the opposing optical force enables significant slowdown of the actual translocation through the underlying nanopore that lasts for tens of milliseconds [2]. This is in contrast to traditional nanopores where molecules will translocate, primarily impeded by barrier forces and molecular crowding while electroosmosis impedes or facilitates translocation depending on its direction. The typical nanopore translocation event is tens of microseconds [3]. Molecular trapping and slower translocation times with the SANE sensor enable bimodal optical-electrical data acquisition, which we show to be favorable for separating individual proteins and ligands from the complexes they form and significantly enhancing the bound fraction relative to free solution. In addition, we show that the optical and electrical data types quantified concurrently by SANE sensing can potentially improve separation of specific from non-specific interactions.

13.1.1 The SIBA Effect

Optical trapping of nanoparticles at low laser powers (mW range) can be attained in the immediate vicinity of metallic nanoapertures through a SIBA mechanism [1]. In SIBA, when a dielectric nanoparticle has a slightly different refractive index than its surrounding medium, a photon-mediated feedback force is actuated due to conservation of momentum against diffusion forces near the nanoaperture. The resulting coupling of light to the far field via the dielectric nanoparticle results in increased light transmission through the plasmonic nanoaperture and therefore enables label-free detection [4]. Double nanohole (DNH) nanoapertures have been reported as SIBA-mediated optical traps by Gordon et al. for high local field enhancement at the intersection of the nanoholes [5]. The Gordon group has reported a series of studies on the design characteristics of the DNH structure [6–8] and their use in many applications, including the trapping of nanoparticles [9–11] and single protein molecules [12–15].

It should be emphasized that the SIBA effect is qualitatively different from optical tweezing. In the latter case, a high numerical aperture objective lens creates a strong electric field gradient near the beam waist. Dielectric particles are attracted to the region of strongest gradient and displaced slightly by scattering, which places them just downstream of the beam waist. Because optical tweezing relies the gradient of light intensity to attract the particle, it is dependent on laser power [16]. This can cause problems with heat dissipation when trying to trap smaller particles. The DNH structure is more efficient than tweezers in heat dissipation due to the thermal conductivity of the surrounding gold (Au) layer. Polarized illumination of the DNH induces plasmonic enhancement of light intensity parallel to its narrow waist to induce the SIBA effect that enables trapping smaller particles, including single proteins and protein complexes.

In opposition to the trapping force from the SIBA effect are four forces of note: electrophoretic, electroosmotic, passive diffusive and thermophoretic forces. The electrophoretic force is the prevalent pulling force in the SANE sensor. It arises from the bias voltage applied to the experimental setup which exerts a force on all charged particles pulling them based on the direction of the electric field and the zeta potential of the particles [17]. The electroosmotic force is the result of a double layer of charges that forms on all the dielectric surfaces in and around the nanopore. This double layer will function as a pump to pull or push ions in solution through or away from the nanopore. The magnitude and direction of this force is primarily dependent on concentration and ionic properties of the bulk ionic solution while the charge of analytes has a minor impact [18, 19]. Passive diffusion, which is driven by concentration gradients, is a relatively small force with respect to the electrophoretic or electroosmotic forces [20]. Thermophoretic forces pull or push particles away from hot spots depending on whether they are thermophilic or thermophobic [21]. By altering properties of the solution through the use of surfactants, a particle can be made more thermophilic, which will increase the force toward the optical trap [22]. Thermophilicity will cause two effects: increase trapping duration while also decreasing the time between trapping events. For thermophilic particles inside the optical trap, the increased thermophoretic forces will offset the electrophoretic forces and make it more difficult for a particle to leave the trap. For a particle on the *cis* side of the nanopore, the thermophoretic forces will combine with the electrophoretic forces drawing it more quickly toward the trap.

13.1.2 Concurrent Optical and Electrical Data Capture with the SANE Sensor

The SANE sensor is capable of bimodal, optical-electrical sensing and provides four data types, two of which are optical and two of which are electrical. These four metrics are optical step change, optical trap time, electrical translocation time and translocation current:

i Optical step change is the change in optical transmission as a particle enters the optical trap. The dielectric properties of the molecule act qualitatively as a lens, focusing the light through the plasmonic nanoaperture, increasing optical transmission [23]. For molecules with spherical shapes, optical step change can be used to estimate the volume of a molecule.
ii Optical trap time is the time duration that a molecule spends inside the optical trap, defined as the time between the timepoint of optical step change completion and the beginning of a translocation event for a single molecule or complex. The optical trap time is a function of the balance between optical forces (SIBA), electrical forces (electrophoresis, electroosmosis and barrier force) and analyte-related forces (concentration gradient and thermophilicity). Depending on the detailed experimental conditions, the relative contributions of these forces can vary, making this a rather complicated system to study.
iii Translocation time is the measurement of the time it takes a molecule that has just escaped the optical trap to travel through the nanopore to the *trans* well. The translocation event is usually denoted by a negative electrical current spike and a decrease in the optical step back to baseline. Like the optical trap time, this metric is also dependent on the voltage and molecule polarity where a larger voltage and larger polarity contribute to increased translocation speed [2].
iv Translocation current is the measurement of current change during a pore translocation event. The negative current spike usually seen in translocation events is due to the molecule blocking the baseline ionic current flow. The size of the negative spike is dependent on analyte size due to its mass blocking the ionic current through the nanopore as well as the charge and concentration of the bulk ion solution and the geometry and charge of the nanopore [24].

The ability to perform optical and electrical sensing concurrently with the SANE sensor enables label-free study of molecules and their interactions in ways that are not accessible by use of traditional optical trapping or classical nanopores on their own:

1 It can offer higher throughput than pure optical trapping because: (a) pure optical trapping can last indefinitely, in principle, and (b) at very low analyte concentrations the time interval between successive trapping events can be impractically long as it is diffusion-limited [25]. In the case of the SANE sensor, the externally applied electric field has a focusing effect, pushing analytes from the *cis* chamber towards the optical trap. The focusing effect is driven by the silicon nitride (Si_xN_y) region not covered by Au. The latter acts as a shield to the applied electric field

because it is a conductor. Electric field focusing enables studying ultra-low analyte concentrations. Furthermore, the balance of opposing optical and electrical forces, potentially combined with thermal ones, shortens trapping times compared to pure optical trapping. Appropriate selection of experimental parameters can be used to roughly tune the range of possible optical trapping times and indirectly control throughput. It should be mentioned, however, that the throughput of an optical-electrical trap is much lower than classical nanopores due to the slowing-down effect of the optical trap. This limitation is discussed further in the Conclusions section of this chapter.

2 It enables tether-free quantification of low-affinity (micromolar) interactions that are relevant to protein pathways occurring in vivo: when experimental conditions can be tuned such that optical trapping times are significantly longer than the mean protein-ligand interaction time, it is possible to measure the off-rates (k_{off}) or binding duration of an interaction event [25]. Measurement of binding kinetics is challenging to do with current commercial technologies such as surface plasmon resonance (SPR).

3 It enables a strong enhancement of protein reactant concentrations at the sensor: Reactants are pushed by the applied electric field inside the optical trap. As we discuss further below, this enables studying protein-ligand interactions with significant bound fractions at concentrations down to 1000-fold lower than the equilibrium binding constant (bound fraction would be near-zero at these low concentrations) [26]. High bound fractions at low concentrations offer significant savings in the amount of purified protein needed for ligand screening experiments.

4 It causes significant slowdown of nanopore translocation: The presence of the optical trap immediately over the nanopore significantly slows down translocation of molecules through the nanopore. The translocation time can last for up to tens of milliseconds in this case, which is orders of magnitude longer than translocation times seen in traditional nanopores that last in the tens of microseconds [2]. More work is needed to extract all possible benefits in molecular characterization by use of this translocation slow-down feature.

5 It offers the possibility of distinguishing between molecular size versus effective molecular charge: The concurrent optical and electrical measurements done with the SANE sensor enable molecular size estimates from the optical data, which can then be substituted into equations typically used to estimate the effective charge around the molecules [2]. In contrast classical nanopore signals depend both on analyte size and charge, making this distinction challenging to attain.

6 It allows for the observation of the frequency spectrum of charged molecules bobbing inside the optical trap: The opposing forces in the optical trap are in dynamic equilibrium in the presence of diffusion-driven perturbation of the trapped analyte. The resulting motions of analytes have been found to have characteristic frequency spectra in the current passing through the nanopore before a translocation event occurs. These spectra have shown to be different when one silicon oxide SiO_2 nanoparticle is inside the optical trap versus two particles [2]. More work is needed to explore the feasibility of using these frequency spectra as an additional metric to distinguish between interacting proteins and complexes.

7 It creates both a positive current spike and a negative spike through the nanopore: Some details for the observed positive spike are discussed below. Briefly, the optically slowed-down approach of the particle to the nanopore from the *cis* side creates a counterion current. However, the physical mechanism and any additional information that can be extracted from this positive spike have not been thoroughly explored to date.

13.2 Sensor Chip Fabrication and Experimental Setup

The SANE sensor was produced using a combination of traditional silicon fabrication and Focused Ion Beam (FIB) milling techniques. A flow cell was then added around the sensor to help handle microliter size liquid samples at physiological conditions. Optical focusing and electrode placement in the *cis* and *trans* chambers were then adjusted and concurrent optical-electrical data acquisition took place by use of an Axopatch 200B system (Molecular Devices, San Jose, CA).

13.2.1 Sensor Chip Fabrication

The SANE sensor was fabricated on a (100) intrinsic silicon (Si) wafer. The wafer was first polished on both sides. A 500 nm (SiO_2) was then grown on both sides of the wafer using wet oxidation followed by a 60 nm Low Pressure Chemical Vapor Deposition (LPCVD) of non-stoichiometric low stress Si_xN_y. A positive photoresist (S1813) and a square window dark field mask were used to pattern the bottom side of the wafer. Deep Reactive Ion Etching (DRIE) was used to etch the Si_xN_y layer to expose the SiO_2, which was then etched using 6:1 buffered hydrofluoric acid to expose the Si wafer. The Si was anisotropically etched using a 22% tetramethylammonium hydroxide (TMAH) solution at 90° C to expose the other SiO_2 layer which was kept intact as a sacrificial layer to protect the remaining Si_xN_y during processing.

A 5 nm chromium (Cr) adhesion layer was deposited by e-beam evaporation followed by a 100 nm gold (Au) layer on the front of the wafer. Finally, a positive resist (S1813) and darkfield mask were used to pattern the FIB alignment markers, using a backside aligner (EVG620) to ensure the markers were properly aligned with the etched well. The Au and Cr layers were etched to expose the Si_xN_y using commercially available wet etchants. The resulting structure markers are seen in Fig. 13.1. A photoresist layer was baked on the chip in order to protected it during the next processing steps. The wafer was then diced into 15 × 15 mm chips. The double nanohole (DNH) structure (Fig. 13.1a–c) was milled through the Au layer by Neon ion FIB. Subsequently, Helium ion FIB was used to mill a 25 nm diameter hole though the Si_xN_y layer at the center of the DNH structure (Fig. 13.1d–f) (Carl Zeiss, ORION NanoFab, Peabody, MA). Gallium ion FIB could not be used for this milling process

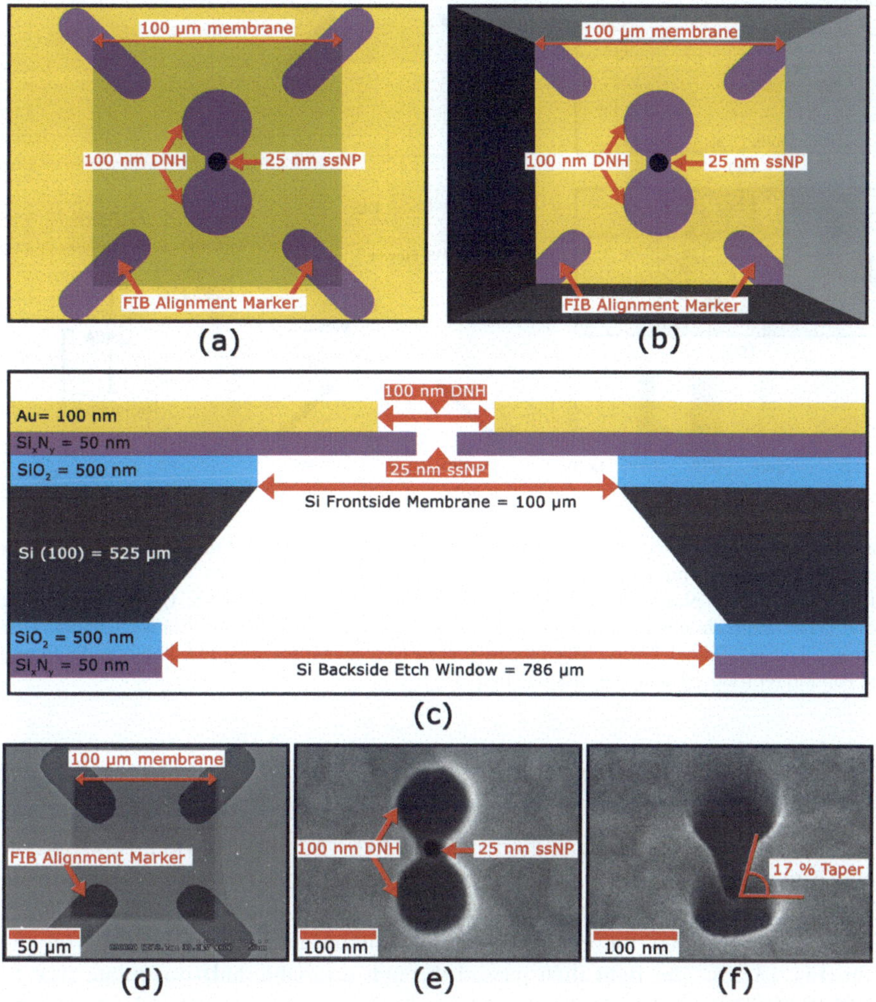

Fig. 13.1 **a** Frontside view of SANE chip. **b** Backside view of SANE chip. **c** Cross-section of the SANE sensor chip. **d** SEM micrograph of front side of the SANE chip before FIB drilling. He ion microscope image of top view (**e**) and tilted view (**f**) of milled DNH with 17% sidewall taper and a 25 nm ssNP drilled at its center. Figure and caption reproduced with permissions by [2]

due to the limitation of its feature resolution. A more detailed fabrication process can be found in Raza et al. [2].

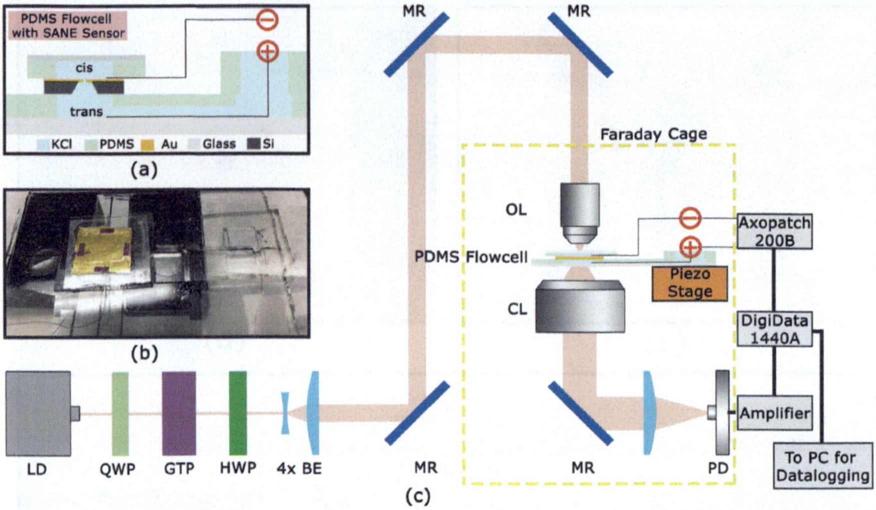

Fig. 13.2 **a** PDMS flow cell cross-sectional view with SANE sensor. **b** Image of prepared PDMS flow cell with SANE chip ready for placement on piezo-controlled stage. **c** Complete optical setup with PDMS flow cell placement and measurement instruments. LD: laser diode, QWP: quarter wave plate, GTP: Glan-Thompson polarizer, HWP: half wave plate, 4× BE: 4× beam expander, MR: mirror, OL: Carl Zeiss 1.3 N.A. 63× objective lens, CL: condenser lens, PD: photodiode. Figure and caption reproduced with permissions by [2]

13.2.2 Optical Measurement Setup

The beam from an 820 nm laser diode (L820P200, Thorlabs, Newton, NJ) was collimated to a 2 mm diameter and circularly polarized through a quarter-waveplate (QWP) (WPQ05M, Thorlabs), followed by a Glan-Thompson linear polarizer (GTH10M, Thorlabs) for controlling the polarization of light incident on the chip (Fig. 13.2c). The light then passed through a tunable half-waveplate (HWP) (WPH05M, Thorlabs) to make the direction of polarization perpendicular to the DNH's long axis to maximally excite wedge plasmons for trapping [10]. A downstream 4× beam expander (Newport) was used in combination with an 8 mm circular aperture (ID.1.0, Newport) to make the intensity profile of the cylindrical beam flatter. The beam then went through a periscope and into the back aperture of a 63× oil immersion objective lens and was focused onto the front Au side of the SANE chip. The 360 μm working distance of the objective lens defined the *cis* flow cell design. A 170 μm thick cover slide was used as the foundation of the *cis* flow cell. That allowed for only 190 μm of height left for the flow cell. Light transmission through the FIB alignment markers was used as a first, coarse step to find the DNH on the chip. The objective's focal spot was aligned with the DNH center by adjusting a piezoelectric stage's controls until polarized light transmission was maximized. The light transmitted through the chip's center and any leakage light scattering through alignment markers was collected by a condenser lens and focused onto a photodiode (PDA36A, Thorlabs) (Fig. 13.2c).

13.2.3 Flow Cell and Electrical Measurement Setup

Flow cells were created from polydimethylsiloxane (PDMS) in three sections. The base was made from a 2 mm thick piece of PDMS that had two 10 mm square holes cut at both ends and then connected by a 2 mm wide trough. The side with the trough was then bonded to a 3 in $\times$ 2 in glass slide using oxygen plasma, sealing the trough and the bottom of the holes. The *trans* side of the SANE sensor was bonded to one well with double-sided tape. A second piece of PDMS, 3 mm thick with a 10 mm hole cut into it, was bonded to the other open hole in the base piece using double-sided tape. This structure serves as a reservoir for ionic solution. A third piece of PDMS was made that was roughly 150 μm thick and bonded to a 170 μm cover slip (the microscope slide intended for use with the objective lens). A 10 mm hole was cut out of the PDMS. This final piece was then bonded to the *cis* side of the SANE sensor using double-sided tape. A 1 mm slit was then cut into the *cis* side of the PDMS to allow the electrode and analytes to be introduced to the *cis* well. The device was attached to the piezoelectric stage for fine positioning. The wells were then filled with an electrolyte solution specific to the experiment. Two silver electrodes were immersed in bleach solution to form a coating of silver chloride (Ag/AgCl). One electrode was inserted in the *cis* and one in the *trans* chamber. The electrode in the *trans* chamber was positioned so that the tip of the electrode was directly below the sensor.

The electrodes were connected to an Axon Axopatch 200B patch clamp amplifier and digitizer equipment (Molecular Devices) through an Axon Headstage (CV 203BU) and operated in voltage clamp mode. A Faraday cage was built with aluminum foil (Reynolds) around the piezo-stage with the PDMS flow cell to eliminate low-frequency electromagnetic noise while recording the ionic current traces. Translocation events were performed at 100 mV voltage bias across the nanopore [25]. The photodiode and Axopatch 200B signals were both sent through an Axon Digidata 1440 ADC to a PC for recording and subsequent data analyses with the Axon pClamp software (version10.6). The electrical signals were filtered by a low 20 Hz 8-pole Bessel filter to remove high frequency noise providing both filtered and unfiltered current data.

13.3 Experiments

SANE sensor measurements were collected and characterized for three different types of analytes. First, experiments were performed with 20 nm SiO_2 and Au nanoparticles to characterize bimodal optical-electrical signatures for rigid nanospheres, representing idealized analytes, and to compare for bimodal signature differences between dielectric versus conducting nanoparticles. Second, a study on high-affinity antibody-ligand interactions (nanomolar range) was performed, mimicking types of interventions relevant to specific antibody mediated cancer immunotherapy. Lastly, we studied low-affinity receptor-ligand interactions (micromolar range) mimicking specific natural killer (NK) cell receptor interactions with cancer-relevant ligands. For the latter two studies, discrimination between specific versus non-specific interactions and a novel effect of bound fraction enhancement were studied as well.

13.3.1 Nanoparticle Sensing with the SANE Sensor

The aim of the nanoparticle experiment was to show that the SANE sensor trapped 20 nm nanoparticles made of SiO_2 (silica) Au for several seconds, while enabling their concurrent optical and electrical sensing. SANE sensing enabled measuring how optical-electrical characteristics differed between SiO_2 and Au nanoparticles due to differences in their physical properties. Both nanoparticle solutions were added to a 1 M, ~7.2 pH KCl solution.

Material	Size	Concentration	Zeta Potential	Model	Company	Note
Silica	20 ± 4 nm	200 pM	±40 mV	MEL0010	NanoComposix	No coating
Gold	20 ± 4 nm	200 pM	±15 mV	C11-20-TM-DIH-50	Nanopartz	2 nm polymer coating

13.3.1.1 Results for Nanoparticles

We were able to demonstrate proof of principle using the SiO_2 and Au nanoparticles [2]. Figure 13.3 shows a clean trapping event of a single SiO_2 nanoparticle. This can be broken down into three regions: the entry of the particle into the trap, Fig. 13.3b; the particle in the trap, Fig. 13.3c; and the translocation of the particle from the *cis* to *trans* chamber, Fig. 13.3d.

We also recorded two-particle trapping events that were inferred from seeing a two-step optical transmission pattern. Interestingly, a broad resonance was detected in the electrical signals while SiO_2 nanoparticles where optically trapped (Fig. 13.3a, Region B) with the peak frequency shifting to lower values for the single particle trapping case [2]. Trapped Au nanoparticles provided similar optical characteristics, but different electrical characteristics relative to the SiO_2 nanoparticles. Specifically, Au nanoparticle measurements did not have the noise that was observed in region B of Fig. 13.3a. We hypothesized that the surface charge of SiO_2 nanoparticles was creating current oscillations while the nanoparticle was bobbing inside the optical

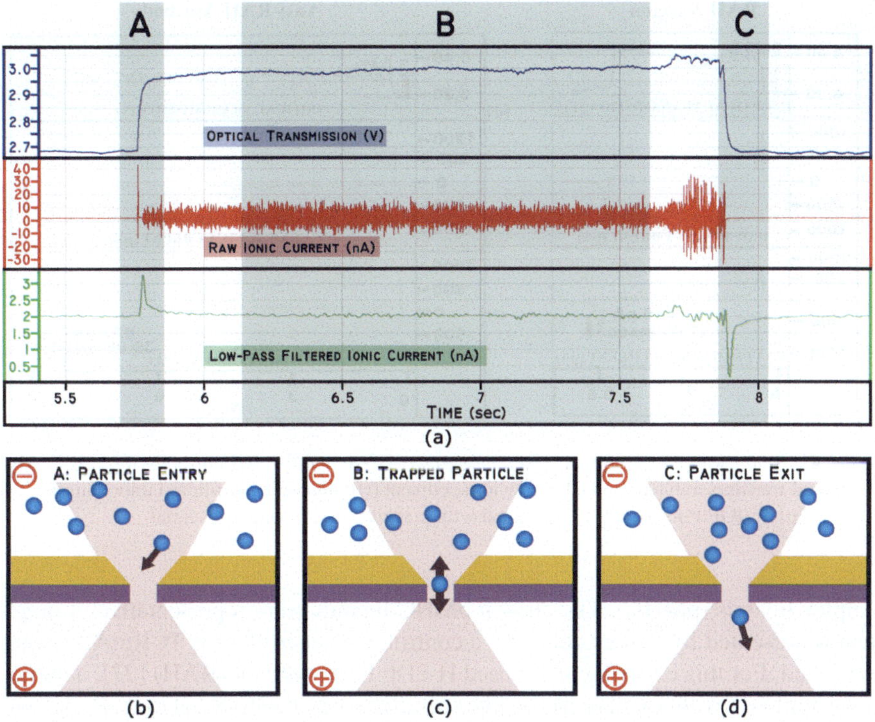

Fig. 13.3 **a** Plots of simultaneously recorded optical transmission (top, blue; V), raw ionic current (middle, red; nA) 20 Hz low-pass filtered ionic current (bottom, green; nA) versus time (sec) for the single 20 nm SiO_2 nanoparticle trapped in the SANE sensor. Physical interpretation schematics for the signals recorded within gray-shaded regions A, B and C are shown in panels (b), (c) and (d), respectively. **b** Region A: negatively charged nanoparticle entering the DNH-ssNP under applied bias. **c** Region B: nanoparticle trapped and bobbing inside the DNH near the ssNP mouth. **d** Region C: nanoparticle exiting the optical trap after the electrophoretic force dominates translocation. Figure and caption reproduced with permissions by [2]

trap, immediately above the underlying nanopore. In contrast, the Au nanoparticles were conducting and likely remained electrically neutral.

13.3.2 High-Affinity (Nanomolar) Interaction Sensing with the SANE Sensor

To expand upon the work of the nanoparticles, we designed an experiment to look at high affinity bonding to determine if we could distinguish between individual proteins and protein complexes. For this experiment we used T-cell receptor-like antibodies (TCRmAbs) that targeted peptide-presenting major histocompatibility

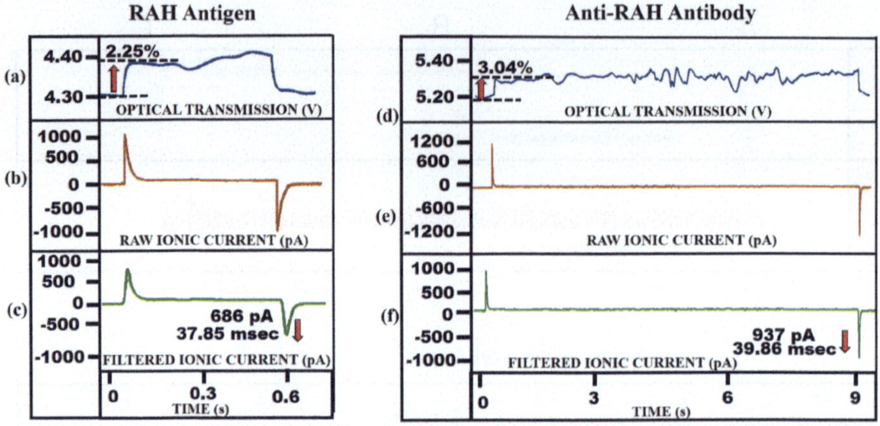

Fig. 13.4 Time traces of individual RAH antigen (**a**)–(**c**) and anti-RAH antibody molecules (**d**)–(**f**). Optical transmission ((**a**) and (**d**)), raw ionic current ((**b**) and (**e**)) and filtered ionic current ((**c**) and (**f**)) [26].

complex ligands (pMHC). The chosen pMHC ligands were representative of target ligands presented by cancer cells. As a control, cancer-irrelevant TCRmABS were also tested. For this experiment we used H-2Db RAH antigen (RAH) [27], a monoclonal antibody with T-cell receptor-like specificity targeted to RAH (anti-RAH) and a non-specific antibody anti-WNV. All three have been prepared as described previously [26]. The experiments were run with analytes at the following concentrations: 200, 20, 2 and 0.2 nM.

13.3.2.1 Results for High-Affinity Interactions

Figure 13.4a–f is a clean example of an optical trapping event in both individual analytes, RAH antigen and anti-RAH antibody. Both traces were taken with 300 nM solutions. The red arrows in Fig. 13.4a, d show a distinct and different optical step change of 2.25% and 3.04% respectively. Similarly, the trapping duration and translocation current were also larger for the antibody. However, the translocation time was similar. A few things can be inferred from this: the larger antibody, anti-RAH would take up more space, which would account for both the larger optical step change and the translocation current, where it would be displace more ions. The translocation time was similar, which is a good indicator that, while larger, the antibody did not have to compress to fit through the pore. The trapping duration would also be affected by the size of the molecule, where according to Neumerier et al., the larger molecule would create a more efficient trap resulting in a longer trapping time [41].

The different optical-electrical parameters were compared and the best data separation of the two analytes was attained by the optical transmission step change metric.

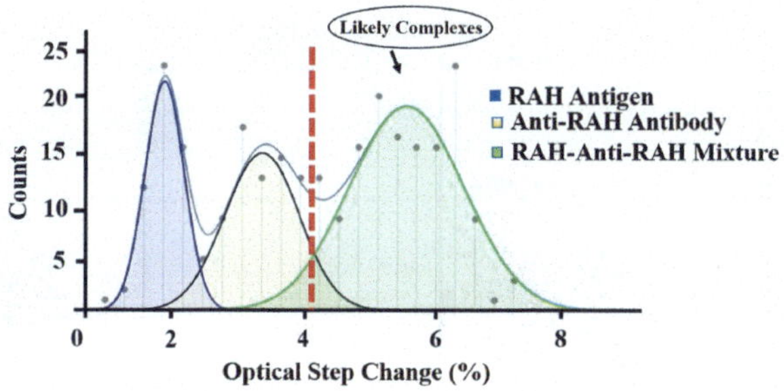

Fig. 13.5 Histogram of optical step change for RAH-anti-RAH equimolar mixtures. A threshold of 4.20% (red dashed line) was defined for classifying events as bound complex (green region) versus unbound RAH (blue region) and anti-RAH (yellow region) [26]. ©IOP Publishing. Reproduced with permission. All rights reserved

Based on the aggregation of data from multiple experiments, a threshold was set to separate between individual analytes and likely complexes (Fig. 13.5) [26].

To look at specific versus non-specific binding event detection with the SANE sensor, we used anti-WNV, an RAH irrelevant antibody. Anti-WNV was chosen because it closely matched the weight and electrical properties of anti-RAH while not having a specific binding site for RAH. In this case, the electrical data alone could not be used to distinguish between the RAH–anti–RAH and RAH–anti–WNV but when combined with the optical data, a clear delineator was present.

Importantly, in all our published work so far, we have focused on separating bound from unbound fraction using 2D projections of these multi-dimensional datasets. However, in future work it may be worthwhile to explore the feasibility of separating interactions, like molecular complexes from unbound ligands, using multi-dimensional data classification approaches such as support vector machine [28] and machine learning [29] algorithms. To elaborate this point, Fig. 13.6a shows a 2D plot of the translocation current versus optical step change for all analytes in that experiment. There is a clear separation of the RAH–anti–RAH complex using the threshold shown in Fig. 13.5. Nevertheless, if a 3rd dimension is added, the optical trap time in this case, one can see a more complex feature (purple plane in Fig. 13.6b) hidden behind a larger mass of data in Fig. 13.6a. This observation suggests that a better threshold for determining bound versus unbound analytes can be developed using a surface derived from multi-dimensional data.

When the bound fraction of a specific complex was compared to estimates based on the K_D value derived by a commercial resonant sensor system (ResoSens, RSI Inc., Arlington, TX, Fig. 13.7a), it was found that the SANE sensor detected considerably higher bound fraction values at corresponding antigen concentrations (Fig. 13.7b). The bound fraction gain with the SANE sensor was lowest at 100 nM possibly due to target saturation and a lower number of available binding sites at that concentration, which was above the K_D value. Bound fraction gains were higher at lower concentrations and followed a pattern that qualitatively mirrored the relative reduction of bound fraction estimated for the bulk solution reactions using the K_D value derived by the commercial resonant sensor system. Interestingly, the resonant sensor did not show appreciable target binding at concentrations $<$10 nM, whereas the SANE sensor was able to detect interactions down to much lower concentrations (10, 1 and

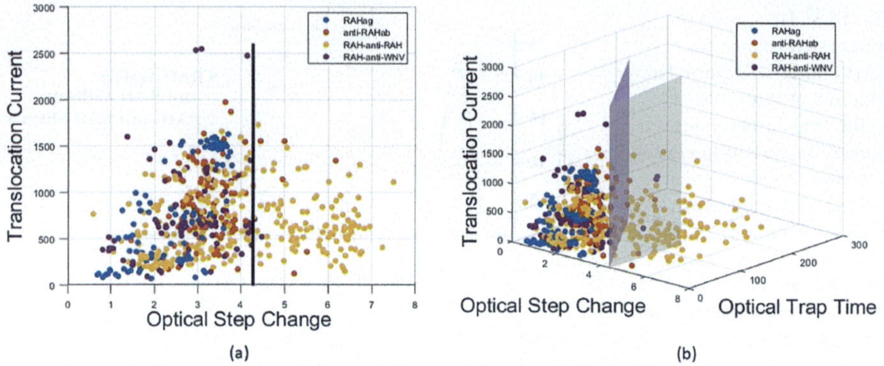

Fig. 13.6 **a** A 2D projection of all experimental data form both specific and non-specific antibody interaction experiments (RAH, anti-RAH, anti-WNV, and their likely complexes). The black line shows the selected threshold of 4.2% separating the likely RAH-anti-RAH complexes (specific binding) from the individual analytes. **b** The same data with an additional data type (optical trap time) added for a third axis and rotated to show a 3D feature that was not noticeable in **a**: The purple plane shows a 2D vector that can be used to separate likely complexes that were not selected using the optical step change 1D vector threshold

0.1 nM, Fig. 13.7b), highlighting its very high sensitivity. The SANE and resonant sensor experiments were not directly analogous as the latter were performed with an excess of antibodies immobilized on a surface whereas the former was performed in solutions of equal antigen-antibody proportions. The fact that fewer antigens should be available in the SANE sensor experiments, since they are not available in excess, yet higher bound fractions were still detected than with the Resonant Sensor, suggests the existence of an underlying mechanism affecting protein interactions at the SANE sensor. We hypothesized that the applied electric field concentrated the protein solution components immediately over the SANE sensor by pushing them inside the optical trapping volume.

13.3.3 Low-Affinity (Micromolar) Interaction Sensing with the SANE Sensor

Further diving into the capabilities of the SANE sensor to identify and characterize protein complexes, this experiment was designed to observe low affinity binding events and delineate between specific and non-specific binding. For this experiment, a simplified model with relevance to cancer immunotherapy was used. A heterodimer presented on Natural Killer lymphocytes was used, CD04 and NKg2a (NK receptor) [30]. This receptor has specific recognition of a pMHC ligand known as QDM/Qa-1 b (QDM) [31]. While the QDM ligand is usually associated with an inhibitory response in NK cells, tumor cells, are known to express QDM ligand as a mech-

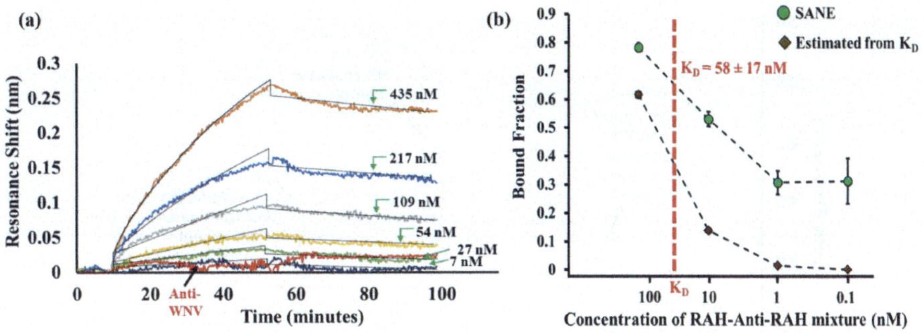

Fig. 13.7 **a** Association-dissociation kinetics for titrations of RAH H-2Db (435 nM, 217 nM, 109 nM, 54 nM, 27 nM, 7 nM) interacting with surface-immobilized TCR-like antibody (anti-RAH TCRmAb at 33 nM) measured by a commercial assay (ResoSens) to derive K_D. Also included, a surface-immobilized anti-WNV TCRmAb (negative control) at 33 nM interacting with RAH H-2Db (7 nM, black arrow). **b** SANE sensor-derived complex bound fraction in equimolar mixture titrations (green dots) bracketing the K_D value derived by the commercial assay (red dashed line) and corresponding free-solution bound fraction estimates for this K_D value [26].

anism to hide from the natural immune system response. As our negative control, the cancer-irrelevant ligand GroeL was used. GroEL is an immunodominant epitope expressed by Salmonella typhimurium and presented by a Qa-1 b MHC molecule. It is specifically recognized by CD8+ cytotoxic T lymphocytes after natural infection in the mouse [33, 34]. Qa-1b Qdm ligand (Qdm) [32], CD94/NKG2A heterodimer (NK receptor) with specificity to Qdm [32] and the non-specific control ligand GroEL [33, 34] were prepared as described previously [25]. The experiments were run with analytes at the following concentrations: 600, 300, 100 and 10 nM.

13.3.3.1 Results for Low-Affinity Interactions

As in previous experiments we started by characterizing the individual components of this simplified ligand-receptor interaction model and once again identified optical step change as the metric allowing the clearest separation between receptor, ligand and ligand-receptor complexes. The optical step threshold for distinguishing the complexes was set empirically as the point of inflection of the joint data histogram (~3.15%, Fig. 13.8). In addition, we identified a smaller subset of step change events that had significantly higher amplitudes, which we identified as likely agglomerates of NK receptors. It is possible that a more accurate threshold for bound complexes may be defined by a multi-dimensional surface when considering more than two data types. This will be examined in future work.

The Qdm-NK data had two step events in transmitted optical intensity, as seen in Fig. 13.9a, that represented the binding of Qdm to the NK receptor. The time duration of the second step feature also served as an indicator of binding duration. The latter

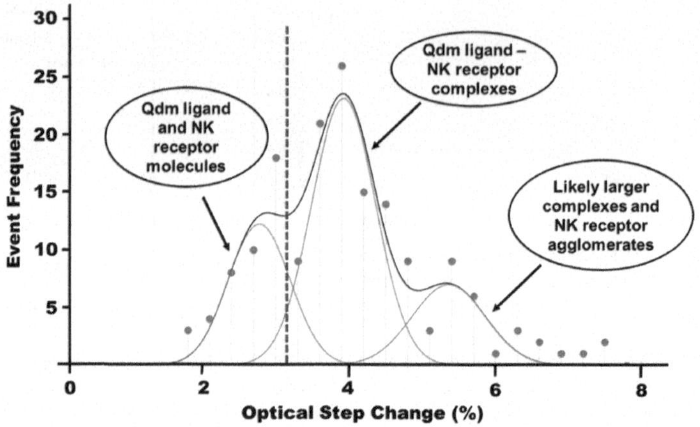

Fig. 13.8 Histogram of optical step change for ligand-receptor (Qdm-NK receptor) equimolar mixtures. A threshold of 3.15% (red dashed line) was defined for classifying events as bound complex versus unbound ligand and receptor [25].

was identified as an important data type for distinguishing between specific Qdm-NK receptor interactions from non-specific GroEL-NK receptor interactions (Fig. 13.9b). Specifically, the GroEL-NK receptor interactions had similar double-step amplitude increases to the specific interactions (Fig. 13.9c), but the binding duration was significantly longer for non-specific interaction (Fig. 13.9d). Furthermore, compilation of multiple binding duration events into a histogram yielded a log-linear dependence, consistent with the known k_{off} value for this interaction [35]. In contrast, the binding time histogram for GroEL-NK receptor interactions looked random (Fig. 13.9e), consistent with non-specific interactions.

Due to single molecule trapping enabled by the SANE nanosensor, low-affinity binding events can be detected which would otherwise require a much more concentrated bulk protein solution. The combination of the electric-field induced movement and the opposition of the optical trap causes a crowding effect at the edge of the double nanohole. This crowding effect creates a localized, high concentration environment that increases the probability that proteins will react within the optical trap volume indicated by the red oval in Fig. 13.10a. Based on our analysis [25] a 1000-fold reduction of purified protein, relative to that required by the commercial assay ELISA, was attained while still being able to study binding events with the SANE sensor. Figure 13.10b depicts the larger than expected bound fraction that was observed at equimolar concentrations of Qdm-NK receptor compared to the calculated bound fraction based on the K_D measured using commercial assays. The nature of this crowding effect can be explored by observing the optical transmission trace from the moment the flowcell/SANE sensor is connected to the experimental setup. We can observe in Fig. 13.10c that there is a long period of time, measured in seconds, in which no events were detected before a rapid succession of trapping

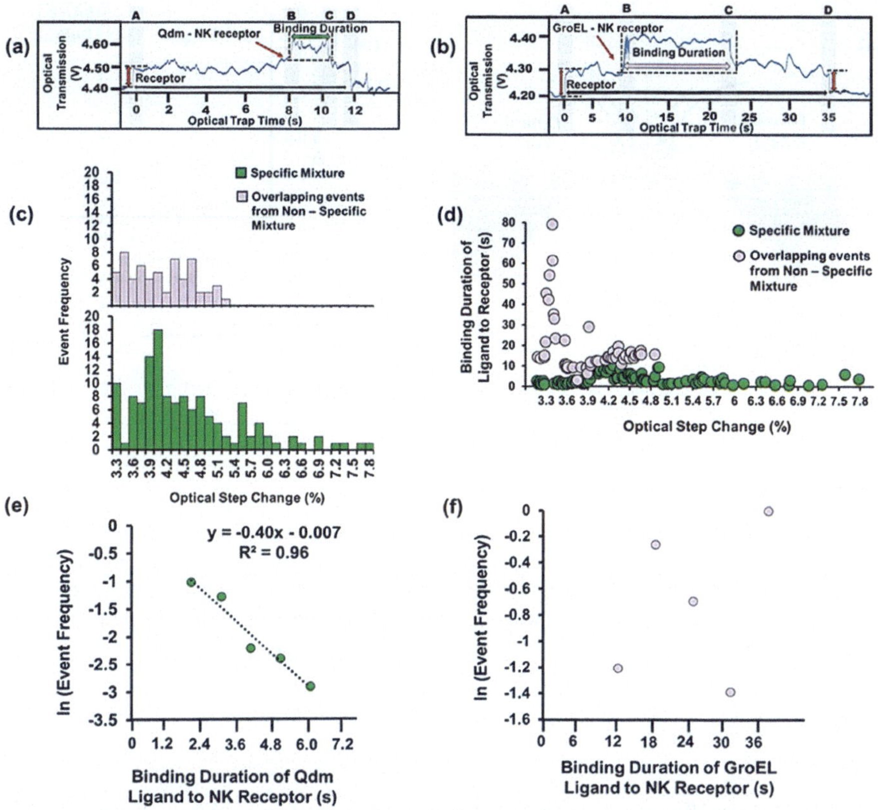

Fig. 13.9 **a** A typical binding duration observed for low-affinity (μM) interactions of Qdm-NK receptor complexes at 10 nM concentration (~2 s). **b** Typical binding interaction of GroEL-NK receptor at 10 nM concentration (~13 s). **c** Event frequency histograms of optical step change for all the specific binding events (green columns) and overlapping events (pink columns) from non-specific mixture. **d** Scatter plot separating the non-specific mixture events (pink circles, n = 55) from specific binding events (green circles, n = 132) based on binding duration and optical step change. Natural log plot of event frequency of ligand-receptor binding interaction duration for **e** specific mixture and **f** non-specific mixture. In contrast to **e**, the data could not be fit to a linear curve in **f**. The slope of the line equation is the k_{off} [25].

events occurred. This time trace indicates the movement of the proteins from the bulk solution to the area near the optical trap where they interacted before translocating through the nanopore.

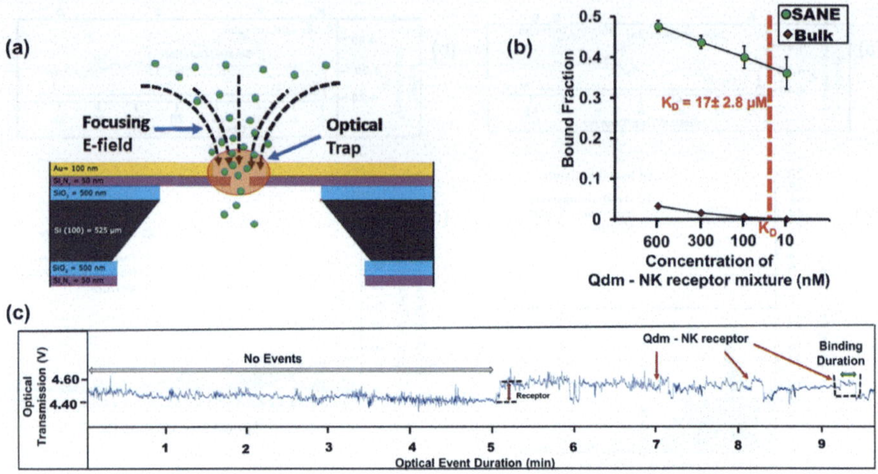

Fig. 13.10 **a** Schematic of the protein crowding effect created by the applied electric field immediately over the SANE sensor. **b** The observed Qdm-NK receptor bound fraction is significantly higher than corresponding values at the same bulk Qdm-NK receptor concentrations, due to the sensor-induced protein crowding. **c** Zoomed out time-series plot showing no trapping events for several seconds at the beginning of an experiment [25].

13.4 Conclusions

13.4.1 Identification of Analytes and Complexes They Form in Simple Mixtures

Using optical-electrical sensing, we were able to collect high quality data that allowed us to successfully identify each individual analyte in all three experiment types performed (nanoparticles, high-affinity, and low-affinity protein-ligand interactions). A set of basic bimodal parameters was acquired that contributed to distinguishing between analytes: optical step change, optical trap time, translocation current, and nanopore translocation time. For the experiments involving protein-ligand interactions, optical data appeared to offer qualitatively better separation between analytes and the complexes that they formed. Optical step change stood out as the best single metric for this purpose in the experiments performed to date.

Interestingly though, using optical step change to separate specific from non-specific interactions works well in one model system (RAH–anti–RAH versus RAH–anti–WNV), but not as well in another (Qdm-NK receptor versus GroEL-NK receptor). An additional data type was identified, namely binding duration, which enabled clear identification of specific interaction in the latter case. Two more data types were identified in our prior studies: bobbing frequency and trapping current. Firstly, the bobbing frequency spectrum of current fluctuations while a charged analyte is in the optical trap. We hypothesized that competing electrophoretic, SIBA and possible other forces felt by analytes inside the optical trap induced bobbing motions inside the optical trap that led to high frequency ionic current oscillations sensed through the nanopore. Secondly, a positive current spike was noted concurrently with the beginning of each trapping event, which is not seen in traditional nanopore measurements. We hypothesized that charge displacement by the analyte as it approaches the nanopore entrance contributes to this spike, although it is possible that mirror charges at the walls of the DNH may have also contribute to driving this counterion current. Both the high-frequency (kHz range) bobbing motions and the positive current spike require additional future studies to clarify their physical underpinnings. In summary, we have identified at least eight bimodal metrics that can be extracted concurrently from optical-electrical data traces enabled by SANE sensing. More data types are likely to be added to this list in the future. The resulting multi-dimensional data sets are inviting of machine learning and other classification that could allowed more sensitive and specific classification of analyte events in the future.

In addition, the bimodal nature of SANE sensor measurements provides opportunities to exploit the synergism between optical and electrical methods. For example, separating the effects of size versus charge on bimodal measurements for a given analyte or molecular complex could be achieved by combining optical (size) and electrical (size-charge) data. Furthermore, optical trapping may offer the opportunity to slow down the translocation of larger molecules, such as DNA through a nanopore [36], which could enhance the signal to noise ratio of electrical measurements. Finally, we speculate that it may be possible to combine SANE sensing with electrophoretic separation to enable study of more complex mixtures in the future.

13.4.2 Enhancement of Protein-Ligand Bound Fractions

A unique advantage of this nanosensor is its ability to detect analyte interactions at much lower concentration than would be required for other assays. In both high-affinity and low-affinity interaction experiments, we observed a much larger bound fraction, upwards of 3 orders of magnitude higher than estimated from the known equilibrium K_D value for these reactions. This concentration effect opens opportunities to screen for low-affinity protein-ligand interactions at much lower concentrations relative to what is required by current label-free technologies like SPR. Being able to screen interactions at lower concentrations could translate to very significant savings in time, effort and reagent costs involved in purifying large amounts of protein. The SANE senor's ultra-high sensitivity for interaction screening could also be valuable for the screening of rare analytes. Interestingly, in recent work, a thermophoretic effect was reported to exist due to plasmon-induced heating in the Au layer of the DNH structure [22]. If analyte complexes are thermophilic, it is possible that their concentration could be enhanced inside the optical trap due to the presence of localized heating. Detailed studies need to be performed in the future to assess the importance of thermal effects on bound-fraction enhancement. Furthermore, rapid advances in the design and fabrication of complex all-dielectric nanostructures that are less prone to heating than plasmonic metal nanostructures could decouple optical from thermal effects. For example, a bullseye all-dielectric structure was developed that has about half the optical trap stiffness of the Au DNH, but had very little Joule heating [37].

13.4.3 Limitations in Sensor Throughput

As a general rule, an acceptable throughput by a classical nanopore is $\sim$1000 events over a 15-min data run. In the case of the SANE sensor, with optical trapping extending individual event times to many seconds, a typical event rate is a few tens of events over the same 15-min period. It is clear therefore that the SANE sensor's throughput is very limited compared to classical nanopores and data acquisition durations adding up to several hours are needed to acquire hundreds of events for robust statistical analysis. In addition, as the objective lens's focal area ($\sim$500 μm) is not much larger than the widest sensor feature size ($\sim$200 μm), optical alignment may need to be adjusted between successive 15-min experimental runs due to piezoelectric stage drift.

Limited SANE sensor throughput could be overcome in the future by multiplexed detection over an array of sensors fabricated on a wafer. On the optical sensing side, a microlens array could enable simultaneous illumination on numerous SANE sensors if the pitch of sensors and lenslets is matched. For further scaling, a vertical cavity surface emitting laser (VCSEL) array is a potential technology that could be used for optical trapping and sensing on SANE sensor array. On the light detection side, a photodiode array, or a scientific complementary metal oxide semiconductor (sCMOS) camera, could be used for parallel light collection from multiple sensors. For nanopore multiplexed sensing, low-noise, high bandwidth on-chip readout electronics have been previously developed [38, 39] and could potentially be integrated with the SANE sensor technology. Recently, Ontera Inc. (Santa Cruz, CA) developed a nanopore array integrated into a CMOS chip to enable readout of 264 separate nanopores simultaneously, recording each one with built-in individual electrodes. Additionally, any future incarnation of a multiplexed platform will require careful engineering to minimize optical and electronic crosstalk between neighboring sensors and the likely addition of microfluidics to enable rapid sample loading and flushing. Finally, it is worth noting that multiplexing of protein-ligand interactions at the single sensor level could be attained if one of the reactants could be barcoded with a unique single strand of DNA or RNA [40]. In that case, the sensor could act as an active nanosieve letting through, say, only bound complexes, that could be identified subsequently by amplification of the nucleotide strands.

13.5 Acronym Glossary

CMOS: complementary metal oxide semiconductor
DNH: double nanohole
DRIE: deep reactive ion etching
FIB: focused ion beam
GFIS: gas field ion source
HWP: half wave plate
K_D: dissociation constant
k_{off}: binding duration
LPCVD: low pressure chemical vapor deposition
NK: natural killer
PDMS: polydimethylsiloxane
pMHC: histocompatibility complex
QDM: Qa-1b Qdm ligand
QWP: quarter wave plate
RAH: H-2Db RAH antigen
SANE: self-induced back-action actuated nanopore electrophoresis
Si_xN_y: silicon nitride
SiO_2: silicon oxide
SIBA: self-induced back action
SPR: surface plasmon resonance
ssNP: solid-state nanopores
TCRmAbs: T-cell receptor-like antibodies
TMAH: tetramethylammonium hydroxide
VCSEL: vertical cavity surface emitting laser
WNV: West Nile virus

Acknowledgements We thank Mr. Soeren Eyhusen and Mr. Chuong Huynh for permitting us with the access and providing technical guidance on focused ion beam milling at the Zeiss ORION NanoFab facility in Peabody, MA. We are also thankful to all the engineers and staff at the Shimadzu Institute Nanotechnology Research Center at the University of Texas at Arlington for their support and guidance. The authors acknowledge the financial support from the National Cancer Institute (1R21CA240220-01A1) and the National Science Foundation (CBET#2022398 and CBET#2022374).

References

1. Mathieu L Juan, Reuven Gordon, Yuanjie Pang, Fatima Eftekhari, and Romain Quidant. Self-induced back-action optical trapping of dielectric nanoparticles. *Nature Physics*, 5(12):915–919, 2009.
2. Muhammad Usman Raza, Sai Santosh Sasank Peri, Liang-Chieh Ma, Samir M Iqbal, and George Alexandrakis. Self-induced back action actuated nanopore electrophoresis (sane). *Nanotechnology*, 29(43):435501, 2018.

3. Benjamin Cressiot, Esther Braselmann, Abdelghani Oukhaled, Adrian H Elcock, Juan Pelta, and Patricia L Clark. Dynamics and energy contributions for transport of unfolded pertactin through a protein nanopore. *ACS nano*, 9(9):9050–9061, 2015.
4. Ahmed A. Al Balushi and Reuven Gordon. Label-free free solution single protein-small molecule binding kinetics: An optical tweezer approach. In *Optics in the Life Sciences*, page OtT2E.3. Optical Society of America, 2015.
5. L Kiran Swaroop Kumar and Reuven Gordon. Overlapping double-hole nanostructure in a metal film for localized field enhancement. *IEEE Journal of selected topics in quantum electronics*, 12(6):1228–1232, 2006.
6. LKS Kumar, A Lesuffleur, MC Hughes, and R Gordon. Double nanohole apex-enhanced transmission in metal films. *Applied Physics B*, 84(1-2):25–28, 2006.
7. Mostafa Ghorbanzadeh, Steven Jones, Mohammad Kazem Moravvej-Farshi, and Reuven Gordon. Improvement of sensing and trapping efficiency of double nanohole apertures via enhancing the wedge plasmon polariton modes with tapered cusps. *ACS Photonics*, 4(5):1108–1113, 2017.
8. Pau Mestres, Johann Berthelot, Srdjan S Aćimović, and Romain Quidant. Unraveling the optomechanical nature of plasmonic trapping. *Light: Science & Applications*, 5(7):e16092–e16092, 2016.
9. Yuanjie Pang and Reuven Gordon. Optical trapping of 12 nm dielectric spheres using double-nanoholes in a gold film. *Nano letters*, 11(9):3763–3767, 2011.
10. Abhay Kotnala, Damon DePaoli, and Reuven Gordon. Sensing nanoparticles using a double nanohole optical trap. *Lab on a Chip*, 13(20):4142–4146, 2013.
11. Abhay Kotnala and Reuven Gordon. Quantification of high-efficiency trapping of nanoparticles in a double nanohole optical tweezer. *Nano letters*, 14(2):853–856, 2014.
12. Yuanjie Pang and Reuven Gordon. Optical trapping of a single protein. *Nano letters*, 12:402–406, 2011.
13. Ahmed A Al Balushi, Ana Zehtabi-Oskuie, and Reuven Gordon. Observing single protein binding by optical transmission through a double nanohole aperture in a metal film. *Biomedical optics express*, 4(9):1504–1511, 2013.
14. Ahmed A Al Balushi and Reuven Gordon. Label-free free-solution single-molecule protein–small molecule interaction observed by double-nanohole plasmonic trapping. *ACS Photonics*, 1(5):389–393, 2014.
15. Ahmed A Al Balushi and Reuven Gordon. A label-free untethered approach to single-molecule protein binding kinetics. *Nano letters*, 14(10):5787–5791, 2014.
16. Jong Min Sung. Dielectrophoresis and optoelectronic tweezers for nanomanipulation. 2007.
17. Mathilde Lepoitevin, Tianji Ma, Mikhael Bechelany, Jean-Marc Janot, and Sebastien Balme. Functionalization of single solid state nanopores to mimic biological ion channels: A review. *Advances in colloid and interface science*, 250:195–213, 2017.
18. Patrick Abgrall and Nam Trung Nguyen. Nanofluidic devices and their applications. *Analytical chemistry*, 80(7):2326–2341, 2008.
19. Eszter Mádai, Bartłomiej Matejczyk, András Dallos, Mónika Valiskó, and Dezső Boda. Controlling ion transport through nanopores: modeling transistor behavior. *Physical Chemistry Chemical Physics*, 20(37):24156–24167, 2018.
20. Matthias Firnkes, Daniel Pedone, Jelena Knezevic, Markus Doblinger, and Ulrich Rant. Electrically facilitated translocations of proteins through silicon nitride nanopores: conjoint and competitive action of diffusion, electrophoresis, and electroosmosis. *Nano letters*, 10(6):2162–2167, 2010.
21. Jan KG Dhont, S Wiegand, S Duhr, and D Braun. Thermodiffusion of charged colloids: Single-particle diffusion. *Langmuir*, 23(4):1674–1683, 2007.
22. Quanbo Jiang, Benoît Rogez, Jean-Benoît Claude, Guillaume Baffou, and Jérôme Wenger. Quantifying the role of the surfactant and the thermophoretic force in plasmonic nano-optical trapping. *Nano Letters*, 2020.
23. Ahmed A Al Balushi, Abhay Kotnala, Skyler Wheaton, Ryan M Gelfand, Yashaswini Rajashekara, and Reuven Gordon. Label-free free-solution nanoaperture optical tweezers for single molecule protein studies. *Analyst*, 140(14):4760–4778, 2015.

24. Junrong Wang, Jian Ma, Zhonghua Ni, Li Zhang, and Guoqing Hu. Effects of access resistance on the resistive-pulse caused by translocating of a nanoparticle through a nanopore. *RSC Advances*, 4(15):7601–7610, 2014.
25. Sai Santosh Sasank Peri, Manoj Kumar Sabnani, Muhammad Usman Raza, Elizabeth L Urquhart, Soroush Ghaffari, Jung Soo Lee, Min Jun Kim, Jon Weidanz, and George Alexandrakis. Quantification of low affinity binding interactions between natural killer cell inhibitory receptors and targeting ligands with a self-induced back-action actuated nanopore electrophoresis (sane) sensor. *Nanotechnology*, 32(4):045501, 2020.
26. Sai Santosh Sasank Peri, Manoj K Sabnani, Muhammad Usman Raza, Soroush Ghaffari, Susanne Gimlin, Debra D Wawro, Jung Soo Lee, Min Jun Kim, Jon Weidanz, and George Alexandrakis. Detection of specific antibody-ligand interactions with a self-induced back-action actuated nanopore electrophoresis sensor. *Nanotechnology*, 31(8):085502, 2019.
27. Mariet CW Feltkamp, Gienke R Vreugdenhil, Michel PM Vierboom, Elisabeth Ras, Sjoerd H van der Burg, Jan Ter Schegget, Cornelis JM Melief, and W Martin Kast. Cytotoxic t lymphocytes raised against a subdominant epitope offered as a synthetic peptide eradicate human papillomavirus type 16-induced tumors. *European journal of immunology*, 25(9):2638–2642, 1995.
28. K SRIVASTAVA Durgesh and B Lekha. Data classification using support vector machine. *Journal of theoretical and applied information technology*, 12(1):1–7, 2010.
29. Howard B Demuth, Mark H Beale, Orlando De Jess, and Martin T Hagan. *Neural network design*. Martin Hagan, 2014.
30. Andrew G Brooks, Phillip E Posch, Christopher J Scorzelli, Francisco Borrego, and John E Coligan. Nkg2a complexed with cd94 defines a novel inhibitory natural killer cell receptor. *The Journal of experimental medicine*, 185(4):795–800, 1997.
31. Russell E Vance, Jennifer R Kraft, John D Altman, Peter E Jensen, and David H Raulet. Mouse cd94/nkg2a is a natural killer cell receptor for the nonclassical major histocompatibility complex (mhc) class i molecule qa-1b. *The Journal of experimental medicine*, 188(10):1841–1848, 1998.
32. Jennifer R Kraft, Russell E Vance, Jan Pohl, Amy M Martin, David H Raulet, and Peter E Jensen. Analysis of qa-1bpeptide binding specificity and the capacity of cd94/nkg2a to discriminate between qa-1–peptide complexes. *The Journal of experimental medicine*, 192(5):613–624, 2000.
33. Wei-Feng Lo, Helena Ong, Eleanor S Metcalf, and Mark J Soloski. T cell responses to gram-negative intracellular bacterial pathogens: a role for cd8+ t cells in immunity to salmonella infection and the involvement of mhc class ib molecules. *The Journal of Immunology*, 162(9):5398–5406, 1999.
34. Wei-Feng Lo, Amina S Woods, Amy DeCloux, Robert J Cotter, Eleanor S Metcalf, and Mark J Soloski. Molecular mimicry mediated by mhc class ib molecules after infection with gram-negative pathogens. *Nature medicine*, 6(2):215–218, 2000.
35. Mar Valés-Gómez, Hugh T Reyburn, Robert A Erskine, Miguel López-Botet, and Jack L Strominger. Kinetics and peptide dependency of the binding of the inhibitory nk receptor cd94/nkg2-a and the activating receptor cd94/nkg2-c to hla-e. *The EMBO journal*, 18(15):4250–4260, 1999.
36. Maxim Belkin, Shu-Han Chao, Magnus P Jonsson, Cees Dekker, and Aleksei Aksimentiev. Plasmonic nanopores for trapping, controlling displacement, and sequencing of dna. *ACS nano*, 9(11):10598–10611, 2015.
37. Wei Zhan, Kui Gao, Chenjun Zhu, and Wuzhou Song. Optical trapping using all silicon nanoantennas with ultra-high electric field enhancement. *Applied Physics Letters*, 117(24):241102, 2020.
38. J Rosenstein, V Ray, M Drndic, and KL Shepard. Nanopore dna sensors in cmos with on-chip low-noise preamplifiers. In *2011 16th International Solid-State Sensors, Actuators and Microsystems Conference*, pages 874–877. IEEE, 2011.
39. Jacob K Rosenstein, Meni Wanunu, Christopher A Merchant, Marija Drndic, and Kenneth L Shepard. Integrated nanopore sensing platform with sub-microsecond temporal resolution. *Nature methods*, 9(5):487, 2012.

40. Marlon Stoeckius, Christoph Hafemeister, William Stephenson, Brian Houck-Loomis, Pratip K Chattopadhyay, Harold Swerdlow, Rahul Satija, and Peter Smibert. Simultaneous epitope and transcriptome measurement in single cells. *Nature methods*, 14(9):865, 2017.
41. Lukas Neumeier, Romain Quidant, Darrick E Chang Self-induced back-action optical trapping in nanophotonic systems *New Journal of Physics*, 17(12):123008, 2015.

Index

A
Achiral, 282
Acoustic
  vibration, 18
Acoustic phonon, 101
Acousto-optic modulator, 9
Actin, 7, 9, 29
Activation energy, 300
Adatoms, 54
Adeno associated virus, 321
Adenosine Diphosphate (ADP), 14
Adenosine Triphosphate (ATP), 14
Ag/AgCl, 330
Aggregate, 9
Alginate, 304
Aliasing, 18
Allan variance, Allan deviation, 73
Allostery, 256
α-Synuclein, 347
Amine, thiol, 53, 54
Aminosilanization, 70, 71
Amphiphilic, 303
Analyte, 299
Angle-resolved nanosphere lithography, 244
Anisotropic nanoparticles, 65
Anti-Stokes, 100
Anti-Stokes Raman Scattering (ASRS), 102
Aperture, 252
Aperture shape, 253
Applied voltage, 343, 347, 350, 352
Aptamers, 371
Asymmetric microsphere, 80
Atomic layer deposition, 243
Atomic polarizability, 13
Atomistic model, 12, 13
ATPase, 7, 14
Attomolar sensitivity, 54, 88
Autocorrelation analysis, 254

B
Backscattered light, 22
Backscattering, 50
Bandwidth, 74, 308
Beam expander, 81
Beam waist, 12, 14
Beat note, 78
Bethe theory, 252
Bias-T, 76
Binding affinity
  high, 389, 398, 399, 406, 407
  low, 389, 390, 393, 398, 402, 404–407
Binding energy, 67
Binding location, 52
Biological nanopores, 343, 345
Biomarker, 304, 345
Biomolecule, 38, 45
Biopolymer, 306
Biorecognition element, 84
Biosensing, 43, 49
Bi-phase ligand exchange, 66
Birefringence, 277
Blue detune, 136
Blue detuning, 131
Bottom-up assembly, 69
Bound fraction, 390, 393, 398, 401, 403, 406, 407
Bovine Serum Albumin (BSA), 9, 12, 13
Brillouin scattering, 100, 101

W. Bowen et al. (eds.), *Single Molecule Sensing Beyond Fluorescence*, Nanostructure Science and Technology,
https://doi.org/10.1007/978-3-030-90339-8

Brownian motion, 208
Bus waveguide, 64

C
Cancer, 258, 345
Capture radius, 300, 327
Carbonic anhydrase, 7
Carrier, sideband, 76
Cas9, 371, 372
Catalytic nanoparticle, 88
Catalytic reaction, 88
Cation, 306
Cavity, 16
Cavity Enhanced Raman Spectroscopy (CERS), 105
Cavity lifetime, 40, 74
Cavity linewidth, 76
Cavity optomechanics, 129, 130
Cavity resonance, 16, 76
Cavity Ring-Up Spectroscopy (CRUS), 75, 77
Centroid, 83
Cetrimonium Bromide (CTAB), 53, 54, 67–69, 71
Chelation, 316
Chemical turnover, 57
Chiral, 272
Chirality, 276
Chromatography, 272
Circulating power, 81
Citrate, 65
Clausius-Mossotti factor, 209, 212
Coherence
  length, 19
Coherent Anti-Stokes Raman Scattering (CARS), 101
Colloidal lithography, 244
Commercialization, 88
Complementary analysis, 262
Complementary Metal-Oxide-Semiconductor (CMOS), 308
Concentration gradient, 86
Conductance, 312
Conformation, 7, 12–14
Conformational change, 17, 57, 85, 86
Conformational dynamics, 84
Conformational dynamics, enzyme, 75
Conformation, conformational change, 52
Conformations, 254
Counterion, 302
Coupling distance, 81
Coupling efficiency, 77
Coupling method, 78
Covalent bond, 53
Cramer-Rao Lower Bound, 43
Critical coupling, 81
Cross-linker, 70
Cross-linking, 316
CW and CCW modes, 50
Cyclodextrin, 273
Cysteamine, 51, 54
Cystein, cysteamine, CTAB, 67

D
Dark solitons, 116
Data Acquisition (DAQ), 78
Debye length, 15, 302
Degenerate FWM, 103
Deoxyribonucleic Acid (DNA), 53, 298, 343, 345, 351, 359, 362
DEP force, 214
DEP trapping volume, 214
Detection limit, 43, 107
Detection volume, 49
Detrending, 83
DFT, 67
Dial-a-reaction, 54
Dichroism, 277
Dielectric breakdown, 299
Dielectric particle, 45
Differential interference contrast microscopy, 5
Diffraction limit, 6, 235
Diffuse layer, 302
Diffusion, 299
Diffusion near surfaces, 260
Dipole moment, 13
Direct frequency comb spectroscopy, 116
"Dirty" samples, 254
Dissipative Kerr Solitons (DKSs), 103
Dissociation constant, 401, 403, 404, 407, 409
Distributed Feedback Laser (DFB), 41, 74, 78
DNA folds, 371, 372
DNA knots, 371
DNA melting temperature, 55
DNA oligonucleotide, 53, 55
DNA origami, 85
DNA-PAINT, 56, 81, 87
DNA, protein, 54
DNA sequence, 55
DNA vibration spectra, 256

dn/dc, refractive index increment, 56
Double nanohole, 389–391, 394–396, 399, 407–409
Drude model, 175
Drug discovery, 257
Drug screen, 86
Dual-comb spectroscopy, 114
Dwell time, 355, 357–359

E
Elastic modes, 256
Elastic network model, 256
Electric double layer, 15, 302
Electric field, 344, 346–348, 351–353
Electrodeformation, 321
Electrolysis, 228
Electrolyte, 344, 347, 349, 350
Electron beam lithography, 243
Electro-optic modulator, 76
Electroosmosis, 298, 390, 392
Electrophoresis, 298, 389, 390, 392, 409
  electrophoretic, 389–391, 399, 407
Electrophoretic mobility, 300
Electrostatic trapping, 54
Electrostriction, 251
Electrostrictive force, 87
Enantiomer, 272
Enantiopure, 280
Endocytosis, 305
Enhanced frequency shift, 47
Enthalpy, entropy, 59
Entropic, 299
Enzymatic rates, 58
Enzyme, 39
Enzyme immobilisation, 57
Enzyme kinetics, 58
Enzyme, MalL, 57, 58
Enzyme-substrate complex, enzyme-transition-state complex, 59
Error signal, 76
Excess polarizability, 13, 56
External Cavity Diode Laser (ECDL), 77
Extraordinary Acoustic Raman (EAR) modes, 88

F
Fabry-Perot, 59, 78
Fabry–Pérot cavity, 98
FDTD, FEM, BEM, 48
Feedback, 18, 19
Feedback control, 19
Femtometer, 39, 44, 83
Femtonewton force, 87
Ferrule, 80
Fiber coupling, prism coupling, 80
Findpeaks algorithm, 84
Finesse, 41
Finite-Difference Time-Domain method (FDTD), 375
Fizeau wavemeter, wavemeter, 78
Flow cell, 394, 396, 397
Fluidic integration, microfluidics, 80
Fluorescence, 81, 249
Fluorescence label, 56
Fluorescence Resonant Energy Transfer Microscopy (FRET), 7
Focused ion beam lithography, 243, 389, 394–396, 409
Fourier transform, 75
Four-Wave Mixing (FWM), 103
Free-energy, 352–354, 361
Free-space coupler, 64
Frequency noise, 75
Frequency resolution, 21
Frequency shift, 76
Frequency shift, wavelength shift, 49
Frequency splitting, 51
Full width half maximum, 308
Functionalization, 71, 84
Fusion splicer, 62

G
Gammadion, 288
Generalized master equation, 192
Gibb's free energy, 59
Glass microsphere, 42, 60, 61, 84
Glass reflow process, 62
Glycan, 306
Glycoprotein, 305
Gold nanoparticles, 64
Gold nanorod, 38
G-Protein Coupled Receptors (GPCR), 86
Gradient force, 234
Graphene, 298, 345, 368
Green function, 168, 173
Grouping threshold, single-molecule event, 84

H
Hamiltonian, 167
Hamiltonian for SERS, 167
Harmonic generation, 102
Head group, 53
Heating, 261

Heisenberg uncertainty, 23, 28
Helicity, 276
Henry's function, 320
Heterodyne
  self, 19, 20
Heterodyne biosensing, 8–11, 29
Heterodyne biosensor, 9, 15, 16
Heterodyne detection, 8, 18, 21, 25, 29
Heterodyne frequency shift, 21
Heterodyne measurement, 77
Hexagonal Boron Nitride (hBN), 368
High-frequency electrical noise, 373
High Q, 73
High throughput, 81
Hinge-bending enzyme, 86
Homodyne detection, 8, 21, 25
Hooge parameter, 322
Hybridization, 55
Hybrid metal-cavity resonators, 175, 185
Hydrophobic, hydrophilic, 66
Hydrophobic ligand, 69

I
Induced dipole, dipole moment, 49
Inelastic scattering process, 100
Influenza A virus, 43
Interference
  beat-note, 9, 20, 21
  contrast, 9
  pattern, 9
Interferometric, 5
Interferometric imaging, 5
Interferometric SCATTering microscopy (iSCAT), 8, 9, 21, 29–31
Interferometric sensing, 6, 8, 11, 14, 17, 18, 25, 26, 28, 29
Interferometric sensor, 5, 6, 17, 32
Intrinsic quality factor, 44
Ion
  counter, 15
Ionic, 298
Ionic conductance, 344–346, 349, 350, 357
Ionic conductivity, 349
Ionic current, 351, 353, 355–362, 368
Ionic current drops, 345, 351–353, 355–359, 361, 362
Ionic current drop speed, 359–361
Ionic current rectification, 313
Isoelectric point, 302

J
Joule heating, 229

K
Kelvin polarization force, 210
Kramer theory, 300

L
Labelling, 253
Labview, 83
Laser
  cavity, 18
  linewidth, 18, 73
Laser relaxation oscillation, 18
Laser scanning, 82
Lasing threshold, 111
Lift-off, 243
Ligand exchange, 66–69
Ligand exchange via intermediate, 66
Ligands, 49
Ligand shell, 65
Linewidth shifts, 51
Linewidth shift, single-molecule linewidth shift, 51
Lipid bilayer, 31
Lipid membrane, 87
Liposomes, 298
Liquid Core Optical Ring Resonators (LCORR), 59
Local oscillator, 18, 20, 21, 25
Lorentzian lineshape, 40
Lowpass filter, 306
Low Pressure Chemical Vapor Deposition (LPCVD), 309
Lysine, 347, 352–355, 357, 361, 362

M
Machine learning, 345, 362
Macromolecular rate theory, 59
Macromolecule, 304
Maltodextrin, 304
Mass distribution, 31
Mass photometry, 9, 30, 31
Mass sensing, 140
Master equations, 169, 170, 173, 184, 192
MATLAB GUI, 83
Maxwell Stress Tensor, 252
Mercaptosilane, 71
Metal nanoparticles, 37
Metasurface, 285
Met-Enkephalin, 347, 352–354
Micelle, 15
Micodisk, 64
Micro-bottle resonator, 62

Micro-bubble, 59
Microcavity, 13, 37
Microcavity frequency combs, 114, 116
Microfluidics, 80, 81, 84, 262
Microlens array, 408
Microring, 64
Microsphere radius, 43
Microspheres, 59
Microtoroid, 59, 63
Microtoroid, microdisk, photonic crystal, 71
Microtubule, 9, 31
Mie, 281
Minimum shift, 73
Mode-locked lasers, 116
Mode splitting, 22, 78
Mode volume, 41, 43, 48
Modulational instability comb, 103
Moiré, 287
Molecular dynamics, 346, 347, 358
Molecular dynamics simulation, 13
Molecular mass, 9
Molecular mass resolution, 9
Molecular optomechanics, 167
Molecular optomechanics in the strong vibrational coupling regime, 181
Molecular polarizability, 12, 13, 44, 56
Molecular weight, 253
Molecules, ions, 45
MoS_2 nanopores, 343–347, 349–353, 355, 357–359, 361, 362
Muscle cell, 7
Myosin, 7, 29

N

Nanoaperture, 239, 368
Nanodisk, 282
Nanofabrication, 261
Nanofibre, 9, 10
Nanofluidic channel, 10, 11
Nano-optical traps, 378
Nanoparticle, 389, 390, 393, 398, 399
Nanoplasmonic, 272, 279
Nanopore, 87, 262, 298, 389–394, 397, 399, 405–409
  solid state nanopore, 389, 395, 399, 409
Nanopore device resolution, 344, 355, 357, 358
Nanorod dimer, 85
Nanorod orientation, 85
Nanorods, nanostars, nanocubes, nanowires, nanoprisms, 65
Nanoscatter coupling, 79
Nanosensor, 407
Nanostar, 65
Nanostructure arrays, 88
Nanostructures, 237
Nanotweezers, 374
Natural killer, 403, 409
  natural killer receptor, 398, 402–406
Near-field, 44
Negative DEP, 213
Noise, 373
  amplitude, 18
  background, 20
  digitisation, 18
  eater, 18
  electronic, 17–21, 25–27
  frequency, 18, 20, 22
  intensity, 18
  laser, 8, 17, 18, 20
  line, 18
  phase, 18, 19
  polarization, 18
  relaxation, 18
  shot, 9, 12, 17, 21–28
  technical, 17, 20, 21, 23, 24, 26, 27
  vacuum, 26
Noise floor, 45
Noise sources, 43
Nondegenerate FWM, 103
Noninvasive, 6
Nonlinear optical effects, 98
Normalized WGM spectrum, 82
Normal mode analysis, 256
NSF11 prism, SF9 prism, 81
Numerical aperture, 9

O

Off-binding constant, 393, 404, 405, 409
Open pore, 351, 353, 357, 358
Optical cavity, 16, 17, 19–23, 45
Optical coherence tomography, 6
Optical fiber, 41
Optical frequency combs, 103
Optical gradients, 374
Optical phonon, 101
Optical potential well, 233
Optical resonance, 22
Optical spring, 129, 138
Optical step change, 392, 401–406
Optical transmission, 374
Optical trap, 390–393, 398, 399, 407, 408
Optical trapping, 233

Optical trap time, 392, 401, 402, 406
Optical tweezer, 15, 28, 87
Optical tweezing, 391
Optomechanical amplification/cooling, 136
Optomechanical Hamiltonian, 173
Optomechanical oscillator, 118
Optomechanical sensing, 148
Optomechanical sensor, 127
Optomechanics, 117
Optomechanics, Biosensing, Single molecule detection, 127
Optoplasmonic, 37
Optoplasmonic sensing, 44, 45, 48, 81
Optoplasmonic sensor, 45
Orientation, 251

P

P53, 258
Particle tracking, 29
Pathlength, 9
PDH, PDH lock, Pound-Drever-Hall, 76
PDMS gasket, sample cell, 80
Peak detection, 84
PECVD, 309
Peptide, 304
Peptide Major Histocompatibility Complex (pMHC), 400
Permittivity, 56
Permutation entropy, 358–362
Perturbation theory, 45
Phase contrast microscopy, 5
Phase matching, 100
Phase shift, 5, 7
Phase space diagram, 24
Photo Activated Localization Microscopy (PALM), 7
Photobleaching, 29
Photodamage, 10, 14
Photolithography, 62
Photonic band-gap, 59
Photonic crystal, 59
Photonic crystal cavity, 98, 175
Photonic crystal resonator, 241
Photoresist AZ6130, 64
Photothermal effect, 117
Physical Vapor Deposition (PVD), 309
Pick-up, 18
Picosecond, 7
PID loop, 76
Piranha, 324
Plasmonic, 279, 389, 390, 392, 408
Plasmonic enhancement, 47, 49
Plasmonic heating, 56
Plasmonic hotspot, 44, 49
Plasmonic nanoantenna, 129
Plasmonic nanoparticle, 38, 44
Plasmonic nanopore, 376
Plasmonic nanostructures, 237
Plasmonic resonance, 6, 38
Plasmonics, 43
Plasmonic tweezer, 87
Plasmon resonance, 49
Plasmons, 374
Poissonian distribution, 23
Poissonian statistics, 23
Polarizability, 13, 14, 29, 30, 56, 209, 251
Polaron transforms, 190
Polycrystalline nanostructures, 64
Polydimethylsiloxane (PDMS), 80, 84, 88, 324
Polyethylenglycol, 54
Polymerase, 13, 58
Polymerase enzyme, enzyme, 87
Polypeptide, 345, 346, 361, 362
Polysaccharide, 298
Positive DEP, 213
Pound-Drever-Hall locking, 19
Pound-Drever-Hall (PDH), 75, 78, 83
Power spectral density, 321
Power spectrum, 18, 19, 27, 40
Principal component, 86, 87
Prism coupling, 79, 81
Probe field, 9–13, 20
Probe photons, 12
Probe power, 12
Probe wavelength, 12
Protein conformations, 254
Protein-DNA interactions, 258
Protein folding, 345, 346
Protein interactions, 258
Proteins, 298, 345, 346, 351, 352, 358, 361, 362
Protein sizing, 253
Protein vibration spectra, 256
Protonate, 301
Purcell effect, 16, 174
Purcell enhancement, 17, 80
Purcell factors, 174
Pure dephasing, 193

Q

Q, 40
Q over V ratio, 42
Quality, 37

Quality factor, 17, 40, 74
Quantum-optical measurement, 88
Quasinormal modes, 173, 185

R
Racemic, 280
Radiation pressure, 233
Raman lasing, 111
Raman scattering, 100, 101
Raman spectra, 177
Raman spectroscopy, 166, 257
Rayleigh, 281
Rayleigh scattering, 50
RC roll-off, 228
Reactive sensing, 128
Receptor, 49
Redshift, 38, 374
Reference interferometer, 128, 142
Refractive index, 5, 12, 15, 25, 30
Refractive index sensing, 6
Resolution, 6, 10
    bandwidth, 27
Resonance shift, linewidth shift, resonance splitting, 45
Resonance shift, mode splitting, mode broadening, 44
Resonance-wavelength, 374
Resonant SERS, 189
Resonant SERS Hamiltonian, 191
Resonant wavelength, 374
Ribonucleic Acid (RNA), 298
Ring-up signal, 77

S
Sampling frequency, 308
SARS, 109
Savitzky-Golay filter, 83
Scan rate, 77
Scattering
    back, 10
    cross-section, 12, 13, 15, 22, 24, 25
    dipole, 5, 7–9, 11, 15, 16, 30
    inelastic, 12
Scattering force, 234
Second Harmonic Generation (SHG), 28, 102, 112
Seed-mediated synthesis, 65
Self-Induced Back-Action (SIBA), 240, 252, 389–392, 407, 409
Self-induced Back-action Actuated Nanopore Electrophoresis (SANE), 389–399, 401–403, 406–409
Sensing sites, 54
Sensitivity, 41, 42, 45, 73, 74
Sensitivity, photon budget, 88
Sensor assembly, 85
Sensor assembly, optoplasmonic sensor, 69, 72
Sequencing, 343–346, 352, 356, 357, 359, 361
SERS spectrum, 171
Shot noise, 73
Shot noise limit, 23–27
Shuriken, 288
Sideband, 76
Signal analysis, 86
Signal enhancement, 47
Signal-to-Noise Ratio (SNR), 11, 345, 351, 353, 355, 357, 362
Signal-to-noise, SNR, detection limit, sensitivity, 71
Silane, 70
Silanization, 70
Silanol, 71, 302
Silica glass, 60
Silica nanoparticles detection, 153
Silicon nitride, 345
Silicon nitride, titanium dioxide, silicon carbide, lithium niobate, chalcogenide, polymers, 64
Single-mode fiber, 80
Single-molecule event, 84
Single-molecule fluorescence, 81
Single-Molecule reaction, 52
Single-molecule scanner, 86
Single-molecule sensing, 43, 44, 49, 59, 74
Single molecule techniques, 249
Single-molecule trapping, 233
Single-phase ligand exchange, 66
Single protein, 156
Slot waveguide, 241
Small molecule interactions with proteins, 257
Sol-gel, 62
Solid-state, 298
Solid-state nanopores, 343–348, 352, 361, 362, 368
Solvent, 345, 346, 349, 352–354
Spatial resolution, 28–30
Spectroscopy, 262
Spike transition, 84
Spin coating, 64
Split modes, mode-split, 50
Spontaneous Raman scattering, 101
Spontaneous scattering, 101

State
coherent, 23, 26–28
oligomeric, 31
squeezed, 27, 28
Step rate, spike rate, surface reaction, 53
Stern layer, 302
Stimulated anti-Stokes Raman spectroscopy, 108
Stimulated Brillouin scattering, 112
Stimulated Emission Depletion Microscopy (STED), 7
Stimulated Raman spectroscopy, 108
Stimulated scattering, 101, 111
Stochastic Optical Reconstruction Microscopy (STORM), 7
Stokes, 100
radius, 9, 12
Stokes and anti-Stokes spectra, 172
Stokes radius, 43
Strong coupling, 189, 195
Substrate turnover, 58
Sum Frequency Generation (SFG), 102
Surface
charges, 15
Surface-Enhanced Raman Spectroscopy (SERS), 104, 105, 166
Surface functionalization, 66
Surface interactions, 259
Surface plasmon resonance, 98, 393, 407, 409
Surface plasmons, 237
Surface reaction, 53
Surface scattering, 62

T

Tapered fiber, 81
Tapered-fiber coupling, 79
Tapered optical fiber, 41
T-cell
T-cell receptor-like antibodies, 399, 409
Temperature tuning, 75
Template-stripping, 243
Temporal resolution, 30
Tethering, 250, 253
Thermal fluctuations, 13
Thermal-optical effect, 117
Thermodynamics, 58
Thermophoresis, 236, 239, 261
Thermophoretic, 391, 407
Thermorefractive noise, 73, 74
Thiol, 66
Third Harmonic Generation (THG)), 102, 113
Threading of biomolecule, 347, 352, 357, 361
3-Aminopropyltriethoxysilane (APTES), 70
Time resolution, 49, 74
TIR objective, 81
Top-down assembly, 71
Toroidal microcavity, microtoroid, 62
Total internal reflection, 59, 79, 81
Toxin, drinking water, 88
Transferrin, 305
Transition metal dichalcogenides, 345, 346, 357
Translocation, 313
Translocation current, 392, 401, 406
Translocation experiments, 347, 350–352, 355, 357–360
Translocation time, 392, 393, 406
Transmission spectrum, 77
Transport phenomena, 299
Trapping potential, 250
Tubulin, 31
Tunable laser, ECDL, external-cavity diode laser, 75
2D materials, 343–345, 357, 374
2D slit, 374

U

Ultimate detection limit, 73
Unprocessed samples, 254
Unresolved mode-split, unresolved split-mode, 51
UV curable adhesive, 80

V

Van der Waals (VdW) forces, 66
Vapor phase, 70
Vernier comb spectroscopy, 115
Vertical Cavity Surface Emitting Laser (VCSEL), 408, 409
Viability, 14

W

Waveguide, 241
Wavelength shift, 53, 55, 57
Wavelet, 87
Wavemeter, 82
Wave vector, 12
Weak coupling, 45
WGM resonances, 43
WGM spectra, 82

WGM standing wave, 52

WGM time trace, 84

WGM transmission spectrum, 41

Whispering gallery microcavity, 127

Whispering gallery mode, 38, 242

X
Xenon difluoride, 62
x-ray crystallography, 57

Z
Zero-mode waveguide, 380
Zeta potential, 302

The manufacturer's authorised representative in the EU is Springer
Nature Customer Service Centre GmbH, Europaplatz 3, 69115 Heidelberg,
Germany. If you have any concerns regarding our products, please
contact ProductSafety@springernature.com

Printed and bound by CPI Group (UK) Ltd, Croydon, CR0 4YY
30/09/2026
02225699-0015